S0-ABE-899

Multiphase Flow in Polymer Processing

Multiphase Flow in Polymer Processing

CHANG DAE HAN

Department of Chemical Engineering
Polytechnic Institute of New York
Brooklyn, New York

1981

ACADEMIC PRESS
A Subsidiary of Harcourt Brace Jovanovich, Publishers
New York London Toronto Sydney San Francisco

ACADEMIC PRESS, INC.
111 Fifth Avenue, New York, New York 10003

United Kingdom Edition published by
ACADEMIC PRESS, INC. (LONDON) LTD.
24/28 Oval Road, London NW1 7DX

Library of Congress Cataloging in Publication Data

Han, Chang Dae.
Multiphase flow in polymer processing.

Includes bibliographical references and index.
1. Polymers and polymerization. 2. Multiphase flow.
I. Title.
QD381.8H36 668.9 80-70598
ISBN 0-12-322460-8 AACR2

PRINTED IN THE UNITED STATES OF AMERICA

81 82 83 84 9 8 7 6 5 4 3 2 1

To my wife, Keyoung,
and my children, Peter, Cecilia, and John

Contents

Preface

Multiphase (or multicomponent) flow arises frequently in chemical processes dealing with chemical reactions and separation processes. In the past, fairly extensive studies of multiphase flow have been reported in the literature dealing with mixtures of Newtonian liquids (e.g., an emulsion of water and oil), suspensions of solid particles in a Newtonian liquid (e.g., a slurry of bentonite and water), and the motion of gas bubbles in a Newtonian liquid. However, relatively less fundamental study has been reported in the literature dealing with mixtures of two or more non-Newtonian viscoelastic liquids or suspensions in a non-Newtonian viscoelastic liquid of solid particles, gas bubbles, or both. Multiphase systems containing at least one non-Newtonian viscoelastic liquid arise in many polymer processing operations, such as in the processing of reinforced plastics, in structural foam processing, and in the processing of polyblends and copolymers. Because of the rheologically complex nature of polymeric liquids, the rheological behavior of multiphase polymeric systems is very complicated compared to that of multiphase Newtonian liquids.

Today there are hundreds of homopolymers commercially available. Since one type of polymer does not possess all the physical/mechanical (or optical or thermal) properties desired in a finished product, it is natural to try to use two or more polymers, or a polymer together with nonpolymeric materials (e.g., solid filler as a reinforcing agent), in order to meet the needed requirements. For instance, many homopolymers are not strong enough to be useful for structural purposes. However, the reinforcement of homopolymers with nonpolymeric materials (e.g., glass fiber) produces new materials very attractive for structural purposes. When two or more homopolymers are to be mixed or a homopolymer is to be mixed with nonpolymeric material, there are a number of fundamental questions that must be answered. Some important questions are (1) Will the mixture of two or more polymers be compatible? (2) How can one be sure of obtaining uniform dispersion? (3) What type of mixing equipment will perform the most effective job?

When two or more polymers are to be processed together, there are two ways of handling them. One is to preblend them by means of a tumbling operation and then to transport the mixture into a shaping device (e.g., extrusion die or mold cavity). More often than not, a mixture of two or more polymers forms two or more distinct phases in the molten state, giving rise to a dispersion in which one component, the *discrete* phase, is suspended in the other component,which forms the *continuous* phase.

Another way of obtaining fabricated products of improved physical/mechanical properties with two or more polymers is to transport them into separate feeding systems and then into a single shaping device where a composite material is produced (e.g., conjugate fiber, multilayer film). This processing technique is widely known as "coextrusion" if it is involved with the extrusion process, or "coinjection molding" if it is involved with the injection molding process. Of course, compatibility is of primary importance in producing a composite material in this manner. When different polymers have poor compatibility, one may use a third component, which acts as an adhesive.

In this monograph, we shall discuss primarily the rheological behavior of multiphase (or multicomponent) polymeric systems as they are involved in various fabrication operations. We shall also point out the importance of the morphological states of multiphase polymeric systems, because they are needed to explain the systems, rheological behavior in the fluid state, and mechanical behavior in the solid state. This monograph is divided into two parts: In Part I we discuss dispersed multiphase flow in polymer processing, and in Part II we discuss stratified multiphase flow (i.e., coextrusion) in polymer processing.

In Part I, we shall discuss the rheological behavior of particulate-filled polymeric systems (in Chapter 3), the rheological behavior of heterogeneous polymeric systems (in Chapter 4), the phenomenon of droplet breakup in dispersed flow (in Chapter 5), and gas-charged polymeric systems (in Chapter 6). In each of these four chapters, we shall first discuss the role of the discrete phase (i.e., solid particles, liquid droplets, gas bubbles) in determining the bulk rheological properties of the multiphase system and then some representative polymer processing operations (namely, fiber spinning and injection molding) of the multiphase (or multicomponent) polymeric systems. In Part II, we shall discuss coextrusion in cylindrical, rectangular, and annular dies (in Chapter 7) and the phenomenon of interfacial instability in coextrusion (in Chapter 8). Each of these flow geometries has specific applications to polymer processing operations. Emphasis will be placed on presenting a unified approach to the study of coextrusion processes from the rheological point of view.

It is fair to say that our efforts to achieve a better understanding of the multiphase (or multicomponent) flow of polymeric systems have just begun. I shall be very satisfied if this monograph contributes to the stimulation of further research on the subject, which is very important from both the fundamental and technological points of view. It is my earnest hope that the materials presented here, when correctly interpreted and properly applied, will help the polymer processing industry to be better prepared for future challenges.

Acknowledgments

I should like to express my sincere gratitude to my colleagues in industry and the National Science Foundation who supported my research, and to my former students, Drs. H. B. Chin, A. A. Khan, Y. W. Kim, D. Rao, R. Shetty, C. A. Villamizar, and H. J. Yoo, who patiently carried out many difficult and time-consuming experiments and theoretical studies during their association with me. Without the financial support of industry and the National Science Foundation and the participation of my former students, this volume would never have materialized.

It is my great pleasure to acknowledge that Professor James L. White of the University of Tennessee kindly provided me with several of his manuscripts before their publication, enabling me to add luster to Chapter 3, and made a number of constructive comments on the original manuscript. I am also very grateful to Dr. Israel Wilenitz, who most unselfishly spent much of his time editing the manuscript, and to publishers who granted permission to reproduce material originally appearing in their journals. My sincere thanks are also extended to Mrs. Ruth Drucker, who typed the entire manuscript with great patience, and to my wife, who skillfully reproduced all the figures appearing in this monograph.

1

Introduction

1.1 CLASSIFICATION OF MULTIPHASE FLOW ENCOUNTERED IN POLYMER PROCESSING

Two types of multiphase flow may be distinguished on the basis of their degree of phase separation. One type is *dispersed* multiphase flow, in which one or more components exist as a discrete phase dispersed in another component forming the continuous phase. The other type is *stratified* multiphase flow, in which two or more components form continuous phases separated from each other by continuous boundaries.

Depending on the nature of the discrete phase, the bulk flow properties of a dispersed multiphase system may vary. That is, the bulk flow properties of a dispersion can be different, depending on whether they are rigid particles, deformable droplets, or gas bubbles that are suspended in the continuous polymeric medium. All three kinds of dispersed systems are widely used in the polymer processing industry. As summarized in Fig. 1.1, the polymer processing industry uses not only two-phase (liquid–liquid, liquid–solid, liquid–gas) systems, but also three-phase (liquid–liquid–solid, liquid–solid–gas) systems and four-phase (liquid–liquid–solid–gas) systems. In Chapters 3–6, we shall discuss the various flow problems involved in the processing of dispersed multiphase polymeric systems.

Depending on the geometry of the die through which a system of fluids may be forced to flow, stratification may occur in a rectangular duct, in a circular tube, and in an annular space. Figure 1.2 summarizes various applications of stratified multiphase flow, often referred to as a "coextrusion" in

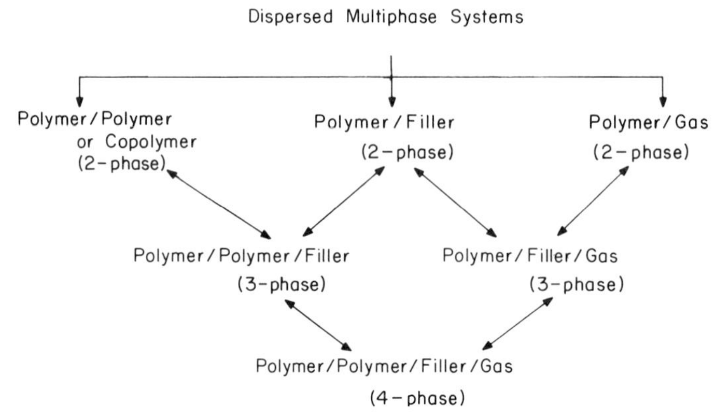

Fig. 1.1 Schematic showing various ways of producing dispersed multiphase systems encountered in the polymer processing industry.

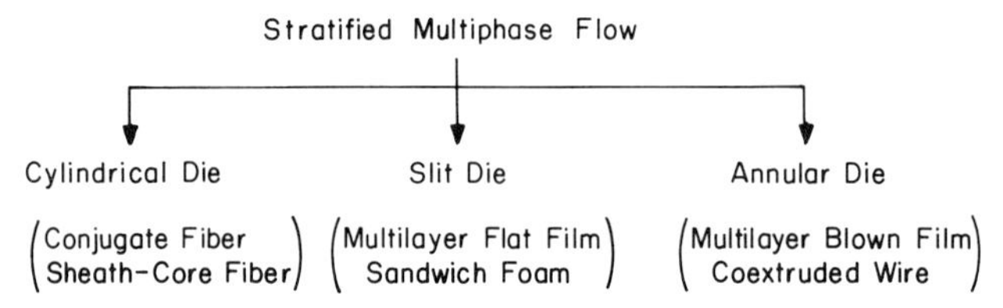

Fig. 1.2 Schematic showing various applications of stratified multiphase flow encountered in the polymer processing industry.

the polymer processing industry. The number of layers, the layer thicknesses, and the manner in which the individual components are arranged, control the bulk flow properties and the ultimate physical/mechanical properties of a stratified system. There are situations where one may have a combination of dispersed and stratified flow, a practice frequently encountered in the polymer processing industry. For instance, a polymer stream containing gas bubbles or additives may be coextruded with another polymer. There are numerous other examples of industrial importance. In Chapters 7 and 8 we shall discuss the various flow problems involved in stratified multiphase flow (i.e., coextrusion).

1.2 DISPERSED MULTIPHASE FLOW IN POLYMER PROCESSING

A rapidly growing interest in engineering thermoplastics has stimulated the polymer industry to develop dispersed multiphase polymeric systems. They comprise many polymeric materials being used in industry, including reinforced plastics, mechanically blended polymers and copolymers, and foamed thermoplastics. When considering the wide choice of base materials available, namely, polymers, solid reinforcing agents, and foaming agents, an enormous number of combinations are possible for making dispersed polymeric materials. In all such cases, one hopes to improve the mechanical/physical properties of the finished product, using some additional component as a modifying agent.

Applications of dispersed multiphase systems to polymer processing operations include: (1) reinforced thermoplastics or thermosets; (2) rubber modified plastics [e.g., high-impact polystyrene (HIPS) and acrylonitrile–butadiene–styrene (ABS) resins], and mechanically blended thermoplastics and elastomers [e.g., blends of polyvinyl chloride (PVC) with impact modifier, blends of styrene–butadiene rubber (SBR) with natural rubbers]; and (3) foamed plastics or rubbers as well as reinforced structural foams.

Rubber modified polymers such as HIPS and ABS differ significantly from mechanically blended polymers such as blends of polyolefin and polystyrene. This is because the dispersed rubber particles are crosslinked and there is substantial graft copolymer between the rubber and matrix. This graft plays a major role in determining the mechanical properties of the products [1,2]. Sometimes, chemical reactions occur between components in blends due to stress-induced degradation during mixing, producing graft structures at the interface [3,4].

When a reinforcing component (e.g., an inorganic filler) is used, improved mechanical/physical properties of the finished product can only be obtained by achieving good adhesion between the reinforcing component and the suspending polymeric medium. In order to obtain good adhesion, multifunctional silanes or titanates are often used as coupling agents.

In the processing of, for instance, heterogeneous polymer blends and copolymers, both phases are usually non-Newtonian and viscoelastic. The polymer processing industry sometimes makes use of blends of two or more polymers of the same molecular structure, but of different molecular weight distributions. It is usually the case that when two polymers of different structure are melt-blended, one polymer forms the discrete phase dispersed in the other (continuous phase), thus giving rise to a two-phase system. Various aspects of polymer blends and copolymers are discussed in several recent monographs [5–11].

In recent years, however, a number of miscible polymer blends have been reported [12], including polystyrene/poly(2,5 dimethyl-1,4, phenylene oxide); polyvinyl chloride/butadiene–acrylonitrile copolymer; polyvinyl chloride/styrene–acrylonitrile copolymers; nylon 6/ethylene–acrylic acid copolymers; polyvinyl chloride/ethylene–ethyl acrylate–carbon monoxide terpolymer; polyacrylic acid/polyethylene oxide; polymethacrylic acid/polyvinyl pyrrolidone; polymethylmethacrylate/polyvinylidene fluoride.

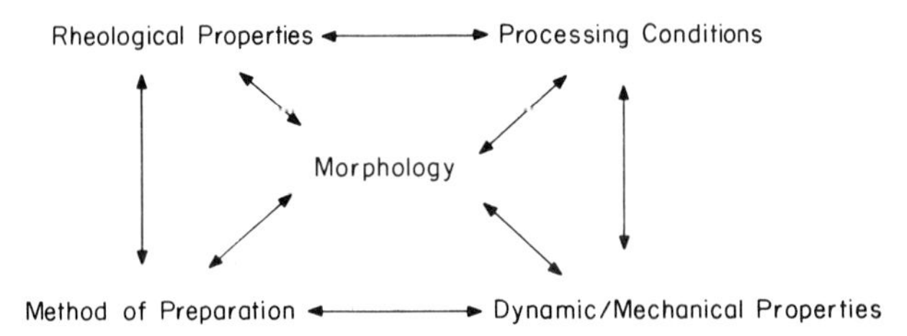

Fig. 1.3 Schematic pointing out processing–morphology–property relationships in dispersed multiphase polymeric systems.

Thermodynamic principles of polymer–polymer miscibility and methods for determining polymer–polymer miscibility are discussed in recent monographs [12, 13]. It is worth pointing out that the homogeneity of the polymer–polymer mixture, because of its very high viscosity in the molten state and its very slow rate of diffusion, will depend very much on the methods of preparation and the time and temperature to which the mixture is subjected.

In dispersed multiphase polymeric systems, there are many variables interrelated to each other, which affect the ultimate mechanical/physical properties of the finished product. Such relationships are pointed out schematically in Fig. 1.3. For instance, the method of preparation (e.g., the method of mixing the polymer blends, the intensity of mixing during polymerization of copolymers) controls the morphology (e.g., the state of dispersion, particle size, and its distribution) of the mixture, which in turn controls the rheological properties of the mixture. On the other hand, the rheological properties strongly dictate the choice of processing conditions (e.g., temperature, shear stress), which in turn strongly influence the morphology, and therefore the ultimate mechanical/physical properties of the finished product. We shall now consider a few specific examples to illustrate the processing–morphology–property relationship in dispersed multiphase polymeric systems.

Today it is well known that a high-impact polystyrene (HIPS) is formed by dissolving rubber in a styrene monomer and then polymerizing. As the polymerization proceeds, the polystyrene in the styrene phase grows at the expense of the rubber–styrene phase, until the latter becomes too small to be the continuous phase. At this point, a phase inversion occurs, and the polystyrene in the styrene phase becomes the continuous phase, with the rubber–styrene phase making up the discrete phase (i.e., droplets). As the polymerization reaches its final stages, the rubber-phase droplets shrink as they continue losing styrene to the continuous phase, resulting in a final product having rubber particles in a polystyrene matrix [14].

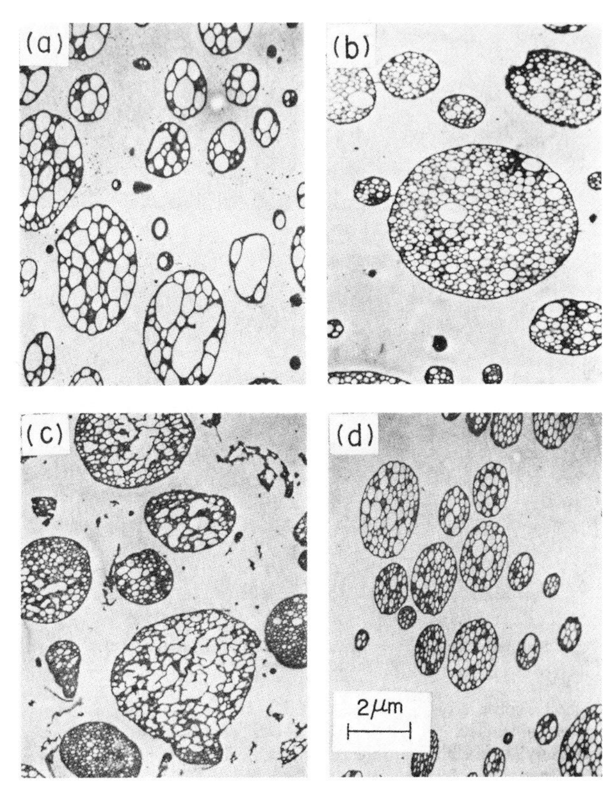

Fig. 1.4 Photomicrographs of commercially available high-impact polystyrenes [15]: (a) Dow Chemical Co., STYRON 456; (b) Monsanto Co., Hi-Test 88; (c) Union Carbide Corp., TGD 6600; (d) Rexall Chemical Co., Elrex 422. From H. Keskkula *et al.*, *J. Appl. Polym. Sci.* **15**, 351. Copyright © 1971. Reprinted by permission of John Wiley & Sons, Inc.

Figure 1.4 gives photomicrographs showing the morphology of commercially available high-impact polystyrenes [15]. It is seen that the HIPS resin forms two phases; the rubber phase is dark and the polystyrene phase is light. Visible in the photomicrographs are the polystyrene occlusions within the rubber particles. It is the large deformable particles that are mainly responsible for reinforcement, since they collectively contain by far the larger amount of rubber. They contain an even higher percentage than is apparent from their larger diameter, since the amount of rubber present is a function of the cube of the diameter.

Table 1.1 gives the range of rubber particle size obtained under the same time–temperature conditions (with varying rates of agitation) in the polymerization of high-impact polystyrene [16]. It is seen that particle size decreases with increasing agitation. Also given in Table 1.1 are the average

TABLE 1.1

Variations of Rubber Particle Size and Mechanical Properties with Intensity of Agitation in the Polymerization of High-Impact Polystyrene[a]

Sample identification	Agitation rate[b] (RPM)	Range of particle size[c] (μm)	Izod impact strength (Nm/m Notch)	Tensile strength at yield (N/m^2)	Tensile strength at break (N/m^2)	Tensile modulus (N/m^2)	Elongation at break (%)
a	50	50 ~ 100	54.36	0.620×10^7	0.758×10^7	0.761×10^9	48
b	100	15 ~ 40	53.08	0.706×10^7	0.758×10^7	0.792×10^9	52
c	200	5 ~ 20	50.53	0.733×10^7	0.794×10^7	0.992×10^9	62
d	400	less than 1 ~ 5	23.66	0.712×10^7	0.675×10^7	0.868×10^9	61

[a] From J. Siberberg and C. D. Han, *J. Appl. Polym. Sci.* **22**, 599. Copyright © 1978. Reprinted by permission of John Wiley & Sons, Inc.
[b] Turbine type of agitator.
[c] Type of rubber used is styrene–butadiene rubber (SBR) by 5 wt. %.

Izod impact strength values found for bars injection molded out of the resins prepared. It is seen that when the particle size is very small, the impact reinforcement becomes small, approaching the impact strength of unmodified polystyrene homopolymer. Once a certain threshold level of particle size is reached, impact reinforcement has achieved most of its full potential. It is also seen in Table 1.1 that, except for sample (d), tensile strength both at yield and at break, as well as tensile modulus, decrease with increasing particle size. This is consistent with the generally more "rubbery" behavior accompanying increasing particle size. Note that with the smallest particle size (i.e., the highest RPM), the tensile strength at yield exceeds its value at break. This further emphasizes the close resemblance this sample has to unmodified polystyrene. When examining the percent elongation data, the values are seen to show a decrease in elongation with increasing particle size and increasing impact reinforcement.

In preparing polymer blends, the method of blending and the mixing conditions (e.g., mixing time and mixing temperature) have a profound influence on the processing characteristics of a blend, its rheological properties in the molten state, and the mechanical/physical properties of the finished product.

Figure 1.5 gives plots of the torque values measured versus the composition of the blends at various times of mixing for blends of natural rubber (NR) and *trans*-polypentenamer (TPP), and for blends of natural rubber (NR) and polybutadiene (BR) [17]. It is seen that the torque values are nonadditive functions of the blend compositions, and that the torque goes through a maximum. The torque values in Fig. 1.5 may be interpreted as

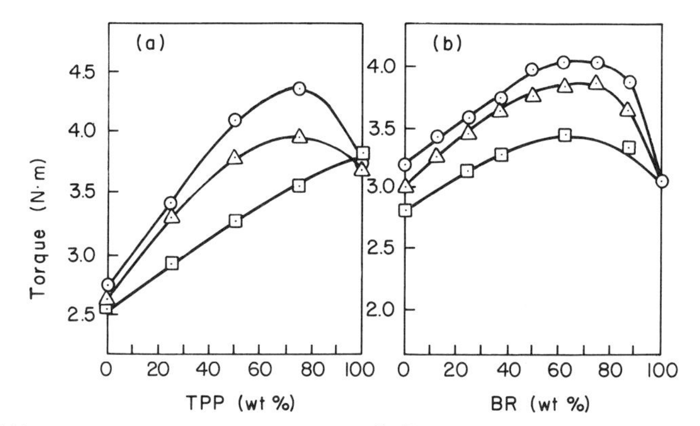

Fig. 1.5 Torque versus blend composition [17]: (a) Blends of natural rubber (NR) and *trans*-polypentenamer (TPP) at various mixing times (min): (○) 2.0; (△) 4.0; (□) 10.0. (b) Blends of natural rubber (NR) and polybutadiene (BR) at various mixing times (min); (○) 2.0; (△) 4.0; (□) 10.0.

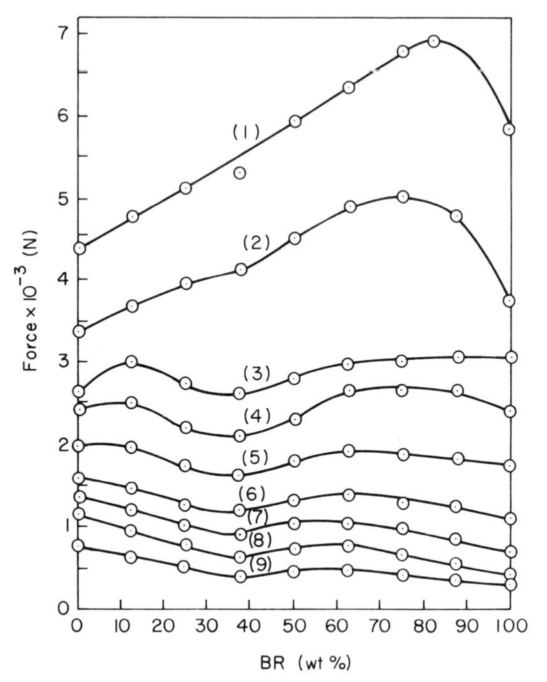

Fig. 1.6 Force required for extrusion versus blend composition for blends of natural rubber (NR) and polybutadiene (BR) at various apparent shear rates (s^{-1}) [17]: (1) 3000; (2) 300; (3) 30; (4) 15; (5) 7.5; (6) 3.0; (7) 1.5; (8) 0.75; (9) 0.3.

equivalent to blend viscosities, and therefore it may be concluded that the viscosity-composition curves for the two blends, NR/TPP and NR/BR, go through a maximum at certain blending ratios. It is of interest to note in Fig. 1.5 that the time of mixing affects the degree of mixing and, consequently, the apparent viscosity of the blends investigated.

Figure 1.6 gives plots of the force required for extrusion versus blend composition for blends of natural rubber (NR) and polybutadiene (BR) at various shear rates [17]. Since, for a given capillary, the measured force is proportional to the shear stress, the plots given in Fig. 1.6 may be interpreted as equivalent to the *apparent* viscosity-blend composition curves. It is seen that at low shear rates (0.3–30 sec^{-1}), the apparent viscosity goes through a maximum and/or a minimum, and that at high shear rates (300–3000 sec^{-1}) a minimum in the viscosity disappears and the apparent viscosity goes through a maximum. It can be concluded therefore that the processing history influences the rheological response of the blends investigated.

Difficulties arise in analyzing the rheological data of multiphase systems in general. This is because their rheological properties are influenced by

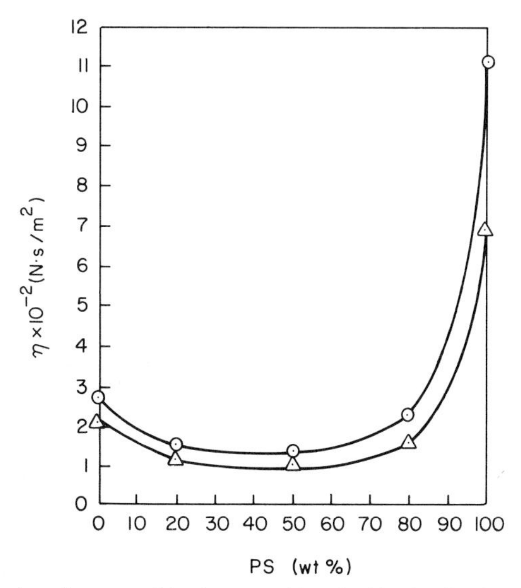

Fig. 1.7 Melt viscosity versus blend composition for blends of polystyrene (PS) and polypropylene (PP) at 200°C, at various shear stresses (N/m^2) [19]: (⊙) 0.41×10^5; (△) 0.48×10^5.

many factors, such as the particle size, its shape, and the volume fraction of the dispersed phase. In other words, the state of dispersion may influence the rheological behavior of multiphase systems [18]. The rheological properties that are important to the design of equipment for polymer processing operations are *viscosity* and *elasticity*. Information concerning these properties is essential, for instance, for determining pressure drops (and thus production rates), viscous heating, stress relaxation, etc.

Figure 1.7 gives plots of bulk viscosity versus blend ratio for blends of polystyrene (PS) and polypropylene (PP) in the molten state at 200°C. It is seen that at a certain blend ratio, the bulk viscosity goes through a minimum. This seemingly peculiar behavior of the bulk flow property can be explained only when one has information on the morphological state of the materials at different blending ratios [18–22].

Figure 1.8 gives photomicrographs of longitudinal sections of polystyrene/polypropylene blend extrudates, in which the *dark* areas represent the polystyrene (PS) phase and the *white* areas represent the polypropylene (PP) phase. Thus it is seen that the blend forms two phases, and that the blend composition influences the extent of dispersion. Note that the mode of dispersion of two incompatible polymers depends, among many factors, on the viscosity and elasticity ratios of the two polymers, and on the extrusion conditions (i.e., melt temperature and shear stress) [18, 21].

In general, the mechanical properties of a finished product consisting of two incompatible polymers are poorer than that of the individual polymers,

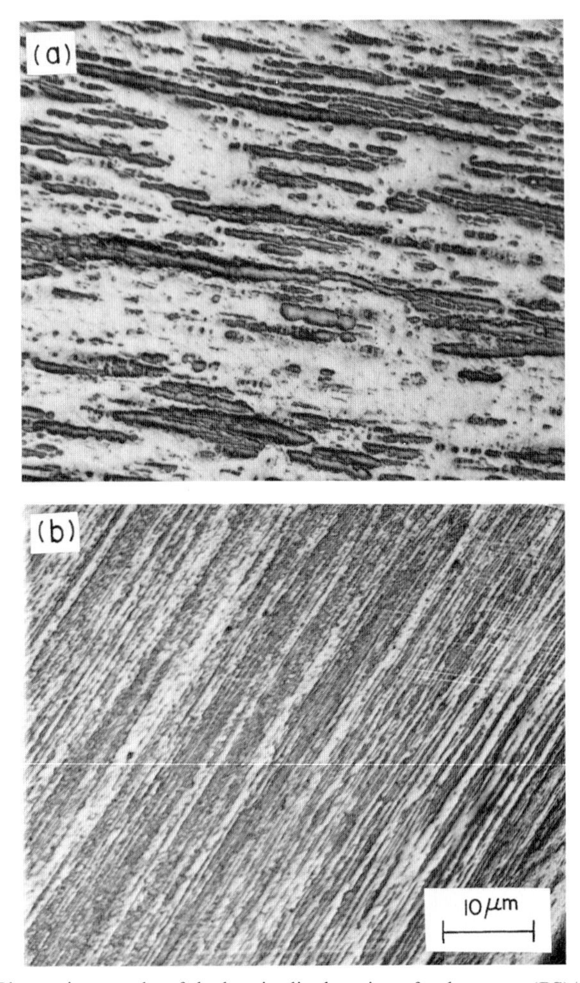

Fig. 1.8 Photomicrographs of the longitudinal section of polystyrene (PS)/polypropylene (PP) blend extrudates: (a) PS/PP = 20/80 (by wt); (b) PS/PP = 80/20 (by wt).

as shown in Fig. 1.9. However, there are situations where improved mechanical properties may be obtained. Figure 1.10 gives plots of tensile strength versus blend composition for blends of low-density polyethylene (LDPE) with partially *crystalline* ethylene–propylene–diene monomer (EPDM) rubber and for blends of LDPE with *amorphous* EPDM rubber. It is seen that the blends of LDPE with *crystalline* EPDM exhibit tensile strengths greater than that of either pure component, and far greater than that of the blends of LDPE with *amorphous* EPDM. The difference in the observed

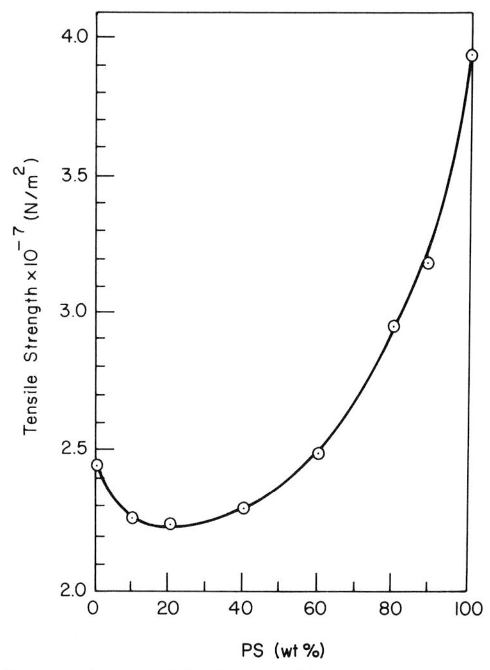

Fig. 1.9 Tensile strength versus blend composition for blends of polystyrene (PS) and polypropylene (PP). The specimens used for the measurement were prepared by injection molding the blends at 220°C.

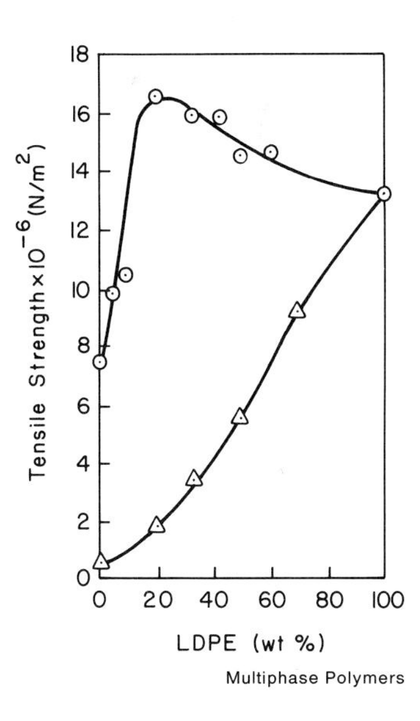

Fig. 1.10 Tensile strength versus blend composition for blends of low-density polyethylene (LDPE) and ethylene–propylene–diene monomer (EPDM) rubber [23]: (⊙) LDPE with partially crystalline EPDM rubber; (△) LDPE with amorphous EPDM rubber.

Multiphase Polymers

tensile strengths between the two blend systems is attributed to the difference in their microphase morphologies [23].

The effect of mixing conditions (e.g., mixing temperature, mixing time) on the mechanical properties of blends of polyvinyl chloride (PVC) and acrylonitrile–butadiene copolymer rubber (NBR) are given in Figs. 1.11–1.14 [24]. It is seen in Fig. 1.11 that both the yield stress and the modulus go

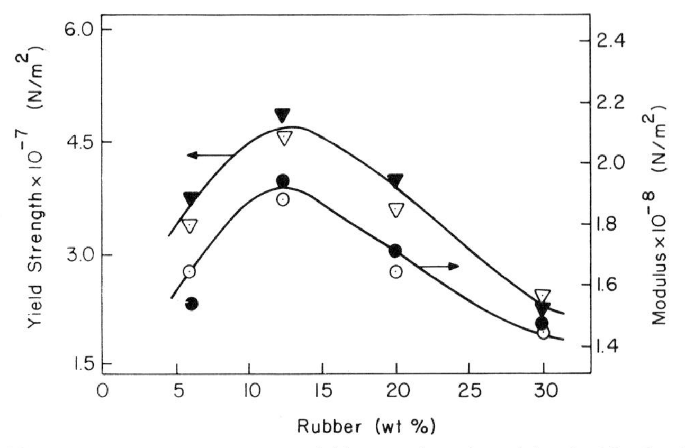

Fig. 1.11 Effect of rubber content on yield strength and modulus for blends of polyvinyl chloride (PVC) and acrylonitrile–butadiene copolymer rubber (NBR) [24]: Open symbols are for 5 min mixing time, and closed symbols for 10 min mixing time. Mixing temperature is 148°C. From C. C. Lee *et al.*, *J. Appl. Polym Sci.* **9**, 2047. Copyright © 1965. Reprinted by permission of John Wiley & Sons, Inc.

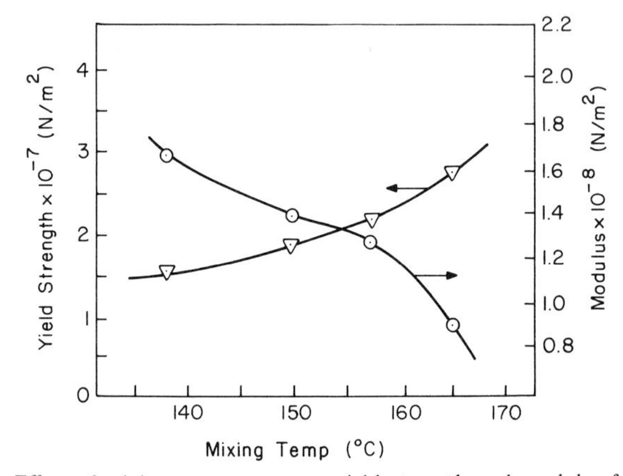

Fig. 1.12 Effect of mixing temperature on yield strength and modulus for the blend of PVC/NBR = 70/30 (by wt.) [24]. The mixing time employed is 10 min. From C. C. Lee *et al.*, *J. Appl. Polym. Sci.* **9**, 2047. Copyright © 1965. Reprinted by permission of John Wiley & Sons, Inc.

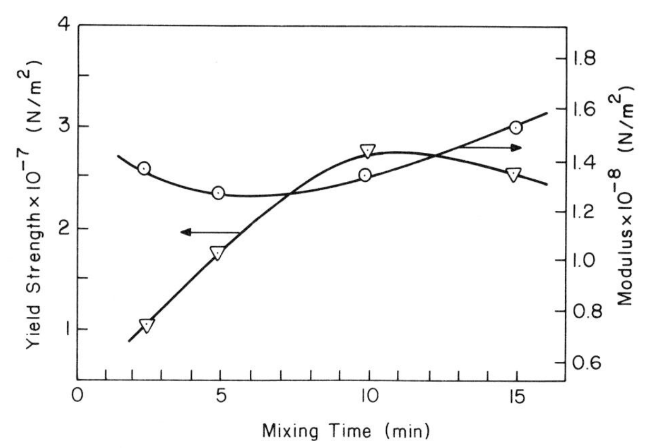

Fig. 1.13 Effect of mixing time on yield strength and modulus for the blend of PVC/NBR = 70/30 (by wt.) at 148°C [24]. From C. C. Lee *et al.*, *J. Appl. Polym. Sci.* **9**, 2047. Copyright © 1965. Reprinted by permission of John Wiley & Sons, Inc.

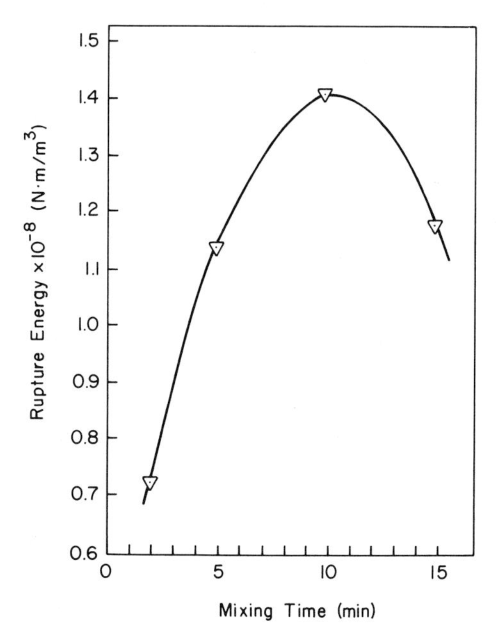

Fig. 1.14 Effect of mixing time on rupture energy for the blend of PVC/NBR = 70/30 (by wt.) at 148°C [24]. From C. C. Lee *et al.*, *J. Appl. Polym. Sci.* **9**, 2047. Copyright © 1965. Reprinted by permission of John Wiley & Sons, Inc.

through a maximum at a certain blending ratio, and in Figs. 1.12 and 1.13 that the efficiency of mixing, judged by the attainment of optimum properties, depends not only on mixing temperature, but also on mixing time. It is of particular interest to note in Fig. 1.14 that undermixed and overmixed blends exhibit very low impact resistance, judged by the index of rupture energy. It is not difficult to surmise that undermixed blends would contain large rubber particles, while overmixed blends would contain too many small particles. This appears to suggest that there exists an optimal range of particle size which gives rise to the greatest achievable impact resistance.

In dealing with either particle-filled polymers (composites) or gas-charged polymers (foamed plastics), processing–morphology–property relationships may be as complex as those for polymer blends and copolymers of heterogeneous nature. For instance, in processing highly filled thermoplastics or thermoset resins, the choice of particle size and its distribution, the shape of particles, and the wettability of the particles by the base polymer matrix are important in controlling the bulk rheological properties of the mixture, and in obtaining a finished product with consistent quality in terms of mechanical/physical properties.

Figure 1.15 shows the effect of filler particles on the shear viscosity of a non-Newtonian viscoelastic molten high-density polyethylene [25]. It is seen that the solid particles increase the melt viscosity very rapidly as the shear rate is decreased, giving rise to a yield stress as the shear rate approaches zero.

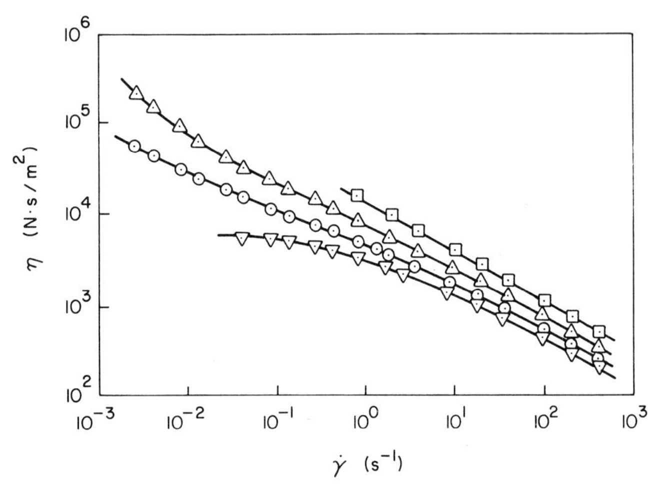

Fig. 1.15 Viscosity versus shear rate of high-density polyethylene ($T = 180°C$) filled with titanium dioxide (vol. %) [25]: (▽) 0.0; (⊙) 12.7; (△) 21.6; (⊡) 35.5. From N. Minagawa and J. L. White, *J. Appl. Polym. Sci.* **20**, 501. Copyright © 1976. Reprinted by permission of John Wiley & Sons, Inc.

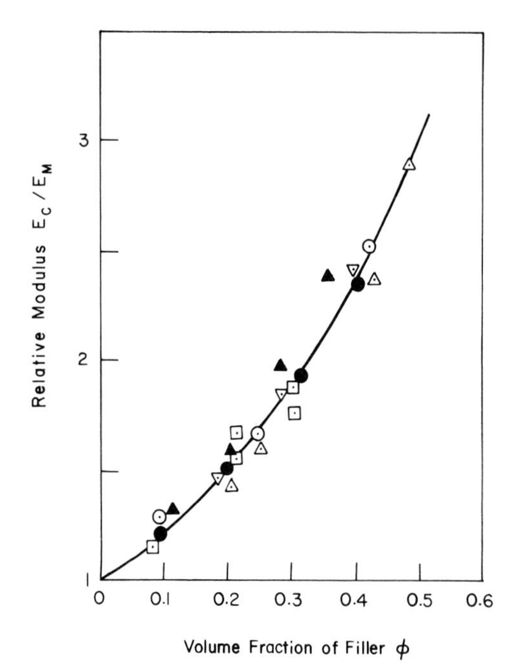

Fig. 1.16 Relative modulus versus volume fraction of glass beads for various composites [26]: (△) SAN; (⊡) PS; (▲) ABS; (⊙) PPO; (●) Epoxy; (▽) Polyester. The solid line is the plot of the Kerner equation.

The primary objective for using solid particles in polymeric materials forming composite materials, is to improve their mechanical/physical properties. Figure 1.16 illustrates the example where the moduli (E_c) of various composites are improved over the moduli (E_m) of polymer matrices, by having glass beads as filler [26]. The solid line in Fig. 1.16 is the theoretical prediction from the Kerner theory [27], which is given by

$$E_c/E_m = (1 + AB\phi)/(1 - B\phi) \tag{1.1}$$

where

$$A = (7 - 5v_m)/(8 - 10v_m) \qquad B = (E_f/E_m - 1)/(E_f/E_m + A) \tag{1.2}$$

v_m is the Poisson's ratio of the polymer matrix, ϕ is the volume fraction of filler, and E_f and E_m are the Young's moduli of the filler and polymer matrix, respectively.

When fillers are used in polymers, the mechanical properties of the composite materials are strongly influenced by the nature of the interface between the phases. Therefore, surface modification of the filler particles

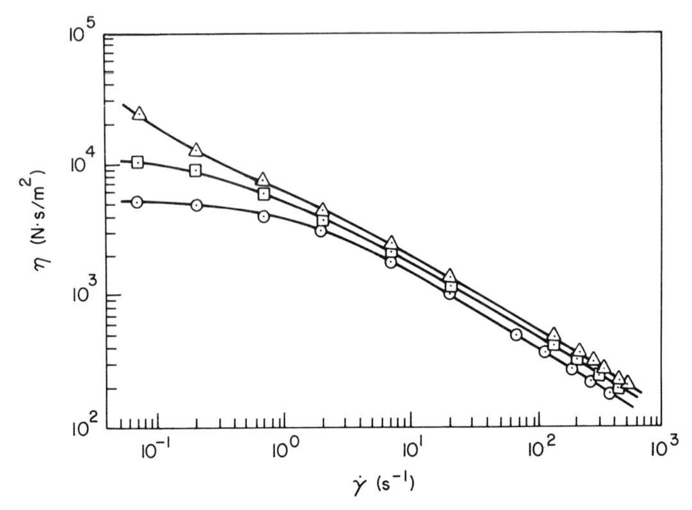

Fig. 1.17 Viscosity versus shear rate of polypropylene (PP) ($T = 200°C$) filled with calcium carbonate ($CaCO_3$), with and without a silane coupling agent: (⊙) Pure PP; (△) PP/$CaCO_3$ = 50/50 (by wt.) *without* coupling agent; (⊡) PP/$CaCO_3$ = 50/50 (by wt.) *with* coupling agent (1 wt. %).

can influence the ultimate mechanical properties of the composite materials. Such practice has indeed been used in the polymer processing industry.

Figure 1.17 shows that the addition of a chemical agent, commonly referred to as "coupling agent," to a calcium carbonate-filled polypropylene reduces its viscosity, implying that the use of the particular coupling agent (Union Carbide Corp., N-octyltriethoxysilane) improves the processability of the filled polymer. Table 1.2 gives measurements of mechanical properties

TABLE 1.2

Effect of Coupling Agent[a] on the Mechanical Properties of Filled Polypropylene

Material	Tensile modulus (N/m^2)	Elongation at break (%)
Polypropylene (PP)	0.62×10^9	300
PP–$CaCO_3$	1.41×10^9	85
PP–$CaCO_3$–Y9187[a]	1.58×10^9	400
PP–Glass beads	0.99×10^9	350
PP–Glass beads–Y9187[a]	1.24×10^9	640

[a] N-octyltriethoxysilane (Union Carbide Corp., Y9187)

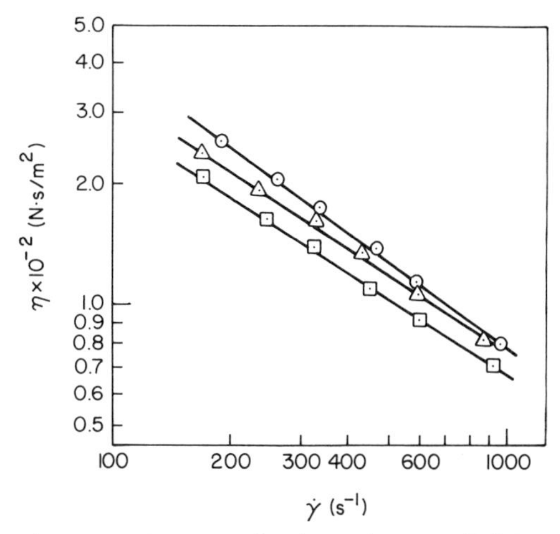

Fig. 1.18 Viscosity versus shear rate of molten polystyrene (PS) ($T = 200°C$) containing sodium bicarbonate ($NaHCO_3$) (wt.%): (⊙) Pure PS; (△) PS with 0.75 wt.% of $NaHCO_3$; (⊡) PS with 1.50 wt.% of $NaHCO_3$.

of the filled polymers with and without a coupling agent. It is seen that the tensile modulus and the percent elongation at the break of the filled polypropylene are improved by the use of the silane coupling agent.

Plastic foams (e.g., polyolefin foam, PVC foam, polycarbonate structural foam) are widely used in many novel applications. In a foam processing operation, gas bubbles, resulting either from the decomposition of a chemical blowing agent or from the injection of a gas, remain in the molten polymer while it is being processed. Foam products are obtained by means of either extrusion or injection molding.

In foam extrusion, as in conventional extrusion operations, the measurement of the rheological properties of the polymer melt, in this case containing dissolved gas, is of fundamental importance to the design of the processing equipment (e.g., screw and die designs). Figure 1.18 shows the effect of the blowing agent on the viscosity of molten polystyrene. It is seen that the gas dissolved in the polymer melt decreases its viscosity, apparently suggesting that the dissolved gas may function in the same way as a plasticizer. As one may surmise, the rheological properties of a polymeric material in foam extrusion vary with the type of foaming agent, the amount of foaming agent, and also with the type and amount of other additives (e.g., plasticizers, modifiers, nucleators, etc.). Also important in foam extrusion is a better understanding of the mechanism of foam formation. In particular, the control of the nucleation and growth of gas bubbles is vitally important to the success of the process.

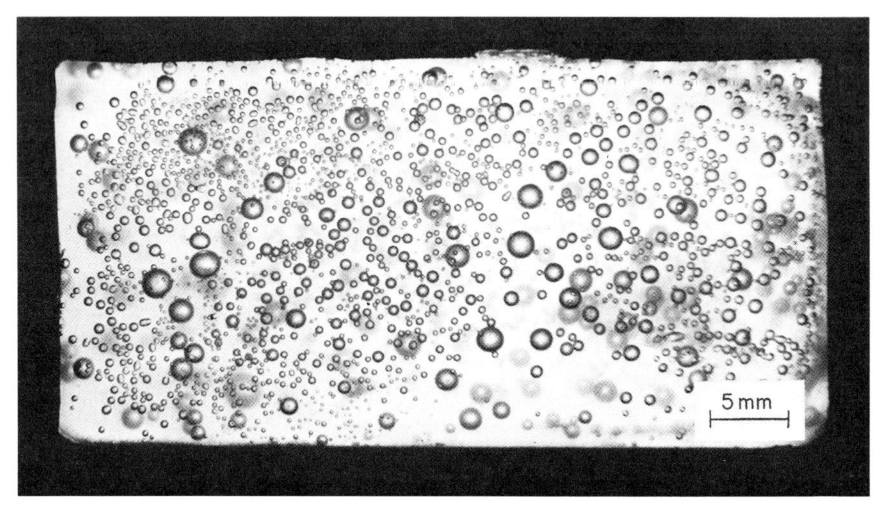

Fig. 1.19 Photograph displaying the bubble size distribution in an injection-molded specimen of high-density polyethylene [28]. The blowing agent employed is Celogen CB which, upon decomposition at elevated temperature, generates nitrogen.

In a foamed plastic, the shape and size of the cells determine the physical properties to a large degree. Figure 1.19 gives a photograph displaying the distribution of cell size in an injection-molded foamed specimen [28]. It may be surmised that the cell size and its distribution would depend on the choice of melt temperature, injection pressure, injection speed, the rate of cooling of the mold, and the amount of blowing agent used. Consequently, all these processing variables are expected to affect the foam density as well as the mechanical properties of the finished foam product. Figure 1.20 gives the mechanical properties of a hydrochloric acid–catalyzed phenolic foam

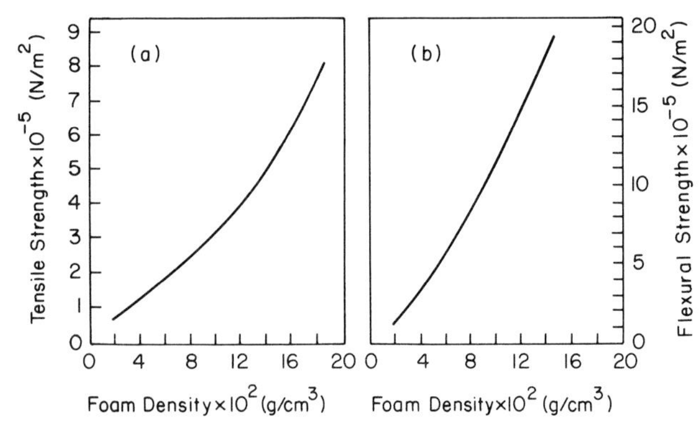

Fig. 1.20 Mechanical properties of phenolic foam as a function of foam density [29].

as a function of foam density [29]. It is seen that both the tensile strength and flexural strength are decreased as the foam density is decreased.

In Chapters 3–6, we shall consider dispersed multiphase flow as applied to polymer processing operations, with emphasis on elucidating the mechanisms of the complex rheological behavior encountered, and on establishing relationships among the processing conditions, morphology, and mechanical properties.

1.3 STRATIFIED MULTIPHASE FLOW IN POLYMER PROCESSING

Recent developments in polymer processing technology have stimulated researchers to obtain a better understanding of the various problems involved in stratified multiphase flow, often referred to as "coextrusion." Representative commercial products produced by coextrusion are conjugate (bicomponent) fiber, multilayer flat film, multilayer blown film, coextruded cables and wires, sandwiched foam composites, etc. Figure 1.21 gives schematics of the cross sections of some representative coextruded products.

In order to operate coextrusion processes successfully, it is important to investigate the processability of given combinations of polymeric materials

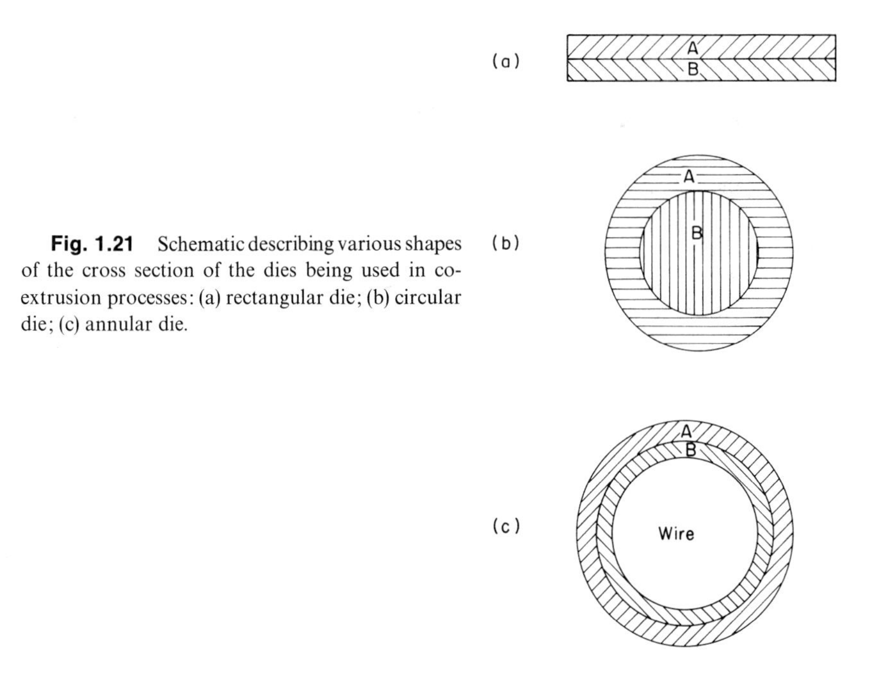

Fig. 1.21 Schematic describing various shapes of the cross section of the dies being used in coextrusion processes: (a) rectangular die; (b) circular die; (c) annular die.

over a wide range of processing variables, and to evaluate the physical/mechanical properties (e.g., adhesion, tensile properties, permeability) of coextruded products as affected by processing conditions. It is also important to develop mathematical models simulating the various coextrusion processes (e.g., flat- and blown-film coextrusion, wire coating coextrusion), in terms of the rheological properties of the individual polymers being coextruded, the variables involved with die design, and the processing variables (i.e., melt extrusion temperature, extrusion rate).

One of the most fundamental and important problems in coextrusion is to obtain an interface of the desired shape between the individual components in the final product. It is reported that under certain situations, the position of the interface migrates as the fluids flow through an extrusion die, and that the direction of interface migration depends on the rheological properties of the polymer system involved [18, 30–33]. Figure 1.22 gives photographs showing the shape of interface of extrudate cross sections in which the initial interface was flat. It is seen that the shape of the interface is quite different from that at the die entrance where the two polymers were first brought into contact side by side. Figure 1.23 gives a schematic suggesting a possible way in which interface deformation could evolve in the coextrusion of two polymers through either a rectangular die or a circular die. The final shape of interface suggested assumes a very long duration flow (i.e., a long die), so that equilibrium in interface shape is achieved. It is also based on the assumption that the viscosity of polymer A is lower than that of polymer B.

It is also reported in the literature [34, 35] that under certain flow conditions or with certain combinations of polymers, the interface between

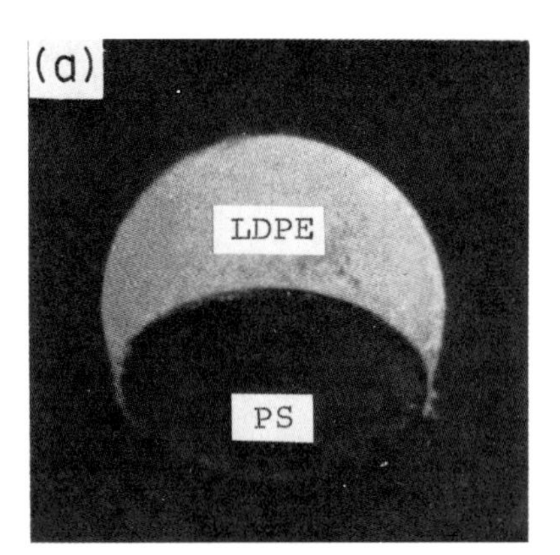

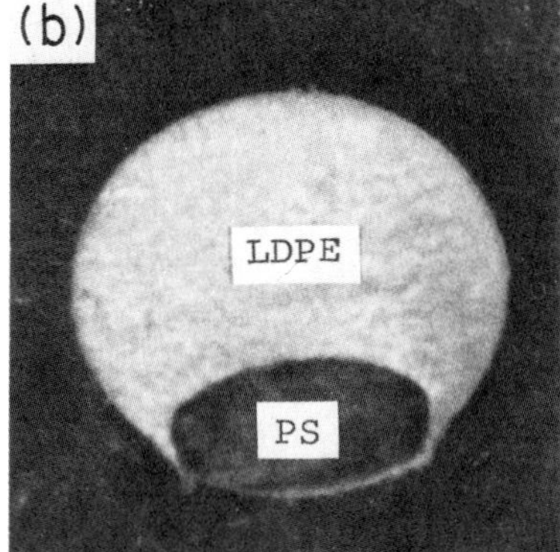

Fig. 1.22 Photograph showing the interface shape of two polymers coextruded through a circular die ($L/D = 4$), with the flat interface at die inlet [18]: (a) $Q_{PS} = 10.9$ cm^3/min, $Q_{LDPE} = 9.9$ cm^3/min; (b) $Q_{PS} = 8.3$ cm^3/min; $Q_{LDPE} = 53.8$ cm^3/min.

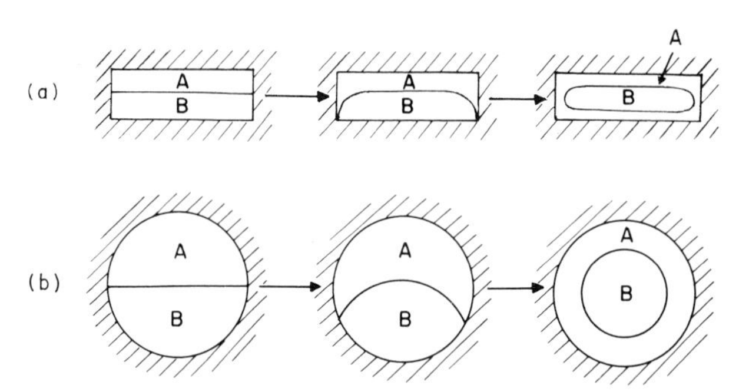

Fig. 1.23 Schematic showing the evolution of interface deformation in the coextrusion of two polymers through the die having: (a) rectangular cross section; (b) circular cross section. It is assumed that the viscosity of polymer A is lower than that of polymer B.

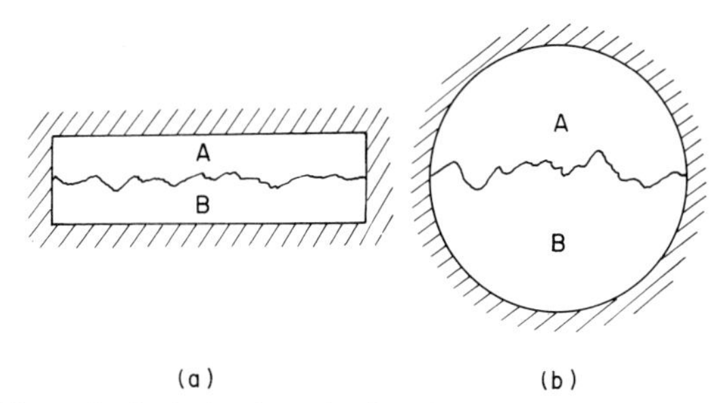

Fig. 1.24 Schematic displaying irregular interfaces between the phases in the coextrusion through the die having: (a) rectangular cross section; (b) circular cross section.

the phases becomes *irregular*, resulting in interfacial instability that must be avoided in obtaining products of acceptable quality. Figure 1.24 shows a schematic illustrating irregular interfaces that may be observed in coextrusion, and Fig. 1.25 gives a photograph displaying a kind of interfacial instability occurring in a two-layer sheet coextruded using the same base polymer, but with one layer pigmented (dark area) and the other not (white area). On the left-hand side of Fig. 1.25, a fairly uniform layer of the pigmented polymer covers the nonpigmented polymer layer, whereas on the right-hand side, the pigmented polymer layer has a very irregular interface, resembling the patterns often observed in a wood panel. The layer thicknesses are *not* uniform across the width of the coextruded sheet.

It is desirable to correlate the processing conditions, at which interfacial instability begins, to the rheological properties of the individual polymers being coextruded, for each coextrusion process (e.g., flat- and blown-film coextrusion, wire coating coextrusion). A better understanding of the phenomenon of interfacial instability occurring in each coextrusion process will

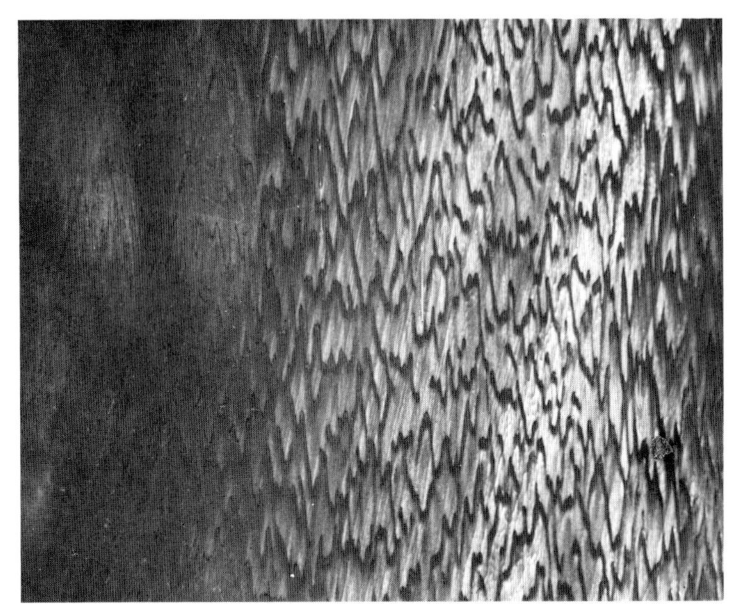

Fig. 1.25 Photograph showing irregular interfaces between the phases in the coextruded sheet. The white area represents the PVC and the dark area represents the pigmented PVC.

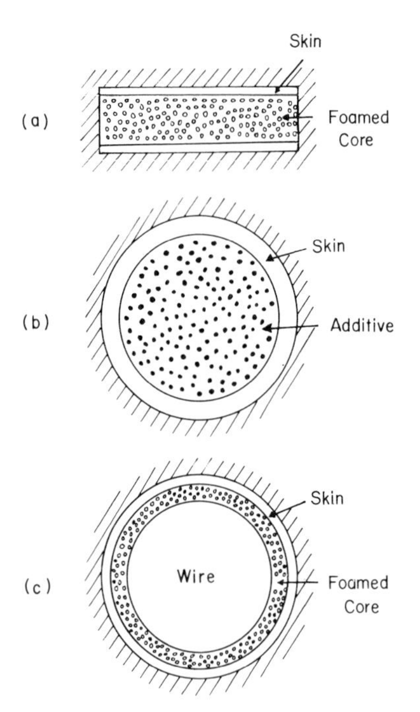

Fig. 1.26 Schematic illustrating the cross sections of the products obtained by a combination of dispersed and stratified flows: (a) sandwiched foam sheet; (b) sheath-core rod with an additive in the core; (c) coextruded wire with the foamed core.

help the industry concerned produce coextruded composites having consistent quality.

A combination of dispersed and stratified multiphase flow has also been used to produce commercial products. For instance, in producing a sandwiched foam product, the gas-charged core component is either coextruded with the skin component through a sheet-forming die or is coinjection molded in a mold cavity, giving rise to the product, as schematically shown in Fig. 1.26. Instead of using a foaming agent, one may use a reinforcing agent (e.g., an inorganic filler) in the core component.

The "sandwich foam coextrusion process" is a relatively new polymer processing technique that combines the film coextrusion process and the foam extrusion process. In producing sandwiched foam products, the core-forming polymer *B* containing a blowing agent, is coextruded with the skin-forming polymer *A*. A large number of combinations of polymer systems may be used for the skin and core components of a sandwiched foam. In the selection of materials, both the core-forming polymer *B* and the skin-forming polymer *A* can be the same (except that *B* contains a blowing agent), or they can be different polymers. If desired, *A* may contain an additive or additives, such as an antistatic or a flame-retardant agent.

In Chapters 7 and 8 we shall consider stratified multiphase flow as applied to polymer processing operations, with emphasis on elucidating the process characteristics of coextrusion operations, which make use of cylindrical or rectangular, or annular dies.

REFERENCES

[1] E. A. Wagner and L. M. Robeson, *Rubber Chem. Technol.* **43**, 1129 (1970).
[2] G. Cigna, S. Matarrese, and G. F. Biglione, *J. Appl. Polym. Sci.* **20**, 2285 (1976).
[3] D. J. Angier and W. F. Watson, *J. Polym. Sci.* **18**, 129 (1955).
[4] R. J. Ceresa, *Polymer* **1**, 72 (1960).
[5] H. H. Keskkula, ed., "Polymer Modification of Rubbers and Plastics," Appl. Polym. Symp. No. 7. Wiley (Interscience), New York, 1969.
[6] P. F. Bruins, ed., "Polyblends and Composites," Appl. Polym. Symp. No. 15. Wiley (Interscience), New York, 1970.
[7] N. A. J. Platzer, ed., "Multicomonent Polymer Systems," Adv. Chem. Ser. No. 99. Am. Chem. Soc., Washington, D.C., 1971.
[8] L. H. Sperling, ed., "Recent Advances in Polymer Blends, Grafts, and Blocks." Plenum, New York, 1974.
[9] N. A. J. Platzer, ed., "Copolymers, Polyblends, and Composites," Adv. Chem. Ser. No. 142. Am. Chem. Soc., Washington, D.C., 1975.
[10] D. R. Paul and S. Newman, ed., "Polymer Blends." Academic Press, New York, 1978.
[11] S. L. Cooper and G. M. Estes eds., "Multiphase Polymers," Adv. Chem. Ser. No. 176. Am. Chem. Soc., Washington, D.C., 1979.

[12] O. Olabisi, L. M. Robeson, and M. T. Shaw, "Polymer-Polymer Miscibility." Academic Press, New York, 1979.
[13] S. Krause, *in* "Polymer Blends" (D. R. Paul and S. Newman, eds.), Vol. 1, Chap. 2. Academic Press, New York, 1978.
[14] G. E. Molau and H. Keskkula, *J. Polym. Sci., Part A-1*, **4**, 1595 (1966).
[15] H. Keskkula, S. G. Turley, and R. F. Boyer, *J. Appl. Polym. Sci.* **15**, 351 (1971).
[16] J. Silberberg and C. D. Han, *J. Appl. Polym. Sci.* **22**, 599 (1978).
[17] V. L. Folt and R. W. Smith, *Rubber Chem. Technol.* **46**, 1193 (1973).
[18] C. D. Han, "Rheology in Polymer Processing." Academic Press, New York, 1976.
[19] C. D. Han, Y. W. Kim, and S. J. Chen, *J. Appl. Polym. Sci.* **19**, 2831 (1975).
[20] C. D. Han and T. C. Yu, *J. Appl. Polym. Sci.* **15**, 1163 (1971).
[21] H. Van Oene, *J. Colloid Interface Sci.* **40**, 448 (1972).
[22] C. D. Han and Y. W. Kim, *Trans. Soc. Rheol.* **19**, 245 (1975).
[23] G. A. Lindsay, C. J. Singleton, C. J. Carman, and R. W. Smith, *in* "Multiphase Polymers" (S. L. Cooper and G. M. Estes, eds.), Adv. Chem. Ser. No. 176, p. 367. Am. Chem. Soc., Washington, D.C., 1979.
[24] C. C. Lee, W. Rovatti, S. M. Skinner, and E. G. Bobalek, *J. Appl. Polym. Sci.* **9**, 2047 (1965).
[25] N. Minagawa and J. L. White, *J. Appl. Polym. Sci.* **20**, 501 (1976).
[26] L. Nicolais, *Polym. Eng. Sci.* **15**, 137 (1975).
[27] E. H. Kerner, *Proc. Phys. Soc., London*, **69**, *Sect. B* 808 (1956).
[28] C. A. Villamizar and C. D. Han, *Polym. Eng. Sci.* **18**, 699 (1978).
[29] R. J. Bender, ed., "Handbook of Foamed Plastics." Lake, Libertyville, Illinois, 1965.
[30] J. H. Southern and R. L. Ballman, *Appl. Polym. Symp.* No. 20, 1234 (1973).
[31] C. D. Han, *J. Appl. Polym. Sci.* **17**, 1289 (1973); **19**, 1875 (1975).
[32] A. E. Everage, *Trans. Soc. Rheol.* **17**, 629 (1973).
[33] B. L. Lee and J. L. White, *Trans. Soc. Rheol.* **18**, 469 (1974).
[34] A. A. Khan and C. D. Han, *Trans. Soc. Rheol.* **21**, 101 (1977).
[35] C. D. Han and R. Shetty, *Polym. Eng. Sci.* **18**, 180 (1978).

2

Fundamentals of Rheology

2.1 INTRODUCTION

Rheology is the science dealing with the deformation and flow of matter. Hence depending on the types of matter one deals with, different branches of rheology may be considered; for instance, polymer rheology, biorheology, lubricant rheology, suspension rheology, and emulsion rheology.

In studying the rheology of a specific type of material, one needs to consider the following three basic steps: (1) define the flow field in terms of the velocity components and the coordinates that are most appropriate; (2) choose a rheological equation of state for the description of the material under deformation; and (3) decide which of the experimental techniques available are most suitable for determining the rheological properties of the material under consideration.

There are two reasons for seeking a precise mathematical description of the rheological models, which relate the state of stress to the state of deformation. The first is that such expressions can be used to identify the significant rheological parameters characteristic of the materials, and to suggest the experimental procedure for measuring them. One would then like to correlate the rheological parameters with the molecular weight, molecular weight distribution, and molecular structure. The second reason is that such an expression can be used, together with the equation of continuity, to solve the equations of motion, in order to relate the rheological parameters to flow conditions and die geometries.

In principle, there are six components of stress that must be specified to completely define the state of stress.† However, in steady shearing flow, a smaller number of stress components is sufficient and the extra stress tensor τ may be written

$$\|\tau\| = \begin{Vmatrix} \tau_{11} & \tau_{12} & 0 \\ \tau_{12} & \tau_{22} & 0 \\ 0 & 0 & \tau_{33} \end{Vmatrix} \tag{2.1}$$

in which τ_{12} is called the shear stress, and τ_{11}, τ_{22}, and τ_{33} are called the normal stresses. Here the subscript 1 denotes the direction of flow, the subscript 2 denotes the direction perpendicular to the flow (i.e., the direction pointing to the velocity gradient), and the subscript 3 denotes the neutral direction.

In an incompressible fluid, the state of stress is determined by the strain (or the rate of strain) or strain history, and, whereas the absolute value of any particular component of normal stress is of no rheological significance, the values of differences (say, $\tau_{11} - \tau_{22}$, and $\tau_{22} - \tau_{33}$) of the normal stress components do have rheological significance [1].

In steady shearing flow, there are no more than three independent material functions needed for correlating the stress quantities of rheological significance to the shear rate $\dot{\gamma}$ [2–4]. These are

$$\tau_{12} = \eta(\dot{\gamma})\dot{\gamma}; \qquad \tau_{11} - \tau_{22} = \Psi_1(\dot{\gamma})\dot{\gamma}^2; \qquad \tau_{22} - \tau_{33} = \Psi_2(\dot{\gamma})\dot{\gamma}^2 \tag{2.2}$$

in which $\eta(\dot{\gamma})$, $\Psi_1(\dot{\gamma})$, and $\Psi_2(\dot{\gamma})$ are called "material functions." $\eta(\dot{\gamma})$ defines the viscosity function, $\Psi_1(\dot{\gamma})$ defines the first normal stress function, and $\Psi_2(\dot{\gamma})$ defines the second normal stress function.

Now, a fluid is called *Newtonian*‡ when, over the range of shear rates tested, the value of viscosity is constant,

$$\eta(\dot{\gamma}) = \eta_0 = \text{const} \tag{2.3}$$

and, also, the values of the normal stress differences are identically zero,

$$(\tau_{11} - \tau_{22}) = (\tau_{22} - \tau_{33}) = 0 \tag{2.4}$$

Figure 2.1 gives plots of η versus $\dot{\gamma}$ for low molecular weight polybutenes (Indopols). It is seen that the plots satisfy Eq. (2.3) and therefore the fluids are considered to be Newtonian. Note that the Indopols show *no* measurable normal stress differences and therefore satisfy Eq. (2.4) also.

† $\mathbf{S} = -p\mathbf{I} + \tau$, where $\mathbf{S}$ is the total stress tensor, p is the hydrostatic pressure, and τ is the "extra stress" tensor.

‡ In terms of the state of stress in steady shearing flow, the stress of Newtonian fluids depends linearly on *instantaneous* rate of deformation, whereas the stresses of viscoelastic fluids depend upon deformation *history*.

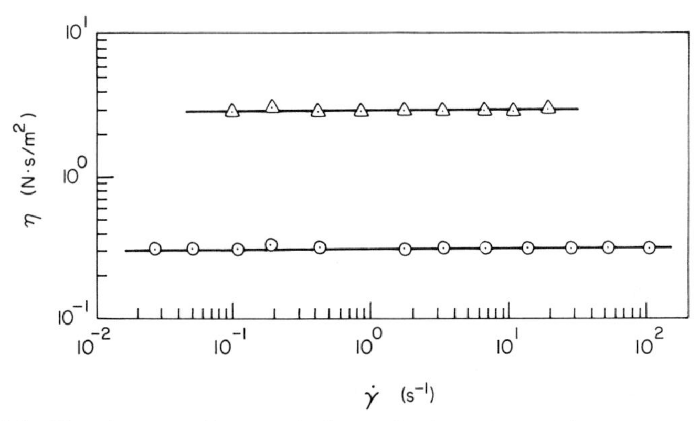

Fig. 2.1 Viscosity versus shear rate for low molecular weight polybutenes (Amoco Chemicals Co., Indopols) of different grades at 25°C: (△) Indopol H25; (⊙) Indopol L50.

A fluid is called *non-Newtonian* and *inelastic* (or *viscoinelastic*) when it gives viscosities, which vary with shear rate, and yet satisfies Eq. (2.4).

A fluid is called *non-Newtonian* and viscoelastic when it satisfies Eq. (2.2), that is, all material functions, $\eta(\dot{\gamma})$, $\Psi_1(\dot{\gamma})$, and $\Psi_2(\dot{\gamma})$, vary with shear rate $\dot{\gamma}$. Plots of $\eta(\dot{\gamma})$ and $\tau_{11} - \tau_{22}$ versus $\dot{\gamma}$ are given in Figs. 2.2, 2.3, and 2.4. Figure 2.2 is for aqueous solutions of polyacrylamide of various concentrations.

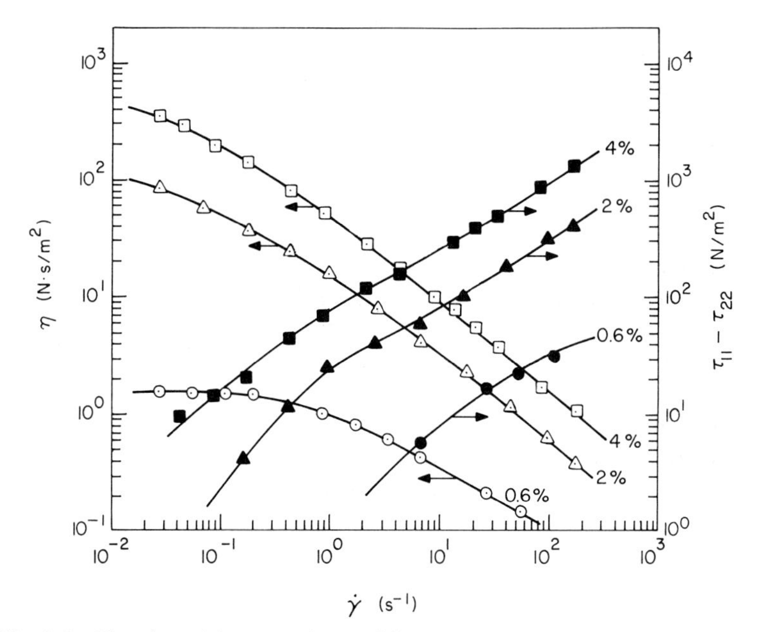

Fig. 2.2 Viscosity and first normal stress difference versus shear rate for aqueous solutions of polyacrylamide at 25°C (Dow Chemical Co., Separan AP 30).

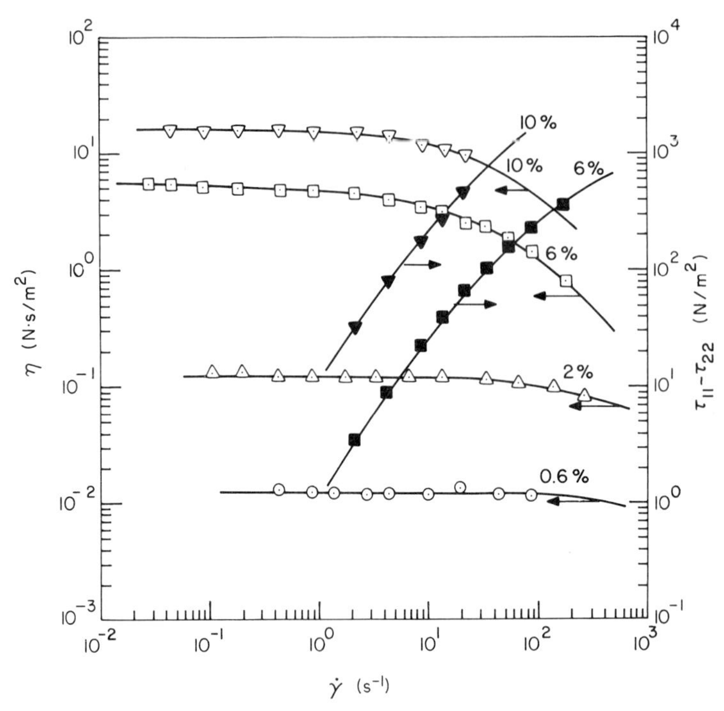

Fig. 2.3 Viscosity and first normal stress difference versus shear rate for solutions of polyisobutylene dissolved in decalin at 25°C.

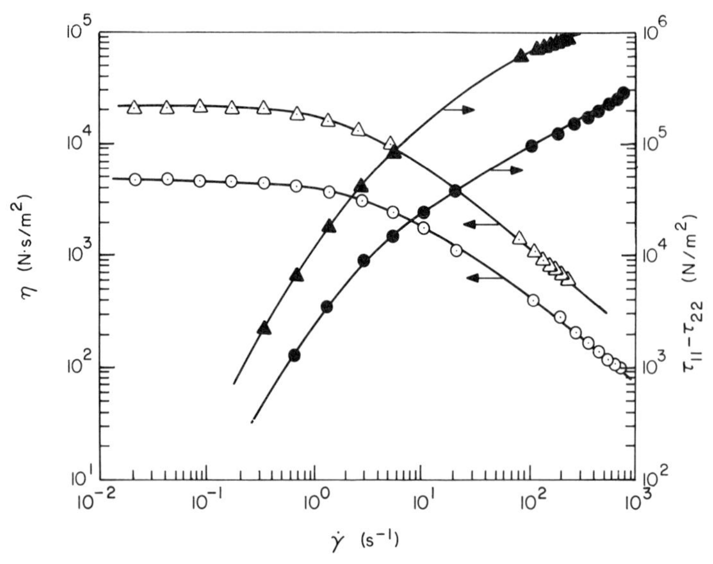

Fig. 2.4 Viscosity and first normal stress difference versus shear rate for polymer melts at 200°C: (⊙, ●) polystyrene; (△, ▲) polymethylmethacrylate.

Figure 2.3 is for polyisobutylene solutions in decalin of various concentrations. Figure 2.4 is for molten polymers. It is seen that the polymer solutions and melts satisfy Eq. (2.2), and therefore they are considered to be *viscoelastic*. The majority of polymeric solutions and melts of industrial importance exhibits *viscoelastic* behavior when subjected to a deformation of reasonably high shear rate.

Another important material function of rheological significance is the elongational viscosity. Whereas the shear viscosity is associated with the shear stress, the elongational viscosity is associated with tensile stress since elongational flows occur when fluid deformation is the result of a stretching motion. Elongational viscosity, unlike the shear viscosity, has meaning only when the type of elongational deformation is specified. There are three types of elongational viscosity: (a) uniaxial elongational viscosity, (b) biaxial elongational viscosity, and (c) pure shear elongational viscosity.

For uniaxial elongational flow, the elongational viscosity† η_E may be defined by the ratio of tensile (axial) stress S_{11} and rate of elongation (or elongation rate) $\dot{\gamma}_E$:

$$\eta_E = S_{11}/\dot{\gamma}_E \tag{2.5}$$

If the surfaces transverse to the direction of principal elongation (i.e., the direction of stretching) are unconstrained, we have

$$S_{22} = S_{33} = 0 \tag{2.6}$$

and, then, Eq. (2.5) may be rewritten

$$\eta_E = (\tau_{11} - \tau_{22})/\dot{\gamma}_E \tag{2.7}$$

Similarly, for uniform (equal) biaxial elongational flow, the corresponding elongational viscosity† η_B may be defined by

$$\eta_B = S_{11}/\dot{\gamma}_B \tag{2.8}$$

in which $\dot{\gamma}_B$ is elongation rate in uniform biaxial stretching. If the surface transverse to the two directions of principal elongation is unconstrained, we have

$$S_{33} = 0 \tag{2.9}$$

and, then, Eq. (2.8) may be rewritten

$$\eta_B = (\tau_{11} - \tau_{33})/\dot{\gamma}_B \tag{2.10}$$

† It is assumed that the elongation rate is constant, yielding *steady* elongational flow.

It can be shown that, for Newtonian fluids, we have

$$\eta_E = 3\eta_0 \tag{2.11}$$

and

$$\eta_B = 6\eta_0 \tag{2.12}$$

in which η_0 is the Newtonian shear viscosity.

In this chapter, we shall briefly review the fundamental concepts of rheology in order to help the reader to understand the main body of this monograph, which follows in Chapters 3–8. The various types of flow fields frequently encountered in polymer processing, and the material functions derived from various rheological equations of state are presented. Also, methods of measurement are discussed, with the aid of some representative experimentally determined values for the rheological properties of homogeneous polymeric liquids.

2.2 SOME REPRESENTATIVE FLOW FIELDS

2.2.1 The Steady Shearing Flow Field

Consider a flow geometry consisting of two infinitely long, parallel planes forming a narrow gap whose distance h is very small compared to the width w of the plane (i.e., $h \ll w$), as schematically shown in Fig. 2.5. The velocity field of the flow at steady state between the two planes is given by

$$v_z = f(y), \qquad v_x = v_y = 0 \tag{2.13}$$

Referring to Fig. 2.5, the lower plane stays stationary (i.e., $v_z = 0$) and the upper plane moves in the z direction at a constant speed $\bar{V}$, i.e., $v_z = \bar{V}$ at $y = h$. We then have a velocity gradient dv_z/dy which is constant, i.e.,

$$\dot{\gamma} = dv_z/dy = \text{const} \tag{2.14}$$

where $\dot{\gamma}$ is called the shear rate. Note in Fig. 2.5 that the channel gap between two parallel planes is filled with a fluid. The flow field defined by Eq. (2.13)

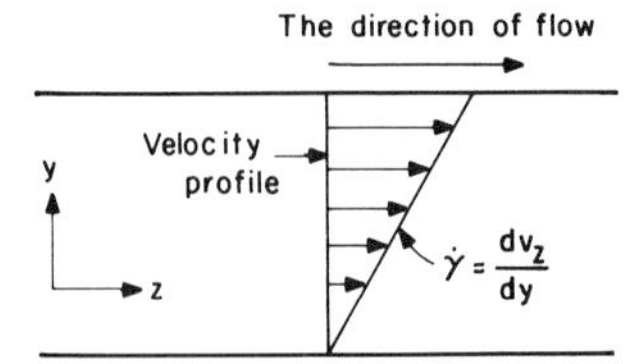

Fig. 2.5 Schematic of simple shearing flow.

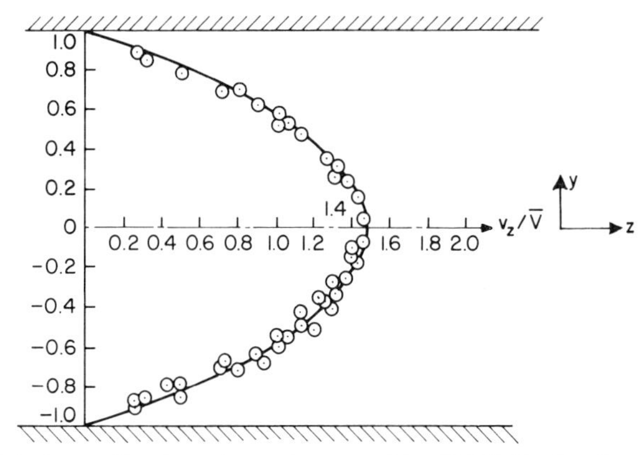

Fig. 2.6 Velocity profile of a molten polystyrene ($T = 225°C$) flowing through a slit die.

is called steady shearing flow, and the rate-of-deformation tensor **d** of shearing flow is given by

$$\|\mathbf{d}\| = \begin{Vmatrix} 0 & \dot{\gamma}/2 & 0 \\ \dot{\gamma}/2 & 0 & 0 \\ 0 & 0 & 0 \end{Vmatrix} \tag{2.15}$$

Consider now the situation where a fluid is forced to flow through two parallel planes, both planes being stationary, i.e., $v_z = 0$ at both $y = 0$ and $y = h$. We then have a velocity gradient dv_z/dy that depends on y, i.e.,

$$dv_z/dy = \dot{\gamma}(y) \tag{2.16}$$

It is seen that Eq. (2.14) is a special case of Eq. (2.16).

Figure 2.6 displays the velocity profile of a molten polystyrene, experimentally determined by means of streak photography, flowing through a thin slit die consisting of two parallel planes, whose length (L) and width (w) are large compared to the channel depth (h). It is seen in Fig. 2.6 that the velocity gradient, dv_z/dy, varies with y.

2.2.2 The Oscillatory Shearing Flow Field

In oscillatory flow, a sinusoidal strain is imposed on the fluid under test. If the viscoelastic behavior of the fluid is linear, the resulting stress will also vary sinusoidally, but will be out of phase with the strain, as schematically shown in Fig. 2.7. Since sinusoidal motion can be represented in the complex

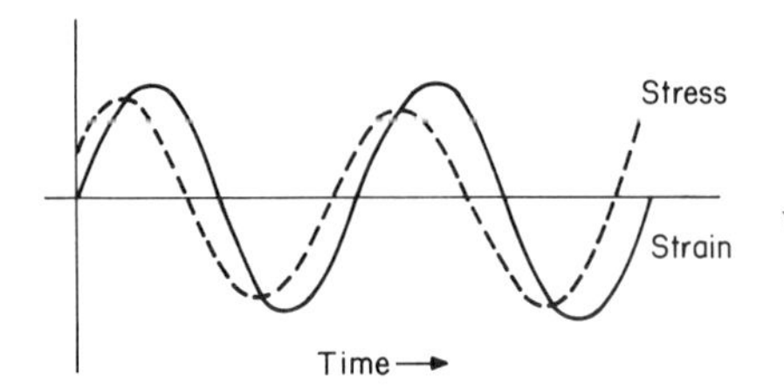

Fig. 2.7 Schematic of the sinusoidally varying stress and strain.

domain, the following complex quantities may be defined:

$$\gamma^*(i\omega) = \gamma_0 e^{i\omega t} = \gamma'(\omega) + i\gamma''(\omega) \tag{2.17}$$

$$\tau^*(i\omega) = \tau_0 e^{i(\omega t + \varphi)} = \tau'(\omega) + i\tau''(\omega) \tag{2.18}$$

where γ_0 and τ_0 are the amplitudes of the complex strain γ^* and the complex stress τ^*, respectively, and φ is the phase angle between them; the quantities with primes (γ' and τ') and double primes (γ'' and τ'') representing the real and imaginary parts of the respective complex quantities. In Eqs. (2.17) and (2.18), the response variable $\tau^*(i\omega)$ is assumed to have the same frequency ω as the input variable $\gamma^*(i\omega)$. This is true only when the system (i.e., the fluid in oscillatory motion) is a linear body.

Using the definitions given in Eqs. (2.17) and (2.18), one can derive material properties of rheological significance, i.e., the complex modulus $G^*(i\omega)$ and the complex viscosity $\eta^*(i\omega)$, which may be defined as

$$G^*(i\omega) = \frac{\tau^*(i\omega)}{\gamma^*(i\omega)} = G'(\omega) + iG''(\omega) \tag{2.19}$$

$$\eta^*(i\omega) = \frac{\tau^*(i\omega)}{\dot{\gamma}^*(i\omega)} = \frac{\tau^*(i\omega)}{i\omega\gamma^*(i\omega)} = \eta'(\omega) - i\eta''(\omega) \tag{2.20}$$

$\eta^*(i\omega)$ in Eq. (2.20) can be expressed in terms of $G^*(i\omega)$ as

$$\eta^*(i\omega) = \frac{G^*(i\omega)}{i\omega} = \frac{G''(\omega)}{\omega} - i\frac{G'(\omega)}{\omega} \tag{2.21}$$

From Eqs. (2.20) and (2.21) we have

$$\eta'(\omega) = G''(\omega)/\omega, \qquad \eta''(\omega) = G'(\omega)/\omega \tag{2.22}$$

Here $G'(\omega)$ is an in-phase elastic modulus associated with energy storage and release in the periodic deformation, and is called the storage modulus. $G''(\omega)$ is an out-of-phase elastic modulus associated with the dissipation of energy as heat, and is called the loss modulus. The real (i.e., in-phase) component of the complex viscosity $\eta'(\omega)$ is called the dynamic viscosity.

Figure 2.8 gives experimentally determined plots of $\eta'(\omega)$ and $G'(\omega)$ versus ω for two bulk polymers, polystyrene (PS) and polymethylmethacrylate (PMMA), in the molten state at 200°C.

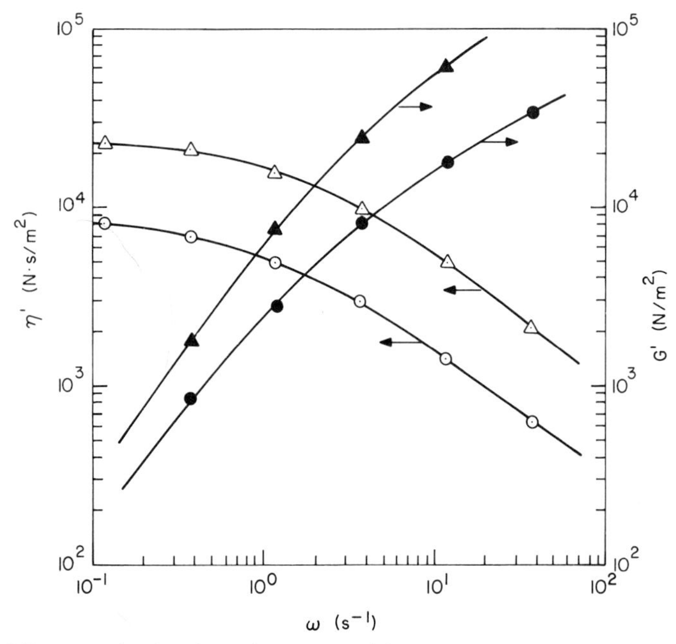

Fig. 2.8 Dynamic viscosity and storage modulus versus frequency for polymer melts at 200°C: (⊙, ●) polystyrene; (△, ▲) polymethylmethacrylate.

2.2.3 The Converging Flow Field

Converging flow is encountered very frequently in many polymer processing operations. For instance, when a fluid flows from a large reservoir into a small channel (or orifice), the fluid forms a converging streamline upstream in the reservoir, as shown in Fig. 2.9. Circulatory flow patterns are seen outside the natural converging stream entering into the small orifice. The converging angle that the melt streamline forms is known to be characteristic of the fluid [5, 6].

The velocity field of converging flow may be written [5]:

(i) Converging flow through two parallel planes†:

$$v_r(r,\theta) = f(\theta)/r, \qquad v_\theta = v_z = 0 \tag{2.23}$$

with cylindrical coordinates (r, θ, z), and

(ii) Converging flow through a conical duct†:

$$v_r(r,\theta) = g(\theta)/r^2, \qquad v_\theta = v_\varphi = 0 \tag{2.24}$$

† It is assumed that there is no secondary flow within the natural converging streamline.

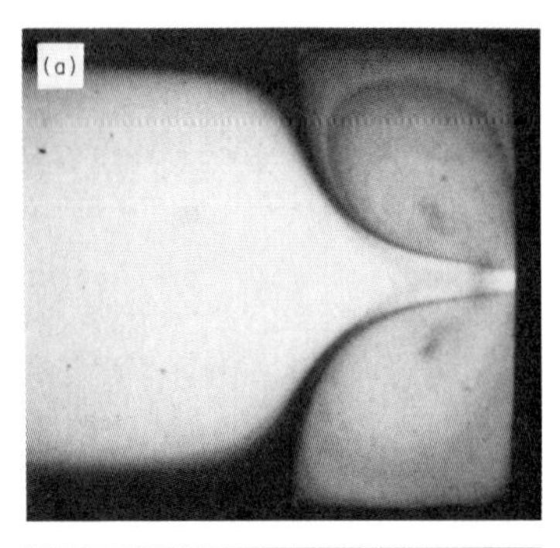

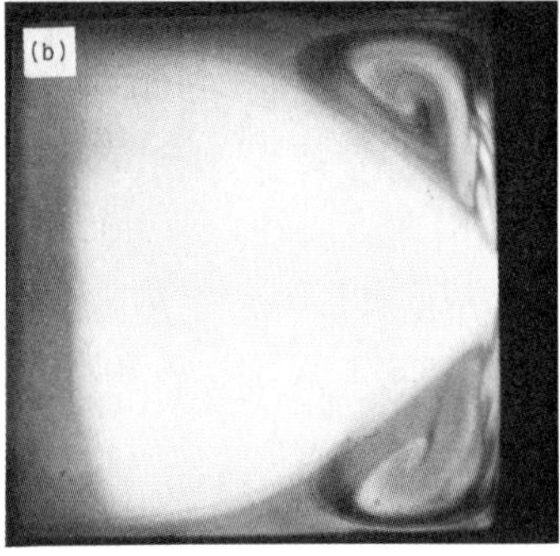

Fig. 2.9 Photographs displaying the flow patterns in the entrance region of a slit die: (a) low-density polyethylene at 200°C; (b) high-density polyethylene at 200°C.

with spherical coordinates (r, θ, φ). Note that $f(\theta)$ in Eq. (2.23) and $g(\theta)$ in Eq. (2.24) are as yet undetermined functions depending on θ only.

For the velocity field given by Eq. (2.23), the rate-of-deformation tensor **d** is

$$\|\mathbf{d}\| = \frac{1}{2r^2}\begin{Vmatrix} -2f & f' & 0 \\ f' & 2f & 0 \\ 0 & 0 & 0 \end{Vmatrix} \tag{2.25}$$

in which $f' = df(\theta)/d\theta$. For the velocity field given by Eq. (2.24) the rate-of-deformation tensor **d** is

$$\|\mathbf{d}\| = \frac{1}{2r^3}\begin{Vmatrix} -4g & g' & 0 \\ g' & 2g & 0 \\ 0 & 0 & 2g \end{Vmatrix} \tag{2.26}$$

in which $g' = dg(\theta)/d\theta$.

Figure 2.10 gives velocity profiles, determined experimentally by means of streak photography, of a molten polystyrene flowing into a thin slit die

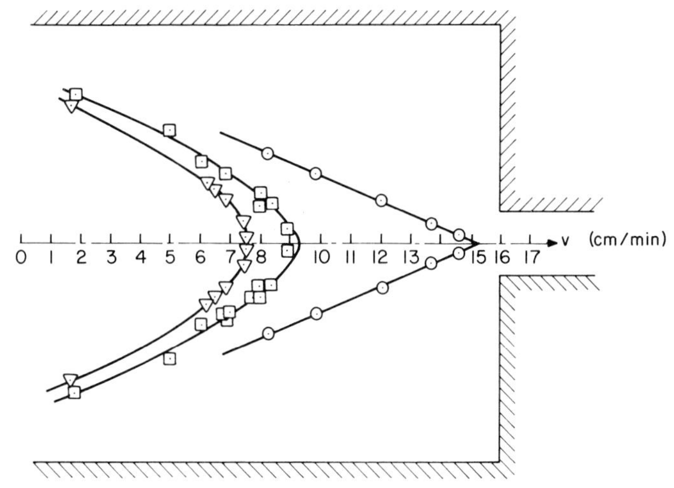

Fig. 2.10 Velocity profile of a molten polystyrene ($T = 200°$) in the entrance region of a slit die.

entrance from a large reservoir. It is seen that the velocity is greatest along the centerline and decreases as one moves from the centerline toward the wall of the reservoir section.

2.2.4 The Elongational Flow Field

Another flow field which is also of very practical importance is elongational (or extensional) flow, that may be found in such polymer processing operations as fiber spinning, flat-film extrusion, and film blowing. For uniaxial stretching (e.g., fiber spinning) the velocity field is given as [5]

$$v_z = f(z), \qquad \frac{\partial v_z}{\partial z} + \frac{\partial v_y}{\partial y} + \frac{\partial v_x}{\partial x} = 0 \tag{2.27}$$

Equation (2.27) assumes a flat velocity profile in the directions perpendicular to the flow direction. For such a flow field, the rate-of-strain tensor **d** is

$$\|\mathbf{d}\| = \begin{Vmatrix} \dot{\gamma}_{\mathrm{E}} & 0 & 0 \\ 0 & -\dot{\gamma}_{\mathrm{E}}/2 & 0 \\ 0 & 0 & -\dot{\gamma}_{\mathrm{E}}/2 \end{Vmatrix} \tag{2.28}$$

in which $\dot{\gamma}_{\mathrm{E}}$ is called the elongation rate, defined as

$$\dot{\gamma}_{\mathrm{E}} = dv_z/dz \tag{2.29}$$

Depending on the type of experiment that may be performed, $\dot{\gamma}_E$ may be either constant or varying with z, the direction of flow. When $\dot{\gamma}_E$ is constant (i.e., the axial velocity is proportional to z), such a flow field gives rise to *steady* uniaxial elongational flow.

Some investigators [7–9] have employed fiber spinning processes for studying uniaxial elongational flow. In such experiments, one measures either the fiber diameter or fiber velocity directly along the direction of stretching. Figure 2.11 gives the velocity profiles of a melt threadline along the direction of stretching, z. It is seen that the axial velocity gradient dv_z/dz varies with z, indicating that steady elongational flow does *not* prevail.

It is of interest to note that elongational flow may be realized, as a special case of converging flow, when one considers only the fluid element flowing along the centerline of the converging streamline. That is, along the centerline (i.e., $\theta = 0$), the wedge flow field defined by Eq. (2.25) reduces to a pure shear elongational flow field,

$$\|\mathbf{d}\| = \frac{1}{r^2}\left\|\begin{matrix} -f(0) & 0 & 0 \\ 0 & f(0) & 0 \\ 0 & 0 & 0 \end{matrix}\right\| \tag{2.30}$$

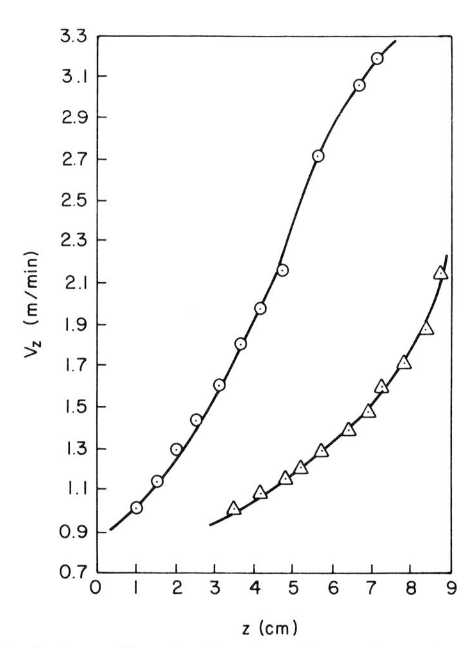

Fig. 2.11 Axial velocity profiles of molten threadlines under an isothermal melt spinning condition: (△) polypropylene ($T = 180°$C) at the draw-down ratio of 212; (⊙) low-density polyethylene ($T = 200°$C) at the draw-down ratio of 8.5.

and, along the centerline (i.e., $\theta = 0$), the conical flow field defined by Eq. (2.26) reduces to a uniaxial elongational flow field,

$$\|\mathbf{d}\| = \frac{1}{r^3}\begin{Vmatrix} -2g(0) & 0 & 0 \\ 0 & g(0) & 0 \\ 0 & 0 & g(0) \end{Vmatrix} \tag{2.31}$$

Clearly, converging flow should *not* be construed as elongational flow.

Figure 2.12 gives a schematic of the axial velocity profile along the centerline of a converging streamline in a conical die, on the basis of a tracer particle experiment [10]. As a matter of fact, such information may also be obtained from the velocity profiles given in Fig. 2.10. It can be said from Fig. 2.12 that a region exists upstream in the conical section where steady elongational flow prevails (that is, a constant axial velocity gradient prevails). However, as the fluid elements approach the die entrance, the centerline velocity increases very rapidly, giving rise to *varying* axial velocity gradients. It can be concluded therefore that *not* all uniaxial elongational flows satisfy the condition for *steady* elongational flow.

There are other types of elongational flow. One is biaxial elongational flow encountered in such polymer processing operations as blown-film extrusion and vacuum thermoforming. When the rates of elongation in two directions are *not* the same (i.e., unequal biaxial stretching) the velocity field is given by

$$v_z = f(z), \qquad v_y = g(y), \qquad \frac{\partial v_z}{\partial z} + \frac{\partial v_y}{\partial y} + \frac{\partial v_x}{\partial x} = 0 \tag{2.32}$$

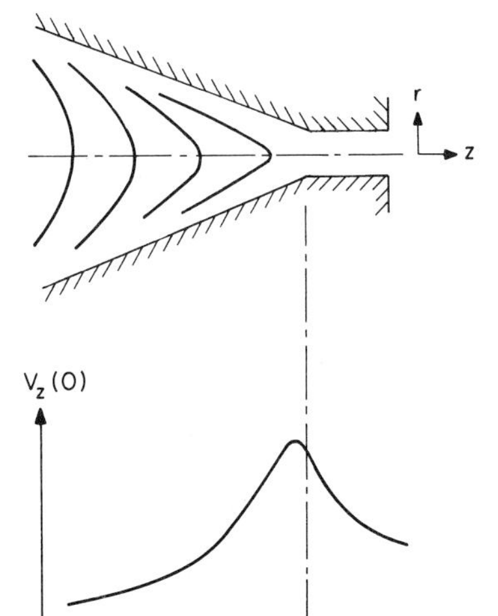

Fig. 2.12 Schematic of the axial velocity profile along the centerline of the converging streamline in a conical die.

and the rate-of-strain tensor **d** is

$$\|\mathbf{d}\| = \begin{Vmatrix} \dot{\gamma}_{B1} & 0 & 0 \\ 0 & \dot{\gamma}_{B2} & 0 \\ 0 & 0 & -(\dot{\gamma}_{B1} + \dot{\gamma}_{B2}) \end{Vmatrix} \tag{2.33}$$

where

$$\dot{\gamma}_{B1} = dv_z/dz, \qquad \dot{\gamma}_{B2} = dv_y/dy, \qquad \dot{\gamma}_{B1} \neq \dot{\gamma}_{B2} \tag{2.34}$$

For equal biaxial elongational flow, the rate-of-strain tensor **d** is

$$\|\mathbf{d}\| = \begin{Vmatrix} \dot{\gamma}_{B} & 0 & 0 \\ 0 & \dot{\gamma}_{B} & 0 \\ 0 & 0 & -2\dot{\gamma}_{B} \end{Vmatrix} \tag{2.35}$$

where

$$\dot{\gamma}_{B} = dv_z/dz = dv_y/dy \tag{2.36}$$

When the elongation rate $\dot{\gamma}_B$ is constant (i.e., independent of the either directions, z and y), such an elongational flow is called *steady*, equal biaxial elongational flow.

2.3 VISCOELASTICITY THEORIES FOR POLYMERIC LIQUIDS

In the past, much effort has been put into developing rheological models in order to predict the rheological properties of viscoelastic fluids. There are two approaches to describing the rheological behavior of a viscoelastic polymeric fluid. One approach is to view the material as a continuum, and then to describe the response of this continuum to stress or strain by a system of mathematical statements having their origin in the theories of continuum mechanics. Another approach is to describe the rheological behavior of the material from molecular structural considerations. In either approach, one basically has to establish a relationship (or relationships), based on either a rigorous mathematical/physical theory or empiricism, or both, which describes the deformation of a fluid in terms of the deformation history or the rate-of-deformation tensor, and the stress tensor. For the development of rheological equations of state for viscoelastic fluids, the reader is referred to standard textbooks [11–14].

2.3.1 Continuum Theories of Viscoelastic Fluids

In the continuum approach, emphasis is put on formulating the relationship between the components of stress and the components of deformation (or the rate of deformation), which should then properly describe the response of the fluid to some specific deformation. The constants involved in a specific model presumably represent the characteristics of the fluid.

Basically, two types of rheological equations of state have been developed, namely, the differential type and the integral type. The differential type is of the form which contains a derivative (or derivatives) of either the stress tensor or the rate-of-strain tensor or both, and the integral type is of the form in which the stress is represented by an integral over the deformation (or strain) history. We shall review below very briefly various forms of rheological model because we will be referring to some of these models in the remaining part of this monograph.

Consider the simple instance in which a spring is attached to a dashpot, as schematically shown in Fig. 2.13. In the spring–dashpot mechanical model, the spring exhibits a purely elastic effect (i.e., like a Hookean solid) and the dashpot exhibits a purely viscous effect (i.e., like a Newtonian fluid). This mechanical model is the basis on which the so-called classical Maxwell model was developed, described in three-dimensional form by

$$\tau + \lambda_1 \, \partial\tau/\partial t = 2\eta_0 \mathbf{d} \tag{2.37}$$

in which τ is the stress tensor, $\mathbf{d}$ is the rate-of-deformation tensor, and $\lambda_1 = \eta_0/G$ is a time constant, where η_0 is the proportionality constant (i.e., fluid viscosity) associated with the dashpot and G is the proportionality constant (i.e., elastic modulus) associated with the spring.

Although the classical Maxwell model is capable of describing many well-known viscoelastic phenomena, such as stress relaxation following a sudden release of imposed stress, it has serious limitations in describing nonlinear viscoelastic phenomena (i.e., shear-dependent viscosity and normal stress effects), in which the rate of deformation is not necessarily very small. It should be pointed out that the use of Eq. (2.37) is limited to infinitesimally small deformations.

In the general case contemplated, the rheological equations of state defining the properties of an element of a viscoelastic material at any instant may involve all the kinematic and dynamic quantities which define the states of the same element during its previous history. Therefore, the form of the

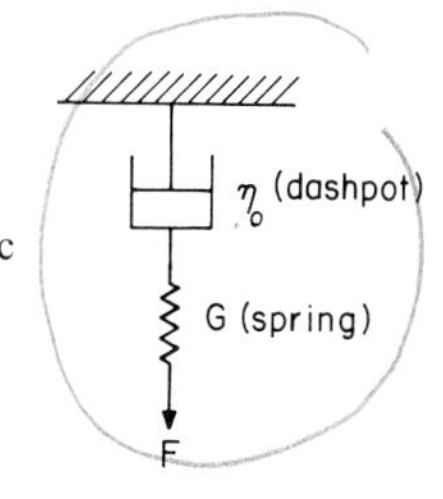

Fig. 2.13 A spring–dashpot mechanical model of a viscoelastic model.

completely general rheological equations must be restricted by the requirement that the equations describe rheological properties independent of the frame of reference. For the purpose of discussing the flow of material in bulk, it is necessary to express the rheological equations of state finally in terms of the same kinematic and dynamic variables as are used to express the equations of motion and continuity and the boundary conditions, so that all the associated equations can be solved simultaneously.

Therefore, in order to overcome the shortcomings of the classical Maxwell model, Eq. (2.37), attempts were made to introduce different types of derivative of a strain (or a stress) with respect to time. They are the convected derivative due to Oldroyd [15] and the Jaumann derivative [16–19]. These are "time" derivatives that transform a tensor from convected or rotational to fixed coordinates.

For a second-order tensor **a**, the convected derivative, $\mathfrak{d}/\mathfrak{d}t$, is defined as

$$\mathfrak{d}a_{ij}/\mathfrak{d}t = (\partial a_{ij}/\partial t) + v^k(\partial a_{ij}/\partial x^k) + (\partial v^k/\partial x^i)a_{kj} + (\partial v^k/\partial x^j)a_{ik} \tag{2.38}$$

for a covariant component of **a**, and

$$\mathfrak{d}a^{ij}/\mathfrak{d}t = (\partial a^{ij}/\partial t) + v^k(\partial a^{ij}/\partial x^k) - (\partial v^i/\partial x^k)a^{kj} - (\partial v^j/\partial x^k)a^{ik} \tag{2.39}$$

for a contravariant component of **a**. The physical interpretation of the convected derivative, Eq. (2.38) or (2.39), may be given as follows. The first two terms represent the derivative of tensor **a** with time, with the fixed coordinate held constant (i.e., the time rate of change as seen by an observer in a fixed coordinate system). The third and fourth terms represent the deformation and rotational motion of a material element referred to a fixed coordinate system.

For a second-order tensor **a**, the Jaumann derivative, $\mathscr{D}/\mathscr{D}t$, is defined as

$$\mathscr{D}a_{ij}/\mathscr{D}t = (\partial a_{ij}/\partial t) + v^k(\partial a_{ij}/\partial x^k) - \omega_{ik}a_{jk} - \omega_{jk}a_{ik} \tag{2.40}$$

in which $\boldsymbol{\omega}$ is the vorticity tensor. The physical interpretation of Eq. (2.40) may be as follows. The first two terms represent the material derivative of **a**, similar to the situation with the convected derivative [see Eq. (2.38)], and the third and fourth terms represent the rotational motion of a material element referred to a fixed coordinate system.

Using the derivatives introduced above, a number of rheological models have been proposed in the past. For instance, Eq. (2.37) is modified as:

(i) A generalized Maxwell model†:

$$\boldsymbol{\tau} + \lambda_1 \, \mathfrak{d}\boldsymbol{\tau}/\mathfrak{d}t = 2\eta_0 \mathbf{d} \tag{2.41}$$

† The use of the contravariant components of $\boldsymbol{\tau}$ and **d** is suggested in order to predict the correct trend of the experimentally observed normal stress effects in steady shearing flow [19].

(ii) Zaremba–DeWitt model [16, 18]:

$$\boldsymbol{\tau} + \lambda_1\, \mathscr{D}\boldsymbol{\tau}/\mathscr{D}t = 2\eta_0 \mathbf{d} \tag{2.42}$$

in which η_0 and λ_1 represent material constants characteristic of a fluid.

There are other forms of differential-type models suggested in the literature which contain nonlinear terms. Oldroyd [19] suggested the following form of rheological model:

$$\begin{aligned} \boldsymbol{\tau} + \lambda_1\, \mathscr{D}\boldsymbol{\tau}/\mathscr{D}t - \mu_1(\boldsymbol{\tau}\mathbf{d} + \mathbf{d}\boldsymbol{\tau}) + \mu_0[\operatorname{tr}\boldsymbol{\tau}]\mathbf{d} + \nu_1[\operatorname{tr}\boldsymbol{\tau}\mathbf{d}]\mathbf{I} \\ = 2\eta_0(\mathbf{d} + \lambda_2\, \mathscr{D}\mathbf{d}/\mathscr{D}t - 2\mu_2\mathbf{d}^2 + \nu_2[\operatorname{tr}\mathbf{d}^2]\mathbf{I}) \end{aligned} \tag{2.43}$$

This model contains eight material constants whose values cannot be determined uniquely from simple experiments. Hence, the practical use of Eq. (2.43) is very limited. Williams and Bird [20], as a special case of Eq. (2.43), suggested the following form of Oldroyd three-constant model:

$$\begin{aligned} \boldsymbol{\tau} + \lambda_1(\mathscr{D}\boldsymbol{\tau}/\mathscr{D}t - \boldsymbol{\tau}\mathbf{d} - \mathbf{d}\boldsymbol{\tau} + \tfrac{2}{3}[\operatorname{tr}\boldsymbol{\tau}\mathbf{d}]\mathbf{I}) \\ = 2\eta_0[\mathbf{d} + \lambda_2(\mathscr{D}\mathbf{d}/\mathscr{D}t - 2\mathbf{d}^2 + \tfrac{2}{3}[\operatorname{tr}\mathbf{d}^2]\mathbf{I})] \end{aligned} \tag{2.44}$$

which contains only three material constants, η_0, λ_1 and λ_2.

In an attempt to generalize the classical Maxwell model, Eq. (2.37), from the point of view of molecular consideration, Spriggs [21] postulated that a polymeric fluid consists of a very large number of sets of "spring and dashpot," and proposed the following form of rheological model:

$$\boldsymbol{\tau}_p + \lambda_p \mathscr{F}_\varepsilon \boldsymbol{\tau}_p = 2\eta_p \mathbf{d} \tag{2.45}$$

where $\mathscr{F}_\varepsilon$ is a nonlinear differential operator defined by

$$\mathscr{F}_\varepsilon \boldsymbol{\tau}_p = \mathscr{D}\boldsymbol{\tau}_p/\mathscr{D}t - (1+\varepsilon)[\boldsymbol{\tau}_p\mathbf{d} + \mathbf{d}\boldsymbol{\tau}_p - \tfrac{2}{3}[\operatorname{tr}\boldsymbol{\tau}_p\mathbf{d}]\mathbf{I}] \tag{2.46}$$

and $\boldsymbol{\tau}_p$ is the stress tensor of the pth unit (or pth mode of action). Therefore the stress of the material is defined as

$$\boldsymbol{\tau} = \sum_{p=1}^{\infty} \boldsymbol{\tau}_p \tag{2.47}$$

λ_p and η_p are material constants of the corresponding modes ($p = 1, \ldots, \infty$), defined, respectively by,

$$\lambda_p = \lambda/p^\alpha \tag{2.48}$$

and

$$\eta_p = \eta_0\lambda_p \bigg/ \left(\sum_{p=1}^{\infty} \lambda_p\right) = \frac{\eta_0}{p^\alpha Z(\alpha)} \tag{2.49}$$

where

$$Z(\alpha) = \sum_{p=1}^{\infty} \frac{1}{p^{\alpha}} \tag{2.50}$$

It is seen above that the Spriggs model, Eq. (2.45), has four material constants: η_0, λ, α, and ε. Note in Eqs. (2.48) and (2.49) that, as p increases, the contribution of the corresponding material constants λ_p and η_p becomes less important. This is in line with the molecular theory, which will be discussed below.

A theory which takes into account the past history of the stress up to the time t is proposed by Green and Rivlin [22], Coleman and Noll [23], and White [24, 25]. The fluid described by the rheological equation of state based on this theory is called the "fluid with memory." According to this theory, we may express the stress tensor $\boldsymbol{\tau}$ as an isotropic hereditary functional $\mathscr{H}$ of the strain history $\mathbf{E}(s)$:

$$\boldsymbol{\tau} = \mathop{\mathscr{H}}_{s=0}^{\infty} \{\mathbf{E}(s)\} \tag{2.51}$$

where

$$\mathbf{E}(s) = \mathbf{C}_t^{-1}(s) - \mathbf{I} \tag{2.52}$$

Here functional $\mathscr{H}$ may be defined as the representation of a variable which is given by one set of functions when another set of functions is specified, $\mathbf{E}(s)$ is considered to be the history of the *relative* Finger deformation tensor $\mathbf{C}_t^{-1}(s)$, and $\mathbf{I}$ is the unit second-order tensor. Here s is the elapsed time defined as $s = t - t'$, so that small s denotes the recent past, and large s denotes the distant past.

Coleman and Noll [23, 26] expanded the deformation history functional $\mathscr{H}$ into an infinite series involving single, double, and higher integrals of the strain history $\mathbf{E}(s)$. The second-order approximation is represented by

$$\boldsymbol{\tau} = \eta_0 \mathbf{A}_{(1)} + \beta \mathbf{A}_{(1)}^2 + \nu \mathbf{A}_{(2)} \tag{2.53}$$

where η_0, β, and ν are material constants, and $\mathbf{A}_{(1)}$ and $\mathbf{A}_{(2)}$ are the Rivlin–Ericksen tensors defined by

$$A_{ij}^{(1)} = (\partial v_i/\partial x_j) + (\partial v_j/\partial x_i) \tag{2.54}$$

and

$$A_{ij}^{(n)} = \frac{\partial A_{ij}^{(n-1)}}{\partial t} + v_k \frac{\partial A_{ij}^{(n-1)}}{\partial x_k} + \frac{\partial v_k}{\partial x_i} A_{kj}^{(n-1)} + \frac{\partial v_k}{\partial x_j} A_{ik}^{(n-1)} \tag{2.55}$$

respectively. Note that the Coleman–Noll second-order model, Eq. (2.53), is

TABLE 2.1

Expressions for the Viscometric Material Functions for Some Differential and Rate-Type Rheological Models

Model	η	Ψ_1	Ψ_2
Upper convected Maxwell	η_0	$2\eta_0\lambda_1$	0
Zaremba–DeWitt [16, 18]	$\dfrac{\eta_0}{1+\lambda_1^2\dot{\gamma}^2}$	$\dfrac{2\eta_0\lambda_1}{1+\lambda_1^2\dot{\gamma}^2}$	$\dfrac{-\eta_0\lambda_1}{1+\lambda_1^2\dot{\gamma}^2}$
Oldroyd Three-constant [20]	$\dfrac{\eta_0(1+\frac{2}{3}\lambda_1\lambda_2\dot{\gamma}^2)}{1+\frac{2}{3}\lambda_1^2\dot{\gamma}^2}$	$\dfrac{2\eta_0(\lambda_1-\lambda_2)}{1+\frac{2}{3}\lambda_1^2\dot{\gamma}^2}$	0
Spriggs [21]	$\sum_{p=1}^{\infty}\dfrac{\eta_p}{1+c^2\lambda_p^2\dot{\gamma}^2}$	$\sum_{p=1}^{\infty}\dfrac{2\lambda_p\eta_p}{1+c^2\lambda_p^2\dot{\gamma}^2}$	$\varepsilon\sum_{p=1}^{\infty}\dfrac{\lambda_p\eta_p}{1+c^2\lambda_p^2\dot{\gamma}^2}$
	where $c^2=(2-2\varepsilon-\varepsilon^2)/3$; $\lambda_p=\lambda/p^\alpha$; $\eta_p=\eta_0\lambda_p\Big/\sum_{p=1}^{\infty}\lambda_p$		
Coleman–Noll Second-order [23]	η_0	-2ν	$\beta+2\nu$

valid for a "slow flow," because, in the expansion of Eq. (2.51), the strain history tensor $\mathbf{E}(s)$ is expressed in terms of the Rivlin–Ericksen tensor $\mathbf{A}_{(n)}$ which is valid for a "slow flow."

Table 2.1 gives expressions for the three viscometric material functions η, Ψ_1, and Ψ_2 defined in Eq. (2.2), and Table 2.2. gives expressions for the elongational viscosities, η_E and η_B, defined by Eqs. (2.7) and (2.10), respectively, for various differential- and rate-type rheological models.

One can find the origin of the integral representation of rheological equations of state in Boltzmann's superposition principle [27]. The general expression of the rheological equation of state of integral type has the form [28–30]:

$$\tau=\int_{-\infty}^{t} m[t-t',\Pi(t')][(1+\tfrac{1}{2}\varepsilon)\mathbf{C}_t^{-1}(x,t,t')-\tfrac{1}{2}\varepsilon\mathbf{C}_t(x,t,t')]\,dt' \quad (2.56)$$

in which $\mathbf{C}_t^{-1}(x,t,t')$ is the *relative* Finger deformation tensor, $\mathbf{C}_t(x,t,t')$ is the *relative* Cauchy deformation tensor, and $m[t-t',\Pi(t')]$ is the memory function which depends on the elapsed time $t-t'$, and on the second invariant $\Pi(t')$ of the rate-of-deformation tensor. ε in Eq. (2.56) is an arbitrary constant introduced in order to remove the Weissenberg assumption (i.e., $\tau_{22}-\tau_{33}=0$).

It should be noted that several researchers have suggested different forms of the memory function based on somewhat different viewpoints of the

TABLE 2.2

Expressions for the Uniaxial and Equal Biaxial Elongational Viscosities for Some Differential and Rate-Type Rheological Models

Model	η_E	η_B
Upper convected Maxwell	$\dfrac{3\eta_0}{(1 - 2\lambda_1\dot{\gamma}_E)(1 + \lambda_1\dot{\gamma}_E)}$	$\dfrac{6\eta_0}{(1 - 2\lambda_1\dot{\gamma}_B)(1 + 4\lambda_1\dot{\gamma}_B)}$
Zaremba–DeWitt [16, 18]	$3\eta_0$	$6\eta_0$
Oldroyd Three-constant [20]	$3\eta_0\left(\dfrac{1 - \lambda_2\dot{\gamma}_E}{1 - \lambda_1\dot{\gamma}_E}\right)$	$6\eta_0\left(\dfrac{1 + 2\lambda_2\dot{\gamma}_B}{1 + 2\lambda_1\dot{\gamma}_B}\right)$
Spriggs [21]	$\displaystyle\sum_{p=1}^{\infty} \frac{3\eta_p}{1 - (1 + \varepsilon)\lambda_p\dot{\gamma}_E}$	$\displaystyle\sum_{p=1}^{\infty} \frac{6\eta_p}{1 + 2(1 + \varepsilon)\lambda_p\dot{\gamma}_B}$
Coleman–Noll Second-order [23]	$3\eta_0\left[1 + \left(\dfrac{\beta + v}{\eta_0}\right)\dot{\gamma}_E\right]$	$6\eta_0\left[1 - 2\left(\dfrac{\beta + v}{\eta_0}\right)\dot{\gamma}_B\right]$

TABLE 2.3

Expressions for the Memory Function for Some Integral-Type Rheological Models

Model	Memory function, $m[t - t', \mathrm{II}(t')]$
Lodge [31]	$\displaystyle\sum_{p=1}^{\infty} (G_p/\lambda_p)\exp[-(t - t')/\lambda_p]$
Meister[a] [32]	$\displaystyle\sum_{p=1}^{\infty} (G_p/\lambda_p)\exp\left[-\int_{t^1}^{t}\left\{\frac{1 + c(\mathrm{II}(\xi))^{1/2}\lambda_p}{\lambda_p}\right\}d\xi\right]$
Bogue[b] [33]	$\displaystyle\sum_{p=1}^{\infty} (G_p/\lambda_p^{\mathrm{eff}})\exp[-(t - t')/\lambda_p^{\mathrm{eff}}]$ where $1/\lambda_p^{\mathrm{eff}} = 1/\lambda_p + a\int_0^{t-t'} \lvert\mathrm{II}(\xi)\rvert^{1/2}\,d\xi \Big/ \int_0^{t-t'} d\xi$
Bird–Carreau[c] [34]	$\displaystyle\sum_{p=1}^{\infty} \frac{\eta_p\exp[-(t - t')/\lambda_{2p}]}{\lambda_{2p}^2[1 + \frac{1}{2}\lambda_{1p}^2\mathrm{II}(t')]}$ where $\eta_p = \dfrac{\eta_0\lambda_{1p}}{\sum_{p=1}^{\infty}\lambda_{1p}}$; $\lambda_{1p} = \lambda_1\left(\dfrac{1 + n_1}{p + n_1}\right)^{\alpha_1}$; $\lambda_{2p} = \lambda_2\left(\dfrac{1 + n_2}{p + n_2}\right)^{\alpha_2}$

[a] $\boldsymbol{\tau} = \int_{-\infty}^{t} m[t - t', \mathrm{II}(t')][\mathbf{C}_t^{-1}(t') - \mathbf{I}]\,dt'$

[b] The second-order term in the original form of Bogue model is neglected for mathematical convenience, and hence it follows the same expression as the Meister model.

[c] $\boldsymbol{\tau} = \int_{-\infty}^{t} m[t - t', \mathrm{II}(t')][(1 + \frac{1}{2}\varepsilon)\mathbf{C}_t^{-1}(t') - \frac{1}{2}\varepsilon\mathbf{C}_t(t')]\,dt'$ where $\mathbf{C}_t^{-1}(t')$ and $\mathbf{C}_t(t')$ are the *relative* Finger and Cauchy deformation tensors, respectively.

deformation history of a material element [31–36]. Table 2.3 gives expressions of some memory functions suggested by several investigators. Table 2.4 gives expressions for the three viscometric material functions η, Ψ_1, and Ψ_2, and Table 2.5 gives the expressions for the elongational viscosities, η_E and η_B, for various integral-type rheological models.

Up to this point, various forms of viscoelastic rheological models based on the continuum theories have been presented. We shall now comment on both the applicability and the limitations of these models in understanding the rheological behavior of real fluids.

As may be seen in Figs. 2.2–2.4, one of the important flow properties of polymeric liquids of practical interest is the shear-dependent viscosity, that is, a fluid viscosity that decreases with shear rate above a certain value of shear rate. At very low shear rates, all fluids exhibit a constant value of viscosity, called the zero-shear viscosity. It is seen, from Tables 2.1 and 2.4 that some models (e.g., the convected Maxwell fluid, the Coleman–Noll second-order fluid, the Lodge model) do *not* predict the shear-dependent viscosity. Therefore these models are useful for qualitative descriptions of deformation because they predict the fluid viscosity at low shear rates only. Note further that these three models predict constant values for the normal stress functions, Ψ_1 and Ψ_2. Again, the experimental data indicate that, at large deformation rates, Ψ_1 and Ψ_2 vary with shear rate, as may be seen in Figs. 2.14 and 2.15.

TABLE 2.4

Expressions for the Viscometric Material Functions for Some Integral-Type Rheological Models

Model	η	Ψ_1	Ψ_2
Lodge [31]	μ_1 [a]	μ_2 [b]	0
Meister [32]	$\sum_{p=1}^{\infty} \frac{G_p\lambda_p}{(1 + c\lambda_p\dot{\gamma})^2}$	$\sum_{p=1}^{\infty} \frac{2G_p\lambda_p^2}{(1 + c\lambda_p\dot{\gamma})^3}$	0
Bogue [33]	$\sum_{p=1}^{\infty} \frac{G_p\lambda_p^{\text{eff}}}{(1 + a\lambda_p^{\text{eff}}\dot{\gamma})^2}$	$\sum_{p=1}^{\infty} \frac{2G_p(\lambda_p^{\text{eff}})^2}{(1 + a\lambda_p^{\text{eff}}\dot{\gamma})^3}$	0
Bird–Carreau [34]	$\sum_{p=1}^{\infty} \frac{\eta_p}{1 + \lambda_{1p}^2\dot{\gamma}^2}$	$\sum_{p=1}^{\infty} \frac{2\eta_p\lambda_{2p}}{1 + \lambda_{1p}^2\dot{\gamma}^2}$	$-\varepsilon \sum_{p=1}^{\infty} \frac{\eta_p\lambda_{2p}}{1 + \lambda_{1p}^2\dot{\gamma}^2}$

[a] $\mu_1 = \int_0^{\infty} \sum_{p=1}^{N} \frac{G_p}{\lambda_p} \exp\left(\frac{-s}{\lambda_p}\right) s\, ds = \text{const}$

[b] $\mu_2 = \int_0^{\infty} \sum_{p=1}^{N} \frac{G_p}{\lambda_p} \exp\left(\frac{-s}{\lambda_p}\right) s^2\, ds = \text{const}$

TABLE 2.5

Expressions for the Uniaxial and Equal Biaxial Elongational Viscosities for Some Integral-Type Rheological Models

Model	η_E	η_B
Lodge	$\sum_{p=1}^{N} \frac{3\eta_0}{(1-2\lambda_p\dot{\gamma}_E)(1+\lambda_p\dot{\gamma}_E)}$	$\sum_{p=1}^{N} \frac{6\eta_0}{(1-2\lambda_p\dot{\gamma}_B)(1+4\lambda_p\dot{\gamma}_B)}$
Meister[a]	$\sum_{p=1}^{\infty} \frac{3\eta_0}{[1+(c\sqrt{3}-2)\lambda_p\gamma_E][1+(c\sqrt{3}+1)\lambda_p\gamma_E]}$	$\sum_{p=1}^{\infty} \frac{6\eta_0}{[1+2(c\sqrt{3}-1)\lambda_p\dot{\gamma}_B][1+2(c\sqrt{3}+2)\lambda_p\gamma_B]}$
Bogue[a]	$\sum_{p=1}^{\infty} \frac{3\eta_0}{[1+(a\sqrt{3}-2)\lambda_p^{eff}\dot{\gamma}_E][1+(a\sqrt{3}+1)\lambda_p^{eff}\dot{\gamma}_E]}$	$\sum_{p=1}^{\infty} \frac{6\eta_0}{[1+2(a\sqrt{3}-1)\lambda_p^{eff}\dot{\gamma}_B][1+2(a\sqrt{3}+2)\lambda_p^{eff}\dot{\gamma}_B]}$
Bird–Carreau[a]	$\sum_{p=1}^{\infty} \frac{3\eta_p}{(1+\frac{3}{2}\lambda_{1p}^2\dot{\gamma}_E^2)(1-2\lambda_{2p}\dot{\gamma}_E)(1+\lambda_{2p}\dot{\gamma}_E)}$ for $\varepsilon = 0$	$\sum_{p=1}^{\infty} \frac{6\eta_p}{(1+6\lambda_{1p}^2\dot{\gamma}_B^2)(1-2\lambda_{2p}\dot{\gamma}_B)(1+4\lambda_{2p}\dot{\gamma}_B)}$ for $\varepsilon = 0$

[a] The second invariant is defined as : $|\text{II}(t')| = 3\dot{\gamma}_E^2$ for uniaxial elongational flow, and $|\text{II}(t')| = 12\dot{\gamma}_B^2$ for biaxial elongational flow.

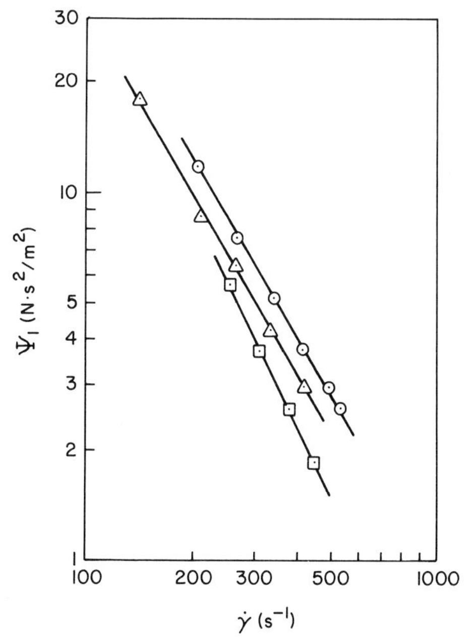

Fig. 2.14 First normal stress function versus shear rate for polymer melts at 200° C: (⊙) low-density polyethylene: (△) high-density polyethylene; (⊡) polystyrene.

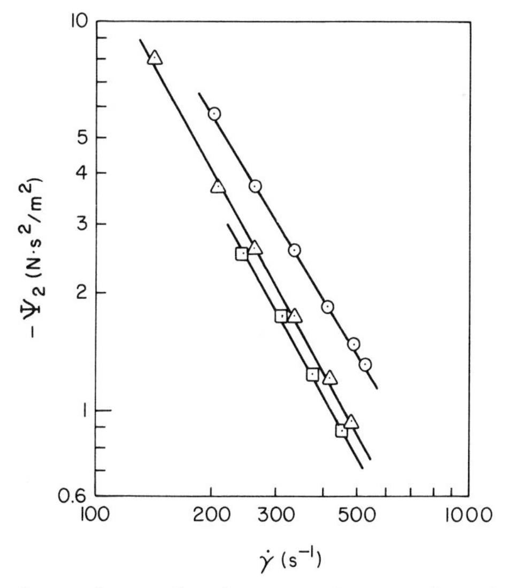

Fig. 2.15 Second normal stress function versus shear rate for polymer melts at 200°C: (⊙) low-density polyethylene; (△) high-density polyethylene; (⊡) polystyrene.

Other models given in Tables 2.1 and 2.4 predict both the shear-dependent viscosity and normal stress behavior. It is fair to say that some of these models can fit steady shearing flow data reasonably well. However, the predictability of nonviscometric flow behavior (e.g., elongational flow) varies from one model to another.

Many flow problems of industrial importance require solutions of the equations of motion in order to either predict the flow patterns or to determine an optimal die geometry. For instance, for flows in the entrance region of a conduit or in a conduit in which there is a sudden change in cross section, a fluid element is subjected to rapidly changing deformations on a time scale (characteristic process time or residence time), which may be small compared to the characteristic time of the fluid. In situations like this, rheological models that portray the deformation history of the fluid are desirable and therefore a rheological model of the integral type (e.g., the Bogue model, the Meister model, the Bird—Carreau model) is preferred. In practice, however, the use of nonlinear rheological models of the integral type will be a formidable task when one has to simultaneously solve the equations of continuity and momentum. It is then clear that, in many practical situations, a compromise is needed between the sophistication of rheological models and the enormous effort required for obtaining solutions of the governing system equations.

As may be seen in Figs. 2.2–2.4, the viscosity of many polymeric solutions and melts may be expressed, over a limited but considerable range of shear rates, by a power-law expression, i.e.,

$$\eta(\dot{\gamma}) = K[\Pi]^{(n-1)/2} \tag{2.57}$$

where Π, the second invariant of the rate-of-deformation tensor, becomes $-\dot{\gamma}^2/4$ for steady shearing flow.

It is of interest to note at this juncture that White and Metzner [37] have suggested that the zero-shear viscosity η_0 in the generalized Maxwell model, Eq. (2.41), be replaced by a variable viscosity function $\eta(\dot{\gamma})$, defined by Eq. (2.57). Such a rheological model, though semiempirical, predicts the shear-dependent viscosity in steady shearing flow.

In later chapters, we shall show how complicated some of the real problems of practical interest are. We shall also demonstrate some successes that we have experienced in dealing with the complicated flow problems encountered in multiphase flow of rheologically complex fluids, by using simplified, though imperfect, rheological models.

2.3.2 Molecular Theories of Viscoelastic Fluids

Today, it is a well-known fact that the rheological properties of polymeric liquids are influenced by the molecular structure, molecular weight, and its distribution. Therefore, a better understanding of the relationship between molecular parameters and rheological properties is very important from the points of view of both polymer preparation and polymer processing. We shall review below very briefly how the molecular parameters influence both the viscous and the elastic properties of polymeric liquids, in the light of the various molecular theories in the literature.

Assuming that a large polymer molecule may be considered to consist of a chain of $N + 1$ identical beads joined together by N completely flexible spring segments, as schematically shown in Fig. 2.16. Rouse [38] first developed a molecular viscoelastic theory for dilute polymer solutions, which relates rheological properties, dynamic viscosity $\eta'(\omega)$, and storage modulus $G'(\omega)$, with molecular parameters in oscillatory shearing flow as follows:

$$\eta'(\omega) = n_s + ckT \sum_{p=1}^{N} \frac{\lambda_p}{1 + \omega^2 \lambda_p^2} \tag{2.58}$$

$$G'(\omega) = ckT \sum_{p=1}^{N} \frac{\omega^2 \lambda_p^2}{1 + \omega^2 \lambda_p^2} \tag{2.59}$$

where ω is the frequency of oscillation, η_s the solvent viscosity, c the concentration, k the Boltzmann constant, T absolute temperature, and λ_p is the relaxation time of the pth segment defined by

$$\lambda_p = [6(\eta_0 - \eta_s)M]/\pi^2 p^2 cRT, \qquad p = 1, 2, \ldots, N \tag{2.60}$$

in which η_0 is the zero-shear viscosity, M is the molecular weight, and R is the gas constant.

It should be pointed out that the Rouse theory assumes vanishing hydrodynamic interaction (i.e., free-draining chains). Zimm [39] has subsequently taken into account the hydrodynamic interaction, and has derived theoretical expressions for $\eta'(\omega)$ and $G'(\omega)$ similar to those of the Rouse theory, but with a relaxation time λ_p slightly different from that of the Rouse theory. Figure 2.17 shows the general behavior of $\eta'(\omega)$ and $G'(\omega)$ predicted by both theories. If the Rouse and Zimm theories were to be used for bulk polymers (although this would not be correct from the rigorously theoretical point of view), one could set η_s equal to zero and replace c with density ρ.

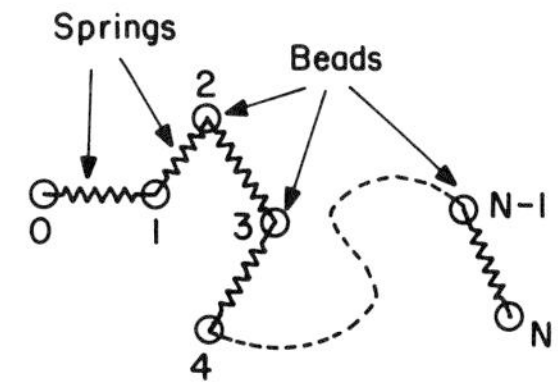

Fig. 2.16 The bead and spring model for linear polymer molecules.

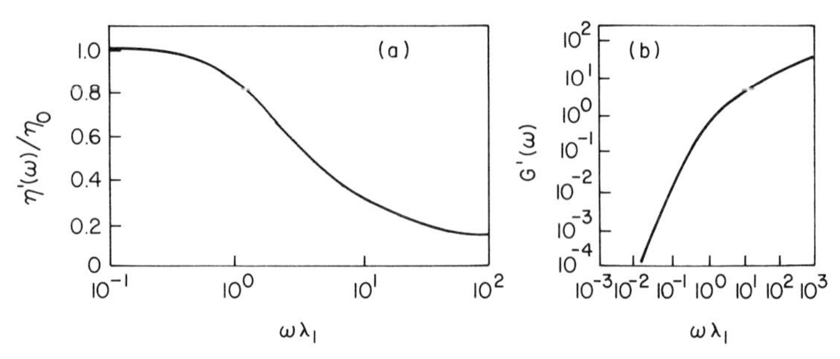

Fig. 2.17 Typical shape of the dynamic viscosity and storage modulus curves of polymeric liquids.

Thus, for polymer melts we have

$$\eta'(\omega) = \rho kT \sum_{p=1}^{N} \frac{\lambda_p}{1 + \omega^2 \lambda_p^2} \tag{2.61}$$

$$G'(\omega) = \rho kT \sum_{p=1}^{N} \frac{\omega^2 \lambda_p^2}{1 + \omega^2 \lambda_p^2} \tag{2.62}$$

where

$$\lambda_p = 6\eta_0 M / \pi^2 p^2 \rho RT, \qquad p = 1, 2, \ldots, N \tag{2.63}$$

It is worth mentioning that although both the Rouse and Zimm theories are generally believed to be strictly applicable only to infinitely dilute polymer solutions, most polymeric liquids, irrespective of whether they are solutions or melts, show curves very similar in shape to those shown in Fig. 2.8 (see Fig. 2.17).

In steady shearing flow, the Rouse and the Zimm theories predict [40]

$$\eta - \eta_s = ckT \sum_{p=1}^{N} \lambda_p \tag{2.64}$$

$$\tau_{11} - \tau_{22} = \frac{2cRT}{M} \dot{\gamma}^2 \sum_{p=1}^{N} \lambda_p^2 \tag{2.65}$$

It is seen that η is constant (i.e., the theory does not predict the shear-dependent viscosity) and that $\tau_{11} - \tau_{22}$ is proportional to the square of shear rate (i.e., Ψ_1 is constant). Note, however, that the molecular theories of Rouse and Zimm provide information on the dependencies of $\tau_{11} - \tau_{22}$ on the molecular weight, the concentration c, and the temperature T of a fluid, whereas the continuum viscoelastic theory does not.

On the basis of the same molecular consideration as that of Rouse and Zimm, Bueche [41] derived a theoretical expression for the shear-dependent viscosity:

$$\frac{\eta - \eta_s}{\eta_0 - \eta_s} = 1 - \frac{6}{\pi^2} \sum_{p=1}^{N} \frac{\dot{\gamma}^2 \lambda_1^2}{p^2(p^4 + \dot{\gamma}^2 \lambda_1^2)} \left(2 - \frac{\dot{\gamma}^2 \lambda_1^2}{p^4 + \dot{\gamma}^2 \lambda_1^2}\right) \tag{2.66}$$

where η_0 is the zero-shear viscosity, η_s is the viscosity of the solvent, $\dot{\gamma}$ is the shear rate, and λ_1 is Bueche's time constant given by

$$\lambda_1 = [12(\eta_0 - \eta_s)M]/\pi^2 cRT \tag{2.67}$$

Graessley [42] also derived a theoretical expression for the shear-dependent viscosity:

$$\frac{\eta}{\eta_0} = \frac{2}{\pi} \left\{ \cot^{-1} \theta + \frac{\theta(1 - \theta^2)}{(1 + \theta^2)^2} \right\} \tag{2.68}$$

where

$$\theta = (\lambda_0 \dot{\gamma}/2)(\eta/\eta_0) \tag{2.69}$$

in which λ_0 is the relaxation time at zero shear rate and is approximately equal to that of the Rouse theory. In developing his theory, Graessley adopted the view that for an entanglement to exist, two molecules must first be within a certain distance of each other, and must remain within this distance for a finite time, or else no entanglement occurs. The entanglement theory of Graessley was introduced for concentrated solutions and polymer melts.

Other molecular theories [43–52] have also been developed. Limitations of space and complexity of the mathematical expressions do not permit us to present the details of these molecular theories. The reader may consult the original papers [43–52].

In view of the fact that almost all polymers are polydispersed in the molecular weight, the Rouse and Zimm theories, which assume monodispersity of materials, need to be modified if the theories are to be useful in dealing with polydispersed materials. In the past, there have been some attempts [53, 54] made to relate the viscosity determined in steady shearing flow to the molecular weight distribution (MWD). It is predicted that the greater the molecular weight distribution of a polymer, the lower its shear viscosity at high shear rates.

Figure 2.18 shows MWD curves for two high-density polyethylenes (HDPE), determined by use of gel permeation chromatography (GPC). Figure 2.19 gives plots of shear viscosity η versus shear rate $\dot{\gamma}$ for the two HDPE resins. It is seen that the resin *A*, having a narrow MWD, is more

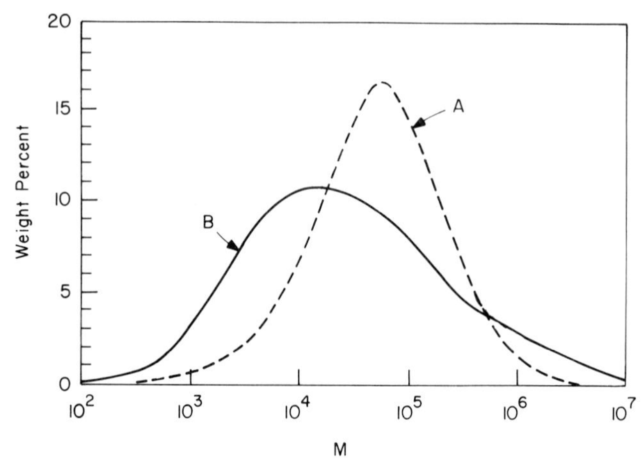

Fig. 2.18 Molecular weight distributions of two high-density polyethylenes: (a) polymer A has $\bar{M}_n = 1.51 \times 10^4$ and $\bar{M}_w = 1.51 \times 10^5$; (b) polymer B has $\bar{M}_n = 0.48 \times 10^4$ and $\bar{M}_w = 2.26 \times 10^5$.

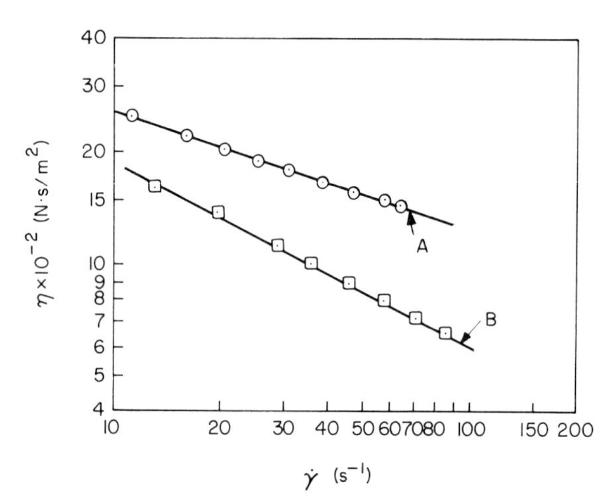

Fig. 2.19 Viscosity versus shear rate for the two high-density polyethylenes whose molecular characteristics are the same as in Fig. 2.18.

viscous than the resin B, having a broad MWD, which is consistent with the theoretical prediction [53].

Figure 2.20 gives plots of first normal stress difference $\tau_{11} - \tau_{22}$ versus shear rate $\dot{\gamma}$, and Fig. 2.21 gives plots of first normal stress difference $\tau_{11} - \tau_{22}$ versus shear stress τ_w for the two HDPE resins A and B. It is of interest

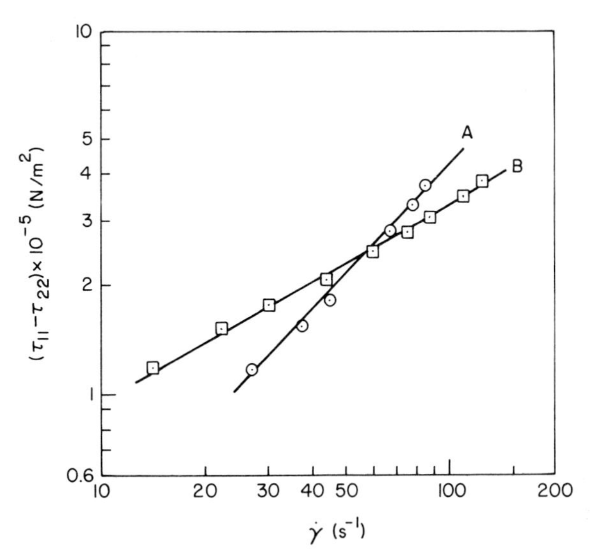

Fig. 2.20 First normal stress difference versus shear rate for the two high-density polyethylenes whose molecular characteristics are the same as in Fig. 2.18.

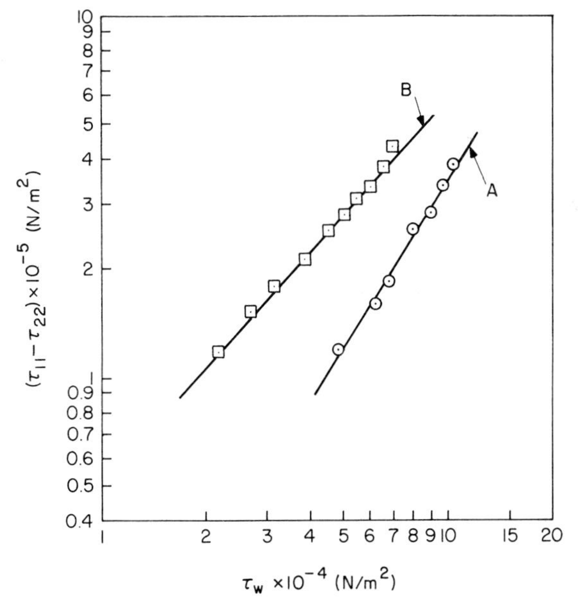

Fig. 2.21 First normal stress difference versus shear stress for the two high-density polyethylenes whose molecular characteristics are the same as in Fig. 2.18.

to note that the two different ways of plotting the $\tau_{11} - \tau_{22}$ data show different responses to the variation in MWD. We therefore have to resolve the question as to which of the two correlations we must use in determining the elasticity of one fluid relative to another.

For this, let us look for another type of correlation, which can provide a basis for molecular interpretation. Elastic recovery (recoil) has long been considered a useful parameter for determining the fluid elasticity. It is often referred to as a measure of stored energy, and is characterized by the steady-state shear compliance J_e defined as [55]

$$J_e = \int_0^\infty sG(s)\,ds \bigg/ \left[\int_0^\infty G(s)\,ds\right]^2 \tag{2.70}$$

where $G(s)$ is the relaxation modulus. However, the experimental determination of $G(s)$ is not straightforward. When information on $G(s)$ is not available, J_e may be represented by [55]

$$J_e = \text{recoil}/\tau_w = (\tau_{11} - \tau_{22})/2\tau_w^2 \tag{2.71}$$

According to Ferry *et al.* [56], J_e for a polydisperse polymer may be represented by

$$J_e = \frac{2}{5\rho RT}\,\frac{\bar{M}_z \bar{M}_{z+1}}{\bar{M}_w} \tag{2.72}$$

where $\bar{M}_z$ and $\bar{M}_{z+1}$ are the z and $z + 1$ average molecular weights. The rheological significance of Eq. (2.72) is that the steady-state shear compliance J_e should increase markedly with spread of MWD.

Figure 2.22 gives plots of steady-state shear compliance J_e versus shear rate $\dot{\gamma}$ for two HDPE resins. Note that J_e was calculated using Eq. (2.71) and the data given in Figs. 2.20 and 2.21. It is seen that J_e is greater for resin B (broad MWD) than for resin A (narrow MWD); behavior similar to that in the plots of $\tau_{11} - \tau_{22}$ versus τ_w given in Fig. 2.21. Therefore, it can be concluded from Figs. 2.21 and 2.22 that resin B, having a broad MWD, is more elastic than resin A, having a narrow MWD.

There is little doubt that polymers with different chemical structures would give different responses to deformation. Of particular interest is the influence of long chain branching (LCB) on the viscoelastic properties of polymers. In the past, on the basis of molecular consideration, several researchers [57–63] attempted to develop theories which would permit one to predict the viscoelastic properties of branched polymers in terms of the extent of long chain branching.

Figure 2.23 gives MWD curves for three low-density polyethylenes (LDPE) determined by use of gel permeation chromatography (GPC). Figure

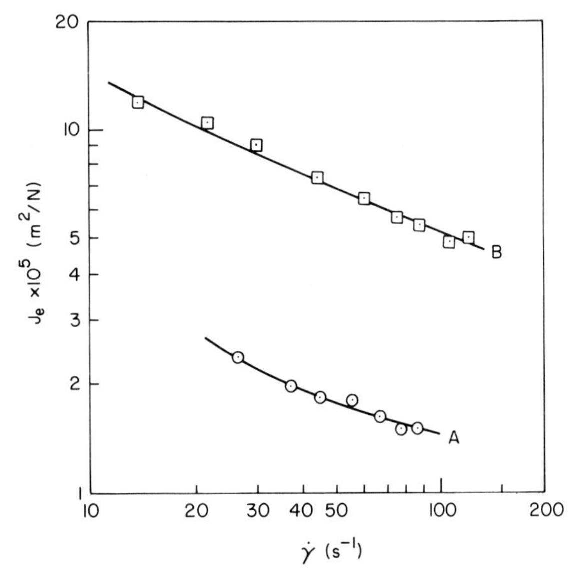

Fig. 2.22 Steady-state shear compliance versus shear rate for the two high-density polyethylenes whose molecular characteristic are the same as in Fig. 2.18.

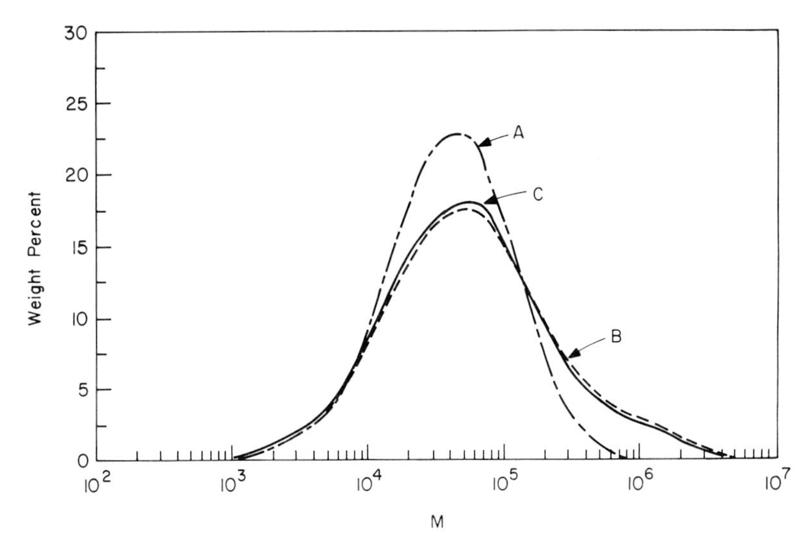

Fig. 2.23 Molecular weight distributions of three low-density polyethylenes: (a) Polymer A has $\bar{M}_n = 1.82 \times 10^4$, $\bar{M}_w = 0.56 \times 10^5$, long chain branching frequency (LCBF) = low; (b) Polymer B has $\bar{M}_n = 2.07 \times 10^4$, $\bar{M}_w = 1.59 \times 10^5$, LCBF = high; (c) Polymer C has $\bar{M}_n = 2.03 \times 10^4$, $\bar{M}_w = 1.59 \times 10^5$, LCBF = high. The LCBF between the polymers B and C is not very different.

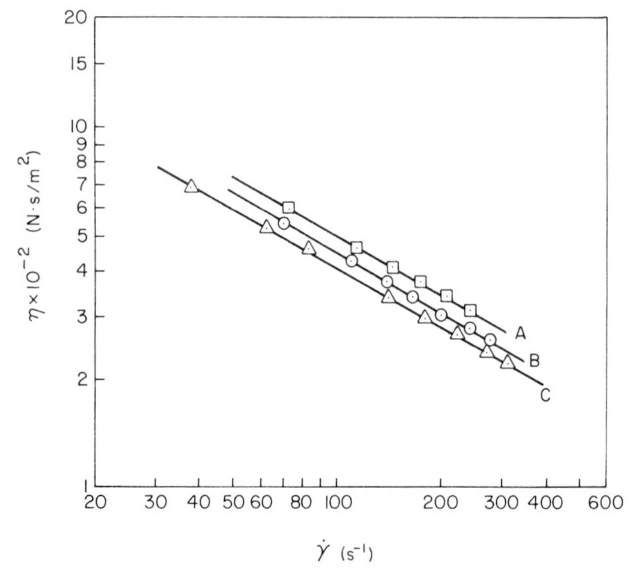

Fig. 2.24 Viscosity versus shear rate for the three low-density polyethylenes whose molecular weight distributions are the same as in Fig. 2.23.

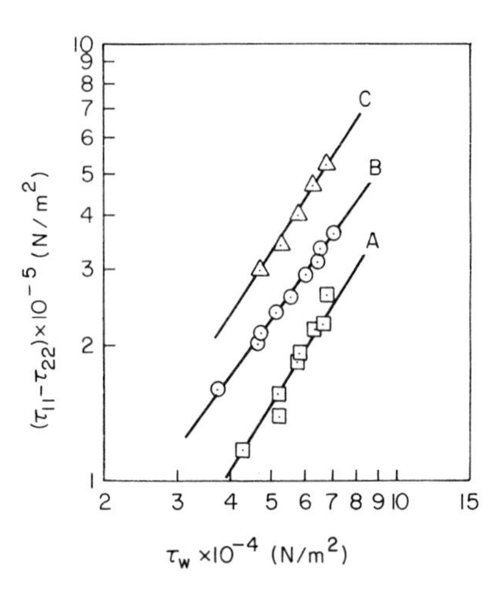

Fig. 2.25 First normal stress difference versus shear stress for the three low-density polyethylenes whose molecular weight distributions are the same as in Fig. 2.23.

2.24 gives plots of shear viscosity η versus shear rate $\dot{\gamma}$, Fig. 2.25 gives plots of first normal stress difference $\tau_{11} - \tau_{22}$ versus shear stress τ_w, and Fig. 2.26 gives plots of elastic compliance J_e versus shear rate $\dot{\gamma}$ for the three LDPE resins. It is seen that resin C, having the most long chain branching, is the least viscous and the most elastic of the three resins. In other words, the presence of long chain branching in a polymer tends to give rise to lower

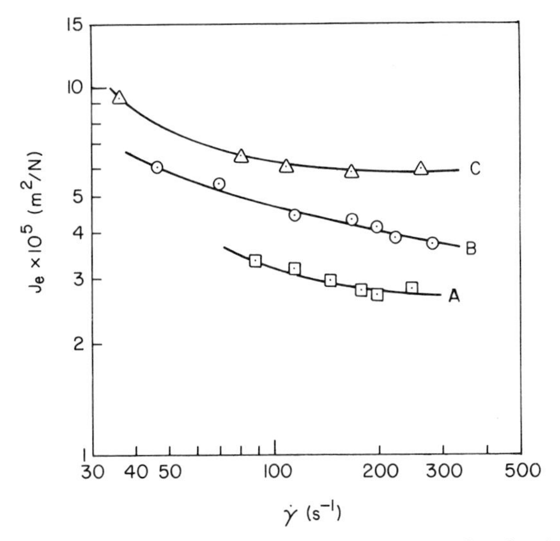

Fig. 2.26 Steady-state shear compliance versus shear rate for the three low-density polyethylenes whose molecular weight distributions are the same as in Fig. 2.23.

viscosity and higher elasticity when the polymer is under reasonably high deformation.

2.3.3 The Dimensionless Material Parameters for Viscoelastic Fluids

The use of the dimensionless material parameters in studying the dynamics of viscoelastic fluids was suggested by White [25] and White and Tokita [64, 65]. White [25] has introduced the Weissenberg number N_{We}

$$N_{\mathrm{We}} = \frac{\lambda_1 V}{L} = \frac{\text{elastic forces}}{\text{viscous forces}} \tag{2.73}$$

in which V and L are a characteristic velocity and a characteristic length, respectively, and λ_1 is a characteristic time for the fluid, and then used N_{We} to investigate the problems of melt fracture in melt extrusion and fiber spinning.

If one has available shear stress and first normal stress difference data as functions of shear rate, the stress ratio S_{R}, often referred to as the recoverable shear strain, may be defined as

$$S_{\mathrm{R}} = (\tau_{11} - \tau_{22})/2\tau_{\mathrm{w}} \tag{2.74}$$

According to Bogue and White [30], S_R plays the same role as N_{We}. S_R may be used to correlate the fluid elasticity of different viscoelastic fluids when plotted against shear stress. Such plots for polymer melts are given in Fig. 2.27. It is seen that the magnitude of S_R for the nylon 6 and PET is about 1 or 2 orders smaller than that for the typical thermoplastics, HDPE, PP, and PS. It can therefore be concluded that the nylon 6 and PET are far less elastic than the HDPE, PP, and PS.

The theory of nonlinear viscoelasticity derives relationships, which connect the normal stress coefficient Ψ_1 with the viscosity η through a relaxation time λ_1. These expressions have the form

$$\Psi_1 = 2\lambda_1\eta \tag{2.75}$$

According to Coleman and Noll [23], we have, for the second-order fluid

$$\lim_{\dot{\gamma}\to 0} (\Psi_1/2\eta) = \lambda_1 = \int_0^\infty sG(s)\,ds \Big/ \int_0^\infty G(s)\,ds \tag{2.76}$$

where $G(s)$ is the relaxation modulus of linear viscoelasticity. Using the material functions presented in Tables 2.1 and 2.4 for various rheological

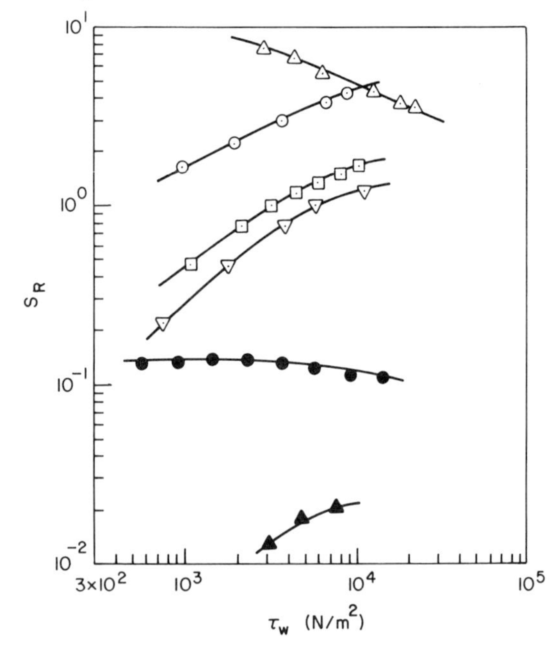

Fig. 2.27 Recoverable shear strain versus shear stress for polymer melts: (△) high-density polyethylene at 200°C; (⊙) low-density polyethylene ($T = 200$°C); (⊡) polypropylene ($T = 200$°C); (▽) polystyrene ($T = 200$°C); (●) nylon 6 ($T = 280$°C); (▲) polyethylene terephthalate ($T = 300$°C).

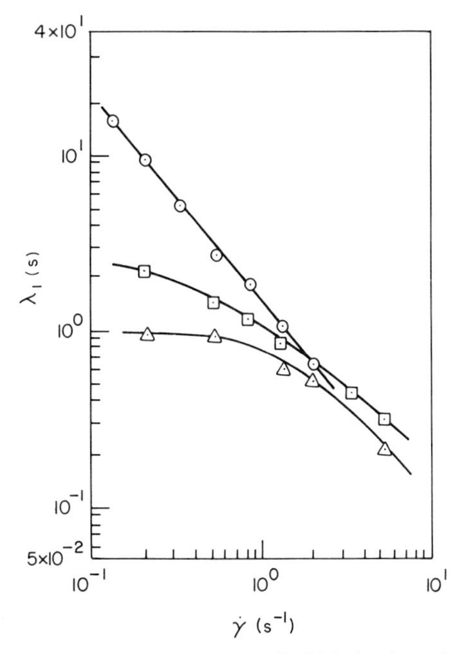

Fig. 2.28 Relaxation time versus shear rate: (⊙) high-density polyethylene ($T = 240°C$); (⊡) polypropylene ($T = 200°C$); (△) polystyrene ($T = 200°C$).

models, one can easily determine the relaxation time λ_1, using Eq. (2.75). Often λ_1 is used to determine the elasticity of one fluid relative to another.

Figures 2.28 and 2.29 give plots of the characteristic relaxation time λ_1 versus shear rate $\dot{\gamma}$ for polymer melts. Note that the values of λ_1 are determined with the aid of Eq. (2.75), using measurements of the first normal stress difference $\tau_{11} - \tau_{22}$ and viscosity η. It is seen that the values of λ_1 for nylon 6 and PET are about 1 or 2 orders of magnitude smaller than the values of λ_1 for the typical thermoplastics, HDPE, PP, and PS. It can be concluded therefore that nylon 6 and PET are far less elastic than HDPE, PP, and PS. This observation is consistent with the conclusion that is drawn from the plot of S_R versus τ_w, given in Fig. 2.27.

It is worth noting, however, that for most polymeric liquids a single value of relaxation time is not adequate and one must turn to a spectrum of relaxation times; especially for the polydisperse polymeric liquids. Methods for determining the relaxation spectrum are presented in the literature [66].

Another important dimensionless material parameter for viscoelastic fluids, first suggested by Reiner [67], is the Deborah number N_{De}

$$N_{De} = \text{characteristic material time/characteristic process time} \quad (2.77)$$

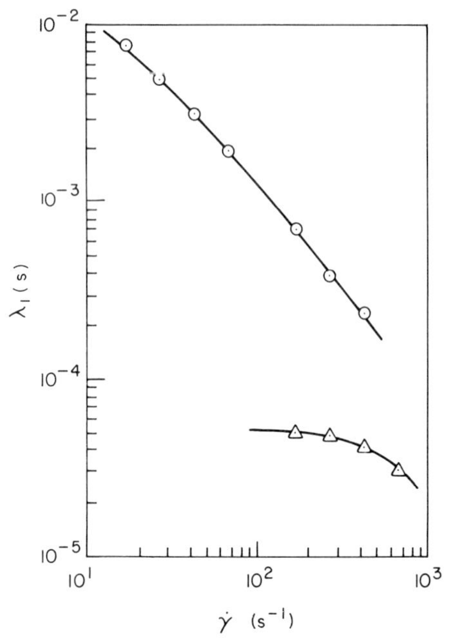

Fig. 2.29 Relaxation time versus shear rate: (⊙) nylon 6 (= 280°C); (△) polyethylene terephthalate ($T = 300°C$).

The Deborah number may be used to interprete the fluidlike or solidlike behavior of a material. If N_{De} is small, the material follows the fluidlike behavior; if large, the solidlike behavior. It is worth noting that Tanner [68] discussed various forms of Deborah number definitions and applied them to the problems of an entrance flow and to the flow through porous media.

2.3.4 The Effect of Temperature on the Rheological Properties

The rheological properties are strong functions of temperature. The effect of temperature on the viscosity and first normal stress difference is given in Fig. 2.30 for a polystyrene melt, and in Fig. 2.31 for a polypropylene melt. An empirical method for reducing viscosity data taken at various temperatures into one master curve has been suggested [69, 70].

It is worth mentioning that the viscosity $\eta(T)$ at temperature T may be evaluated from the viscosity $\eta(T_R)$ at a reference temperature T_R, by using the following relationship:

$$\eta(T) = a_T \eta(T_R) \tag{2.78}$$

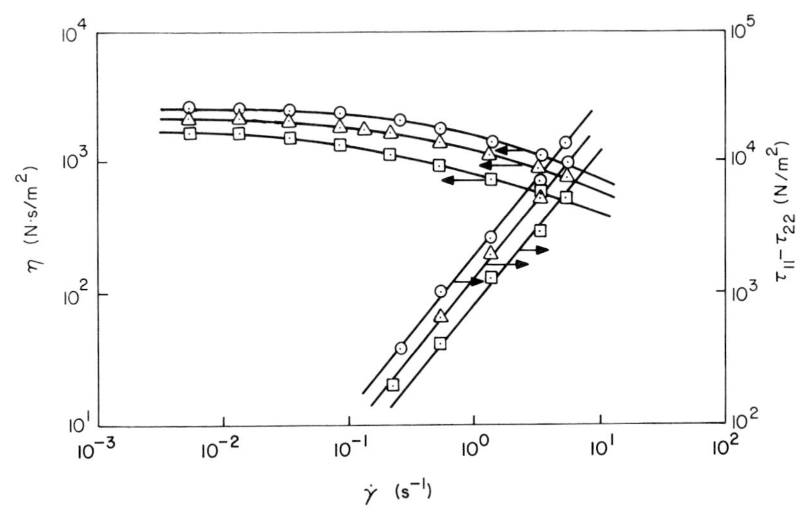

Fig. 2.30 Viscosity and first normal stress difference versus shear rate for polystyrene melt: (⊙) 200°C; (△) 200°C; (⊡) 240°C.

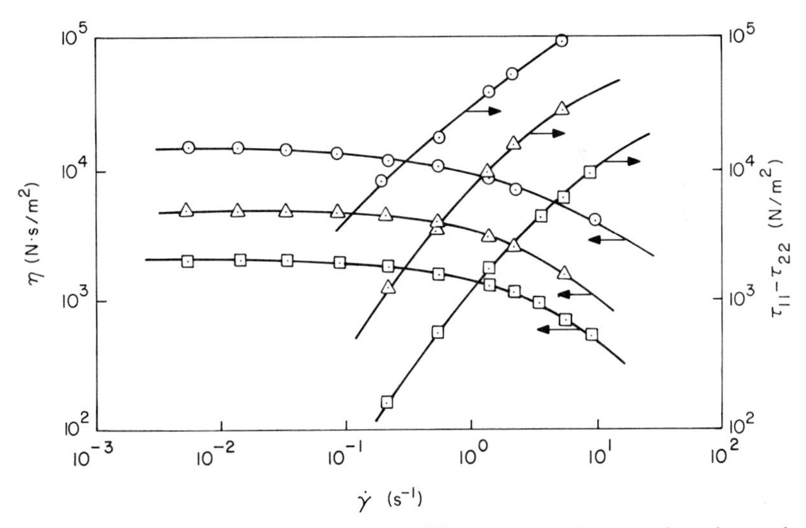

Fig. 2.31 Viscosity and first normal stress difference versus shear rate for polypropylene melt: (⊙) 180°C; (△) 200°C; (⊡) 220°C.

where a_T is the shift factor defined by

$$a_T(T) = \exp\left[\frac{E}{R_g}\left(\frac{1}{T} - \frac{1}{T_R}\right)\right] \tag{2.79}$$

in which E is the flow activation energy and R_g is the universal gas constant.

Polymers are known to possess one or more thermodynamic transition. In practice, the glass transition temperature T_g plays an important role in the temperature variation of physical properties of polymers [66, 71]. Williams *et al.*, (69) have proposed a universal relationship

$$\log a_T = [-17.44(T - T_g)]/[51.6 + (T - T_g)] \tag{2.80}$$

where a_T represents the temperature variation of the segmental friction coefficient for mechanical relaxation. This empirical relationship, known as the WLF equation, has been applied successfully to describe the relaxation or viscosity variation of polymers in the temperature range of

$$T_g < T < T_g + 100^\circ\text{C}.$$

2.4 EXPERIMENTAL TECHNIQUES FOR DETERMINING THE VISCOELASTIC PROPERTIES OF POLYMERIC LIQUIDS

There has been a continuing interest in developing experimental techniques for the measurement of the rheological properties of viscoelastic fluids, and rheologists have spent much effort on it. Reliable experimental data are needed in order to evaluate how good a rheological model is in its capability of predicting the rheological properties of viscoelastic fluids, and also in determining an optimum processing condition or a particular desired set of physical properties in the finished product. In other words, reliable measurement of the rheological properties can be used as a means of quality control.

There are two types of measuring apparatus which give rise to viscometric flow fields (e.g., simple shearing flow, tubular and plane Poiseuille flow). They are: (a) the rotational type (cone and plate, parallel plate, and coaxial cylinder), and (b) the capillary type. In the measurement of the viscometric rheological properties of polymer melts, the use of the rotational type is limited to low shear rates because of flow instability [72, 73]. On the other hand, the use of the capillary type has no such limitation, until melt fracture starts to occur at a critical wall shear stress.

We shall discuss below very briefly methods for determining steady and oscillatory shearing flow properties with the cone-and-plate instrument, and steady shearing flow properties with the capillary/slit instrument. We shall also review briefly methods for determining steady elongational viscosities. Space limitations do not permit us to describe various experimental techniques in great detail, and the reader may consult other monographs [66, 74].

2.4.1 The Measurement of Steady Shearing Flow Properties

Let us consider the flow of an incompressible fluid placed in a cone-and-plate instrument (e.g., the Weissenberg rheogoniometer, or Rheometrics Mechanical Spectrometer), in which a cone with a wide vertical angle is placed on a horizontal flat plate as schematically shown in Fig. 2.32. The wedgelike space between the cone and plate is filled with the liquid under test. One of the surfaces is fixed and the other rotates around the axis of the cone. It is desired to find the relationships between the torque and angular velocity, and between the net thrust acting on the cone (or plate) and angular velocity.

If we neglect inertia forces and edge effects for the flow geometry given in Fig. 2.32, the equations of motion, when integrated with the use of appropriate boundary conditions, yield [74]:

$$\tau_{\phi\theta} = \tau_{12} = 3\mathcal{T}/2\pi R^3 \tag{2.81}$$

in which $\mathcal{T}$ is the torque measured on the surface of the cone and R is the radius of the cone. Note that in the derivation of Eq. (2.81), the value of θ_c (see Fig. 2.32) is assumed to be very small (say, $\theta_c \leq 2°$).

For the geometry shown in Fig. 2.32, the flow field may be given as

$$v_\phi = v_\phi(\theta, r) \qquad v_\theta = v_r = 0 \tag{2.82}$$

and therefore the shear rate $\dot{\gamma}$ is defined as

$$\dot{\gamma} = \frac{\sin\theta}{r}\frac{\partial}{\partial r}\left(\frac{v_\phi}{\sin\theta}\right) \cong \frac{1}{r}\frac{\partial v_\phi}{\partial\theta} \tag{2.83}$$

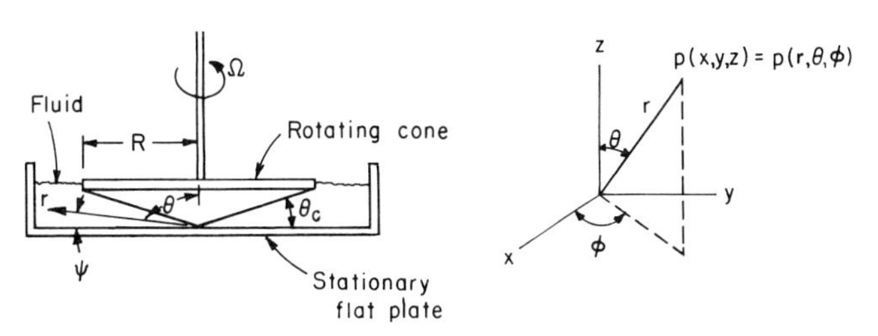

Fig. 2.32 Cone-and-plate geometry and spherical coordinate systems.

Using appropriate boundary conditions, it can be shown that Eq. (2.83) may be represented by

$$\dot{\gamma} = -\Omega/\theta_c \tag{2.84}$$

in which Ω denotes the angular velocity of the rotating cone (see Fig. 2.32).

It is seen from Eqs. (2.81) and (2.84) that for small enough gaps between the cone and plate, both the shear stress τ_{12} and shear $\dot{\gamma}$ are uniquely determined from measurements of the torque $\mathscr{T}$ and angular velocity Ω, thus permitting one to determine the viscosity function $\eta(\dot{\gamma})$. Let us now discuss what other quantities one should measure in order to determine the normal stress functions $\Psi_1(\dot{\gamma})$ and $\Psi_2(\dot{\gamma})$.

The total force F exerted normal to the cone (or plate) may be calculated from

$$F = -\int_0^R 2\pi r S_{\theta\theta}\, dr \tag{2.85}$$

Integrating by parts, Eq. (2.85) becomes

$$F = -\pi\left[R^2 S_{\theta\theta}(R) - \int_0^R r^2 \frac{\partial S_{\theta\theta}}{\partial r}\, dr\right] \tag{2.86}$$

It can be shown that if all stresses are evaluated at the surface of the plate ($\psi = 0$), we have

$$dS_{\theta\theta}/d\ln r = N = \text{const} \tag{2.87}$$

where

$$N = \tau_{\phi\phi} + \tau_{\theta\theta} - 2\tau_{rr} = \tau_{11} + \tau_{22} - 2\tau_{33} \tag{2.88}$$

Introducing Eq. (2.87) into the right-hand side of Eq. (2.86), we have

$$F = -\pi R^2[S_{\theta\theta}(R) - \tfrac{1}{2}(\tau_{\phi\phi} + \tau_{\theta\theta} - 2\tau_{rr})_R] \tag{2.89}$$

Noting that

$$S_{\theta\theta}(R) = -p(R) + \tau_{\theta\theta}(R) \tag{2.90}$$

Eq. (2.89) may be rewritten

$$F = -\pi R^2[-p(R) + \tau_{rr}(R) - \tfrac{1}{2}(\tau_{\phi\phi} - \tau_{\theta\theta})_R] \tag{2.91}$$

If the edge of the gap is in equilibrium with the atmosphere, one has

$$S_{rr}(R) = -p(R) + \tau_{rr}(R) = -p_0 \tag{2.92}$$

Use of Eq. (2.92) in Eq. (2.91) gives

$$\tau_{\phi\phi} - \tau_{\theta\theta} = \tau_{11} - \tau_{22} = 2F'/\pi R^2 \tag{2.93}$$

where $F' = F - \pi R^2 p_0$ is the net thrust measured on the cone (or plate), in excess of that due to ambient pressure.

It is seen, in Eq. (2.93), that the first normal stress difference $\tau_{11} - \tau_{22}$ can be determined from measurement of the net thrust F', which varies with angular velocity Ω. Since the measurement of angular velocity Ω enables one to determine the shear rate $\dot{\gamma}$ [see Eq. (2.84)], one can then establish the relationship between $\tau_{11} - \tau_{22}$ and $\dot{\gamma}$, and therefore determine the first normal stress function $\Psi_1(\dot{\gamma})$ [see Eq. (2.2)].

Figures 2.30 and 2.31 give typical plots of η and $\tau_{11} - \tau_{22}$ versus $\dot{\gamma}$ for molten polymers. It should be pointed out at this juncture that the cone-and-plate instrument has a major drawback, in that its use is limited to low shear stresses (or low shear rates), because the test fluid exudes from the space between the cone and plate as the speed of the plate (or cone) exceeds a certain critical value [72, 73]. This behavior is inherent in the elasticity of the test fluid. In the use of typical molten thermoplastics as test liquid, the upper limit of operable shear stress of the cone-and-plate instrument is about 10^4 N/m^2 (10^5 dyn/cm^2). Translating this shear stress into the upper limit of operable shear rate depends on the level of viscosity of the particular fluid under the test. For instance, it could be as high as 1000 sec^{-1} for dilute polymer solutions (see Figs. 2.2 and 2.3), it could be less than 10 sec^{-1} for typical molten thermoplastics (see Figs. 2.30 and 2.31), and it could be less than 0.1 sec^{-1} for very viscous elastomers. However, the use of shear stress can be applied uniformly, regardless of the type of test fluid (i.e., regardless of the level of fluid viscosity). In practical polymer processing operations, the range of shear stresses one encounters is

$$10^4 \sim 10^6 \text{ N/m}^2 \ (10^5 \sim 10^7 \text{ dyn/cm}^2).$$

The capillary-type instrument has been used for determining fluid viscosity for a long time. However, its use for determining fluid elasticity, in terms of normal stress differences, is relatively new. Han and co-workers [75–77] have made an extensive investigation, both experimentally and theoretically, of the determination of the three material functions $\eta(\dot{\gamma})$, $\Psi_1(\dot{\gamma})$, and $\Psi_2(\dot{\gamma})$, from measurements of the wall normal stress distribution along a circular tube or slit die. We shall present below, in brief, the basic principles of slit- and capillary-die rheometry. The reader may consult the monograph of Han [5] for details of the experimental and theoretical background.

Let us consider the flow of an incompressible viscoelastic fluid entering a capillary tube from a large reservoir. The velocity profile starts to develop in the entrance region and continues to change in the capillary until it reaches a certain distance, beyond which the flow is said to be fully developed. Now, when pressures are measured in the reservoir and in the capillary tube with pressure transducers, one may obtain pressure profiles, similar to those shown

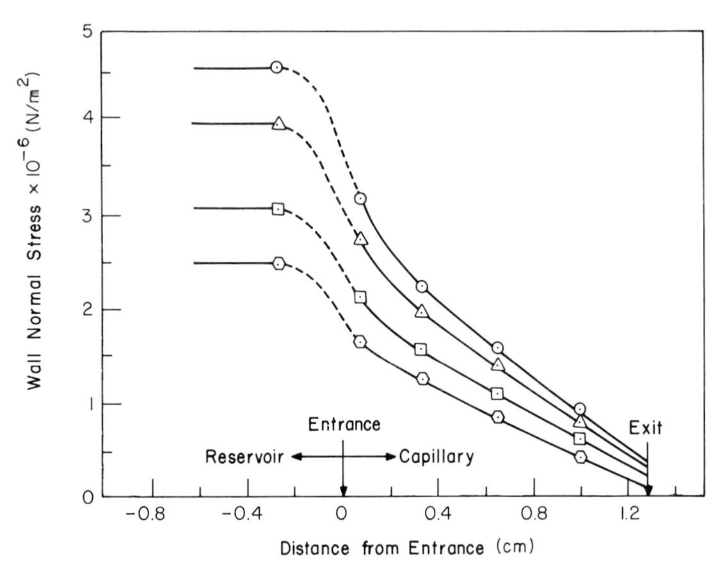

Fig. 2.33 Wall normal stress distributions of high-density polyethylene ($T = 180°C$) in a cylindrical tube having L/D ratio of 4, at various shear rates (s^{-1}) [75]: (⊙) 790; (△) 616; (⊡) 313; (⬡) 160. From C. D. Han *et al.*, *Trans. Soc. Rheol* **14**, 393. Copyright © 1970. Reprinted by permission of John Wiley & Sons, Inc.

in Fig. 2.33. Note that what is measured is not just the pressure p, but the outward-acting total wall normal stress $S_{rr}(R, z)$ which is defined as

$$S_{rr}(R, z) = -p(R, z) + \tau_{rr}(R, z) \tag{2.94}$$

in which $\tau_{rr}(R, z)$ is the extra stress which is generated due to the deformation of the fluid.

Two facts are of particular interest in Fig. 2.33: (1) Beyond a certain distance downstream in the tube, a constant value of pressure gradient develops, which increases with flow rate (or shear rate). (2) The extrapolation of the pressure readings to the exit of the die yields a nonzero gauge pressure, termed the exit pressure P_{exit}. This is higher than the ambient pressure, and it increases with flow rate (or shear rate). In the past, Han and co-workers [75–77] have extensively discussed the rheological significance of the exit pressure, some precautions necessary in the measurement of pressure, as well as some implications of the extrapolation procedure.

For the capillary die, the following equations may be used to calculate the shear rate $\dot{\gamma}$, shear stress τ_w, and normal stress differences, $\tau_{11} - \tau_{22}$ and $\tau_{22} - \tau_{33}$ [5, 77]:

$$\dot{\gamma} = [(3n + 1)/4n]\dot{\gamma}_{\text{app}} \tag{2.95}$$

$$\tau_w = (-\partial p/\partial z)R/2 \tag{2.96}$$

$$\tau_{11} - \tau_{22} = P_{\text{exit}} + \tau_{\text{w}}\, dP_{\text{exit}}/d\tau_{\text{w}} \tag{2.97}$$

$$\tau_{22} - \tau_{33} = -\tau_{\text{w}}\, dP_{\text{exit}}/d\tau_{\text{w}} \tag{2.98}$$

in which $\dot{\gamma}_{\text{app}}$ and n are defined by

$$\dot{\gamma}_{\text{app}} = 4Q/\pi R^3 \tag{2.99}$$

$$n = d\ln\tau_{\text{w}}/d\ln\dot{\gamma}_{\text{app}} \tag{2.100}$$

Q denotes the volumetric flow rate, R denotes the capillary radius, $-\partial p/\partial z$ is the pressure gradient, and P_{exit} is the exit pressure.

For the thin slit die, the working equations for shear rate $\dot{\gamma}$, shear stress τ_{w}, and normal stress difference $\tau_{11} - \tau_{22}$ are given by [5, 77]

$$\dot{\gamma} = [(2n + 1)/3n]\dot{\gamma}_{\text{app}} \tag{2.101}$$

$$\tau_{\text{w}} = (-\partial p/\partial z)h/2 \tag{2.102}$$

$$\tau_{11} - \tau_{22} = P_{\text{exit}} + \tau_{\text{w}}\, dP_{\text{exit}}/d\tau_{\text{w}} \tag{2.103}$$

where n is as defined in Eq. (2.100) and $\dot{\gamma}_{\text{app}}$ is given by

$$\dot{\gamma}_{\text{app}} = 6Q/wh^2 \tag{2.104}$$

w being the slit width, and h the thickness of the slit.

Figure 2.34 gives schematic layouts for the two different die geometries; slit and capillary. Although one can have more than three transducers in one die, only three, the minimum recommended, are shown in the figure.

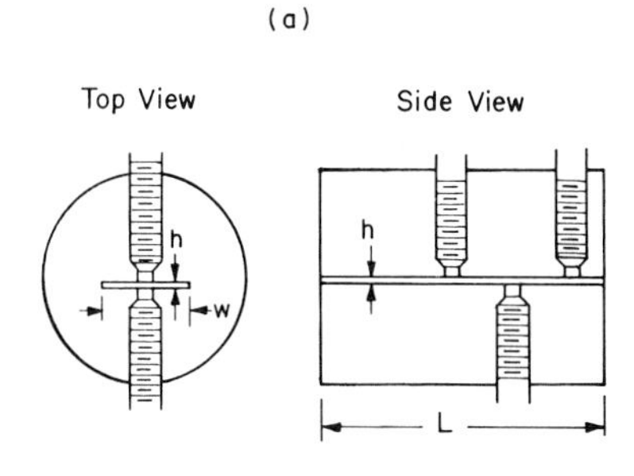

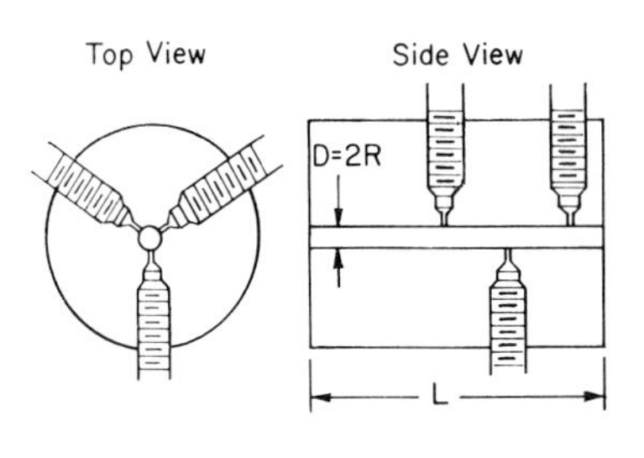

Fig. 2.34 Schematic of die design: (a) slit die; (b) circular die.

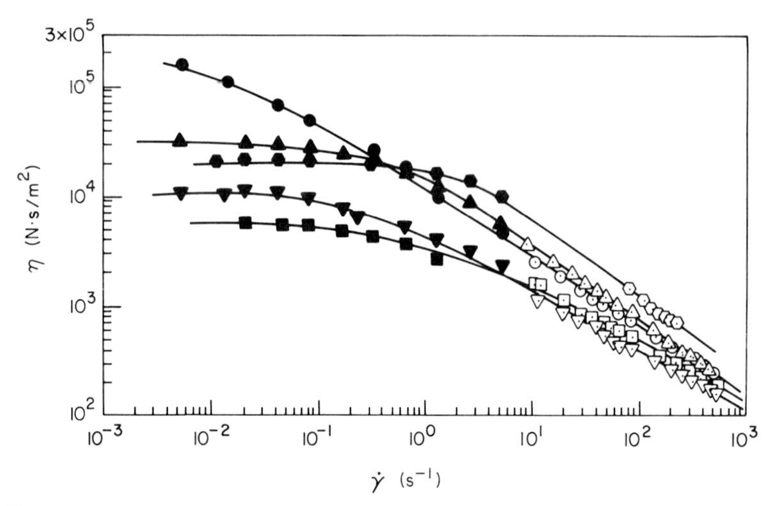

Fig. 2.35 Viscosity versus shear stress for polymer melts at 200°C, where the *closed* symbols represent the data obtained with a cone-and-plate instrument, and the *open* symbols represent the data obtained with a slit/capillary rheometer [77]: (⊙, ●) high-density polyethylene; (△, ▲) polystyrene; (⬡, ⬢) polymethylmethacrylate; (▽, ▼) low-density polyethylene; (⊡, ■) polypropylene. From C. D. Han, *Trans. Soc. Rheol.* **18**, 163. Copyright © 1974. Reprinted by permission of John Wiley & Sons, Inc

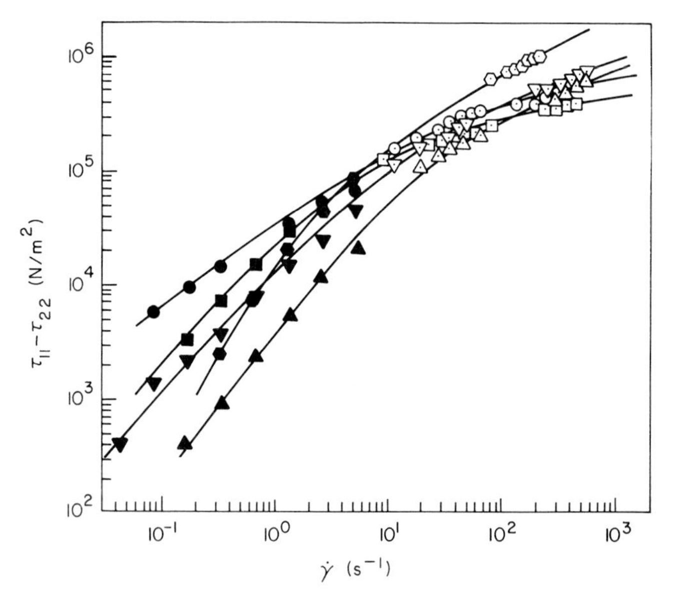

Fig. 2.36 First normal stress difference versus shear rate for polymer melts at 200°C [77]. Symbols are the same as in Fig. 2.35. From C. D. Han, *Trans. Soc. Rheol.* **18**, 163. Copyright © 1974. Reprinted by permission of John Wiley & Sons, Inc.

There is one important aspect, which one has to be aware of, in mounting pressure transducers on a die wall. As schematically shown in Fig. 2.34, the capillary die cannot avoid having small "pressure holes" (i.e., dead spaces) formed by the curved die wall and the flat tip of the transducer. (In general, transducers with curved tips are not available.) On the other hand, the slit die permits flush mounting of pressure transducers, and the elimination of the dead space. Thus the slit die has an advantage over the others, since there is no dead space in which the polymers can reside and either degrade or form crosslinkages.

Some representative plots of viscosity versus shear rate are given in Fig. 2.35, and plots of first normal stress difference versus shear rate in Fig. 2.36 for molten polymers, determined by the use of a cone-and-plate rheometer at low shear rates and the slit/capillary rheometer at high shear rates. It is seen that the slit/capillary rheometric data are in good agreement with the cone-and-plate rheometric data. It should be noted that the range of shear rates in the use of the slit (or capillary) rheometer can be varied by using different sizes of slit opening (or capillary diameter) (e.g., the Seiscor/Han Rheometer).

2.4.2 The Measurement of Oscillatory Shearing Flow Properties

Oscillatory flow measurement, often referred to as dynamic measurement, has been widely used in studying the viscoelastic properties of polymeric materials. Dynamic measurement requires an instrument which can generate sinusoidal strain as an input to the fluid under test and record the stress resulting from the fluid deformed as an output (see Fig. 2.7). Frequently used for dynamic measurement are the cone-and-plate instrument, concentric cylinders instrument, and eccentric parallel plates instrument.

Using the one-dimensional Maxwell model [see Eq. (2.37)], the following expression may be obtained:

$$\tau^* + \lambda_1 i\omega\tau^* = \eta i\omega\gamma^* \tag{2.105}$$

or

$$G^*(i\omega) = \tau^*/\gamma^* = \eta i\omega/(1 + \lambda_1 i\omega) = G'(\omega) + iG''(\omega) \tag{2.106}$$

Therefore, from Eq. (2.106) one has

$$G'(\omega) = G\lambda_1^2\omega^2/(1 + \lambda_1^2\omega^2); \qquad G''(\omega) = G\lambda_1\omega/(1 + \lambda_1^2\omega^2) \tag{2.107}$$

in which use is made of $\lambda_1 = \eta/G$, G being the elastic modulus. Hence, using Eqs. (2.22) and (2.107) one has

$$\eta'(\omega) = G\lambda_1/(1 + \lambda_1^2\omega^2), \qquad \eta''(\omega) = G\lambda_1^2\omega/(1 + \lambda_1^2\omega^2) \tag{2.108}$$

TABLE 2.6

Expressions of the Dynamic Viscosity and Storage Modulus for Some Representative Rheological Models

Model	$\eta'(\omega)$	$G'(\omega)$
Oldroyd Three-constant	$\dfrac{\eta_0(1+\lambda_1\lambda_2\omega^2)}{1+\lambda_1^2\omega^2}$	$\dfrac{\eta_0(\lambda_1-\lambda_2)\omega^2}{1+\lambda_1^2\omega^2}$
Spriggs	$\dfrac{\eta_0}{Z(\alpha)}\sum_{p=1}^{\infty}\dfrac{p^\alpha}{p^{2\alpha}+\lambda^2\omega^2}$	$\dfrac{\eta_0}{Z(\alpha)}\sum_{p=1}^{\infty}\dfrac{\lambda\omega^2}{p^{2\alpha}+\lambda^2\omega^2}$
Bird–Carreau	$\sum_{p=1}^{\infty}\dfrac{\eta_p}{1+\lambda_{2p}^2\omega^2}$	$\sum_{p=1}^{\infty}\dfrac{\eta_p\lambda_{2p}\omega^2}{1+\lambda_{2p}^2\omega^2}$
Bogue	$\sum_{p=1}^{\infty}\dfrac{G_p\lambda_p^{\text{eff}}}{1+(\lambda_p^{\text{eff}})^2\omega^2}$	$\sum_{p=1}^{\infty}\dfrac{G_p(\lambda_p^{\text{eff}})^2\omega^2}{1+(\lambda_p^{\text{eff}})^2\omega^2}$
Meister	$\sum_{p=1}^{\infty}\dfrac{G_p\lambda_p}{1+\lambda_p^2\omega^2}$	$\sum_{p=1}^{\infty}\dfrac{G_p\lambda_p^2\omega^2}{1+\lambda_p^2\omega^2}$

One can apply the same analysis to other rheological equations of state in order to derive the dynamic viscosity $\eta'(\omega)$ and complex moduli $G'(\omega)$ and $G''(\omega)$. Table 2.6 gives material functions $\eta'(\omega)$ and $G'(\omega)$ for some representative rheological models for oscillatory shearing flow.

The salient feature of dynamic measurement lies in that it yields information of both the viscous property $\eta'(\omega)$ and the elastic property $G'(\omega)$ of a fluid. Figure 2.8 gives plots of $\eta'(\omega)$ and $G'(\omega)$ versus ω for two polymer melts. Even when the data is obtainable only over one or two decades of the logarithmic frequency scale, at any one time, the viscoelastic functions can be traced out over a much larger effective range by making measurements at different temperatures and using the superposition principle [69]. When properly applied, it yields plots in terms of reduced variables which can be used with considerable confidence to deduce molecular parameters as well as to predict viscoelastic behavior in regions of the time or frequency scale not readily accessible experimentally.

2.4.3 The Measurement of Elongational Flow Properties

It should be remembered that the theoretical predictions of steady elongational viscosities from various rheological equations of state, for both uniaxial and uniform biaxial elongational flows, are given in Tables 2.2 and 2.5. It is of interest to note, in Table 2.2, that the relaxation time λ_1 in the convected Maxwell model brings about a dependency of the elongational viscosity on the elongation rate, whereas the relaxation time λ_1 in the Zaremba–DeWitt model has no influence on its predicted elongational viscosity. This observation is quite opposite to what these models predict with regard to shear viscosity (see Table 2.1). It can be concluded, therefore, that there is no general relationship between the viscometric material functions and elongational viscosity, and thus there is no way of predicting elongational viscosities from the viscometric material functions, except for Newtonian fluids [see Eqs. (2.11) and (2.12)].

A number of investigators [78–91] have reported measurements of steady elongational viscosities. Unfortunately, however, there is no single experimental technique that would be useful for all types of fluids. Some experimental techniques are better suited to solutions, whereas others are better suited to very viscous liquids such as polymer melts. Some experimental techniques are better suited to low rates of strain (elongation rates), whereas others are better suited to high elongation rates. Here we will confine our discussion to the measurement of the elongational viscosity of very viscous liquids (namely, polymer melts), which are of practical interest in polymer processing operations.

The various experimental techniques may broadly be classified [74], in accordance with the degree of control of the elongational deformation, into two: (1) the controlled flow type of experiment, and (2) the uncontrolled flow type. The former refers to the experiment where the sample is subjected to a constant strain rate, and the latter to a nonconstant strain rate.

To find the elongational strain as a function of time for steady uniaxial elongational flow, consider an initially unstrained, rod-shaped, specimen, fixed at one end and stretched in the uniaxial direction. In order to generate a constant rate of strain (elongation rate) $\dot{\gamma}_E$, it is necessary to have the velocity of the free end such that the following relationship is maintained:

$$\varepsilon = \ln(L/L_0) = \dot{\gamma}_E t \tag{2.109}$$

in which ε is the total Hencky strain and L_0 is the initial length of the specimen. In this type of experiment, difficulties are often encountered in avoiding the slippage of the specimen from the clamp. Precise control of the free end velocity is also of utmost importance.

To overcome this difficulty, Meisner [79] suggests an alternative experimental technique, in which two sets of gripping wheels are used, instead of end loading. In this way, the specimen is stretched between the distance L fixed in space and the elongation rate $\dot{\gamma}_E$ is determined by

$$\dot{\gamma}_E = dv_z/dz = V/L = R\Omega/L \tag{2.110}$$

in which R is the radius of the rotating wheel with angular velocity Ω, and V is the linear wheel velocity. In this type of apparatus, the tensile stress is measured through the deflection of a spring associated with one of the rotating wheels. A slightly different experimental technique has been used by Ide and White [85], in which the test specimen, clamped to an Instron load cell at one end, is stretched and wrapped around a rotating roller at the other end. In this type of apparatus, the tension (total force) is measured by the Instron load cell, and the elongation is determined by Eq. (2.110).

In determining the elongational viscosity η_E [see Eq. (2.5)] with the methods described above, one may have to wait a long time for the stress to build up the level where a steady state (in both the Lagrangian and Eulerian senses) is attained, because the specimen used is strain free before the test begins.

Figure 2.37 gives some representative results of uniaxial, steady elongational viscosities for commercially available thermoplastics, namely, low-density polyethylene (LDPE), polystyrene (PS), high-density polyethylene (HDPE), and polypropylene (PP). It is seen that the elongational viscosities of the LDPE and PS increase with elongation rate (sometimes referred to as "elongation hardening"), whereas those of the HDPE and PP decrease with elongation rate ("elongation softening"). It is clear that the theoretical predictions presented above (see Tables 2.2 and 2.5) cannot describe the elongation softening behavior of the HDPE and PP given in Fig. 2.37. Note that some models predict an increase in η_E without bound at some critical value of $\dot{\gamma}_E$. White and co-workers [92–94] have discussed the rheological significance of the elongation softening behavior on flow instabilities in elongational flow.

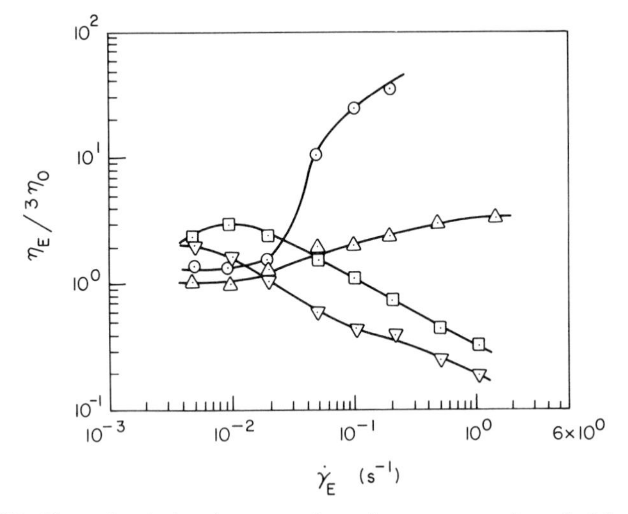

Fig. 2.37 Elongational viscosity versus elongation rate at steady, uniaxial elongational flow for polymer melts [85]: (⊙) low-density polyethylene at 160°C; (△) polystyrene at 160°C; (□) polypropylene at 180°C; (▽) high-density polyethylene at 160°C. From Y. Ide and J. L. White, *J. Appl. Polym. Sci.* **22**, 1061. Copyright © 1978. Reprinted by permission of John Wiley & Sons, Inc.

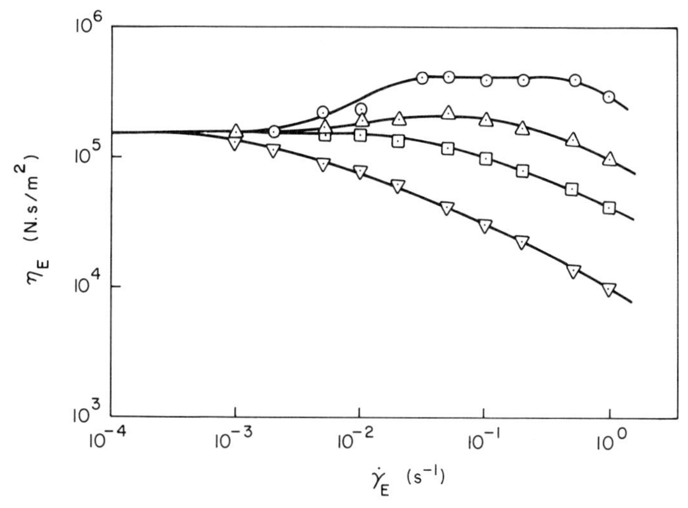

Fig. 2.38 Uniaxial elongational viscosity versus elongation rate for low-density polyethylene ($T = 150°C$) at various Hensky strains [91]: (⊙) 3.0; (△) 2.0; (⊡) 1.0; (▽) 0.1. From A. E. Everage and R. L. Ballman, *J. Appl. Polym. Sci.* **21**, 841. Copyright © 1977. Reprinted by permission of John Wiley & Sons, Inc.

On the basis of various experimental observations, it appears that elongational responses vary with the structure of macromolecules and, also, with the molecular weight distribution. There is experimental evidence that broadening distributions of molecular weight tend to cause a greater tendency to deformation rate softening in elongational flow [94, 95].

Figures 2.38 and 2.39 give plots of η_E versus $\dot{\gamma}_E$, with the total Hencky strain ε as parameter [see Eq. (2.109) for the definition of ε]. It is seen that,

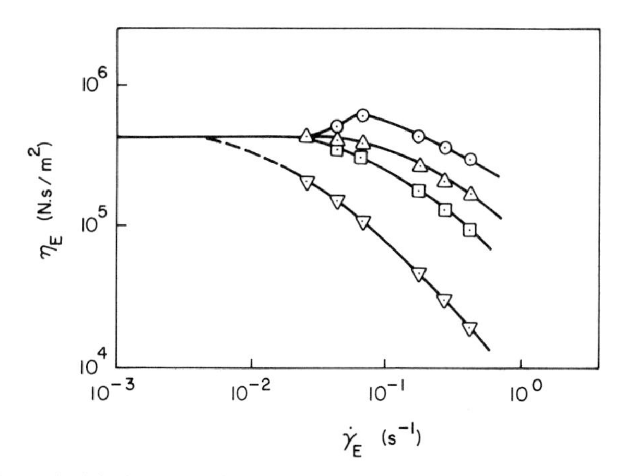

Fig. 2.39 Uniaxial elongational viscosity versus elongation rate for polystyrene ($T = 155°C$) at various Hencky strains [91]: (⊙) 1.5; (△) 0.9; (⊡) 0.5; (▽) 0.1. From A. E. Everage and R. L. Ballman, *J. Appl. Polym. Sci.* **21**, 841. Copyright © 1977. Reprinted by permission of John Wiley & Sons, Inc.

for very high $\dot{\gamma}_E$, η_E decreases with $\dot{\gamma}_E$ for all strain levels and that, at high level of strain, η_E first increases and then decreases as $\dot{\gamma}_E$ increases.

Everage and Ballman [91] have tried to explain the wide variety of material response observed in these elongational flow experiments with a single unifying theory. They used the convected Maxwell model to obtain the expression for elongational viscosity η_E:

$$\eta_E = \frac{3\eta_0}{(1 - 2\lambda_1\dot{\gamma}_E)(1 + \lambda_1\dot{\gamma}_E)} - \frac{2\eta_0 \exp[-(1 - 2\lambda_1\dot{\gamma}_E)\varepsilon/\lambda_1\dot{\gamma}_E]}{1 - 2\lambda_1\dot{\gamma}_E} - \frac{\eta_0 \exp[-(1 + \lambda_1\dot{\gamma}_E)\varepsilon/\lambda_1\dot{\gamma}_E]}{1 + \lambda_1\dot{\gamma}_E} \tag{2.111}$$

in which the relationship $\varepsilon = \dot{\gamma}_E t$ given by Eq. (2.109) is employed. They used Eq. (2.111) to plot η_E/η_0 against $\lambda_1\dot{\gamma}_E$ with ε as a parameter, as given in Fig. 2.40. It is revealing to see that η_E goes through a maximum at high values of ε, behavior very similar to that observed by experiment. Everage and Ballman postulated, in terms of the spring–dashpot model, that at high values of $\dot{\gamma}_E$, the total deformation occurs over a short time and primarily involves a stretching of the spring with little movement of the dashpot, giving rise to a constant value of the spring stress. Therefore a constant stress divided by an increasing rate of strain (elongation rate) leads to a decrease in η_E with $\dot{\gamma}_E$. On the other hand, at the other extreme of small

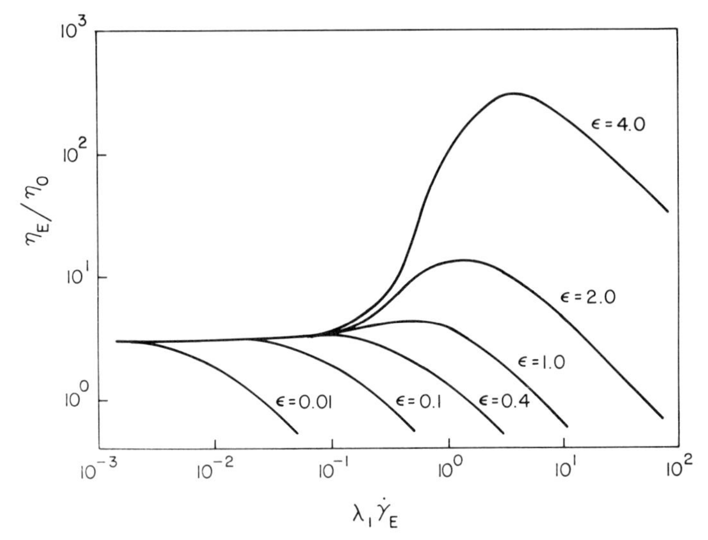

Fig. 2.40 Theoretical predictions of uniaxial elongational viscosity curve with Hencky strain as parameter [91]. From A. E. Everage and R. L. Ballman, *J. Appl. Polym. Sci.* **21**, 841. Copyright © 1977. Reprinted by permission of John Wiley & Sons, Inc.

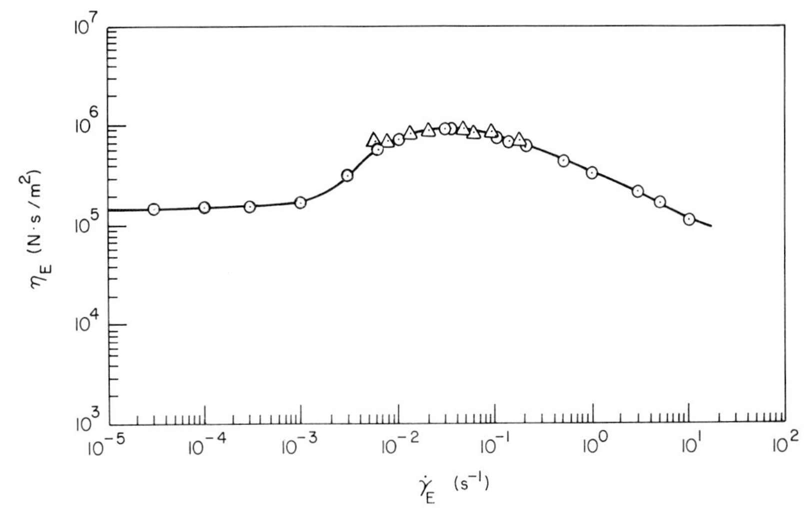

Fig. 2.41 Uniaxial elongational viscosity versus elongation rate for low-density polyethylene ($T = 150°C$) [89]: (⊙) with the measurements of constant elongation rate; (△) with the measurements of tensile creep.

values of $\dot{\gamma}_E$, the total deformation occurs over a long period and involves primarily dashpot motion, giving rise to the elongational response of a Newtonian fluid (i.e., $\eta_E = 3\eta_0$).

Figure 2.41 gives plots of η_E versus $\dot{\gamma}_E$ for low-density polyethylene at 150°C, over several decades of $\dot{\gamma}_E$, in which both a constant stretching rate apparatus and a tensile creep apparatus were employed [89]. It is seen that the elongational viscosity goes through a maximum, decreasing with elongation rate at high values of $\dot{\gamma}_E$.

Denson and co-workers [96, 97] measured the uniform equal biaxial elongational viscosity for polyisobutylene by inflating a flat sheet of polymer into a contour approximating a spherical cap using an inert gas. In this type of experiment, strain is determined as a function of time by measuring the radius of curvature R of the bubble, and the uniform biaxial strain rate $\dot{\gamma}_B$ is determined by

$$\dot{\gamma}_B = (1/R)(dR/dt) \tag{2.112}$$

The biaxial stress S_{11} is determined by using the hoop stress equation for a thin spherical shell:

$$S_{11} = PR/2b \tag{2.113}$$

where P is the differential pressure applied to generate an expandable bubble of polymer and b is the thickness of the bubble wall. Figure 2.42 gives plots of η_B versus $\dot{\gamma}_B$ determined by Joye and co-workers [97], using Eqs. (2.112)

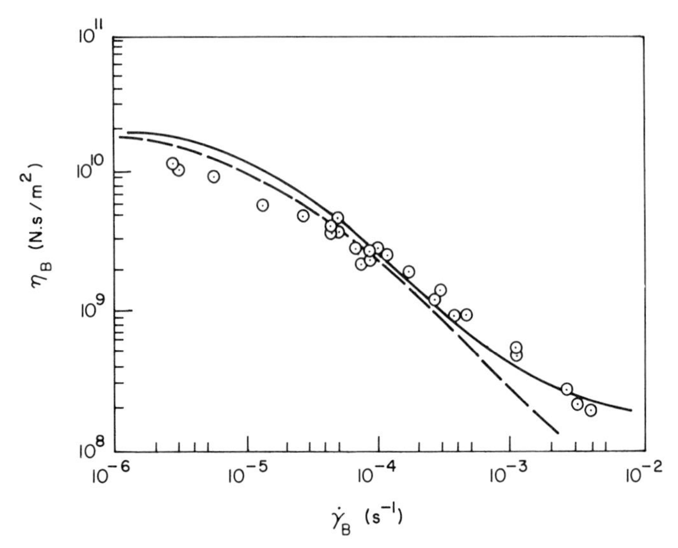

Fig. 2.42 Comparison of theoretical prediction with experimental data of uniform (equal) biaxial elongational viscosity of polyisobutylene [97]. The data were obtained with the bubble inflation technique, and the theoretical predictions were made from the Oldroyd three-constant model (solid line) and, also, from the Bogue model (broken line). From D. D. Joye *et al.*, *Trans. Soc. Rheol.* **17**, 287. Copyright © 1973. Reprinted by permission of John Wiley & Sons, Inc.

and (2.113) [see Eq. (2.8)]. It is seen that η_B decreases with increasing $\dot{\gamma}_B$. This is, apparently, as predicted by the rheological models given in Tables 2.2 and 2.5, except for the Lodge model, which predicts an increase in η_B without bound at some critical value of $\dot{\gamma}_B$.

In spite of the fact that in the past much effort, both experimental and theoretical, has been spent on determining elongational viscosity, relatively little has been discussed about the usefulness of this parameter in predicting the processability of polymeric materials. In this connection, it seems most useful to take experimental data on the type of elongational viscosity relevant to the particular process of interest, to select a rheological model describing the data, and to obtain solutions of the momentum and energy balance equations describing the process under investigation.

REFERENCES

[1] B. D. Coleman and W. Noll, *Arch. Ration Mech. Anal.* **3**, 289 (1959); **6**, 355 (1960).

[2] B. D. Coleman and W. Noll, *Ann. N.Y. Acad. Sci.* **89**, 762 (1961).

[3] C. Truesdell and W. Noll, "The nonlinear field theories of mechanics." Springer-Verlag, Berlin and New York, 1965.

[4] B. D. Coleman, H. Markovitz, and W. Noll, "Viscometric Flow of Non-Newtonian Fluids." Springer-Verlag, Berlin and New York, 1966.
[5] C. D. Han, "Rheology in Polymer Processing." Academic Press, New York, 1976.
[6] T. F. Ballenger and J. L. White, *J. Appl. Polym. Sci.* **15**, 1949 (1971).
[7] D. Acierno, J. N. Dalton, J. M. Rodriguez, and J. L. White, *J. Appl. Polym. Sci.* **15**, 2395 (1971).
[8] C. D. Han and R. R. Lamonte, *Trans. Soc. Rheol.* **16**, 447 (1972).
[9] I. Chen, G. E. Hagler, L. E. Abbott, J. N. Dalton, D. C. Bogue, and J. L. White, *Trans. Soc. Rheol.* **16**, 473 (1972).
[10] C. D. Han, unpublished research (1977).
[11] A. S. Lodge, "Elastic Liquids." Academic Press, New York, 1964.
[12] A. G. Fredrickson, "Principles and Applications of Rheology." Prentice-Hall, Englewood Cliffs, New Jersey, 1964.
[13] S. Middleman, "The Flow of High Polymers." Wiley (Interscience), New York, 1968.
[14] R. B. Bird, R. C. Armstrong, and O. Hassager, "Dynamics of Polymeric Liquids," Vol. 1. Wiley, New York, 1977.
[15] J. G. Oldroyd, *Proc. R. Soc. London, Ser. A* **200**, 523 (1950).
[16] S. Zaremba, *Bull. Acad. Cracovie* p. 594 (1903).
[17] H. Fromm, *Z. Angew. Math. Mech.* **25**, 146 (1947).
[18] T. W. DeWitt, *J. Appl. Phys.* **26**, 889 (1955).
[19] J. G. Oldroyd, *Proc. R. Soc. London, Ser. A* **245**, 278 (1958).
[20] M. C. Williams and R. B. Bird, *Phys. Fluids* **5**, 1126 (1962).
[21] T. W. Spriggs, *Chem. Eng. Sci.* **20**, 931 (1965).
[22] A. E. Green and R. S. Rivlin, *Arch. Ration Mech. Anal.* **1**, 1 (1957).
[23] B. D. Coleman and W. Noll, *Rev. Mod. Phys.* **33**, 239 (1961).
[24] J. L. White, *J. Appl. Polym. Sci.* **8**, 1129 (1964).
[25] J. L. White, *J. Appl. Polym. Sci.* **8**, 2339 (1964).
[26] B. D. Coleman and W. Noll, *Trans. Soc. Rheol.* **5**, 41 (1961).
[27] L. Boltzmann, *Pogg. Ann. Phys.* **7**, 624 (1876).
[28] I. F. MacDonald and R. B. Bird, *J. Phys. Chem.* **70**, 2068 (1966).
[29] M. Yamamoto, *Trans. Soc. Rheol.* **15**, 331 (1971).
[30] D. C. Bogue and J. L. White "Engineering Analysis of Non-Newtonian Fluids." NATO Agardograph No. 144 (1970).
[31] A. S. Lodge, *Trans. Faraday Soc.* **52**, 120 (1956).
[32] B. J. Meister, *Trans. Soc. Rheol.* **15**, 63 (1971).
[33] D. C. Bogue, *Ind. Eng. Chem. Fundam.* **5**, 253 (1966).
[34] R. B. Bird and P. Carreau, *Chem. Eng. Sci.* **23**, 427 (1968).
[35] M. H. Wagner, *Rheol. Acta* **15**, 136 (1976).
[36] B. Bernstein, E. Kearsley, and L. Zapas, *Trans. Soc. Rheol.* **7**, 391 (1963); **9**, 27 (1965).
[37] J. L. White and A. B. Metzner, *J. Appl. Polym. Sci.* **7**, 1867 (1963).
[38] P. E. Rouse, Jr., *J. Chem. Phys.* **21**, 1272 (1953); **22**, 1570 (1954).
[39] B. H. Zimm, *J. Chem. Phys.* **24**, 269 (1956).
[40] M. C. Williams, *J. Chem. Phys.* **42**, 2988 (1965).
[41] F. Bueche, *J. Chem. Phys.* **22**, 1570 (1954).
[42] W. W. Graessley, *J. Chem. Phys.* **43**, 2696 (1965); **47**, 1942 (1967).
[43] A. S. Lodge, *Trans. Faraday Soc.* **52**, 120 (1956).
[44] M. Yamamoto, *J. Phys. Soc. J.* **11**, 413 (1956); **12**, 1148 (1957).
[45] F. W. Wiegel, *Physica (Amsterdam)* **42**, 156 (1969).
[46] F. W. Wiegel and F. T. De Bats, *Physica (Amsterdam)* **43**, 33 (1969).
[47] P. G. De Gennes, *Macromolecules* **9**, 587 (1976).
[48] S. F. Edwards and J. W. V. Grant, *J. Phys. A* **6**, 1186 (1973).
[49] M. Doi, *Chem. Phys. Lett.* **26**, 269 (1974).

[50] M. Doi, *J. Phys. A* **8**, 417, 959 (1975).
[51] M. Doi and S. F. Edwards, *J. C. S. Faraday Trans. II* **74**, 1789, 1802, 1818 (1978).
[52] M. Doi, *J. Polym. Sci.* **18**, 1005 (1980).
[53] S. Middleman, *J. Appl. Polym. Sci.* **11**, 417 (1967).
[54] W. W. Graessley and L. Segal, *AIChE J.* **16**, 261 (1970).
[55] B. D. Coleman and H. Markovitz, *J. Appl. Phys.* **35**, 1 (1964).
[56] J. D. Ferry, M. L. Williams, and D. M. Stern, *J. Chem. Phys.* **58**, 987 (1954).
[57] B. H. Zimm and R. W. Kilb, *J. Polym. Sci.* **37**, 19 (1959).
[58] W. L. Peticolas, *J. Polym. Sci.* **58**, 1405 (1962).
[59] F. Bueche, *J. Chem. Phys.* **40**, 484 (1964).
[60] G. Kraus and J. T. Gruver, *J. Appl. Polym. Sci.* **9**, 739 (1965).
[61] D. P. Wyman, L. J. Elyash, and W. J. Frazer, *J. Polym. Sci., Part A* **3**, 681 (1965).
[62] L. H. Tung, *J. Polym. Sci.* **46**, 409 (1960).
[63] W. W. Graessley and J. S. Prentice, *J. Polym. Sci., Part A-2* **6**, 1887 (1968).
[64] N. Tokita and J. L. White, *J. Appl. Poly. Sci.* **10**, 1011 (1966).
[65] J. L. White and N. Tokita, *J. Appl. Polym. Sci.* **11**, 321 (1967).
[66] J. D. Ferry, "Viscoelastic Properties of Polymers." Wiley, New York, 1961.
[67] M. Reiner, *Phys. Today* **17**(1), 62 (1964).
[68] R. I. Tanner, *AIChE J.* **22**, 910 (1976).
[69] M. L. Williams, R. F. Landel, and J. D. Ferry, *J. Am. Chem. Soc.* **77**, 3701 (1955).
[70] G. V. Vinogradov and A. Y. Malkin, *J. Polym. Sci., Part A* **21**, 2357 (1964).
[71] C. D. Armeniades and E. Baer, *in* "Introduction to Polymer Science and Technology" (H. S. Kaufman and J. J. Falcetta, eds.), Chap. 6, Wiley, New York, 1977.
[72] J. F. Hutton, *Nature* (*London*) **200**, 646 (1963).
[73] R. G. King, *Rheol. Acta* **5**, 35 (1966).
[74] K. Walters, "Rheometry." Chapman & Hall, London, 1975.
[75] C. D. Han, M. Charles, and W. Philippoff, *Trans. Soc. Rheol.* **14**, 393 (1970).
[76] C. D. Han and C. A. Villamizer, *J. Appl. Polym. Sci.* **22**, 1677 (1978).
[77] C. D. Han, *Trans. Soc. Rheol.* **18**, 163 (1974).
[78] R. L. Ballman, *Rheol. Acta* **4**, 137 (1965).
[79] J. Meissner, *Rheol. Acta* **8**, 78 (1969); *ibid.* **10**, 230 (1971).
[80] F. N. Cogswell, *Plast. Polym.* **36**, 109 (1968).
[81] J. F. Stevenson, *AIChE J.* **18**, 540 (1972).
[82] C. W. Macosko and J. M. Lorntson, SPE, 31st ANTEC, Prepr. p. 461 (1973).
[83] G. V. Vinogradov, V. D. Fikhman, and B. V. Radushkevich, *Rheol. Acta* **11**, 286 (1972).
[84] J. Rhi-Sausi and J. M. Dealy, *Polym. Eng. Sci.* **16**, 799 (1976).
[85] Y. Ide and J. L. White, *J. Appl. Polym. Sci.* **22**, 1061 (1978).
[86] A. E. Everage and R. L. Ballman, *Nature* (*London*) **273**, 213 (1978).
[87] H. Münstedt, *Rheol. Acta* **14**, 1077 (1975).
[88] H. M. Laun and H. Münstedt, *Rheol. Acta* **15**, 517 (1976).
[89] H. M. Laun and H. Münstedt, *Rheol. Acta* **17**, 415 (1978).
[90] A. E. Everage and R. L. Ballman, *J. Appl. Polym. Sci.* **20**, 1137 (1976).
[91] A. E. Everage and R. L. Ballman, *J. Appl. Polym. Sci.* **21**, 841 (1977).
[92] Y. Ide and J. L. White, *J. Non-Newtonian Fluid Mech.* **2**, 281 (1977).
[93] J. L. White and Y Ide, *J. Appl. Polym. Sci.* **22**, 3057 (1978).
[94] W. Minoshima, J. L. White, and J. E. Spruiell, *Polym. Eng. Sci.* **20**, 1166 (1980).
[95] W. Minoshima, J. L. White, and J. E. Spruiell, *J. Appl. Polym. Sci.* **25**, 287 (1980).
[96] C. D. Denson and R. J. Gallo, *Polym. Eng. Sci.* **11**, 174 (1971).
[97] D. D. Joye, G. W. Poehlein, and C. D. Denson, *Trans. Soc. Rheol.* **16**, 412 (1972); **17**, 287 (1973).

Part I

Dispersed Multiphase Flow in Polymer Processing

When a mixture of two incompatible polymers, or polymer and filler, or polymer and blowing agent, is processed, it forms two distinct phases, the continuous phase and the dispersed phase. For instance, an extruded foam has gas bubbles, a reinforced plastic has solid particles, and a blend of incompatible polymers has the discrete phase dispersed in the continuous phase. It is well understood today that the rheological behavior of heterogeneous polymeric systems is influenced by many factors such as the particle size, particle shape, volume percentage and relative deformability of the dispersed phase, and the state of dispersion. Moreover, different processing conditions (e.g., pressure and temperature) can bring about a substantial change in the mechanical/physical properties of the final products. Basically, the processing of heterogeneous polymeric systems requires the following considerations: (1) control of the rheological properties; (2) an efficient method of mixing; (3) control of the mechanical/physical properties resulting from mixing; (4) control of the microstructure in the solid state.

We encounter flow instabilities in dispersed multiphase flows. When dispersing one liquid in another to form emulsions or polymer blends, breakup of droplets (the discrete phase) occurs. It is well understood today that breakup of droplets is preceded by the elongation of spherical droplets. Therefore, the process of generating small droplets from a long threadlike liquid cylinder may be considered to be an unstable flow phenomenon. From the processing point of view, the extent of breakup of droplets depends on the rheological properties of the two liquids being dispersed, the flow field chosen, and the extent of the rates of deformation. Therefore a better understanding of the mechanism(s) of droplet breakup is very important to the control of the *bulk* rheological properties of liquid–liquid dispersed systems and to the processing of such materials.

In the next four chapters, we shall discuss the rheology and processing of dispersed multiphase polymeric systems. Chapter 3 deals with particulate-filled polymeric systems, Chapter 4 deals with heterogeneous polymeric systems (e.g., blends or emulsions), Chapter 5 deals with the phenomenon of droplet breakup in dispersed flow, and Chapter 6 deals with gas-charged polymeric systems. Emphasis will be placed on the role that the discrete phase plays in determining the *bulk* rheological properties of dispersed multiphase polymeric systems.

3

Dispersed Flow of Particulate-Filled Polymeric Systems

3.1 INTRODUCTION

Composites of polymeric materials and inorganic (or metallic) particles are very common in the plastics and elastomer industries. A composite may be defined as a combination of several distinct materials, designed to achieve a set of properties not possessed by any of the components alone. In practice, composites most often consist of two component materials, one of them forming a continuous phase (matrix) and the other one forming discrete phase dispersed in the matrix.

The continuous phase is most often a polymeric material. Polymeric matrices can be thermoplastic resins, which soften and behave as viscous liquids when heated above their glass transition temperatures (in the case of amorphous thermoplastic resins) or above their melting temperatures (in the case of crystalline thermoplastic resins). Polymeric matrices can also be thermoset resins, which undergo a transformation from a viscous resinous liquid to a hard or rubbery solid in the presence of heat and/or chemicals. Polymeric matrices of either type allow the relatively easy incorporation of a discrete phase (particles or glass fibers) and the processing of the mixture. Composites based on a polymeric matrix are very numerous [1–3]; for example, reinforced thermoplastics containing mica, asbestos; fiber-reinforced thermoset resins containing glass fibers, carbon–graphite, boron, steel; carbon black-reinforced elastomers; and particulate mineral-filled thermoplastics.

The ultimate objective of using extending and reinforcing fillers is to improve the mechanical properties of polymeric materials. However, fillers themselves *usually* supply little or no reinforcement, since there is little interaction between the resin and filler surfaces. This has led to the development of "coupling agents," chemical additives capable of improving the interfacial bond between the filler and the resin.

The use of coupling agents for the surface modification of fillers and reinforcements in polymers has generally been directed towards improving the mechanical strength and chemical resistance of composites by improving adhesion across the interface.

When reinforcing and extending fillers are added to polymers, the resulting material is a complex rheological fluid which is practically impossible to characterize in terms of classic ideal fluids. Polymer processing operations (e.g., injection and compression molding) of filled polymers generally employ between 5 and 60 wt. % of solid particles, and from the point of view of flow properties, particulate-filled polymers may be considered as *concentrated suspensions* of rigid particles. It is then easily surmised that a better understanding of the rheological behavior of concentrated suspensions would help in the choice of optimal processing conditions.

Today, it is well known in the literature [4–9] that, in general, the addition of inert solid particles to a polymer increases the melt viscosity, and decreases the melt elasticity. Therefore the measurement (and prediction) of the rheological properties of concentrated suspensions is of practical interest in controlling polymer processing operations.

The volume fraction of the particles in a suspension (volume occupied by particles per unit volume of suspension) as well as other factors, such as the shape of the particles, the particle size and its distribution, and the state of dispersion of the particles (well-dispersed state or agglomerated state) are known to influence the rheological properties of filled polymer melts [10–13].

The role of the glass fiber in the flow of the suspension may be compared to the situation where, in a polymer solution, some of the coiled molecules are oriented during flow. Figure 3.1 describes the extent of fiber orientation as the polymeric liquid flows into the entrance of a rectangular channel [14]. It is seen that alignment of the long threadlike glass fiber parallel to the axis of extension occurs in the entrance region of the die, where tensile stresses are high compared to shearing stresses at the same deformation rates. For a polymer solution, this may be considered as equivalent to the uncoiling of coiled macromolecules into elongated conformations.

In this context, it is worth mentioning that when adding a small amount (0–5%) of very long, flexible fibrils of polytetrafluoroethylene (Teflon) to linear polyethylene, Busse [15] observed an increase in the normal stress effect of the polyethylene, in terms of the extrudate swell. On the other hand,

Fig. 3.1 Photograph displaying the orientation of long glass fibers in the die entrance [14]. Reprinted by permission of *Modern Plastics Magazine*, McGraw-Hill, Inc.

he did *not* observe an increase in extrudate swell when glass beads and asbestos were added to the same resin. Busse postulated that the very long thin filaments of Teflon penetrate through and entangle many randomly coiled polymer molecules and act as an elastic coupling between them.

There are situations where solid particles in a polymeric material influence the processing characteristics. For example, the presence of carbon black in unvulcanized rubber reduces the degree of extrudate distortion, as shown in Fig. 3.2 [16]. In view of the fact that the phenomenon of melt fracture, causing extrudate distortion, is attributable to the normal stress effect [5], the enhancement of the extrusion characteristics may be due to a decrease in the normal stress effect caused by the presence of the carbon black in the elastomer [11, 12, 17–19].

The surface modification of the filler particles also affects the rheology of the polymer by changing the dispersion of the particles, and thereby the viscosity, normal stresses, and flow during polymer processing operations. Coupling agents, designed to improve the coupling of the fillers to the resins, often improve the processability of filled polymeric materials [20–22].

Fig. 3.2 Photographs showing the effect of carbon black on the shape of extrudate of oil extended styrene–butadiene rubber (SBR): (a) without carbon black; (b) with carbon black (40 phr). Extrusion conditions: $T = 100°C$; $L/D = 4$ ($D = 0.152$ cm).

In this chapter, we shall first discuss the rheological behavior of concentrated suspensions with Newtonian fluids as the suspending medium, and then the rheological behavior of concentrated suspensions with viscoelastic fluids as the suspending medium. Along this line, we shall review some of the semiempirical expressions, suggested in the literature, for correlating experimental data of the rheological properties of suspensions of rigid particles dispersed in Newtonian fluids. Although our primary interest is in particles dispersed in *polymeric* materials, which in most cases are viscoelastic, there is little theoretical work published in this important area. Therefore we shall review the theoretical work published on suspensions of rigid particles in

Newtonian fluids with the hope that such a review will shed some light on the important role that rigid particles play in influencing the *bulk* (macroscopic) rheological properties of concentrated suspensions with viscoelastic fluids. Finally, we shall discuss some polymer processing operations involving highly filled thermoplastic resins and reinforced thermoset resins. Emphasis will be placed on demonstrating the effects of processing condition on the mechanical properties of the composites produced.

3.2 THE RHEOLOGICAL BEHAVIOR OF CONCENTRATED SUSPENSIONS OF RIGID PARTICLES IN NEWTONIAN FLUIDS

When the particles are spherical, and no external force or couple acts on them, the suspension has a wholly isotropic structure and so behaves as a Newtonian fluid for sufficiently small rates of strain. In these circumstances, the effect of the presence of the particles is simply equivalent to an increase in the shear viscosity of the suspension.

In recent years, research on suspension rheology has been directed toward the extensions of these well-known results, in particular for the cases of (a) more concentrated suspensions of spherical particles, (b) deformable particles with both viscous and elastic properties, and (c) nonspherical rigid particles. None of these extensions is yet complete, and there are also other problems which have not yet been given much consideration, such as the effects of inertia forces in the relative motion near a particle, the effect of an externally imposed couple on the particles, and the effect of the polydispersity of particles. Our understanding of the rheological properties of a suspension of rigid (and deformable) particles is still in its early stages [23–26].

The reader should be reminded that in this chapter we are primarily interested in the rheological behavior of highly filled polymer systems, insofar as it is relevant to the processing of such materials. As will be shown below, the rheological behavior of highly filled polymers is indeed very complicated in that: (1) the suspending medium is, more often than not, a viscoelastic liquid, and (2) the level of concentration (the loading of filler particles) is usually very high. Realizing that very few rigorous theories for predicting the *bulk* rheological properties of suspensions in viscoelastic fluids are available in the literature, we shall briefly review the existing theories of suspension rheology, which deal, almost invariably, with suspensions having Newtonian liquids as the suspending medium. We believe, however, that such a review will serve a useful purpose in providing a better understanding of the rheological behavior of more complex fluid systems, namely, highly filled polymeric materials.

3.2.1 Shearing Flow of Concentrated Suspensions in Newtonian Fluids

(a) Theoretical Considerations

The ultimate objective of theoretical suspension rheology is to develop a theory (or theories), whereby the *bulk* (macroscopic) rheological properties of a suspension may be predicted from a knowledge of the properties of the particles (e.g., shape, size, and its distribution) and the suspending fluid. For this, one needs to determine the relation between the *macroscopic* rheological properties of the suspension and its *microscopic* structure on the particle scale.

It has been shown theoretically that when the Reynolds number of relative motion near one particle is small (i.e., inertia is negligible), a suspension of couple-free particles (i.e., at a dilute concentration) of spherical shape is *Newtonian* when such particles are suspended in a Newtonian fluid. The presence of rigid particles in a liquid raises the *bulk* viscosity which is generally higher than the viscosity of the suspending medium.

Einstein [27] was the first to develop a theory for predicting the viscosity of a dilute suspension of rigid spheres and obtained the following expression for the *effective* viscosity η of a suspension:

$$\eta = \eta_0(1 + 2.5\phi) \tag{3.1}$$

where η_0 is the Newtonian viscosity of the suspending medium and ϕ is the volume fraction of the spheres. Equation (3.1) is valid only for extremely *dilute* suspensions, in which interactions between neighboring particles are negligible (i.e., in the absence of hydrodynamic interactions), and for a *Newtonian* fluid as the suspending medium. Einstein's derivation of Eq. (3.1) is based on the energy dissipation, i.e., the energy dissipated in the suspension is equated to the energy dissipated in a liquid having the effective viscosity η.

Following Einstein [27], Jeffery [28] investigated the motion of *non-spherical* particles (rigid ellipsoidal particles) in a shear field of Newtonian liquid, on the basis of the creeping flow equations, and derived an expression for the *effective* viscosity

$$\eta = \eta_0(1 + \overline{v}\phi) \tag{3.2}$$

where $\overline{v}$ is a parameter which depends on the geometry of ellipsoidal particles. Jeffery reports that $\overline{v}$ is less than 2.5, which is the value for spherical particles derived earlier by Einstein [27] [see Eq. (3.1)]. The significance of the Jeffery study lies in that the presence of *nonspherical* particles in a Newtonian liquid can give rise to *non-Newtonian* flow behavior.

For very *dilute* suspensions, one may neglect the interactions among particles suspended in the medium, which simplifies the theoretical treatment considerably. However, for *concentrated* suspensions, one has to give consideration to particle–particle interactions, which could influence the flow properties. A number of semiempirical equations have been developed to predict the bulk viscosity of suspensions. One such equation may be represented by a power series in volume fraction ϕ:

$$\eta/\eta_0 = 1 + a_1\phi + a_2\phi^2 + a_3\phi^3 \tag{3.3}$$

where a_1 is 2.5 for spherical particles, and a_2 allows for hydrodynamic or other interactions between the particles at higher values of ϕ than originally proposed by Einstein [27]. The reported values of a_2 range between 2.5 and 15.0 [29, 30]. It has been reported that a_2 increases with decreasing particle size, but the exact nature of the interrelation has not been established. Few values have been reported for the constant a_3.

Simha [31] has developed a "cell model" and obtained the following theoretical expression for the viscosity of concentrated suspensions of spheres:

$$\frac{\eta}{\eta_0} = 1 + \frac{5}{2}\phi\left\{1 + \frac{25}{4f^3}\phi + \frac{75}{4f^4}\phi^{4/3} + \frac{27}{f^5}\phi^{5/3} + \frac{785}{16f^6}\phi^2 + O\left(\frac{\phi^{7/3}}{f^7}\right)\right\} \tag{3.4}$$

in which f is regarded as a semiempirical quantity that depends on the particle size and volume fraction ϕ. According to Simha, f increases with ϕ and approaches a limit given by

$$f^3 = 8\phi_m \tag{3.5}$$

in which ϕ_m is the maximum concentration corresponding to close packing. For a monodisperse suspension of spherical particles, the value of ϕ_m is bounded by the volume fractions for touching spheres in cube and hexagonal packing (0.52 and 0.74) and will be close to the volume fraction for random packing (0.62). According to Simha, as ϕ approaches ϕ_m, Eq. (3.4) reduces to

$$\lim_{\phi \to \phi_m}\left(\frac{\eta}{\eta_0}\right) = \frac{54}{5f^3}\left[\frac{\phi^2}{(1 - \phi/\phi_m)^3}\right], \qquad 1 < f < 2 \tag{3.6}$$

Equation (3.6) may be compared with other similar semiempirical expressions obtained by various investigators, namely:

Eilers [32]:

$$\eta/\eta_0 = \tfrac{25}{16}[\phi^2/(1 - \phi/\phi_m)^2] \tag{3.7}$$

Mooney [33]:

$$\ln(\eta/\eta_0) = 2.5\phi/(1 - C_1\phi), \qquad 1.35 < C_1 < 1.91 \tag{3.8}$$

It should be noted that the above classical equations defining the viscosity of concentrated suspensions contain no terms for the influence of particle size. Sherman [34] obtained the following empirical correlation, which does contain the influence of particle size,

$$\ln\left(\frac{\eta}{\eta_0}\right) = \left[\frac{C_1 a}{(\phi_m/\phi)^{1/3} - 1}\right] + C_2 \tag{3.9}$$

where C_1 and C_2 are constants which depend on particle size a, and ϕ_m is about 0.74 on the basis of experiment. Presumably, the form of Eq. (3.9) varies with the degree of polydispersity of the system.

Without recourse to empiricisms, Frankel and Acrivos [35] derived a theoretical expression for relating the effective viscosity of concentrated suspensions to volume fraction for rigid spheres. In their analysis, the hydrodynamic interactions among particles in relative motion, and in close proximity to one another, were considered, and the suspension was assumed to behave as a Newtonian continuum on a macroscopic scale. In developing the asymptotic solution, they adopted the point of view that the viscous dissipation of energy in highly concentrated suspension, which, due to Einstein [27], is related to the effective viscosity, arises primarily from the flow within the narrow gaps separating the various rigid spheres from one another. Their asymptotic solution yields [35]:

$$\lim_{\phi \to \phi_m} \left(\frac{\eta}{\eta_0}\right) = \frac{9}{8}\left[\frac{(\phi/\phi_m)^{1/3}}{1 - (\phi/\phi_m)^{1/3}}\right] \tag{3.10}$$

Goddard [36] also derived an equation very similar to Eq. (3.10), with a somewhat different numerical coefficient ($\pi/8$ instead of 9/8).

In the past, therefore, efforts have been spent on developing rheological models by considering suspensions as *continuous media* (homogeneous fluids), instead of mixtures of particles and fluid. Such an effort is justifiable when the length scales describing the motion of the suspension as a whole are much larger than the average size or average separation of the particles. In such an approach, as described in Chapter 2, one develops a constitutive equation which relates the stress tensor to the tensors that measure the rate of deformation (or strain).

Batchelor [37] has provided a new assessment of the means by which one proceeds from a knowledge of microscopic flow behavior to an expression of "bulk stress" in a suspension. In crude terms, we wish to know what stress is generated in the suspension when a prescribed bulk motion is imposed on it. However, the velocity, pressure, and stress all vary with position in the suspension, depending on proximity to a particle, and the terms "bulk stress" and "bulk velocity gradient" in the suspension have meaning only in some integral or average sense, and the method of averaging must now be specified. According to Batchelor [37], the bulk stress and bulk velocity gradient in the suspension may be defined as ensemble averages or volume averages, yielding the full constitutive equation, in contrast to the Einstein "viscous dissipation" approach [27]. Averaging processes for rheological properties have also been studied by Hashin [38], Roscoe [39], and Goddard and Miller [40]. Details of these averaging processes are beyond the scope of our interest here, and therefore the reader may refer to the original papers.

Using an averaging technique, Lin *et al.* [41] investigated the effect of inertia on the rheological behavior of a *dilute* suspension and obtained the following expressions for the viscometric functions:

$$\eta/\eta_0 = 1 + \phi(2.5 + 1.34\,\mathrm{Re}^{3/2}) \tag{3.11}$$

$$\tau_{11} - \tau_{22} = \eta_0\dot{\gamma}\phi\,\mathrm{Re}(-\tfrac{4}{3} + 0.287\,\mathrm{Re}^{1/2}) \tag{3.12}$$

$$\tau_{22} - \tau_{33} = \eta_0\dot{\gamma}\phi\,\mathrm{Re}(\tfrac{2}{3} - 0.252\,\mathrm{Re}^{1/2}) \tag{3.13}$$

where Re is the particle shear Reynolds number $\mathrm{Re} = \dot{\gamma}a^2\rho/\eta_0$, in which $\dot{\gamma}$ is shear rate, a is the particle radius, ρ is the fluid density, and η_0 is the fluid viscosity.

Note that Eqs. (3.11)–(3.13) are derived on the premise that uniform, rigid spheres are suspended in a Newtonian fluid and hydrodynamic interactions among particles are negligible. It is seen from these equations that the effective viscosity of the suspension increases with an increase in Re, that the first normal stress difference is negative at low Re and becomes positive at high Re, and that second normal stress difference is positive at low Re and becomes negative at high Re. Experimental verification of these predictions is not available in the literature. It appears that these predictions of normal stress differences run in a direction opposite to what has been observed in some limited experiments. The significance of this theoretical study, however, lies in that the inertia in the suspension can give rise to nonisotropic normal stresses, even when a Newtonian fluid is used as the suspending medium.

The non-Newtonian behavior of suspensions can also be predicted by considering the nonsphericity of particles, as pointed out by Jeffery [28] as early as in 1922. Recently, Hinch and Leal [42] performed a theoretical analysis of a viscometric flow of a dilute suspension by including Brownian forces acting on nonspherical particles, and obtained the following expressions:

$$\frac{\eta}{\eta_0} = 1 + 2.5\phi + \varepsilon^2\left\{k_1 + \frac{k_2}{1 + (\dot{\gamma}/6D_r)^2}\right\}\phi + O(\phi\varepsilon^3) \tag{3.14}$$

$$\tau_{11} - \tau_{22} = \left\{\frac{\eta_0 D_r k_3 \varepsilon^2}{1 + (6D_r/\dot{\gamma})^2}\right\}\phi + O(\phi\varepsilon^3) \tag{3.15}$$

$$\tau_{22} - \tau_{33} = -\tfrac{1}{6}(\tau_{11} - \tau_{22}) + O(\phi\varepsilon^3) \tag{3.16}$$

where

$$D_r = k_B T/8\pi\eta_0 a^3 \tag{3.17}$$

In Eqs. (3.14)–(3.17), k_1, k_2, k_3 are known constants; D_r is the rotational Brownian diffusion constant for a dilute suspension of spheres, where k_B is the Boltzmann constant; T is the temperature; η_0 is the fluid viscosity (Newtonian); a is the effective particle radius equivalent to a sphere; and ε describes a measure of deviation from sphericity. It is seen in Eqs. (3.14)–(3.16) that for $\varepsilon = 0$ (perfectly spherical particles), the non-Newtonian effects disappear, and that the first normal stress difference is positive and the second normal stress difference is negative. The significance of this theoretical study lies in that the nonsphericity of particles suspended in a Newtonian fluid gives rise to normal stress effects in the suspension.

There are situations where interparticle effects (i.e., London–van der Waals attraction and electrostatic repulsion) must be taken into account in predicting theoretically the rheological properties of suspensions [43–50]. In certain colloidal systems, flocculation of particles may take place due to London–van der Waals attractions, influencing the rheological properties of suspensions. However, when electrostatic repulsions become appreciable, the rate of flocculation is reduced. Consideration of interparticle effects is very important in describing the stability of colloidal dispersions.

(b) Experimental Observations

It has been known for some time that concentrated suspensions of rigid particles in a Newtonian fluid can exhibit non-Newtonian behavior, such as yield stress, shear-thinning behavior, and shear-thickening behavior [51–63]. It has been found that the flow behavior of various suspensions in a Newtonian fluid, over the range of shear rates investigated, may be fitted satisfactorily by the Bingham plastic model [64]:

$$\begin{aligned} \dot{\gamma} &= 0, \qquad \text{for} \quad \tau_w < Y \\ \tau_w &= Y + \eta_0 \dot{\gamma}, \qquad \text{for} \quad \tau_w > Y \end{aligned} \tag{3.18}$$

in which τ_w is the wall shear stress, Y denotes the yield stress, η_0 denotes the zero-shear viscosity of the medium, and $\dot{\gamma}$ denotes apparent shear rate.

A tensorial formulation of a Bingham plastic fluid was first introduced by Hohenemser and Prager [65] and by Oldroyd [66], using the von Mises criterion [67]:

$$\operatorname{tr} \boldsymbol{\tau}^2 = 2Y^2 \tag{3.19}$$

in which $\boldsymbol{\tau}$ is the stress tensor and Y is a *yield* value. A three-dimensional form of Eq. (3.18) may then be written [66]

$$\mathbf{d} = 0, \qquad \text{for} \quad \operatorname{tr} \boldsymbol{\tau}^2 < 2Y^2 \tag{3.20}$$

$$\boldsymbol{\tau}\{1 - Y(\mathrm{II}_{\tau})^{-1/2}\} = 2\eta_0 \mathbf{d}, \qquad \text{for} \quad \operatorname{tr} \boldsymbol{\tau}^2 > 2Y^2 \tag{3.21}$$

or

$$\boldsymbol{\tau} = 2\{\eta_0 + Y(4\mathrm{II}_{\mathbf{d}})^{-1/2}\}\mathbf{d}, \qquad \text{for} \quad \mathrm{tr}\,\boldsymbol{\tau}^2 > 2Y^2 \tag{3.22}$$

in which $\mathrm{II}_{\boldsymbol{\tau}}$ is the second invariant of the "extra" stress tensor $\boldsymbol{\tau}$ (i.e., $\mathrm{II}_{\boldsymbol{\tau}} = \frac{1}{2}\mathrm{tr}\,\boldsymbol{\tau}^2$), $\mathrm{II}_{\mathbf{d}}$ is the second invariant of the rate-of-deformation tensor $\mathbf{d}$ (i.e., $\mathrm{II}_{\mathbf{d}} = \frac{1}{2}\mathrm{tr}\,\mathbf{d}^2$), and η_0 is the viscosity of the medium. Note that Eqs. (3.20)–(3.22) hold only for small strain rates (or small rates of deformation) and are limited to the von Mises yield criterion [67]. Other yield criteria (e.g., Tresca yield criterion) have also been suggested in the literature [68, 69].

Einstein's theory, Eq. (3.1), has been tested experimentally by a number of investigators. Some investigators report that the coefficient of ϕ in Eq. (3.1) ranges somewhere between 1.5 and 5.0. This deviation may be attributable to one or more of the following factors: (1) the particles in suspension might have formed aggregates [violating one of the assumptions made in the derivation of Eq. (3.1)], which can give rise to a Einstein coefficient larger than 2.5; (2) the particle size is not uniform, (3) the state of dispersion is nonuniform, so that the medium cannot be considered to be isotropic.

Figure 3.3 gives plots of *relative* viscosity, $\eta_r = \eta(\phi)/\eta(0)$, versus volume fraction of particles ϕ for two sets of experimental data, in which Eq. (3.8) was curve fitted with an adjustable parameter C_1. It is not surprising to

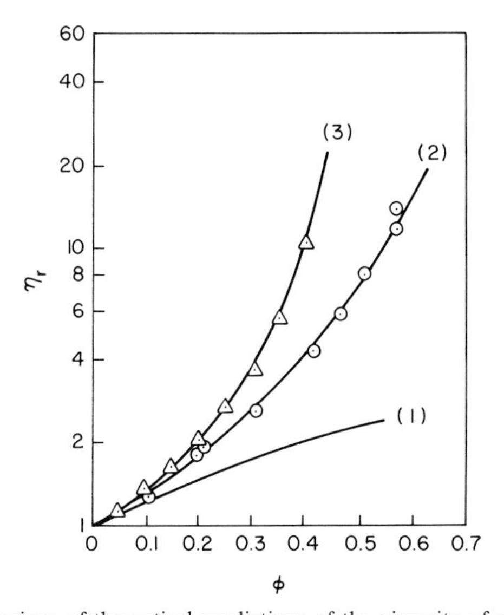

Fig. 3.3 Comparison of theoretical predictions of the viscosity of concentrated suspensions with experimental data [33]: curve (1) represents the Einstein theory, Eq. (3.1); curve (2) represents the Eilers' data [32] fitted to Eq. (3.8) with $C_1 = 0.75$; curve (3) represents the Vand's data [29] fitted to Eq. (3.8) with $C_1 = 1.43$.

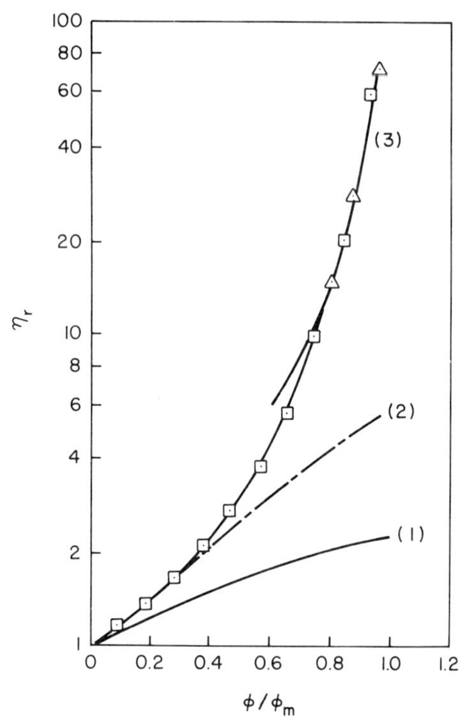

Fig. 3.4 Comparison of theoretical predictions of the viscosity of concentrated suspensions with experimental data [35]: curve (1) represents the Einstein theory, Eq. (3.1); curve (2) represents the Simha theory, Eq. (3.4); curve (3) represents the asymptotic theory of Frankel and Acrivos, Eq. (3.10). The experimental data are: (⊡) $\phi_m = 0.535$ due to Rutgers [23]; (△) $\phi_m = 0.625$ due to Thomas [59]. Reprinted with permission from *Chemical Engineering Science* **22**, N. A. Frankel and A. Acrivos, Copyright 1967, Pergamon Press, Ltd.

see that Eq. (3.8), having an adjustable parameter, fits data better than Einstein's equation, Eq. (3.1), which is valid only for dilute suspensions (i.e., for small values of ϕ).

Figure 3.4 gives a comparison between the experimental data and the theoretical expressions. It is seen that Eq. (3.10) is in good agreement with the data at values of $0.75 < \phi/\phi_m < 1.0$, whereas other theories are in reasonable agreement with the data at values of ϕ/ϕ_m less than 0.4.

Figure 3.5 gives plots of wall shear stress versus apparent shear rate for suspensions of glass beads ($5 \sim 45$ μm in diameter) in Indopol L100 (a low molecular weight polybutene, which follows Newtonian behavior). These data were obtained by use of a Weissenberg rheogoniometer. It is seen that the suspensions give rise to yield stress Y, which increases with the volume fraction of glass beads ϕ, as given in Fig. 3.6.

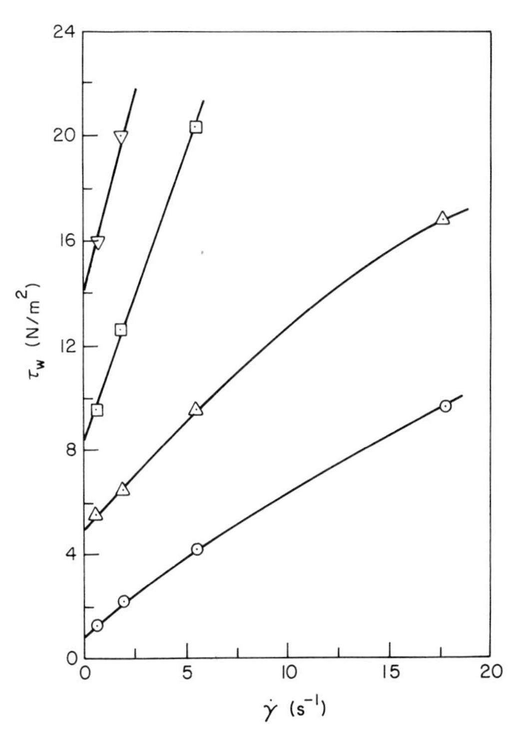

Fig. 3.5 Shear stress versus shear rate for a low molecular weight polybutene (Indopol L100) suspended with glass bead (vol.%): (⊙) 5; (△) 10; (⊡) 15; (▽) 20.

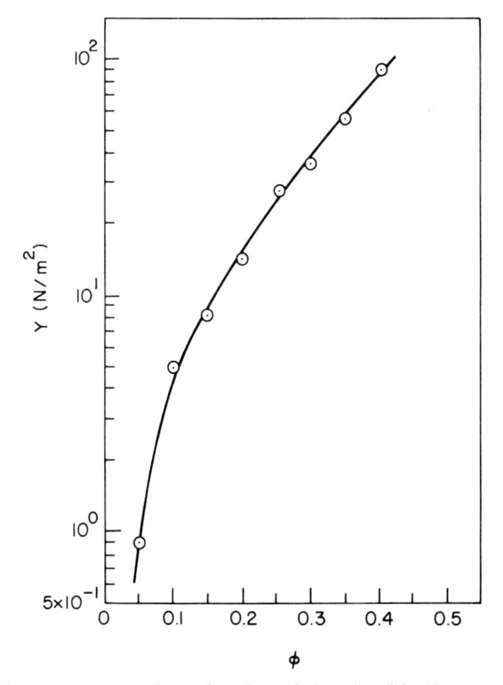

Fig. 3.6 Yield stress versus volume fraction of glass bead in the suspensions of Indopol (L100).

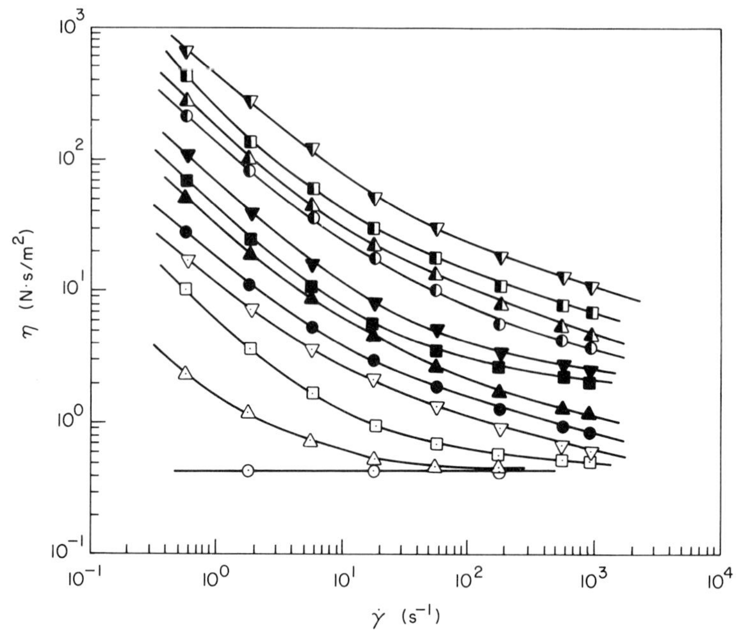

Fig. 3.7 Viscosity versus shear rate for low molecular weight polybutene (Indopol L100) suspended with glass bead (vol.%): (⊙) 0.0; (△) 5; (⊡) 10; (▽) 15; (●) 20; (▲) 25; ( ) 30; (▼) 35; (◐) 40; (◭) 45; (◧) 50; (▼) 55.

Figures 3.7 and 3.8 give plots of *effective* viscosity versus apparent shear rate and effective viscosity versus shear stress, respectively, for suspensions of glass beads in Indopol L100. It is seen that, over the range of shear stresses investigated, as shear stress decreases, the effective viscosity rapidly increases (it becomes especially pronounced as the volume fraction of glass beads increases), and that as shear stress increases, the effective viscosity approaches a constant value whose magnitude increases with the volume fraction of particles. Shear-thinning behavior of concentrated suspensions of rigid particles is attributed to "crowding" [51].

There is experimental evidence that some suspensions exhibit shear-thickening behavior [60–62]. Of particular interest is the study of Hoffman [60, 61], who observed both shear-thinning and shear-thickening behavior in concentrated suspensions, depending on the rates of shear applied, as may be seen in Fig. 3.9. It is seen that for volume fractions ϕ of above 0.5, a discontinuity in the effective viscosity occurs as the apparent shear rate reaches a certain critical value. Using an optical technique to measure diffraction patterns, Hoffman [61] observed that there was a change in flow patterns of particles (from an ordered array of particles to a disordered array) at the shear rate at which the discontinuity in effective viscosity was actually

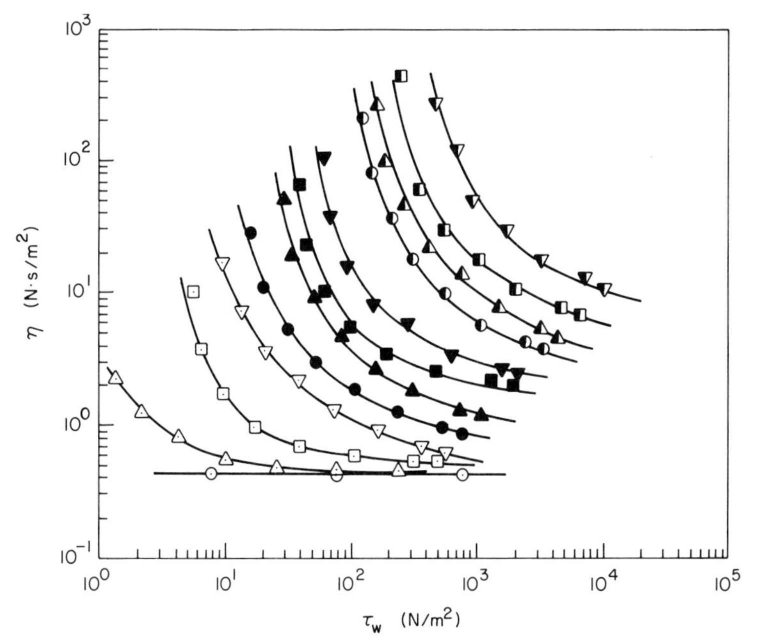

Fig. 3.8 Viscosity versus shear stress for Indopol L50 suspended with glass bead. Symbols are the same as in Fig. 3.7.

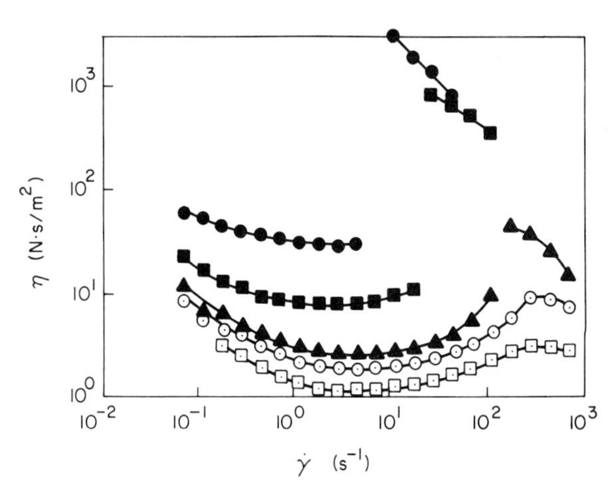

Fig. 3.9 Viscosity versus shear rate for dioctyl phthalate suspended with PVC bead (vol. %) [60]: (⊡) 47; (⊙) 49; (▲) 51; (■) 55; (●) 57. From R. L. Hoffman, *Trans. Soc. Rheol.* **16**, 155. Copyright © 1972. Reprinted by permission of John Wiley & Sons, Inc.

observed. This appears to indicate that the rheological properties of concentrated suspensions depend on the state of dispersion of the rigid particles suspended in the continuous medium. Shear-thickening behavior of concentrated suspensions appears to depend on certain features of the microstructure of rigid particles. For instance, it is reported [62] that shear-thickening (dilatant) behavior was observed in suspensions of titanium dioxide particles (in water or aqueous sucrose), but not in suspensions of glass beads in Newtonian fluids.

Many colloidal suspensions are also known to exhibit shear-thinning behavior. This non-Newtonian behavior has been attributed to mechanisms in which the shear stress, transmitted through the continuous medium, orients (or distorts) the suspended particles in opposition to the randomizing effects of Brownian motion. Variation of viscosity with shear rate is then a result of the lowered resistance to flow offered by the oriented (or distorted) arrangements. Later in this chapter, the rheological properties of suspensions of nonspherical particles (e.g., glass fiber) will be discussed from this point of view.

Another important variable that may affect the rheological properties of concentrated suspensions is the particle size and its distribution [29, 34, 63, 70–73]. Chong *et al.* [63] investigated the dependence of the effective viscosities of concentrated suspensions of glass beads on particle size distribution. They noted that for monodisperse systems, the effective viscosity was independent of the particle size and was a function only of the solids concentration. Figure 3.10 gives plots of relative viscosity, $\eta_r = \eta(\phi)/\eta(0)$, versus

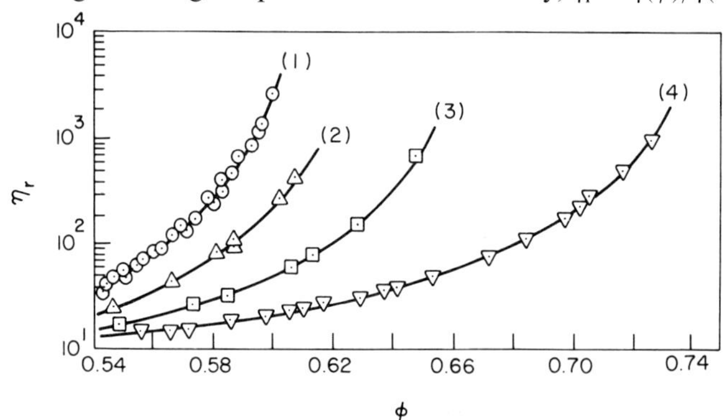

Fig. 3.10 Relative viscosity versus volume fraction for a low molecular weight polyisobutylene (exhibiting the Newtonian behavior, $\eta_0 = 2\text{Ns/m}^2$, at room temperature) suspended with glass beads having bimodal size distribution (small particle diameter, $d_q = 80 \sim 125\ \mu\text{m}$; large particle diameter, $d_l = 210 \sim 250\ \mu\text{m}$) [63]: (⊙) monodispersed system; (△)$\phi_s = 0.25$, $d_s/d_l = 0.477$;(□) $\phi_s = 0.25$, $d_s/d_l = 0.313$;(▽) $\phi_s = 0.25$, $d_s/d_l = 0.138$. ϕ_s denotes the volume fraction of small particles in the suspension, and d_s/d_l is the particle size ratio. From J. S. Chong *et al.*, *J. Appl. Polym. Sci.* **15**, 2007, Copyright © 1971. Reprinted by permission of John Wiley & Sons, Inc.

volume fraction of particle ϕ, in which bimodal distributions of spheres were used, with 25 percent by volume of the total solids as small spheres. Note that in Fig. 3.10 the particle diameter ratio of small to large spheres, d_s/d_l, was used as a measure of particle size distribution. It is seen that the relative viscosity decreases as the particle diameter ratio decreases, indicating that polydisperse systems yield lower viscosities than monodisperse systems. Chong *et al.* [63] noted that when the particle diameter ratio reached about 0.1, the relative viscosity did not decrease to any appreciable amount.

Note that in the suspension of nonspherical particles immersed in a Newtonian fluid, a couple (but no force) may be imposed by external means. The effect of exerting a coupling on the particle (Brownian motion) gives rise to *non-Newtonian* behavior [74, 75]. Non-Newtonian behavior of a dilute suspension of rigid spherical particles suspended in a Newtonian fluid is reported by Segré and Silberberg [76]. They attributed the non-Newtonian behavior observed to radial migration of solid particles in a cylindrical tube.

It has been observed that suspended particles sometimes migrate across the streamlines during flow. Goldsmith and Mason [77] studied the suspension of spheres, rods, and discs flowing through tubes, reporting that at very low Reynolds numbers ($\mathrm{Re} < 10^{-6}$), the particles do not migrate. Karnis *et al.* [78, 79] observed however that at a somewhat higher Reynolds number ($\mathrm{Re} = 10^{-3}$–0.36), the particles migrate towards the tube axis and a particle-free layer develops near the tube wall. Gauthier *et al.* [80] reported that in tubular flow, there were some migrations of neutrally buoyant particles, suspended in a dilute viscoelastic polymer solution, away from and towards the tube wall. Theoretical studies on the migration of rigid particles suspended in a Newtonian fluid have been reported by Cox and Brenner [81] and Ho and Leal [82].

Figure 3.11 gives plots of viscosity and first normal stress difference versus shear stress for a Newtonian fluid (Indopol L100) containing glass fiber, having a radius of 1.32×10^{-3} cm and a length of 1.27 cm. It is seen that the viscosity of the suspension first decreases and then levels off as the shear stress is increased, while exhibiting a yield value at a low shear stress. This behavior is typical of concentrated suspensions, as shown in Fig. 3.8. What is, however, of great interest in Fig. 3.11 is the appearance of a first normal stress difference. According to Han [83], when glass beads ($0.5 \times 10^{-3} \sim 4.0 \times 10^{-3}$ cm in diameter) were suspended in the same medium (Indopol L100), the suspension gives rise to *no* measurable first normal stress difference, while exhibiting viscosity behavior very similar to that of glass fiber-suspended Indopol.

Figure 3.12 gives photographs demonstrating the difference in the "rod climbup" effects between the glass bead-suspended Indopol and the glass fiber-suspended Indopol. It is quite clear that only the glass fiber-suspended Indopol exhibits the Weissenberg rod climbup effects [84], giving credence

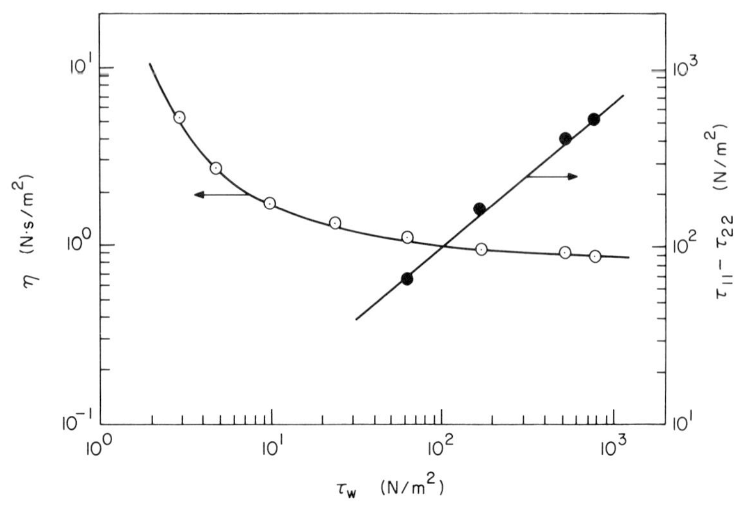

Fig. 3.11 Viscosity and first normal stress difference versus shear stress for a low molecular weight polybutene (Indopol L100) suspended with glass fiber (0.3 vol.%) at 25°C.

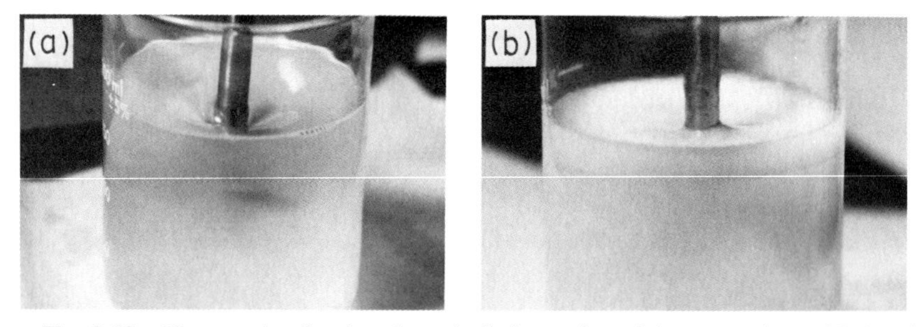

Fig. 3.12 Photographs showing the rod climb-up effect of the suspensions of Indopol L100: (a) suspended with glass bead (5 ~ 40 μm); (b) suspended with glass fiber (length = 1.27 cm; aspect ratio = 962).

to the measurement of the first normal stress difference shown in Fig. 3.11. Other investigators [85–87] also observed the rod climbup effects with Newtonian fluids suspended with long, flexible fibers.

3.2.2 Extensional Flow of Suspensions

The extensional flow of suspensions, in particular suspensions containing long fibrils, is of theoretical interest in that the suspended fibrils may align along the direction of stretching. Thus one is tempted to apply the so-called slender-body theory [88–92] for predicting the rheological behavior of such suspensions, subjected to extensional flow.

Batchelor [92] has given a fairly comprehensive analysis of the extensional flow behavior of elongated particles, oriented in a uniaxial direction, and suspended in a Newtonian fluid. In his analysis, Batchelor assumed that interactions among particles are negligible, and that the shape of particles is rodlike, justifying the use of the slender-body theory. He then obtained the following expression for an *effective* elongational viscosity η_E:

$$\eta_E = 3\eta_0 \left[\frac{(4\pi/9)Nl^3}{\ln(2l/R_0) - \ln(1 + 2l/h) - \frac{3}{2}} \right] \tag{3.23}$$

where η_0 is the viscosity of the Newtonian suspending medium, N is the number of particles per unit volume, $2l$ and R_0 are the particle length and radius, respectively, and

$$h = 1/(2Nl)^{1/2} \tag{3.24}$$

is a measure of the distance between particles. Equation (3.23) has the appropriate asymptotic forms, approaching either the dilute-suspension or close-particle formula as the ratio h/l becomes large or small, respectively.

Batchelor [92] obtained the following expression as an asymptotic form of Eq. (3.23).

$$\eta_E = 3\eta_0 \left[1 + \frac{4}{9} \frac{(l/R_0)^2 \phi}{\ln(\pi/\phi)} \right] \tag{3.25}$$

where l/R_0 is the aspect ratio of the rodlike particles and $\phi = \pi R_0^2 l N$ is the volume fraction of particles in the suspension. It is seen in Eq. (3.25) that the second term on the right-hand side represents the contribution from the presence of particles in the suspension, and that even though ϕ is small, the factor l/R_0 can produce substantial increases in η_E/η_0 over the Newtonian value of 3. The reader may refer to Chapter 2 for the discussion of elongational viscosity.

The significance of Batchelor's work is that even small concentrations of slender rodlike particles, suspended in a Newtonian fluid, can give rise to *non-Newtonian* behavior in extensional flow. There are some experimental results [87, 93, 94], though not many, which support the theoretical predictions of Eq. (3.25).

Weinberg and Goddard [94] carried out extensional flow experiments using glass fiber suspensions in Newtonian fluids, and found that the tensile stresses of the suspensions were 9 to 10 times those of the suspending fluids, as shown in Fig. 3.13, while the tensile stress calculated from Eq. (3.23) was 8.4 times that of the suspending fluid. Since the glass fiber particles used in the Weinberg–Goddard experiment may not have been of uniform size and Batchelor's theory has some simplifying assumptions, the agreement between the two is considered good. Other investigators [87, 93] also reported measurements of tensile stresses of glass fiber suspensions, exhibiting non-

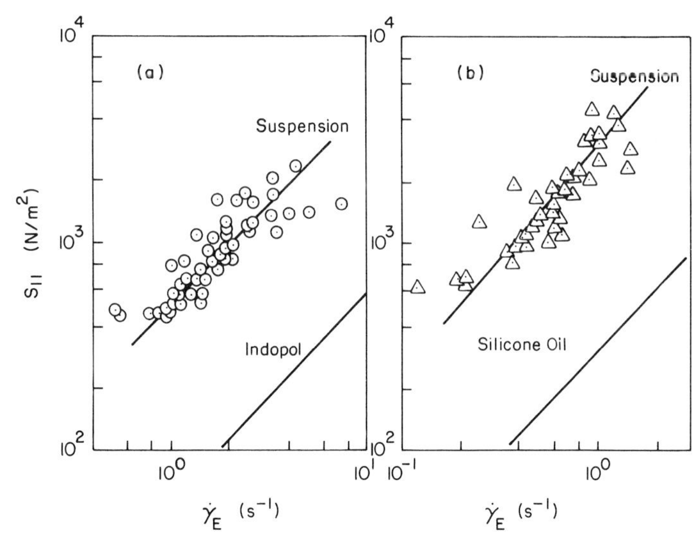

Fig. 3.13 Tensile stress versus elongation rate for Newtonian liquids suspended with glass fiber [94]: (a) Indopol ($\eta_0 = 20.5$ N s/m^2) suspended with glass fiber; (b) Silicone oil ($\eta_0 = 102.5$N s/m^2) suspended with glass fiber. The glass fibers employed have 3.5 μm in average diameter and 200 μm in average length. The lower solid lines represent the theoretical stresses for the pure liquids ($S_{11} = 3\eta_0\dot{\gamma}_E$). Reprinted with permission from *International Journal of Multiphase Flow* **1**, C. B. Weinberg and J. D. Goddard, Copyright 1975, Pergamon Press, Ltd.

Newtonian behavior in extensional flow. The marked increase in tensile stress observed with glass fiber suspension may be attributable to alignment of the rodlike particles parallel to the direction of stretching.

It should be mentioned that for dilute suspensions containing spherical particles, the Einstein theory implies the following expression:

$$\eta_E = 3\eta_0(1 + 2.5\phi) \tag{3.26}$$

which reduces to Eq. (2.11) when $\phi = 0$.

3.3 THE RHEOLOGICAL BEHAVIOR OF HIGHLY FILLED POLYMERIC SYSTEMS

Fillers and reinforcements are high-modulus particles and fibers, which are dispersed in polymer matrices to improve processing and the mechanical/physical (or optical) properties of the final products. In particulate fillers, fine particle size and high surface area generally favor reinforcement. In fibrous reinforcement, the fiber length-to-diameter (aspect) ratio is of primary importance.

From the polymer processing point of view, one is interested in information on the *bulk* rheological properties, either experimentally determined or theoretically predicted, of molten polymers containing particulate fillers or reinforcements. In recent years, some research efforts [4–13, 95–102] have been directed toward a fundamental understanding of the rheological behavior of highly filled polymeric systems. Such systems may be considered as *concentrated* suspensions of rigid particles in *viscoelastic* fluids because, except for certain liquid resins (e.g., urethane or epoxy precursors), almost all thermoplastic resins exhibit non-Newtonian, viscoelastic behavior over the range of practical processing conditions (see Chapter 2). We shall now discuss the rheological behavior of molten polymers as influenced by the presence of particulate fillers or fibrous reinforcements.

3.3.1 Experimental Observations of the Rheological Behavior of Highly Filled Polymer Melts

Figure 3.14 gives plots of melt viscosity versus apparent shear rate for low-density polyethylene (LDPE) melts containing various loadings (filler concentration) of titanium dioxide particles, and Fig. 3.15 gives similar plots for polystyrene (PS) melt containing various loadings of carbon black. It is seen that as the loading of filler particles is increased, the melt viscosity

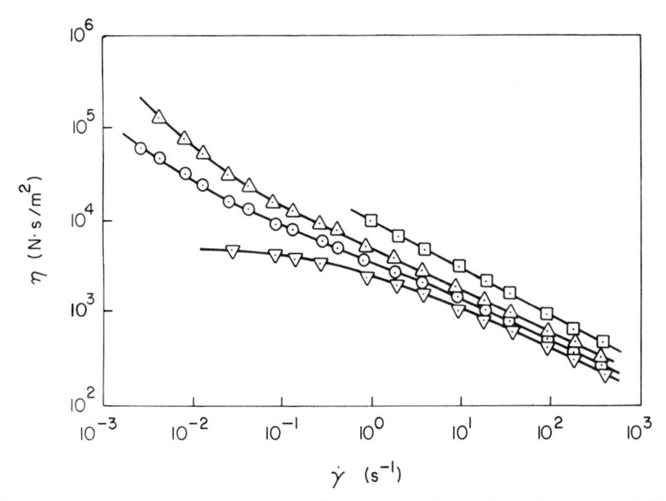

Fig. 3.14 Viscosity versus shear rate for low-density polyethylene ($T = 180°C$) filled with titanium dioxide (vol. %) [6]: (▽) 0.0; (⊙) 13; (△) 22; (⊡) 36. From N. Minagawa and J. L. White, *J. Appl. Polym. Sci.* **20**, 501. Copyright © 1976. Reprinted by permission of John Wiley & Sons, Inc.

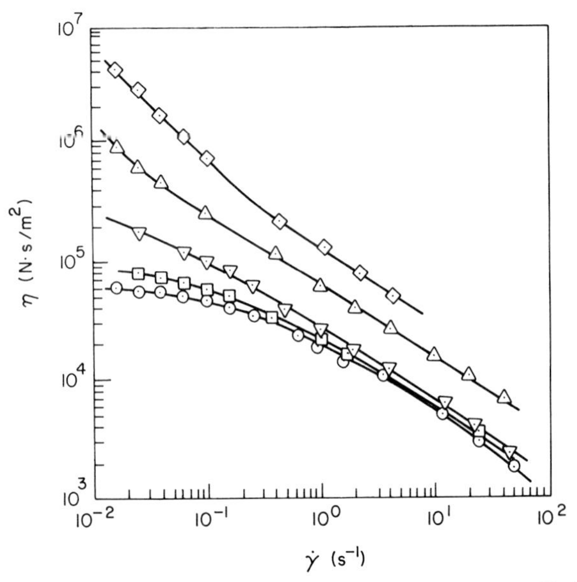

Fig. 3.15 Viscosity versus shear rate for polystyrene ($T = 170°C$) filled with carbon black (vol. %) [8]: (⊙) 0.0; (⊡) 5; (▽) 10; (△) 20; (◇) 25.

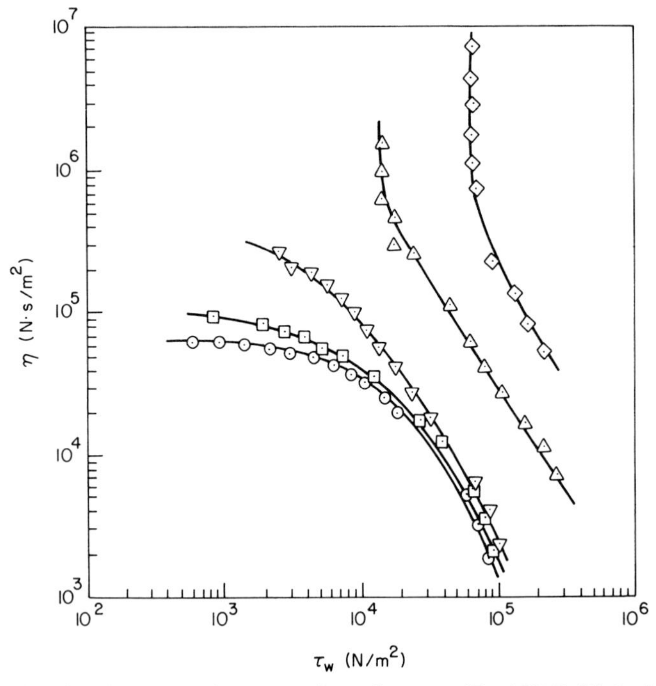

Fig. 3.16 Viscosity versus shear stress for polystyrene ($T = 170°C$) filled with carbon black [8]. Symbols are the same as in Fig. 3.15.

increases rapidly, in particular at low shear rates. Such behavior becomes more pronounced when the viscosity is plotted against shear stress, instead of against shear rate, as may be seen in Fig. 3.16.

Figure 3.17 gives plots of viscosity versus shear stress for calcium carbonate-filled polypropylene melts at various filler concentrations. It is seen that at a fixed shear stress, the viscosity increases with filler concentration.

Figure 3.18 gives plots of relative viscosity, $\eta_r = \eta(\phi)/\eta(0)$, versus concentration for TiO_2-filled molten polymers. Note that these plots are given

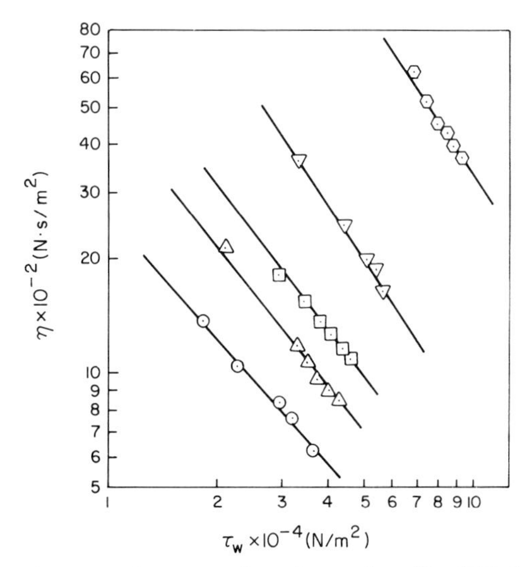

Fig. 3.17 Viscosity versus shear stress for polypropylene ($T = 200°C$) filled with calcium carbonate (vol. %) [5]: (⊙) 0.0; (△) 2.9 (⊡) 6.4 (▽) 15.4 (⬡) 38.9.

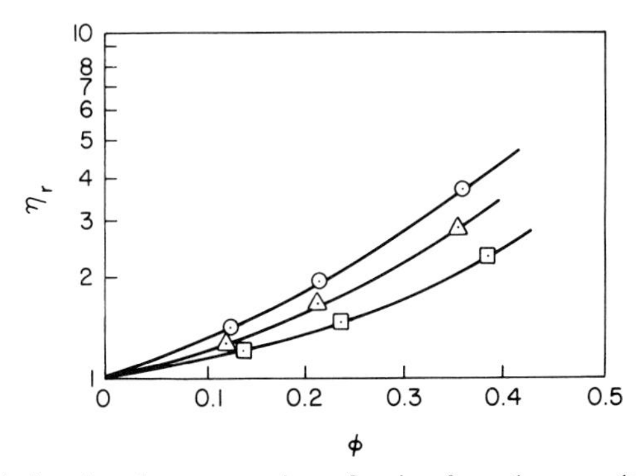

Fig. 3.18 Relative viscosity versus volume fraction for polymer melts ($T = 180°C$) filled with titanium dioxide (TiO_2) [6]: (⊙) low-density polyethylene with TiO_2; (△) high-density polyethylene with TiO_2; (⊡) polystyrene with TiO_2. The shear rate used is 10 s^{-1}. From N. Minagawa and J. L. White, *J. Appl. Polym. Sci.* **20**, 501. Copyright © 1976. Reprinted by permission of John Wiley & Sons, Inc.

at a fixed value of apparent shear rate because, as shown in Figs. 3.14 and 3.15, the filled polymers exhibit shear-thinning non-Newtonian behavior. It is seen in Fig. 3.18 that the general trend of relative viscosity follows the same pattern as that of concentrated suspensions of Newtonian fluids (see Figs. 3.3 and 3.4).

That polymer melts filled with particulates follow a power law at high shear rates (or shear stresses), while exhibiting yield stresses at low shear rates (or shear stresses), suggest the following empirical equation:

$$\tau_w = Y + K\dot{\gamma}^n \tag{3.27}$$

As a matter of fact, such flow behavior was first observed by Herschel and Bulkley [103, 104] and later by other investigators [105–107]. Skelland [108] has discussed the use of Eq. (3.27) to solve problems associated with flows in cylindrical tubes and between flat parallel plates. Equation (3.27) represents the nonlinear Bingham plastic fluids and is sometimes referred to as the Herschel–Bulkley model.

Kambe and Takano [13] made a study of the dynamic viscosity of molten polymers (uncrosslinked polyethylene) containing high percentage of fillers (glass spheres, barium sulfate powder, and calcium carbonate powder) of various particle sizes, and reported that dynamic viscosities at low frequencies were very sensitive to the structural change of the network formed by the particles. They reported that the dynamic viscosity increased with filler concentration, as may be seen in Fig. 3.19, and that in suspensions of finer particles, the dynamic viscosity of the filled molten polymer increased abruptly beyond a critical value of filler concentration. On the other hand, in suspensions of larger particles, such a critical concentration was not observed up to the highest filler concentration used in their experiments. They concluded, therefore, that the critical concentration may depend on

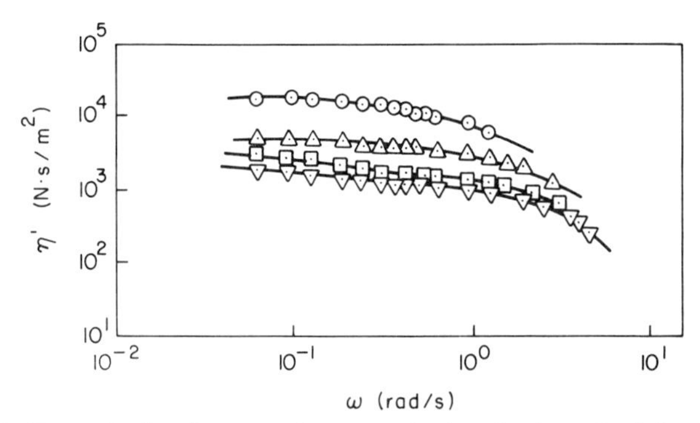

Fig. 3.19 Dynamic viscosity versus frequency for low-density polyethylene ($T = 200°C$) filled with glass bead (wt. %) [13]: (⊙) 0.0; (△) 28.6; (⊡) 60.0; (▽) 75.0. From H. Kambe and M. Takano, *Proc. Int. Cong. Rheol., 4th*, p. 557. Copyright © 1965. Reprinted by permission of John Wiley & Sons, Inc.

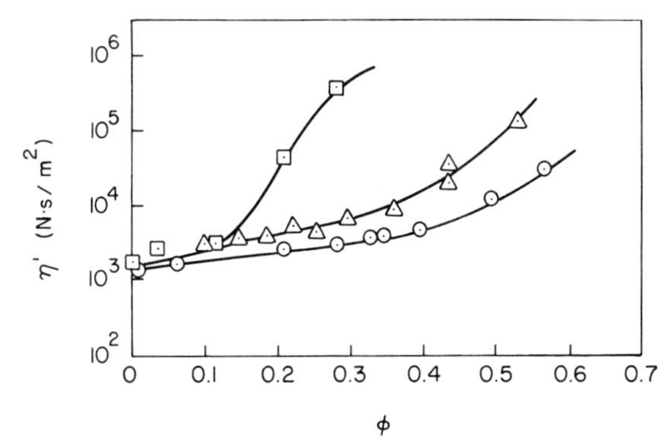

Fig. 3.20 Dynamic viscosity versus volume fraction for low-density polyethylene ($T = 200°C$) filled with various particulates [13]: (⊙) glass bead; (△) calcium carbonate; (⊡) barium sulfate. The frequency used is 0.02 rad/s. From *Proc. Int. Cong. Rheol., 4th*, p. 557. Copyright © 1965. Reprinted by permission of John Wiley & Sons, Inc.

the particle size. It should be noted that the exact concentration at which a continuous structural network forms depends on the nature of the filler and on the interaction between the filler and the suspending medium. In suspensions of glass spheres, they have found that the concentration dependence of the dynamic viscosity $\eta'(\omega)$ follows the Mooney equation, Eq. (3.8), with the value of $C_1 = 1.351$. Figure 3.20 gives plots of $\eta'(\omega)$ versus filler concentration ϕ at a fixed frequency, $\omega = 0.02$ cycles/sec.

Figure 3.21 gives the effect of filler particle size on the viscosity of talc-filled polypropylene melts. It is seen that the effect of particle size is negligible

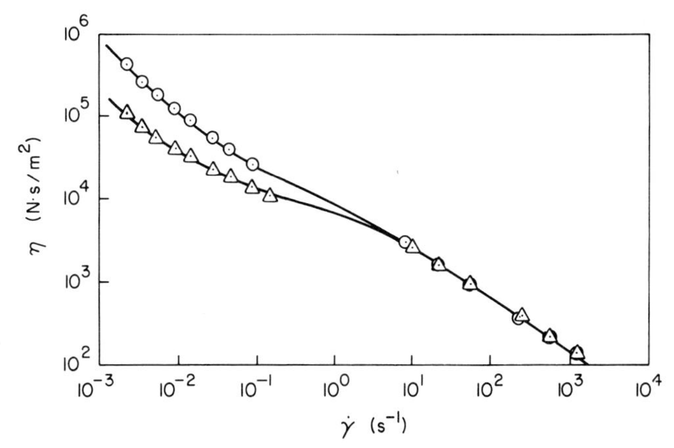

Fig. 3.21 Effect of filler particle size on the viscosity of polypropylene ($T = 200°C$) having talc particles (40 wt.%) [10]: (⊙) talc I; (△) talc II. The average size of talc I is smaller than that of talc II.

at shear rates greater than 1.0 sec^{-1}, whereas, at low shear rates, the material with the larger particles gives rise to lower viscosities than the material with smaller ones. This is expected since low-shear viscosity behavior is governed more by the filler particles than by the suspending medium, whereas the opposite is expected of high-shear viscosity behavior.

Figure 3.22 gives the effect of filler particle shape on the viscosity of filled polypropylene melts, in which the glass bead fillers used have the same density as the talc materials (2.4 g/cm^3) and similar particle size distribution (44 μm or less). It is seen that the material with spherical glass beads gives rise to lower viscosities than the material with talc particles of irregular shape. This is not surprising, in view of the fact that glass beads have less surface activity and the spherical shape minimizes the surface contact between particles, leading to weaker interaction for the glass beads than for the talc particles.

It is seen above that the viscosity of highly filled polymer melts shows shear-thinning non-Newtonian behavior, and depends on the particle size and particle shape. From their study with carbon black-filled elastomeric materials, White and Crowder [11] have suggested the following form of dimensionless relationship for the bulk viscosity $\eta(\phi, \dot{\gamma})$ of concentrated suspensions:

$$[\eta(\phi, \dot{\gamma}) - \eta(0, \dot{\gamma})]/\eta(0, \dot{\gamma}) = F(\phi, d_p, \alpha, \dot{\gamma}) \tag{3.28}$$

in which $\eta(0, \dot{\gamma})$ is the viscosity of the suspending medium, ϕ is the volume fraction of the particles in the suspension, d_p is the average particle diameter, α is a constant characteristic of the filler used (e.g., a shape factor), and $\dot{\gamma}$ is

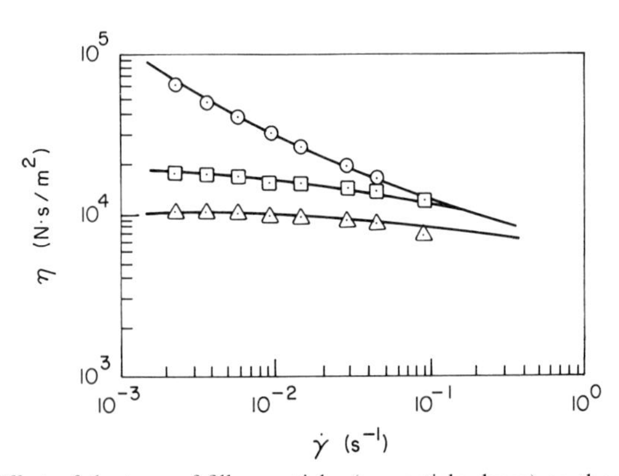

Fig. 3.22 Effect of the type of filler particles (or particle shape) on the viscosity of polypropylene ($T = 200°C$) [10]: (△) pure polypropylene (PP); (□) PP with 40 wt.% glass beads; (⊙) PP with 40 wt.% of talc. Particle size distributions in both cases are similar, having sizes 44 μm or less.

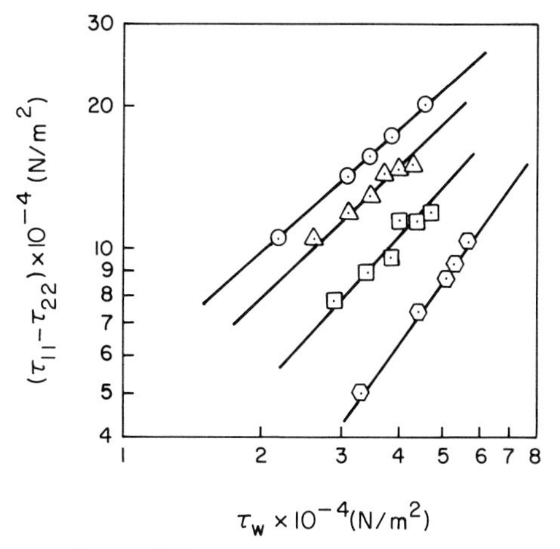

Fig. 3.23 First normal stress difference versus shear stress for polypropylene ($T = 200°$C) filled with calcium carbonate (vol. %) [5]: (⊙) 0.0; (△) 2.9 (⊡) 6.4 (⊙) 15.4.

the shear rate. Using polyisobutylenes filled with carbon blacks, White and Crowder [11] have found that the following empirical equation:

$$[\eta\phi, \dot{\gamma}) - \eta(0, \dot{\gamma})]/\eta(0, \dot{\gamma}) = A(\dot{\gamma})(\phi\alpha/d_p) + B(\dot{\gamma})(\phi\alpha/d_p)^2 \tag{3.29}$$

describes their experimental data reasonably well. In Eq. (3.29), $A(\dot{\gamma})$ and $B(\dot{\gamma})$ are functions which depend on shear rate $\dot{\gamma}$. Equation (3.29) predicts that viscosity increases with decreasing particle size (or increasing surface area) and with filler concentration.

Figure 3.23 gives plots of first normal stress difference versus shear stress for calcium carbonate-filled polypropylene melts at 200°C at various filler concentrations. It is seen that, as would be expected intuitively, the normal stresses decrease as the filler concentration increases. That polymers with high filler concentrations have smaller normal stresses than those with low concentrations can be explained by the fact that the filler itself increases the rigidity of the polymer.

Figure 3.24 gives plots of first normal stress difference versus shear rate for carbon black-filled polystyrene melts at 180°C at various filler concentrations. Figure 3.24 appears to indicate that the normal stresses increase with filler concentration. However, when first normal stress difference is plotted against stress, instead of against shear rate, as given in Fig. 3.25, the effect of filler concentration on normal stresses is seen to indicate a trend opposite to that shown in Fig. 3.24. It should be remembered that in Chapter 2 we have already discussed the practical significance of plots of first normal stress difference versus shear stress. Moreover, in correlating the rheological properties

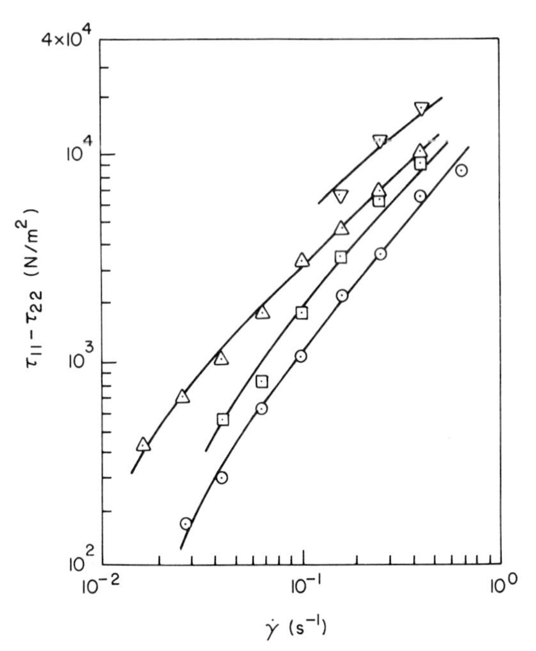

Fig. 3.24 First normal stress difference versus shear rate for polystyrene ($T = 180°C$) filled with carbon black (vol. %) [9]: (⊙) 0.0; (⊡) 5; (△) 20; (▽) 30.

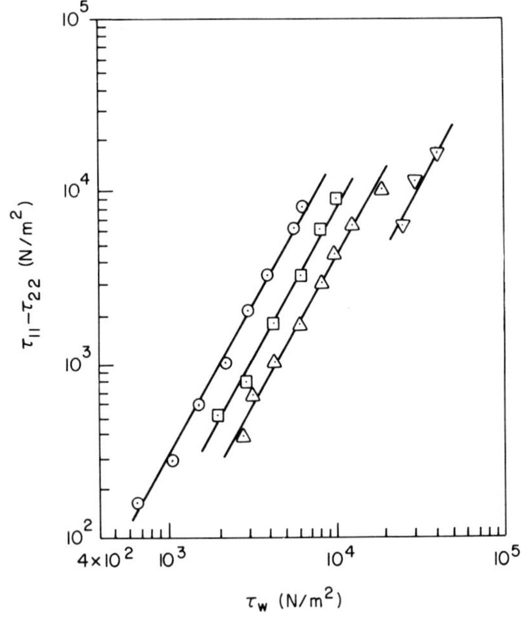

Fig. 3.25 First normal stress difference versus shear stress for polystyrene ($T = 180°C$) filled with carbon black [9]. Symbols are the same as in Fig. 3.24.

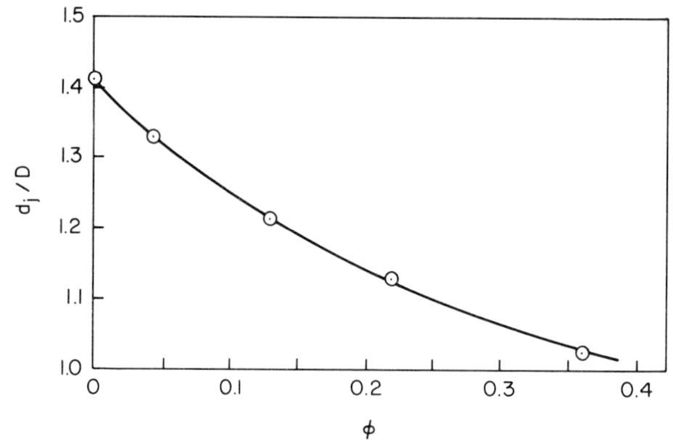

Fig. 3.26 Extrudate swell ratio versus volume fraction for high-density polyethylene ($T = 180°C$) filled with titanium dioxide [6]. The capillary die used has an $L/D = 28$ and the shear rate employed is 10 s^{-1}. From N. Minagawa and J. L. White, *J. Appl. Polym. Sci.* **20**, 501. Copyright © 1976. Reprinted by permission of John Wiley & Sons, Inc.

of multiphase systems, the shear rate (i.e., velocity gradient) may not be continuous at the interface between the phases, whereas shear stress may be continuous at the interface when no slippage occurs between the phases.

Attempts have been made to show that the extrudate swell ratio, d_j/D, defined as the extrudate-to-capillary diameter ratio, depends on the die entrance angle, the capillary length-to-diameter (L/D) ratio, shear rate (or shear stress), and the characteristic time of the fluid under test [5]. Figure 3.26 gives a plot of extrudate swell ratio versus filler concentration for high-density polyethylene melts filled with titanium dioxide at a constant shear rate (10 sec^{-1}), with the capillary having an L/D ratio of 28. It is seen that the extrudate swell decreases with an increase in filler concentration, supporting the correlation of the first normal stress difference and shear stress, given in Figs. 3.23 and 3.25.

Figure 3.27 gives plots of first normal stress difference versus shear stress for polystyrene melts filled with glass fiber. It is interesting to note that the normal stresses increase with filler concentration, showing a trend opposite to the one observed when carbon black was used as filler (see Fig. 3.25). Czarnecki and White [102] reported measurements of the first normal stress difference for polystyrene melts containing cellulose fibers and organic aromatic polyamide fibers (Kevlar®). They noted that the normal stresses of melts containing long, flexible particulates depend on the fiber length-to-diameter ratio. The increase in normal stresses in the presence of long, flexible particulates should not be a surprise, because we have already noted that even Newtonian liquids can exhibit normal stress effects when glass fibers are suspended in them as shown in Figs. 3.11 and 3.12.

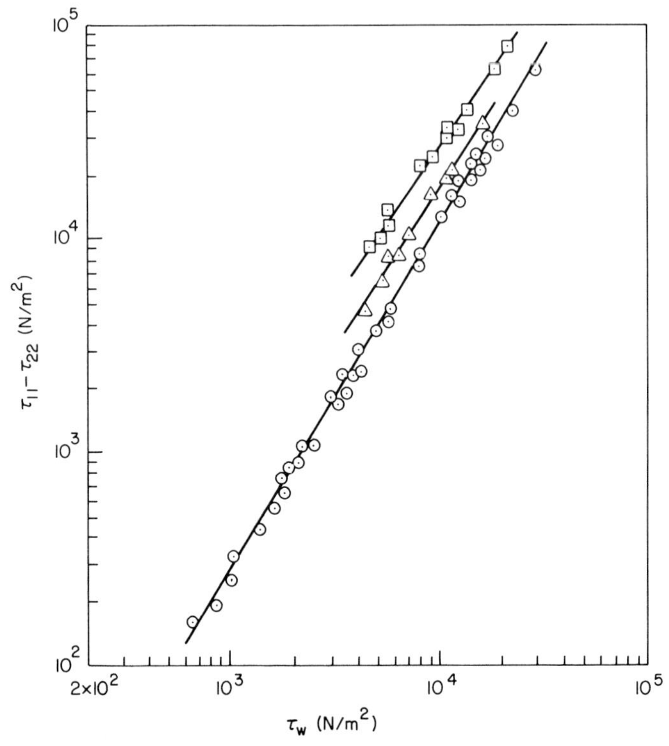

Fig. 3.27 First normal stress difference versus shear stress for polystyrene ($T = 180°C$) filled with glass fiber (vol. %) [102]: (⊙) 0.0; (△) 10; (⊡) 22. From L. Czarnecki and J. L. White, *J. Appl. Polym. Sci.* **25**, 1217. Copyright © 1980. Reprinted by permission of John Wiley & Sons, Inc.

In Chapter 2, we have suggested the use of the characteristic time of a fluid in correlating the normal stress effects of viscoelastic fluids. Figure 3.28 gives plots of relaxation time versus filler concentration for various polymer melts containing small particles (titanium dioxide), and Fig. 3.29 gives plots of relaxation time versus shear rate for high-density polyethylene melts containing long, flexible particulates (glass fiber). It is seen that, when the filler consists of *small* particles of TiO_2, the relaxation time *decreases* with an increase in filler concentration, whereas, when long threadlike particulates of glass fiber are used, the relaxation time *increases* with an increase in filler concentration.

It is then fairly well established that both the size and shape of the particulates and the filler loading, have a profound influence on the rheological properties of polymer melts subjected to steady shearing flow.

Relatively little experimental study has been reported in the literature dealing with the elongational viscosity of highly filled polymeric materials.

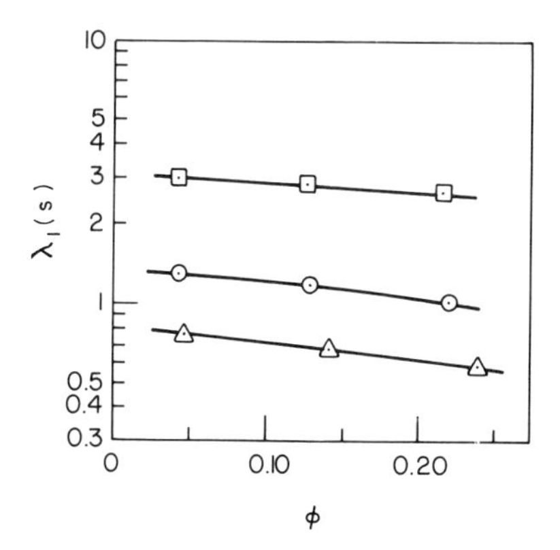

Fig. 3.28 Characteristic time versus volume fraction for polymer melts ($T = 180°C$) filled with titanium dioxide [6]: (□) high-density polyethylene; (⊙) low-density polyethylene; (△) polystyrene. The shear rate employed is $0.5\ s^{-1}$. From N. Minagawa and J. L. White, *J. Appl. Polym. Sci.* **20**, 501. Copyright © 1976. Reprinted by permission of John Wiley & Sons, Inc.

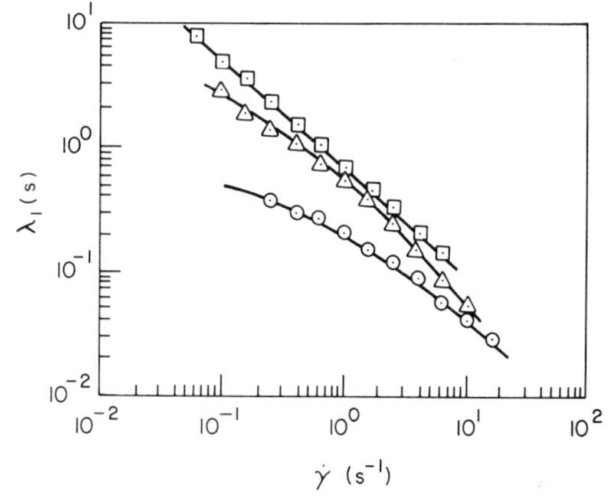

Fig. 3.29 Characteristic time versus shear rate for polystyrene ($T = 180°C$) filled with glass fiber (wt. %) [7]: (⊙) 0.0; (△) 20; (□) 40. From Y. Chan *et al.*, *J. Rheol.* **22**, 507. Copyright © 1978. Reprinted by permission of John Wiley & Sons, Inc.

White and co-workers [7–9] have carried out measurements of the elongational properties of polymer melts filled variously with calcium carbonate, titanium dioxide, carbon black, and glass fibers.

Figure 3.30 gives plots of elongational viscosity η_E versus elongation rate $\dot{\gamma}_E$ for carbon black-filled polystyrene (PS) melts. It is seen that at low elongation rates the pure PS melt asymptotically achieves a constant value of η_E (three times the zero-shear viscosity η_0), but that at higher elongation rates, η_E increases somewhat. On the other hand, in the filled systems, η_E decreases rapidly with increasing $\dot{\gamma}_E$, and increases with the amount of filler (carbon black) added. Tanaka and White [9] report similar elongational flow behavior using other types of fillers (titanium dioxide, calcium carbonate).

Using a melt spinning apparatus, Han and Kim [109] also studied the elongational flow behavior of calcium carbonate-filled polypropylene (PP)

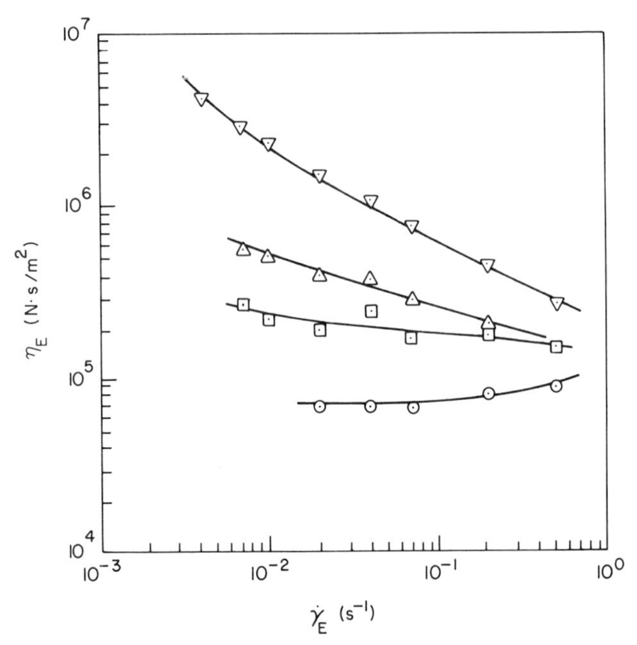

Fig. 3.30 Elongational viscosity versus elongation rate for polystyrene ($T = 180°C$) filled with carbon black (vol. %) [9]: (⊙) 0.0; (⊡) 10; (△) 20; (▽) 30.

melts. Their results, as given in Fig. 3.31, show a trend similar to that reported by White and co-workers [8,9]. Note, however, that the melt-spinning apparatus does not provide steady elongational flow and thus the elongational responses determined from such an instrument are often referred to as "spinning" viscosity (see Chapter 2).

Figure 3.32 gives plots of $\eta_E/3\eta_0$ versus $\dot{\gamma}_E$ for high-density polyethylene (HDPE) filled with glass fibers. It is seen that, for the pure HDPE melt, η_E/η_0 is about 3 at low elongation rates as the theory predicts (see Chapter 2), and for the filled melts the values of η_E/η_0 are much larger than 3 at low elongation rates (heading upward as $\dot{\gamma}_E$ decreases). It seems somewhat peculiar that the values of $\eta_E/3\eta_0$ are much higher for the melt containing 20 wt. % of glass fibers than for the one containing 40 wt. %. Chan *et al.* [7] attribute this peculiarity to the smaller aspect ratios of the particles employed. It has also been reported that the highly filled polymer melts exhibit yield values in elongational flow, very similar to that in shear flow [9].

The large values of η_E/η_0 for glass fiber-filled polymers may be explained qualitatively by the theoretical framework, Eq. (3.23), of Batchelor [92], although he considered a Newtonian fluid as the suspending medium. Note, however, that according to Chan *et al.* [7], the application of Eq. (3.23) to

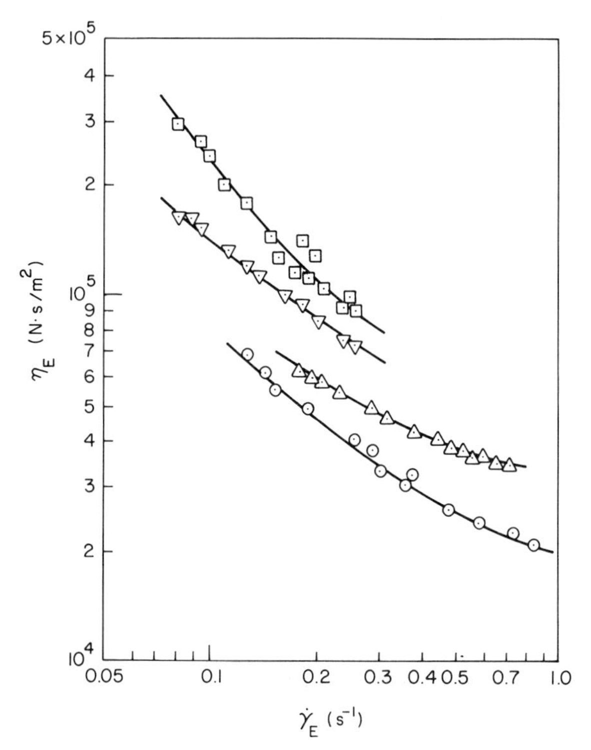

Fig. 3.31 Apparent elongational viscosity versus elongation rate for polypropylene ($T =$ 200°C) filled with calcium carbonate (vol. %): (⊙) 0.0; (△) 2.9; (▽) 6.4; (⊡) 15.4.

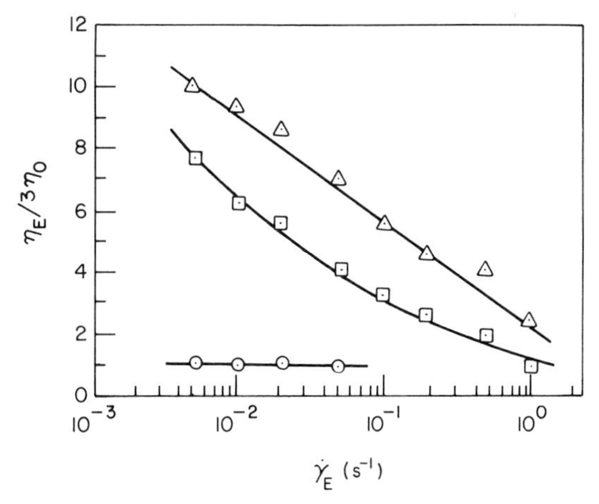

Fig. 3.32 Relative elongational viscosity versus elongation rate for high-density polyethylene ($T = 180$°C) filled with glass fiber (wt. %) [7]: (⊙) 0.0; (△) 20; (⊡) 40. From Y. Chan *et al.*, *J. Rheol.* **22**, 507. Copyright © 1978. Reprinted by permission of John Wiley & Sons, Inc.

their data (see Fig. 3.32) gave rise to values of $\eta_E/3\eta_0$ of the order of several hundred, which certainly was far too large, compared to their experimental results. Note further that Eq. (3.23) does not predict a decreasing trend for the elongational viscosity as the elongation rate is increased.

As will be discussed below, Goddard [110, 111] has advanced a theory of elongational flow of suspensions of non-Newtonian fluids. According to his theory, a decrease in elongational viscosity is predicted when the suspending medium exhibits shear-thinning behavior, because the shear viscosity between the glass fibers will be much lower than the zero-shear viscosity of the medium without the suspended particles.

3.3.2 Theoretical Consideration of the Rheological Behavior of Highly Filled Polymer Melts

The experimental data presented previously show that highly filled polymer melts exhibit both non-Newtonian viscosity and normal stress effects at large strain rates or large shear stresses, while exhibiting *yield* values at small strain rates or small shear stresses. Therefore it is desirable to develop a three-dimensional rheological model for highly filled polymer melts which is capable of predicting such rheological behavior.

Extending the concept originated by Hohenemser and Prager [65], Oldroyd [66], and others [112–114], White [115] has formulated a rheological model for nonlinear viscoelastic Bingham fluids given by

$$\mathbf{S} = -p\mathbf{I} + G_{11}\mathbf{C}^{-1} - G_{12}\mathbf{C}, \qquad \text{for} \quad \operatorname{tr}\tau^2 < 2Y^2 \tag{3.30}$$

$$\tau = \{1 + Y(\tfrac{1}{2}\operatorname{tr}\mathbf{H}^2)^{-1/2}\}\mathbf{H}, \qquad \text{for} \quad \operatorname{tr}\tau^2 > 2Y^2 \tag{3.31}$$

in which $\mathbf{S}$ is the total stress tensor defined as $\mathbf{S} = -p\mathbf{I} + \tau$, p is the pressure, $\mathbf{C}^{-1}$ and $\mathbf{C}$ are Finger and Cauchy rate-of-deformation tensors, respectively, G_{11} and G_{12} are constants, and $\mathbf{H}$ is a nonlinear memory integral of the form used for isotropic viscoelastic fluids. White [115] notes that $\mathbf{H}$ must be defined so as to make $\operatorname{tr}\mathbf{H}$ and $\operatorname{tr}\tau$ equal to zero. Note that Eqs. (3.30) and (3.31) are limited to the von Mises yield criterion [67]. Other yield criteria (e.g., Tresca yield criterion) have also been suggested in the literature [68, 69].

Using the following form of a single integral,

$$\mathbf{H} = \int_0^\infty m(t)\{\mathbf{C}^{-1} - \tfrac{1}{3}(\operatorname{tr}\mathbf{C}^{-1})\mathbf{I}\}\,dt \tag{3.32}$$

and representing the memory function $m(t)$ as

$$m(t) = (G/\lambda_1)e^{-t/\lambda_1} \tag{3.33}$$

in which G is the modulus and λ_1 is a relaxation time, White [115] has shown that for steady simple shearing flow:

$$\tau_{12} = \{1 + Y[G^2(\lambda_1^2\dot{\gamma}^2 + \tfrac{4}{3}\lambda_1^4\dot{\gamma}^4)]^{-1/2}\}G\lambda_1\dot{\gamma} \tag{3.34}$$

$$\tau_{11} - \tau_{22} = \{1 + Y[G^2(\lambda_1^2\dot{\gamma}^2 + \tfrac{4}{3}\lambda_1^4\dot{\gamma}^4)]^{-1/2}\}2G\lambda_1^2\dot{\gamma}^2 \tag{3.35}$$

and for steady, uniaxial elongational flow:

$$\eta_E = (3^{1/2}Y/\dot{\gamma}_E) + 3G\lambda_1/(1 + \lambda_1\dot{\gamma}_E)(1 - 2\lambda_1\dot{\gamma}_E) \tag{3.36}$$

It can be seen from Eqs. (3.34) and (3.35) that

(i) At low shear rates:

$$\lim_{\dot{\gamma}\to 0} \tau_{12} = Y \tag{3.37}$$

$$\lim_{\dot{\gamma}\to 0} \tau_{11} - \tau_{22} = 0 \tag{3.38}$$

and

(ii) At high shear rates:

$$\lim_{\dot{\gamma}\to \infty} \tau_{12} = G\lambda_1\dot{\gamma} \tag{3.39}$$

$$\lim_{\dot{\gamma}\to \infty} \tau_{11} - \tau_{22} = 2G\lambda_1^2\dot{\gamma}^2 \tag{3.40}$$

Also, from Eq. (3.36) we have

$$\lim_{\dot{\gamma}_E\to 0} \eta_E = 3^{1/2}Y/\dot{\gamma}_E \tag{3.41}$$

It is of great interest to note that the above theoretical development predicts the qualitative features of the experimentally observed rheological behavior of highly filled polymer melts, namely: (a) Eq. (3.37) predicts a yield values at low shear rates (see Figs. 3.14 and 3.15); (b) Eq. (3.34) predicts non-linear behavior of viscosity at high shear rates (see Figs. 3.14 and 3.15); (c) Eq. (3.35) predicts first normal stress difference at high shear rates (see Figs. 3.23 and 3.25); (d) Eq. (3.41) predicts a yield value of elongational viscosity (see Fig. 3.30). It should be pointed out that the White analysis [115] assumes small solid particles and therefore it is not applicable to the situation where long flexible particles (e.g., glass fibers) are used. An experimental study by White and Tanaka [116] supports the theoretical development described above, Eqs. (3.30)–(3.41).

Let us turn our attention to the situation where glass fibers or the like are suspended in a polymer matrix. At present there exists no theory dealing with the shearing flow of suspensions of long, flexible particles, analogous to the White analysis. However, attempts [110, 111] have been made to predict theoretically the tensile stress of suspensions of long, flexible particles suspended in a medium that obeys a power law when subjected to a shearing flow. Using slender-body theory, Goddard [111] obtained the following form representing the uniaxial elongational viscosity $\eta_E(\phi, \dot{\gamma}_E)$:

$$\eta_E(\phi, \dot{\gamma}_E) = \eta_E(0, \dot{\gamma}_E)\left\{1 + \frac{4}{9}\left(\frac{l}{R_0}\right)^{n+1}\left[\frac{((1-n)/n)\phi}{1-(\pi/\phi)^{(n-1)/n}}\right]B\right\}; \qquad n \neq 1 \tag{3.42}$$

where

$$B = [9/(2+n)]\eta(\dot{\gamma}_E)/\eta_E(\dot{\gamma}_E) \tag{3.43}$$

in which $2l$ and R_0 are particle length and radius, respectively, n is a power law index of the suspending medium, ϕ is the volume fraction of slender particles, $\eta_E(\dot{\gamma}_E)$ is the elongational viscosity of the medium in the absence of particles, and $\eta(\dot{\gamma}_E)$ is the value of shear viscosity of the medium evaluated at the strain rate $\dot{\gamma}_E$ under consideration.

Note that, in the limit as n approaches unity, Eq. (3.42) reduces to Eq. (3.25) for a Newtonian fluid. For the usual run of thermoplastic resins in the molten state, n lies somewhere between 0.3 and 0.6. Under such circumstances, the magnitude of $\eta_E(\phi, \dot{\gamma}_E)$ calculated from Eq. (3.42) is much smaller than that calculated from Eq. (3.25). Goddard [111] attributes this to the possible tensile stiffening behavior of the suspending medium, which is assumed to obey a power law when subjected to a steady shearing flow.

3.4 EFFECTS OF COUPLING AGENTS ON THE RHEOLOGICAL PROPERTIES OF FILLED POLYMERS

The mechanical properties of mineral-filled composites, mineral-filled elastomers, and glass fiber-reinforced plastics are primarily dependent upon three factors: (1) the strength and modulus of the filler or reinforcing particles, (2) the strength and chemical stability of the resin, and (3) the effectiveness of the bond between resin and filler in transferring stress across the interface. Difficulty is often encountered in producing new polymer composites with improved mechanical properties, due mainly to the fact that bonding is lacking between the polymer and the particulate fillers. In general, mineral fillers as extenders and glass fibers as reinforcing agents have little interfacial interaction with the resin and, therefore, coupling agents have been developed

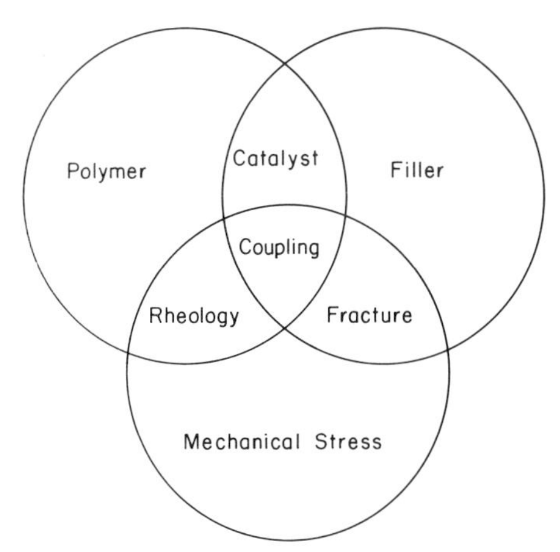

Fig. 3.33 Schematic displaying the role that coupling agents play in the processing of filled polymers and in improvement of the mechanical properties of polymer composites [117].

for improving the bond between rigid particles and the matrix resin. Although adhesion is central to any "coupling" mechanism, it is recognized that many factors are involved in the total performance of a composite system, as is indicated in Fig. 3.33 [117].

Figure 3.34 shows photomicrographs of broken glass fiber-polypropylene composites, with and without silane coupling agent, γ-(β-aminoethyl) aminopropyl-trimethoxysilane (Dow Corning Co., Z-6020) [118]. It is seen that a composite *with* coupling agent shows excellent retention of polymer on the broken glass ends, whereas a composite *without* a coupling agent shows clean holes in the matrix resin and uncoated fiber ends pulled from the polymer. The improvement of adhesion between the particles and the resin must have resulted from interfacial modification brought about by the coupling agent. The adhesion mechanism through coupling agents is a complex subject, and the reader is referred to recent review articles [119, 120].

Today, a variety of organofunctional silanes [121, 122] and titanates [123] are commercially available as coupling agents. Silane coupling agents are monomeric silicon chemicals, and they are believed to be compatible with almost every type of organic polymer, ranging from thermosetting resins through elastomers to thermoplastic resins [124–126]. Titanate coupling agents are monoalkoxy organochemicals, and when added to mixtures of resins and fillers, they are believed to fulfill several functions [127, 128].

Coupling agents usually have a dual reactivity since they contain pendant groups capable of reacting both with the resin and with the filler surface. The stability of the composites appears to be related to the strength of the covalent bonds between resin and filler via the coupling agent. Although the exact

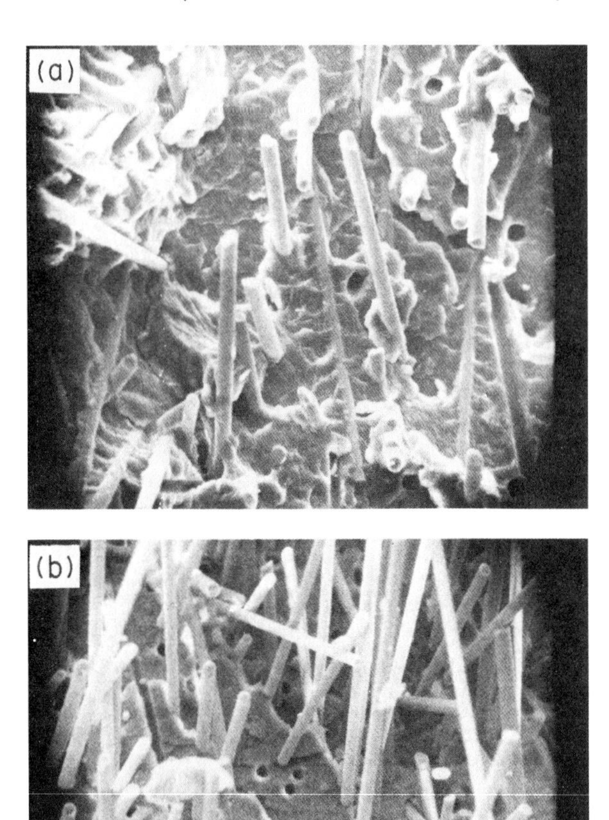

Fig. 3.34 Photomicrographs of broken glass fiber–polypropylene composites [118]: (a) with a coupling agent; (b) without a coupling agent. The coupling agent employed is a silane coupling agent (Dow Corning Co., Z-6020). Reprinted with permission from R. C. Hartlein, *Ind. Eng. Chem. Prod. Res. Dev.* **10**, 92 (1971). Copyright 1971 American Chemical Society.

mechanism of bonding may still be controversial, it is believed that the organofunctional portion of the coupling agent reacts with the resins and becomes covalently bonded to the resin matrix. All coupling agents for thermosetting resins are reactive with the resins.

From the point of view of polymer processing, the mixture of a resin, particulate filler, and a coupling agent may be considered as a three-phase system. Interfacial modification with coupling agents should affect the

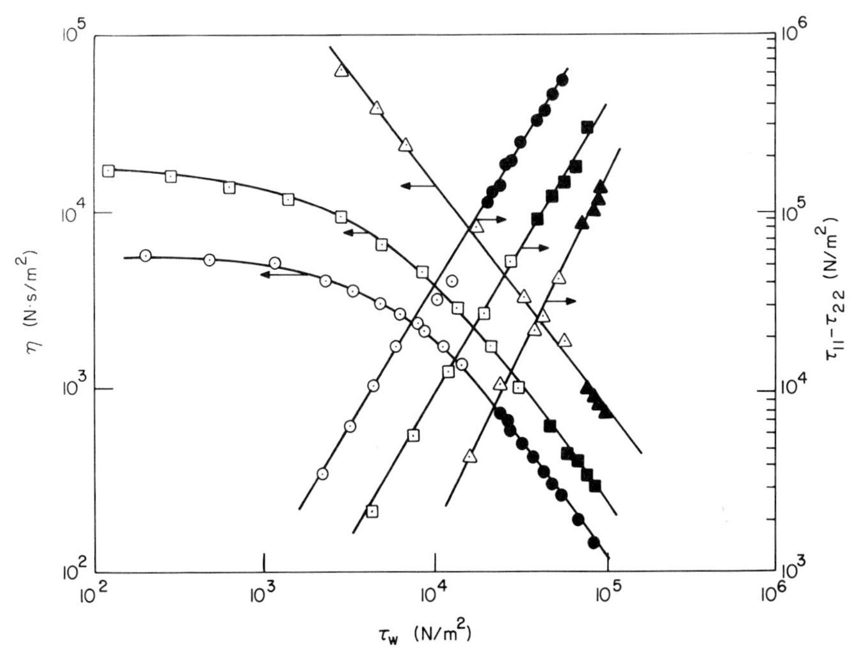

Fig. 3.35 Viscosity and first normal stress difference versus shear stress for polypropylene ($T = 200°C$) filled with calcium carbonate (50 wt.%), with and without a titanate coupling agent (Kenrich Petrochemicals Inc., TTS) [22]: (⊙, ●) pure polypropylene (PP); (△, ▲) PP/$CaCO_3$ = 50/50 (by wt.); (⊡, ■) PP/$CaCO_3$ = 50/50 with TTS (1 wt.%). The open symbols are the data obtained by a cone-and-plate instrument, and the closed symbols are the data obtained by a slit/capillary rheometer.

rheology of the filled polymers during processing. Therefore, a better understanding of the role that a coupling agent may play in controlling the rheological properties of filled molten polymers is of fundamental and practical importance. In the past, some research efforts [20–22] have been spent on understanding this role.

Figure 3.35 gives plots of viscosity and first normal stress difference versus shear stress for calcium carbonate ($CaCO_3$)-filled polypropylene (50% by weight) with and without a titanate coupling agent. The coupling agent used is isopropyl triisostearoyl titanate (Kenrich Petrochemicals, TTS). It is seen that the TTS decreases the melt viscosity of the filled polymer, but increases its normal stress difference over the range of shear stresses investigated.

Figure 3.36 gives plots of viscosity and first normal stress difference versus shear stress for glass fiber-filled polypropylene (50 percent by weight) with and without TTS. Note that at low values of shear stress, the TTS has little effect on the viscosity of the filled polypropylene, but, as shear stress increases,

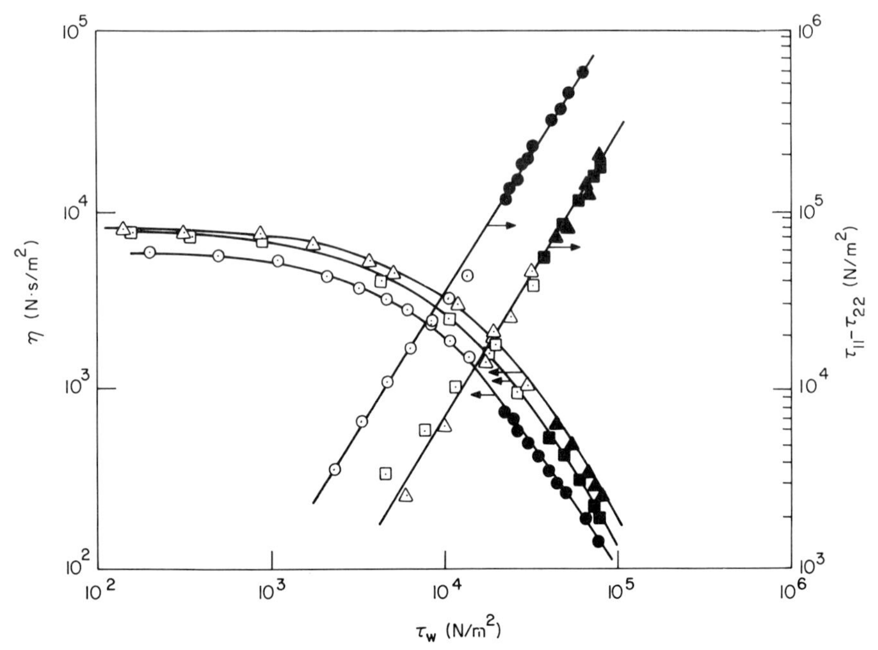

Fig. 3.36 Viscosity and first normal stress difference versus shear stress for polypropylene ($T = 200°C$) filled with glass fiber (50 wt.%), with and without a titanate coupling agent (TTS): (⊙, ●) pure PP; (△, ▲) PP/glass fiber = 50/50 (by wt.); (⊡, ■) PP/glass fiber = 50/50 with TTS (1 wt.%). Open and closed symbols have the same meanings as in Fig. 3.35.

the viscosity *with* the TTS decreases rapidly, yielding values lower than those *without* the TTS. Note also in Fig. 3.36 that the coupling agent has little influence on the normal stress difference of the glass fiber-filled polypropylene.

Figure 3.37 gives plots of viscosity versus shear stress for talc-filled polypropylene and talc-filled high-density polyethylene with and without a coupling agent (Kenrich Petrochemicals, ETDS-201), which is a chelated carboxyl diisostearoyl ethylene titanate. It is seen that the ETDS-201 has little effect on the melt viscosity of the filled polymers investigated.

Figure 3.38 and 3.39 give plots of viscosity versus shear stress and plots of first normal stress difference versus shear stress, respectively, for calcium carbonate-filled and glass bead-filled polypropylene melts with and without silane coupling agents (Union Carbide, Y9187 and A1100). Y9187 is N-octyltriethoxysilane and A1100 is γ-aminopropyltriethoxysilane. It is seen that both Y9187 and A1100 decrease the melt viscosity and increase the normal stresses of the calcium carbonate-filled polypropylene melts. However the effect of these coupling agents on the rheological properties of the glass bead-filled polypropylene melts are somewhat complex. Whereas A1100

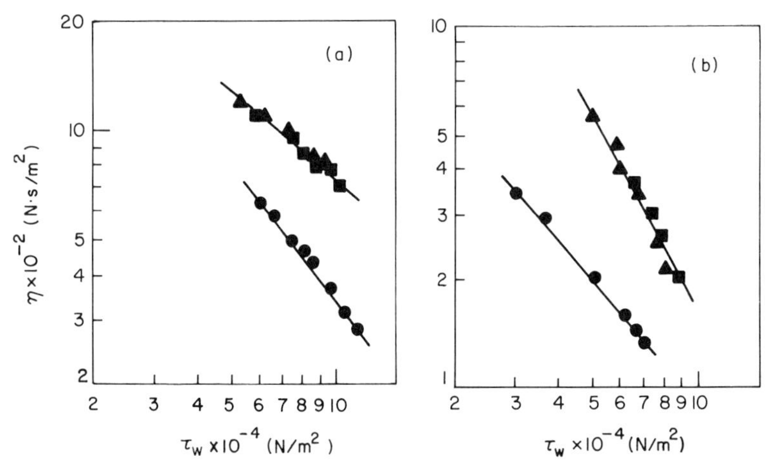

Fig. 3.37 Viscosity versus shear stress for talc-filled molten polymers ($T = 200°C$) with and without a titanate coupling agent (Kenrich Petrochemicals Inc., ETDS 201): (a) high-density polyethylene (HDPE)/talc system, (●) pure HDPE; (▲) HDPE/talc = 50/50 (by wt.); (■) HDPE/talc = 50/50 with ETDS (1 wt.%). (b) polypropylene (PP)/talc system, (●) pure PP; (▲) PP/talc = 50/50 (by wt.); (■) PP/talc = 50/50 with ETDS (1 wt.%).

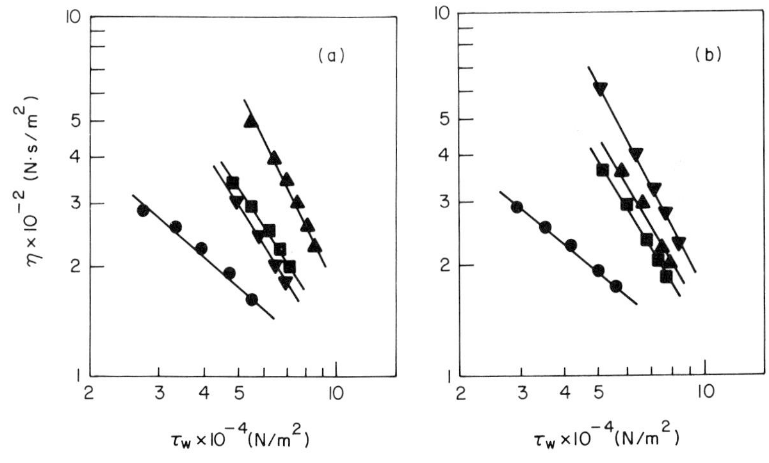

Fig. 3.38 Viscosity versus shear stress for polypropylene ($T = 200°C$) filled with particulates ($CaCO_3$ or glass bead), with and without a silane coupling agent (Union Carbide Corp., either Y9187 or A1100): (a) Polypropylene (PP)/$CaCO_3$ system, (●) pure PP; (▲) PP/$CaCO_3$ = 50/50 (by wt.); (■) PP/$CaCO_3$ = 50/50 with A1100 (1 wt.%); (▼) PP/$CaCO_3$ = 50/50 with Y9187 (1 wt.%). (b) Polypropylene (PP)/glass bead system, (●) pure PP; (▲) PP/glass bead = 50/50 (by wt.); (■) PP/glass bead = 50/50 with A1100 (1 wt.%); (▼) PP/glass bead = 50/50 with Y9187 (1 wt.%).

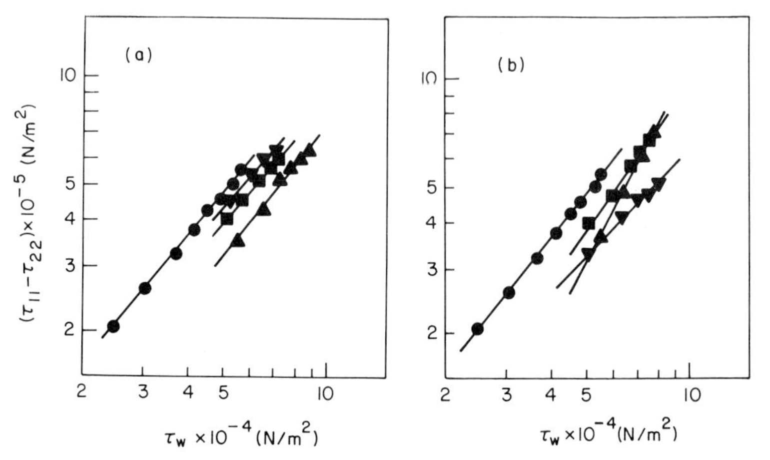

Fig. 3.39 First normal stress difference versus shear stress for polypropylene ($T = 200°C$) filled with particulates [22]: (a) Polypropylene/$CaCO_3$ system; (b) Polypropylene/glass bead system. Symbols are the same as in Fig. 3.38.

decreases the melt viscosity, Y9187 increases it, and whereas the Y9187 decreases the normal stresses, the A1100 increases them over the limited range of shear stresses investigated.

We can then conclude from Figs. 3.35–3.39 that the effects of coupling agents on the rheological properties of filled molten polymers depends on on both the type of coupling agent and the polymer/filler system under consideration.

Figure 3.40 shows the effects of the filler loading on both the viscosity and normal stresses of $CaCO_3$-filled high-density polyethylene, where the titanate coupling agent TTS added is 1.0 percent by weight of the filler. It is of great interest to note in Fig. 3.40 that the viscosities of the HDPE/$CaCO_3$ = 60/40 (wt. %) are lower even than that of the pure HDPE over the entire range of shear stress investigated. Note further that as the filler loading increases, the highly filled system exhibits yield stress behavior. In all the cases discussed, the coupling agents increase the normal stresses of the filled systems.

It appears from the experimental observations presented above that a coupling agent which reduces the melt viscosity of a filled polymer also increases the normal stresses and vice versa. In other words, the change in viscosity due to the presence of a coupling agent is in the direction opposite to the change in normal stresses.

A decrease in melt viscosity of a filled polymer, due to the presence of a coupling agent, may have resulted from the role that the coupling agent played in decreasing interparticle forces and help prevent flocculation of

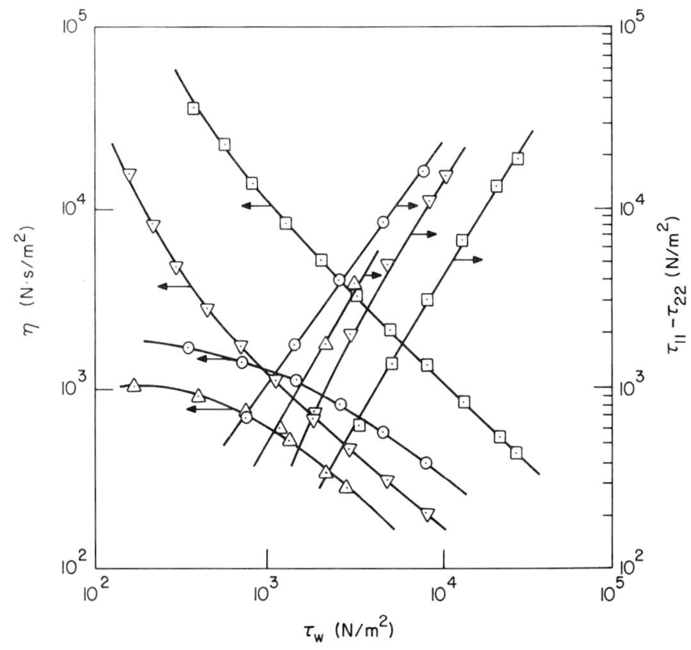

Fig. 3.40 Viscosity and first normal stress difference for high-density polyethylene (HDPE) ($T = 240°C$) filled with calcium carbonate, with and without a titanate coupling agent (Kenrich Petrochemicals Inc., TTS): (⊙) pure HDPE; (⊡) HDPE/$CaCO_3$ = 30/70 (by wt.) with TTS (1 wt.%); (▽) HDPE/$CaCO_3$ = 45/55 with TTS (1 wt.%); (△) HDPE/$CaCO_3$ = 60/40 with TTS (1 wt.%).

particles. If that is the case, under a shearing flow polymer molecules will slip between the filler particles treated with the coupling agent, encountering a frictional resistance far less than that from untreated filler particles. This argument, though speculative, implies that there probably is little "true" coupling between the filler particles and the polymer matrix. It is difficult to imagine how a reduction of melt viscosity can occur if macromolecules are interconnected to each other through filler particles, when a chemical agent acts as a true coupling agent, instead of as a surface modifier.

As shown in Figs. 3.38 and 3.39, there is an instance where the use of a coupling agent increases the melt viscosity, while decreasing the normal stresses of a filled polymeric material. In such a case, it may be conjectured that the particular coupling agent makes the long macromolecules, subjected to a shearing flow, less flexible (or mobile) by connecting (or bridging) them through filler particles. If this is so, one would expect to have stiff, long molecules interconnected to each other (similar to the crosslinking of large molecules). It is not difficult to imagine that such a molecular arrangement would have less capability of storing an elastic energy than flexible macromolecules would have.

It is clear from the above observations that under proper combinations of polymer, filler, and coupling agent, a drastic change in the rheological properties of highly filled polymeric systems is possible. However, at present little basic study has been performed to define the mechanism (or mechanisms) that can convincingly explain the changes.

3.5 PROCESSING OF PARTICULATE-FILLED POLYMERIC SYSTEMS

Filled polymer composites have a continuous matrix phase (i.e., polymer) and a discontinuous filler phase made up of discrete particles. Two examples of this type of system are short fiber-reinforced elastomers and glass fiber-reinforced plastics. The primary objective for using solid particles in polymeric materials is twofold: (1) To improve the mechanical/physical properties of the composite materials; and (2) to reduce the cost of production by replacing expensive resins with inexpensive solid particles.

Common filler types widely used in polymer composites are: (1) mineral silica (sand, quartz), (2) mineral silicate (Kaolin clay, mica, talc, asbestos), (3) calcium carbonate, and (4) metallic oxides (alumina, titania, zinc oxide). For instance, clay is most commonly used, particularly in high-density polyethylene, to improve rigidity and tensile strength for automotive and pipe applications.

Fiber reinforcement of polymeric materials was first developed as a means of increasing the mechanical properties (tensile modulus, dimensional stability, fatigue endurance, deformation under load, hardness, abrasion resistance) of thermoset resins. These fibers were originally in the form of continuous filaments, chopped mat, or woven fabric. However, fibers in these forms have little versatility in newer, faster processing techniques, such as injection molding. This fact, coupled with the development of whisker technology, has generated an interest in injection molding of short fiber-reinforced thermoplastic composites. These fibers may be in the form of chopped strand, milled filaments, or whiskers. As glass is one of the cheapest and most readily available fibrous materials, its use is the most common and best understood.

The mechanical (static or dynamic) properties of composite materials are determined by the properties of the components, the morphology of the system, and the nature of the interface between the phases. Therefore a great variety of properties can be obtained by varying the structure of the system or the interface properties. An important property of the interface is the degree of adhesion bonding between the phases.

In the past, a number of theoretical and experimental studies have been reported in the literature dealing with the mechanical properties of polymer composites containing fillers. The theoretical studies [129–140] attempted to predict the mechanical properties (e.g., tensile or shear modulus) of polymer composites, and the experimental studies [141–152] tried either to verify theoretical predictions or to establish relationships between the mechanical properties and the morphology in reinforced plastics and reinforced elastomers.

The structure–processing–property relationships of such classical solids as metals, oxides, and ceramics, and of organic crystals have been quite extensively explored, but our corresponding knowledge of polymeric materials and their composites is still very incomplete. In dealing with polymeric materials and their composites, the greatest difficulties arise in understanding the effects of processing history (both deformation and thermal histories) on the structure–property relationships.

We shall discuss below the injection molding and fiber spinning of thermoplastic resins containing particulate fillers, and the processing of some thermoset resins. In line with the primary objective of this monograph, emphasis will be placed on demonstrating the relationships between processing conditions and mechanical properties of such materials. Some selected mechanical properties of composite materials will be discussed as relevant to the processing conditions and the choice of materials from the rheological point of view. Detailed discussion of the structure–processing–property relationship in composite materials is beyond the scope of this monograph.

3.5.1 Melt Spinning of Filled Polymers

Melt spinning is basically an extrusion process since the molten polymer is pumped through the spinnerette [153–155]. The fibers are usually solidified by a cross current blast of air (or inert gas) as they proceed to drawing rolls. The drawing step stretches the fiber and orients the molecules in the direction of stretch. Today it is well known to the textile fiber industry that the manner in which the molten threadlines are cooled upon exiting from the spinnerette holes, influences the extent to which the molecules are oriented, and thus the ultimate tensile properties of the fibers.

Han *et al.* [22] performed melt spinning experiments, using $CaCO_3$-filled polypropylene containing coupling agents. The purpose of their study was to determine the effect of coupling agents on spinnability of the PP–$CaCO_3$ system and to evaluate the effectiveness of the coupling agents in improving the tensile properties of the melt-spun fibers. Not only did they spin the fiber into quiescent air, but they also spun the fiber using an isothermal chamber attached beneath the spinnerette. In the former case, cooling occurs as soon as the melt threadlines exit from the spinnerette holes. On the other hand, the use of an isothermal chamber delays the initial cooling of melt threadlines. Such a cooling method, often referred to as *delayed cooling*, has been reported to have a considerable influence on the stability of threadlines.

Table 3.1 gives a summary of spinnability tests. Under identical spinning conditions, the maximum draw-down ratio, $(V_L/V_0)_{max}$, may be used as a means of comparing the spinnability of different materials. In other words, the larger the value of $(V_L/V_0)_{max}$, the more spinnable the material. This is because higher values of $(V_L/V_0)_{max}$ indicate greater elongation of the threadline without breaking.

The following conclusions may be drawn from Table 3.1: (1) The use of an isothermal chamber increases the spinnability; (2) As the length of the isothermal chamber is increased from 7.5 to 30 cm, the spinnability is

TABLE 3.1

Summary of Spinnability Test of Filled Polypropylene with and without Coupling Agent[a]

Material	Velocity at spinnerette, V_0 (m/min)	Maximum speed at takeup device, V_L (m/min)	Maximum draw-down ratio, $(V_L/V_0)_{max}$
	(a) Nonisothermal spinning		
PP	3.59	553.8	154.3
PP/$CaCO_3$[b]	1.69	255.6	157.3
PP/$CaCO_3$/Y9187[c]	1.52	378.2	248.0
PP/$CaCO_3$/A1100[d]	1.52	432.3	284.5
PP/$CaCO_3$/TTS[e]	1.64	310.7	189.3
	(b) With 7.5 cm long isothermal chamber		
PP	3.56	918.6	257.9
PP/$CaCO_3$	1.67	445.8	266.5
PP/$CaCO_3$/Y9187	1.56	729.5	468.3
PP/$CaCO_3$/A1100	1.56	810.5	518.0
PP/$CaCO_3$/TTS	1.64	952.5	580.3
	(c) With 30 cm long isothermal chamber		
PP	3.57	1202.4	336.9
PP/$CaCO_3$	1.61	513.4	318.9
PP/$CaCO_3$/Y9187	1.55	864.6	557.2
PP/$CaCO_3$/A1100	1.62	1026.7	632.8
PP/$CaCO_3$/TTS	1.70	1202.4	707.5

[a] From Han *et al.* [22].
[b] $CaCO_3$ (50 wt.%): Thompson and Wienman, Atomite.
[c] Y9187: Union Carbide, N-octyl triethoxy silane.
[d] A1100: Union Carbide, γ-aminopropyl triethoxy silane.
[e] TTS: Kenrich Petrochemical, isopropyl triisostearoyl titanate.

increased; (3) In the case of nonisothermal spinning, the silane coupling agents (Union Carbide, Y9187 and A1100) yield greater spinnability than the titanate coupling agent (Kenrich Petrochemicals, TTS); (4) In the case of delayed cooling (with an isothermal chamber), the titanate coupling agent (TTS) gives rise to greater spinnability than the silane coupling agents (Y9187 and A1100); (5) For the two silane coupling agents used, the A1100 gives rise to greater spinnability than the Y9187.

It should be remembered that the effectiveness of a coupling agent depends, among other things, on the structure of a polymer, the nature of the filler surface, and the chemistry of the coupling agent. Therefore the conclusions given above may be applicable only to the polymer/filler/coupling agent systems investigated.

Figures 3.41–3.43 display tensile strength measurements made on melt-spun fibers. It is seen that the use of coupling agents improved the tensile strength of the PP-$CaCO_3$ fibers. Figure 3.41 shows a clear trend that as the draw-down ratio is increased (hence as the fiber is stretched more during the melt spinning operation), the titanate coupling agent TTS gives rise to higher fiber tensile strength than the silane coupling agents, Y9187 and A1100.

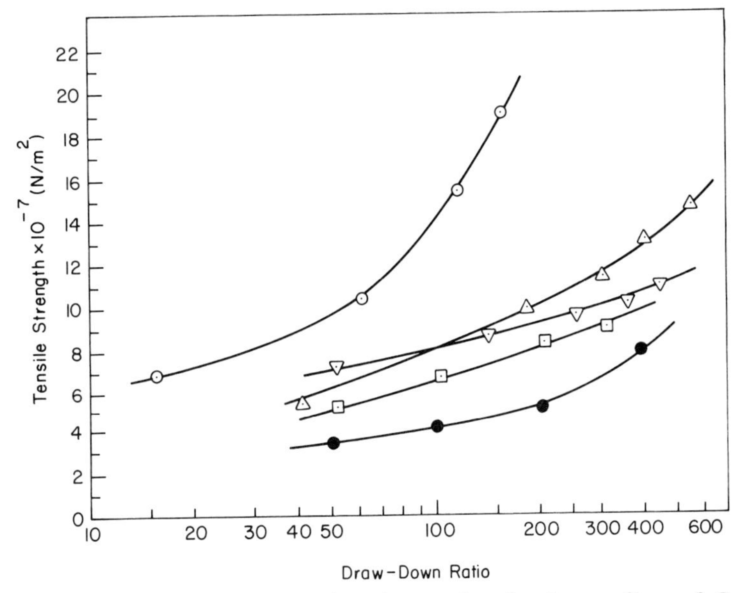

Fig. 3.41 Tensile strength versus draw-down ratio of melt-spun fibers of $CaCO_3$-filled polypropylene (PP) with and without coupling agent (1 wt.%) [22]: (○) pure PP; (●) PP/$CaCO_3$ = 50/50 (by wt.): (△) PP/$CaCO_3$ = 50/50 with a titanate coupling agent, TTS; (▽) PP/$CaCO_3$ = 50/50 with a silane coupling agent, Y9187; (□) PP/$CaCO_3$ = 50/50 with a silane coupling agent, A1100.

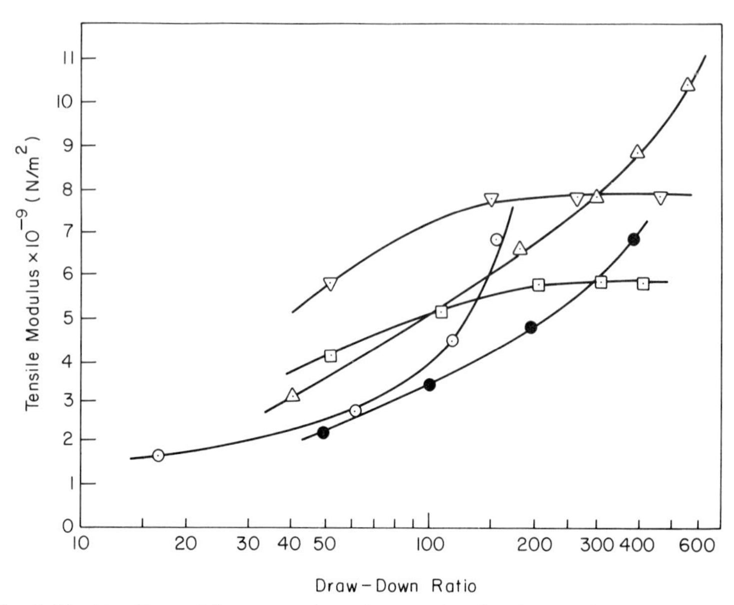

Fig. 3.42 Tensile modulus versus draw-down ratio of melt-spun fibers of $CaCO_3$-filled polypropylene with and without coupling agent [22]. Symbols are the same as in Fig. 3.41.

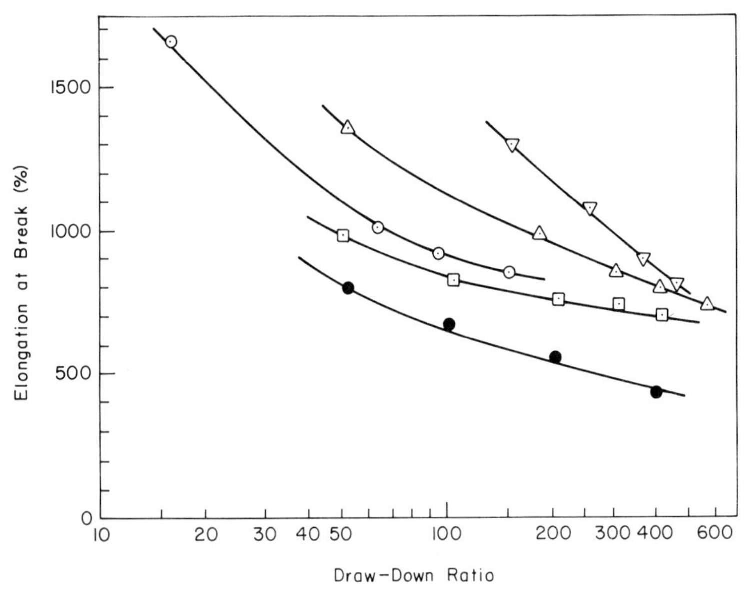

Fig. 3.43 Elongation at break versus draw-down ratio of melt-spun fibers of $CaCO_3$-filled polypropylene with and without coupling agent [22]. Symbols are the same as in Fig. 3.41.

However, the effect of coupling agent on the modulus of the melt-spun fibers of PP-$CaCO_3$ is somewhat complicated. Figure 3.42 shows that the modulus of the PP-$CaCO_3$ fiber with the silane coupling agents, Y9187 and A1100, first increases with draw-down ratio (up to about 200) and then levels off as the draw-down ratio is increased further. However, the modulus of the PP-$CaCO_3$ fiber with the titanate coupling agent TTS increases continuously with draw-down ratio, giving rise to values much greater than the moduli obtained by the use of the silane coupling agents, Y9187 and A1100. Figure 3.43 shows that the *untreated* PP-$CaCO_3$ has the lowest ultimate elongation of all the materials tested, and that the PP-$CaCO_3$ treated with either the silane coupling agent A1100 or the titanate coupling agent TTS has a greater ultimate elongation than the homopolymer PP.

White and Tanaka [156] performed melt spinning experiments, using suspensions of carbon black, titanium dioxide, and calcium carbonate in polystyrene. They found that the presence of particulates gives rise to enhanced instabilities in melt spinning.

3.5.2 Injection Molding of Filled Thermoplastic Resins

The mechanical/physical properties of injection-molded specimens are determined by changes in three controlling variables [157, 158]: (1) melt temperature, (2) injection pressure, and (3) molding time. These variables are dependent upon the machine variables, mold variables and the polymer employed. The temperature of the polymer entering the mold is determined by: (1) the cylinder wall temperature, (2) the contact time (mold cycle time), (3) the polymer thermal diffusivity and heat of fusion (a function of crystallinity and association energies), and (4) the size of the contact surface between the polymer and the cylinder wall. Maximum pressure will eventually be transmitted to the mold (except at very short ram forward times). However, the speed at which maximum pressure is attained is strongly dependent upon other variables (viscosity of the melt, cooling rate, etc.).

Part defects (warpage, sinks, bubbles, scoring, cracks) are mainly a result of strains frozen into the polymer during molding. Frozen strains result when polymer chains, oriented by flow, solidify in a nonequilibrium condition. During filling, the polymer forms a thin, highly oriented layer on the surface of the mold (as a result of contact with the cold mold surface). During compression, more polymer enters the mold. As the mold is being cooled, some of the orientation resulting from this flow is frozen before it can relax.

Particulate-filled polymeric materials, used in injection molding operations, generally contain between 5 and 60 wt.% of solid particles. Optimal loading is reported to be between 20 and 40 wt.%. For instance, in the use of

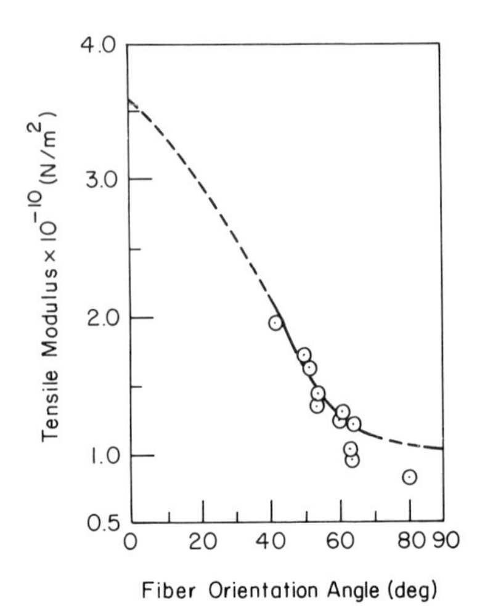

Fig. 3.44 Tensile modulus versus longitudinal fiber orientation for epoxy reinforced with glass fiber (50 vol.%) [14]. Dashed line is extrapolation to limits predicted by Tsai–Halpin theory. Reprinted by permission of *Modern Plastics Magazine*, McGraw-Hill, Inc.

glass fiber, below a certain critical loading, the glass fiber supplies no reinforcement. Above the upper critical loading, breakage of glass fiber occurs during processing and the resin fails to support the glass fiber. In this context, a good understanding of the mold-filling patterns of highly filled molten polymers is very important [159, 160].

Processing conditions can alter the fiber orientation, fiber length, and fiber dispersion, greatly influencing the properties of the composite. A group of fibers entering a constriction (such as an injection nozzle) tends to become oriented in the direction of flow [14, 161, 162]. (Also, see Fig. 3.1.) As a result, the longitudinal properties of reinforced molded parts are frequently improved more than the transverse properties (perpendicular to the flow), as shown in Fig. 3.44. The degree of orientation is a complex function of flow rate, shear rate, mold geometry, resin properties, fiber loading, and fiber size. No quantitative method is available for predicting orientation in injection molded parts.

The mechanical properties of injection-molded specimens of highly filled polypropylene, with and without titanate coupling agent, were investigated by Han *et al.* [21, 22]. Let us examine the roles that particulate fillers and coupling agents play in influencing the mechanical properties of filled polypropylene.

It is seen in Table 3.2 that elongation at break (ultimate tensile elongation) obtained with $CaCO_3$-filled polypropylene (PP), *with* the titanate coupling agent TTS, exceeds 300 percent. This equals that observed for virgin PP, whereas the $CaCO_3$-filled PP *without* coupling agent has an elongation at break of only 44%. On the other hand, the addition of coupling agent TTS to glass fiber-filled PP has little influence on the ultimate tensile elongation

TABLE 3.2

Mechanical Properties of Injection-Molded Filled Polypropylene with and without Titanate Coupling Agent[a]

Material	Injection temp (°C)	Tensile strength $\times 10^{-7}$ (N/m^2)	Elongation at break (%)	Tensile modulus $\times 10^{-9}$ (N/m^2)	Flexural modulus $\times 10^{-9}$ (N/m^2)
PP[b]	243	2.46	300	0.62	1.03
	260	2.46	300	0.55	0.99
PP/$CaCO_3$[c]	243	1.82	48	0.98	2.56
	260	1.79	44	0.97	2.40
PP/$CaCO_3$/TTS	243	1.68	300	0.85	1.98
	260	1.65	300	0.71	1.87
PP/Glass fiber[d]	243	2.04	22	1.38	4.19
	260	1.95	21	1.36	4.11
PP/Glass fiber/TTS[e]	243	1.73	32	1.13	3.73
	260	1.67	25	1.14	3.51

[a] From Han *et al.* [21].
[b] PP: Exxon Chemical Co., E115.
[c] $CaCO_3$ (50 wt.%): Thompson and Weinman, Atomite.
[d] Glass fibers (50 wt.%): Ferro Corp., 661-11-0312.
[e] TTS (1 wt.%): Kenrich Petrochemical, isopropyl triisostearoyl titanate.

of the filled polymers. This appears to indicate that the filler type has a strong influence on the ultimate tensile elongation of filled polymeric systems.

In reference to Table 3.2, all of the samples tested decreased slightly in tensile strength with the addition of the titanate coupling agent. The maximum decrease in strength ($\sim$15%), resulting from the addition of 1.0 wt.% coupling agent, was observed in the PP-glass fiber-TTS systems.

Referring again to Table 3.2, both the tensile and flexural moduli are observed to decrease with the addition of coupling agent. The smallest decrease (14%) occurs in the PP-glass fiber-TTS system. The largest decrease in tensile modulus (27%) occurs in the PP-$CaCO_3$-TTS system and the minimum decrease (18%) occurs with the PP-glass fiber system. Higher injection temperatures result in minor decreases (generally 5% or less) in modulus. This is probably due to greater uniformity or decreased chain length (degradation). This decrease is generally more pronounced for the tensile modulus results than for the flexural modulus ones.

The largest differences in modulus, as with tensile strength, result from the type of filler employed. All samples are stiffer than virgin material by a factor of 2 to 4. As with tensile strength, differences in modulus due to filler type probably result from interfacial effects. Stiffness increases with the addition of filler because the relatively high modulus particles prevent the matrix from flexing as it would if no filler were present. Modulus decreases (in most cases) with the addition of titanate, which is believed due to the fact that the plasticizing effect of the short chain molecule exceeds the effect of the interfacial bond created by van der Waals forces between the polymer and titanate molecules.

To explain the observations made above, it is necessary to first examine the chemical structure of titanate coupling agents [123]. When the titanate is bonded to the inorganic filler (in this case, with the removal of an isopropyl group), the hydrocarbon chains improve the compatibility of the inorganic particle with the polymer. This may tend to improve dispersion and surface wetting. It is apparent, therefore, that there are at least two significant factors, the state of dispersion and the interface characteristics, affecting the change in tensile strength of filled thermoplastics due to the addition of titanates.

Table 3.3 gives a summary of the mechanical property measurements of the injection-molded specimens, with two different silane coupling agents. There is an improvement in tensile strength of the filled polypropylene, compared to the virgin resin. Note also that the tensile strength of the PP/$CaCO_3$ is greater than that of the PP/glass beads. It is of particular interest to note that the coupling agent Y9187 has a relatively small effect on the tensile strength of both PP/$CaCO_3$ and PP/glass beads. (In the case of the PP/$CaCO_3$ the tendency is to decrease.) On the other hand, the coupling agent A1100 improves the tensile strength considerably. On the other hand, the effect of the silane coupling agents on the tensile modulus seems somewhat more complex. The coupling agent Y9187 increases the tensile modulus of both

TABLE 3.3

Mechanical Properties of Injection-Molded Filled Polypropylene with and without Silane Coupling Agent[a]

Material	Injection temp. (°C)	Tensile strength $\times 10^{-7}$ (N/m^2)	Tensile modulus $\times 10^{-9}$ (N/m^2)	Elongation at break (%)
PP/$CaCO_3$[b]	243	5.20	1.41	85
PP/$CaCO_3$/Y9187	243	4.96	1.58	400
PP/$CaCO_3$/A1100	243	6.37	1.34	110
PP/Glass beads[c]	243	4.20	0.99	350
PP/Glass beads/Y9187[d]	243	4.37	1.24	640
PP/Glass beads/A1100[e]	243	4.51	1.72	75

[a] From Han *et al.* [22].
[b] $CaCO_3$ (50 wt. %): Thompson and Weinman, Atomite.
[c] Glass beads (50 wt. %): Ferro Corp., MS-XLX.
[d] Y9187: Union Carbide, N-octyl triethoxy silane.
[e] A1100: Union Carbide, γ-aminopropyl triethoxy silane.

PP/$CaCO_3$ and PP/glass beads, and the coupling agent A1100 has little effect on the tensile modulus of the PP/$CaCO_3$ while increasing the modulus of the PP/glass beads considerably. Table 3.3 clearly shows that the ultimate tensile elongation (percent elongation at break) is very high when the coupling agent Y9187 was used, compared to the situation when the coupling agent A1100 was used.

Figure 3.45 gives photomicrographs of the tensile fracture surface of injection-molded specimens of PP/glass beads, with and without silane

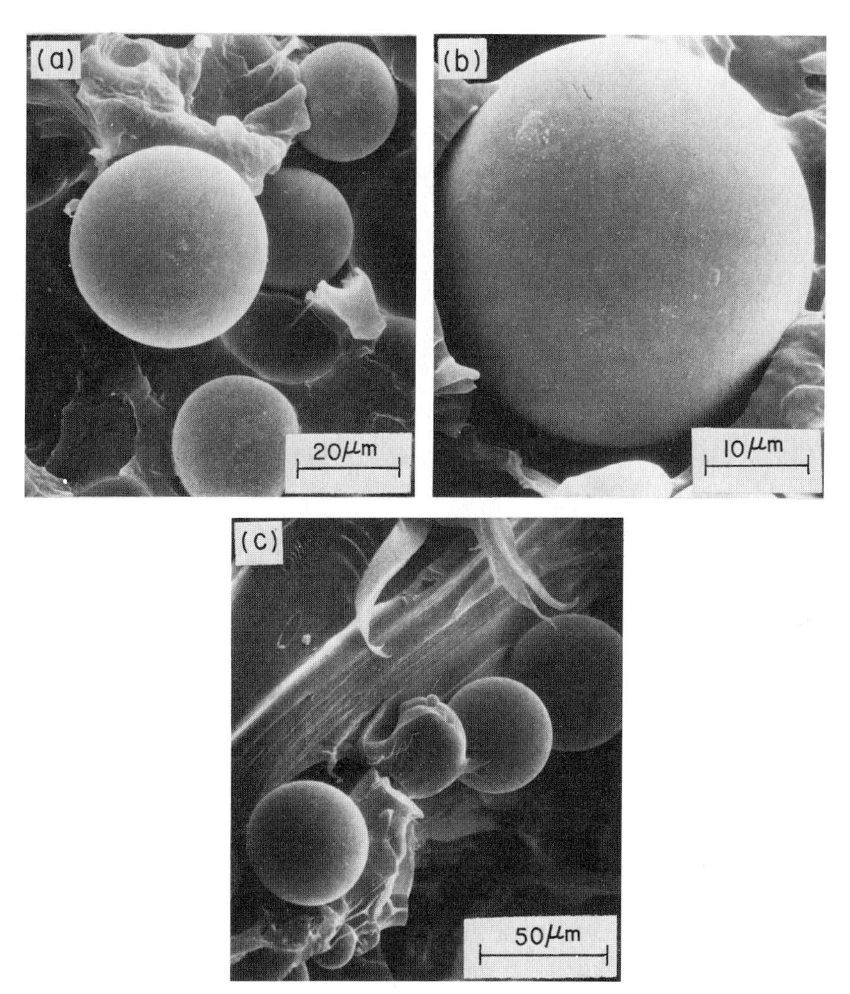

Fig. 3.45 Scanning electronmicrographs of the tensile fracture surface of glass bead-filled polypropylene [22]: (a) without coupling agent; (b) with a silane coupling agent, A1100; (c) with a silane coupling agent, Y9187.

coupling agents. Coupling agent Y9187 appears to have promoted some interactions between the glass beads and polypropylene, whereas coupling agent A1100 has little effect.

Figure 3.46 gives photomicrographs of the tensile fracture surface of injection-molded specimens, with and without coupling agents. It is seen that although these coupling agents appear to have created little bonding between the $CaCO_3$ particles and the polypropylene phase, coupling agents Y9187

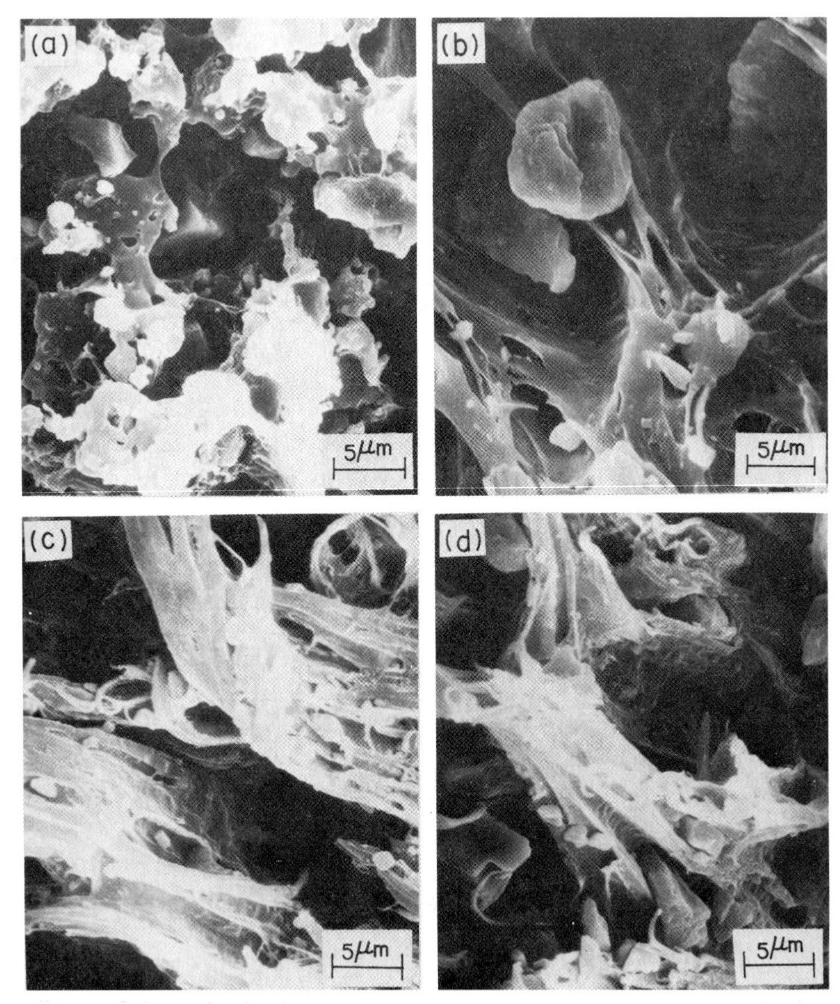

Fig. 3.46 Scanning electronmicrographs of the tensile fracture surface of $CaCO_3$-filled polypropylene [22]: (a) without coupling agent; (b) with a silane coupling agent, A1100; (c) with a silane coupling agent, Y9187; (d) with a titanate coupling agent, TTS.

and TTS have certainly changed the morphology of the virgin polypropylene. Figures 3.46c and 3.46d clearly show that the polypropylene phase exhibits a morphology of long fibrils, whereas such a morphology is not seen in Fig. 3.46a where no coupling agent is added, and in Fig. 3.46b where coupling agent A1100 is added. It is possible that Y9187 and TTS influenced the crystallization kinetics of the partially crystalline polypropylene when the specimens were injection-molded accompanied by cooling.

It should be remembered, as given in Table 3.3, that the silane coupling agent Y9187 considerably increases the percent elongation of injection-molded PP-$CaCO_3$, but has little effect on its tensile strength. On the other hand, the percent elongation of injection-molded PP-$CaCO_3$ is increased little in the presence of the coupling agent A1100. It should also be remembered that, according to Table 3.2, the titanate coupling agent TTS also considerably increases the percent elongation of injection-molded PP-$CaCO_3$, with little effect on its tensile modulus and tensile strength. It appears then that the unusually high percent elongation of injection-molded PP-$CaCO_3$, in the presence of either the silane coupling agent Y9187 or the titanate coupling agent TTS, may be attributable to a change in the morphology of the polypropylene phase, which exhibits long fibrils (see Figs. 3.46c and d) quite different from the morphology of the virgin polypropylene (see Figs. 3.45a and 3.46a). It is now clear why, as shown in Fig. 3.43, the percent elongation of melt-spun fibers of PP-$CaCO_3$ is quite high in the presence of either the silane coupling agent Y9187 or the titanate coupling agent TTS, compared to the situation where either no coupling agent is added or the coupling agent A1100 is added.

The mechanical properties of mica-filled polypropylene with coupling agents were investigated by Okuno and Woodhams [163] who used a silane coupling agent (Dow Corning, Z-6032), and by Boaira and Chaffey [20] who used both a silane coupling agent (Dow Corning, Z-6032) and an amino coupling agent (Rohm and Haas, DMAEMA). They report that when coupling agents were used, improvements in certain mechanical properties (flexural modulus and the tensile and flexural strengths) were observed, as given in Figs. 3.47 and 3.48. Such improvements were attributed to the better bonding between the matrix resin and the fillers.

Boaira and Chaffey [20] noted that different degrees of improvement in the flexural modulus were observed with different types of coupling agents. They reported that an increase of 22% in a mica-filled composite's strength was obtained with the silane coupling agent, whereas an increase of 16% was obtained with the amino coupling agent, compared to the strength of the composite *without* coupling agent. They also reported that the amino coupling agent had a negligible effect on the flexural modulus of the mica-filled composite, whereas the silane coupling agent had a significant effect.

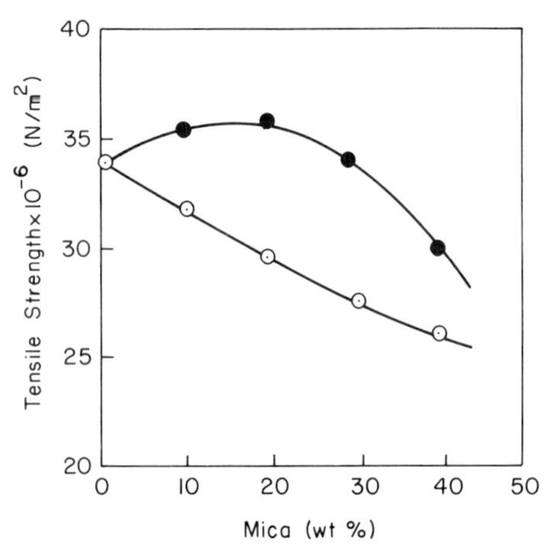

Fig. 3.47 Effect of the mica content on the tensile strength of mica-filled polypropylene [163]: (⊙) without coupling agent; (●) with a silane coupling agent (Dow Corning Co., Z-6032).

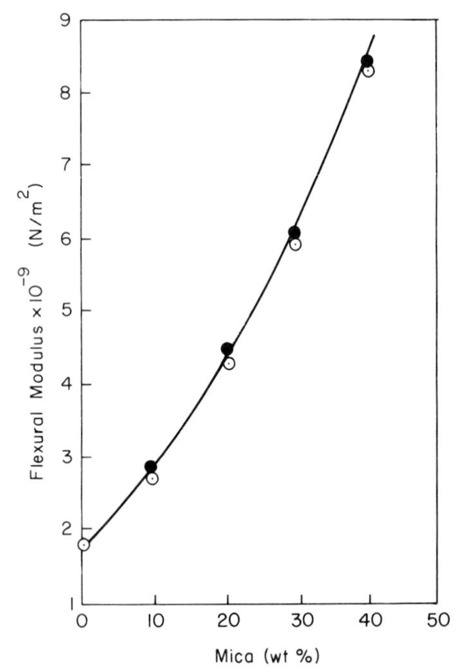

Fig. 3.48 Effect of mica content on the flexural modulus of mica-filled polypropylene [163]. Symbols are the same as in Fig. 3.47.

In view of the fact that minute portions of coupling agents at the interface can have a profound effect on the performance of composites, it is expected that elucidation of the "coupling" mechanism will be helpful in understanding the fundamental nature of the adhesion of organic polymers to mineral (or metallic) surfaces. Further discussion of mechanisms at the interface and of adhesion, when particulate fillers are treated with coupling agents, is beyond the scope of this monograph. The reader is referred to recent review articles [119, 120, 164].

3.5.3 Molding of Reinforced Thermoset Resins

Unlike the processing of particulate-filled thermoplastic resins, the processing of reinforced thermoset resins requires an understanding of the reaction mechanisms in a polymer matrix. It should be pointed out that thermoset resins contain reactive groups, which form crosslinkages at the incipient gel points and release heat from their reactions. Moreover, in the presence of coupling agents, the polymer matrix and the reinforcing agent give rise to strong interfacial bonding, which is the basic requirement for the enhancement of the end-use properties of the finished product. On the other hand, the nature of the processing techniques significantly controls the dimensional stability and mechanical properties of the reinforced thermoset composites.

The term "reinforced thermoset resin" covers a wide range of combinations of thermoset resins and reinforcing agents. Glass fiber-reinforced thermoset resin has one of the best strength characteristics of many materials and is available in many different types, depending on the method of fabrication and the formulation of the glass-resin mixture used. Glass fiber-reinforced thermoset resins have several properties, which are improved in comparison to unreinforced versions of the same polymer, namely: (1) stiffness is increased; (2) the coefficient of thermal expansion is reduced, yielding a better dimensional stability of molded parts; (3) droop and sag of the composite at elevated temperature are reduced.

There are a variety of fabrication processes being used in industry for thermoset resins with reinforcing agents [165, 166]. However, we shall confine our discussion here to the injection molding (or transfer molding or compression molding) of glass fiber-reinforced unsaturated polyesters and to the reaction injection molding (RIM) of glass fiber-reinforced urethanes.

Unsaturated polyester accounts for the greater part of all resin used in glass fiber-reinforced plastics. Polyesters offer the advantages of a balance of good mechanical, chemical, and electrial properties. A typical formulation of a polyester premix molding compound is as follows: Resin: 100; Filler: 300 phr (parts per hundred resin); Lubricant: 5 phr; Catalyst: 2 phr; Glass fiber: 75 phr. Resins used in a premix molding compound can vary from rigid to flexible with viscosities from 3 ~ 60 N sec/m^2 (30 ~ 600 poises). The choice of a rigid or flexible resin would be a function of the part and the type of reactive monomer in the resin. For instance, diallyl phthalate–polyester resins would exhibit significantly higher viscosities than styrene–polyester resins. The two fillers most commonly used are calcium carbonate and clay. The catalyst commonly used is organic peroxide (e.g., benzoyl peroxide, tertiary butyl perbenzoate). Lubricant is used for the purpose of releasing the molded part from the mold, and zinc, calcium, or magnesium stearates are commonly used.

The polyester premix molding compounds of commercial use are supplied as sheet molding compound (SMC), bulk molding compound (BMC), or thick molding compound (TMC) [167, 168]. These molding compounds can be molded on standard compression or transfer molds. The basic problem in molding polyester premix compounds is to get a uniform layer of glass reinforcement held in place in the die cavity while the resin fills the cavity and reaches its gel stage during cure. Temperature, mold closing speed, pressure, and cure time will be a function of the part and its design. Transfer molding does require a proper gate design in the mold. The flow of the mixture through the gate(s) can result in a variation in strength across the part, due to fiber orientation during the flow [14]. The precise level of end-use properties depends on the fiber orientation, fiber distribution, and fiber content in the polymer mixture, these being greatly influenced by the processing conditions. Since the mechanical properties of the molded articles depend strongly upon the orientation of the glass fibers (see Fig. 3.44), it is important to understand how to control fiber orientation during molding [169–174].

Note that the viscosity of the premix compound is a dominant factor, which not only affects the fiber orientation, but also controls the processing conditions. Little has been published dealing with the rheological properties of polyester premix compounds. It is worth mentioning that the viscosity of the polyester molding compound will influence the molding characteristics, and is determined by the level of chemical thickener in the premix and the time-temperature history of the compound. For instance, too low a viscosity can cause the resin to flow ahead of the glass fiber and result in resin-rich areas with low mechanical properties. On the other hand, too high a viscosity may not properly wet-out the glass fibers.

In determining the rheological properties of glass fiber-reinforced resins, the choice of experimental technique is very crucial for obtaining meaningful results. In other words, depending on the size of the glass fibers in a molding compound, the geometry of the rheometer employed can influence the rheological data obtained, especially for a compound having a high glass content. Using a squeeze-flow rheometer (i.e., the test material is squeezed between two parallel discs), Gandhi and Burns [175] took rheological measurements of glass fiber-reinforced polyester molding compounds, and reported that a power law was obeyed. They reported that the length-to-diameter ratio (aspect ratio) of the glass fibers significantly affected the rheological behavior of the test materials.

Although reaction injection molding (RIM) has been used for some time in industry, glass fiber-reinforced RIM is a relatively new processing technique, and has recently attracted the attention of industrial researchers [176–181]. The RIM process involves the impingement mixing of two or more streams of highly reactive components in a small chamber at high pressure, followed by rapid flow of the mixture directly into a mold cavity [181]. The mixture finally crosslinks *in situ*. The process is widely used for manufacturing exterior automotive parts, the most popular polymer used being polyurethane.† When glass fibers are added to RIM urethanes, the mixture produces the composite materials that are stiffer and have a lower coefficient of thermal expansion, than unreinforced urethane.

In reinforced RIM urethane, the first step in the process for incorporating reinforcement involves preparing a slurry of glass fibers in one or both of the individual components. Because of the fast chemical reaction, there is not sufficient time to add the reinforcement after the components have been mixed. The use of milled fibers for reinforcement is attractive because they can be more easily processed in prototype RIM equipment. Milled fibers are continuous glass filaments which have been hammermilled into short lengths. They are available with chemical surface treatments which make them compatible with different resin systems.

Glass fibers increase the viscosity of polyfunctional alcohol (polyol). The increase in viscosity is dependent on both the amount of glass and the average length of the glass filaments. The viscosity of the slurry is reported to have non-Newtonian characteristics [176, 177]. Note that the polyols commercially employed in industry are Newtonian fluids. The non-Newtonian (i.e., shear-thinning) behavior of the slurry is attributable to the orientation of the fiber filaments in the flow direction. Hence the fiber aspect ratio and the fiber content are also of importance for controlling the nature of the shear-thinning behavior. Materials and test variables which affect the rheological properties of glass fiber-polyol slurries are: (1) the viscosity of the unfilled

† Polymer is made *in situ* from polyol and isocyanate.

TABLE 3.4

Mechanical and Thermal Properties of Reinforced RIM Urethanes[a]

Material	Flexural modulus $\times 10^{-9}$ (N/m^2)	Tensile strength $\times 10^{-7}$ (N/m^2)	Elongation at break (%)	Mold shrinkage (%)	Coefficient of thermal expansion $\times 10^6$ (cm/cm/
Unreinforced urethane	0.834	3.11	83	1.2	128
Reinforced urethane with 15 wt.% glass fibers	1.469	3.38	27	0.6	65

[a] From Isham [177].

monomer; (2) the concentration of milled fibers; (3) the length of the milled fibers; (4) the slurry temperature. As mentioned above, in connection with glass fiber-reinforced polyester premix, most of the existing experimental techniques for rheological measurements are designed for pure liquids and produce questionable results, especially when a large quantity of glass fibers is added to the liquid. An accurate and universally accepted method for measuring the rheological properties of glass fiber-polyol slurries is very much needed. It should be pointed out that control of, at least the viscosity of the slurry, through the variation of temperature, glass content, and polyol, is critical to success in the reinforced RIM process.

Isham [177] notes that stiffness, tensile strength, and tensile elongation are influenced by both the glass fibers and the urethane chemistry. Table 3.4 gives typical mechanical and thermal properties of glass fiber-reinforced RIM urethanes. The addition of milled glass fibers does not significantly improve tensile strength and yet part shrinkage values are decreased significantly by their use. Similar observations are also reported by Simpkins [178]. The maximum glass content in RIM urethane is normally 20–25%, and it is limited to this level due to the increased slurry viscosity in the process equipment.

Chemical bonding between the coupling agent and the polymer matrix is necessary for obtaining a high performance in glass fiber-reinforced composites. Schwarz and co-workers [179, 180] investigated the effect of a silane coupling agent on the mechanical properties and morphology of glass fiber-reinforced RIM urethanes. The observations of the mechanical and thermal properties of their test materials are in general agreement with those reported by Isham [176, 177]. Figure 3.49 shows the effect of a silane coupling agent on the flexural modulus of reinforced RIM composites. Both untreated and

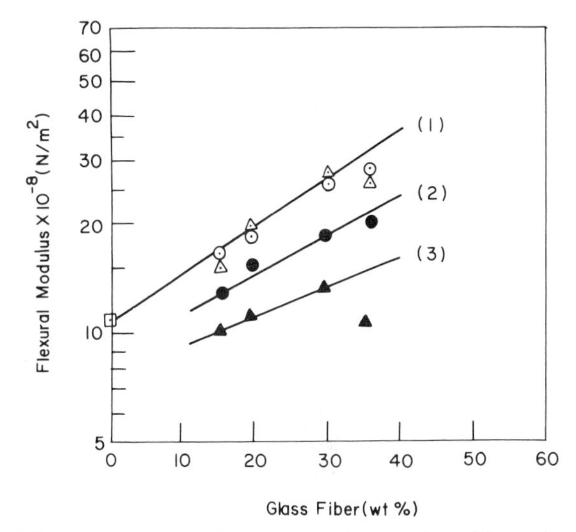

Fig. 3.49 Effect of a silane coupling agent on the retention of flexural modulus of RIM urethane composites reinforced with glass fiber [180]. (1) Initial state after cure: (△) untreated specimen; (⊙) silane treated specimen; (2) Silane treated specimen after 8 hr in boiling water; (3) Untreated specimen after 8 hr in boiling water. From *Journal of Cellular Plastics* **15**, No. 1, 1979, R. M. Gerkin *et al.*, Courtesy of Technomic Publishing Co., Inc., Westport, CT 06880.

silane-treated composites produce essentially the same initial modulus, but after exposure to boiling water for 8 hours, the composite containing silane-treated glass fiber retained a significantly higher fraction of the original flexural modulus. Gerkin *et al.* [180] report that silane coupling agents contribute, also, to the tensile and impact properties of reinforced RIM composites. Figure 3.50 shows scanning electron micrographs of the impact surfaces of untreated and silane-treated composites. It is seen that there appears to be virtually no polymer bonded to the untreated glass surface, whereas, on the treated surface, there is significant polymer adhesion with cohesive failure in the polymer, as well as fracture of glass fiber itself during the fracture process.

In summary, the addition of glass fibers to thermoset resins (e.g., unsaturated polyester, urethane) yields significant improvement in physical/mechanical (and/or thermal) properties, and makes the molding of large lightweight parts possible. However, much has yet to be learned about the preparation of thermoset molding compounds having consistent quality, the efficient ways of processing them, and the quantitative determination of the relationship of chemical bonds to the strength of reinforced composites, especially when reinforcement is realized through chemical bonding at interfaces, i.e., the coupling agent/fiber interface, and the coupling agent/polymer matrix interface.

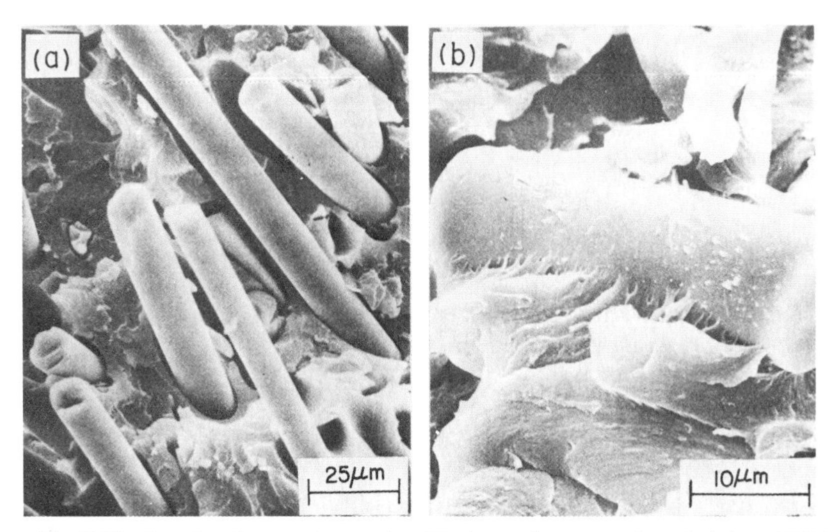

Fig. 3.50 Scanning electronmicrographs of the impact fracture surface of reinforced RIM urethane composites [179]: (a) containing *untreated* milled glass fibers (after 8 hr in boiling water); (b) containing *silane treated* milled glass fibers (after 8 hr in boiling water).

Problems

3.1 A calcium carbonate-filled polypropylene melt is found to obey the relationship given by Eq. (3.27), in which Y is 2.5×10^4 N/m^2, K is 4.5×10^4 N sec^n/m^2, and $n = 0.5$ at 200°C.

(i) Sketch the viscosity versus shear rate curve.

(ii) Sketch velocity and shear stress distributions in a long cylindrical tube through which this material is extruded.

(iii) What will the pressure drop be when this material is extruded in a flat film die (60 cm wide, 0.2 cm thick, and 10 cm long) at a volumetric flow rate of 100 cm^3/sec at 200°C?

3.2 Consider that a string of small, spherical glass beads of uniform size are injected into a position somewhere between the tube wall and the centerline of a long cylindrical tube through which a viscous liquid flows at a constant volumetric flow rate. Assume that the diameter d_p of the glass beads is very small compared to the diameter D of the tube (i.e., $d_p \ll D$). Will there be a radial migration of the glass beads as they move along the tube axis? If so, in which direction will the glass beads migrate (i.e., toward the tube wall or toward the centerline of the tube)? In answering the question, consider the following alternative condi-

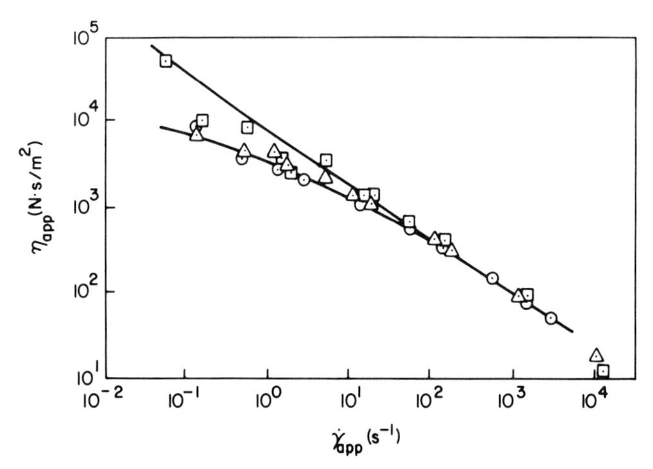

Fig. 3.51 Viscosity versus shear rate for short glass fiber-filled molten polypropylene at various loadings of glass fiber [100]: (⊙) pure PP; (△) 20 wt.% of glass fiber ($d = 0.01$ mm; $l = 2.4$ mm); (⊡) 40 wt.% of glass fiber ($d = 0.01$ mm; $l = 0.635$ mm).

tions: (1) the suspending medium is a Newtonian liquid; (2) the suspending medium may be represented by the Zaremba–DeWitt model; (3) the suspending medium may be represented by the Coleman–Noll second-order fluid. Assume that the viscosity is so large that the inertia effect is negligible compared to the viscous force.

3.3 Figure 3.51 gives the viscosity, determined by the use of the plunger-type viscometer, of polypropylene melts filled with short glass fibers [100]. It is seen that the viscosities of the glass fiber-filled polypropylene are almost the same as that of pure polypropylene at high shear rates (say, at above 10 sec^{-1}). Give your explanation as to why the addition of short glass fibers to polypropylene shows little influence on its viscosities at high shear rates.

REFERENCES

[1] G. Lubin, ed., "Handbook of Fiberglass and Advanced Plastics Composites." Van Nostrand-Reinhold, New York, 1969.
[2] G. Kraus, ed., "Reinforcement of Elastomers." Wiley (Interscience), New York, 1965.
[3] R. D. Deanin and N. R. Schott, eds., "Fillers and Reinforcements for Plastics," Adv. Chem. Ser. No. 134. Am. Chem. Soc., Washington, D.C., 1974.
[4] C. D. Han, *J. Appl. Polym. Sci.* **18**, 821 (1974).
[5] C. D. Han, "Rheology in Polymer Processing," Chap. 7, p. 184. Academic Press, New York, 1976.

[6] N. Minagawa and J. L. White, *J. Appl. Polym. Sci.* **20**, 501 (1976).
[7] Y. Chan, J. L. White, and Y. Oyanagi, *J. Rheol.* **22**, 507 (1978).
[8] V. M. Lobe and J. L. White, *Polym. Eng. Sci.* **19**, 617 (1979).
[9] H. Tanaka and J. L. White, *Polym. Eng. Sci.* **20**, 949 (1980).
[10] F. M. Chapman and T. S. Lee, *SPE J.* **26**(1), 37 (1970).
[11] J. L. White and J. W. Crowder, *J. Appl. Polym. Sci.* **18**, 1013 (1974).
[12] G. V. Vinogradov, A. Y. Malkin, E. P. Plotnikova, O. Y. Sabsai, and N. E. Nikolayeva, *Int. J. Polym. Mater.* **2**, 1 (1972).
[13] H. Kambe and M. Takano, *Proc. Int. Congr. Rheol., 4th* Part 3, p. 557. Wiley (Interscience), New York, 1965.
[14] L. A. Goettler, *Modern Plast.* **47**(4), 140 (1970).
[15] W. F. Busse, *J. Polym. Sci., Part A-2* **5**, 1249 (1967).
[16] B. L. Lee, personal communication (1979).
[17] C. C. McCabe and N. Mueller, *Trans. Soc. Rheol.* **5**, 329 (1961).
[18] J. R. Hopper, *Rubber Chem. Technol.* **40**, 463 (1967).
[19] N. Nakajima and E. A. Collins, *Rubber Chem. Technol.* **48**, 615 (1975).
[20] M. S. Boaira and C. E. Chaffey, *Polym. Eng. Sci.* **17**, 715 (1977).
[21] C. D. Han, C. Sandford, and H. J. Yoo, *Polym. Eng. Sci.* **18**, 849 (1978).
[22] C. D. Han, T. Van den Weghe, P. Shete, and J. R. Haw, *Polym. Eng. Sci.* **21**, 196 (1981).
[23] I. R. Rutgers, *Rheol. Acta* **2**, 202 (1962); **2**, 305 (1962).
[24] J. Happel and H. Brenner, "Low Reynolds Number Hydrodynamics." Prentice-Hall, Englewood Cliffs, New Jersey, 1965.
[25] H. L. Goldsmith and S. G. Mason *in* "Rheology" (F. R. Eirich, ed.), Vol. 4, p. 86. Academic Press, New York, 1967.
[26] D. J. Jeffrey and A. Acrivos, *AIChE J.* **22**, 417 (1976).
[27] A. Einstein, *Ann. Phys.* **19**, 289 (1906); **34**, 591 (1911).
[28] G. B. Jeffery, *Proc. R. Soc. London, Ser. A* **102**, 161 (1922).
[29] V. Vand, *J. Phys. Colloid Chem.* **52**, 277 (1948).
[30] E. Guth and R. Simha, *Kolloid-Z.* **74**, 266 (1936).
[31] R. Simha, *J. Appl. Phys.* **23**, 1020 (1952).
[32] H. Eilers, *Kolloid-Z.* **97**, 313 (1941).
[33] M. Mooney, *J. Colloid Sci.* **6**, 162 (1952).
[34] P. Sherman, *Proc. Int. Congr. Rheol., 4th* Part 3, p. 605. Wiley (Interscience), New York, 1965.
[35] N. A. Frankel and A. Acrivos, *Chem. Eng. Sci.* **22**, 847 (1967).
[36] J. D. Goddard, *J. Non-Newtonian Fluid Mech.* **2**, 169 (1967).
[37] G. K. Batchelor, *J. Fluid Mech.* **41**, 545 (1970).
[38] Z. Hashin, *J. Appl. Mech.* **32**, 630 (1965).
[39] R. Roscoe, *J. Fluid Mech.* **28**, 273 (1967).
[40] J. D. Goddard and C. Miller, *J. Fluid Mech.* **28**, 657 (1967).
[41] C. J. Lin, J. H. Perry, and W. R. Schowalter, *J. Fluid Mech.* **44**, 1 (1970).
[42] E. J. Hinch and L. G. Leal, *J. Fluid Mech.* **52**, 683 (1972).
[43] H. C. Hamaker, *Physica (Amsterdam)* **4**, 1058 (1937).
[44] R. Hogg, T. W. Healy, and D. W. Fuerstenau, *Trans. Faraday Soc.* **62**, 1638 (1966).
[45] T. G. M. Van De Ven and S. G. Mason, *J. Colloid Interface Sci.* **57**, 505 (1976).
[46] P. M. Adler, *Rheol. Acta* **17**, 298 (1978).
[47] W. B. Russel, *J. Fluid Mech.* **85**, 209 (1978).
[48] G. R. Zeichner and W. R. Schowalter, *AIChE J.* **23**, 243 (1977).
[49] G. R. Zeichner and W. R. Schowalter, *J. Colloid Interface Sci.* **71**, 237 (1979).
[50] H. Tanaka and J. L. White, *J. Non-Newtonian Fluid Mech.* **7**, 333 (1980).

[51] I. M. Krieger and J. J. Dougherty, *Trans. Soc. Rheol.* **3**, 137 (1959).
[52] G. Segré and A. Silberberg, *J. Colloid Sci.* **18**, 312 (1963).
[53] S. H. Maron and S. M. Fok, *J. Colloid Sci.* **10**, 482 (1955).
[54] S. H. Maron, B. P. Madow, and I. M. Krieger, *J. Colloid Sci.* **6**, 584 (1951).
[55] I. M. Krieger and S. H. Maron, *J. Colloid Sci.* **6**, 528 (1951).
[56] S. Prager, *Trans. Soc. Rheol.* **1**, 53 (1957).
[57] T. Kotaka, *J. Chem. Phys.* **30**, 1566 (1959).
[58] D. G. Thomas, *AIChE J.* **6**, 631 (1960); **7**, 431 (1961).
[59] D. G. Thomas, *J. Colloid Sci.* **20**, 267 (1965).
[60] R. L. Hoffman, *Trans. Soc. Rheol.* **16**, 155 (1972).
[61] R. L. Hoffman, *J. Colloid Interface Sci.* **46**, 419 (1974).
[62] A. B. Metzner and M. Whitlock, *Trans. Soc. Rheol.* **11**, 239 (1958).
[63] J. S. Chong, E. B. Christiansen, and A. B. Bare, *J. Appl. Polym. Sci.* **15**, 2007 (1971).
[64] E. C. Bingham, "Fluidity and Plasticity." McGraw-Hill, New York, 1922.
[65] V. K. Hohenemser and W. Prager, *ZAMM* **12**, 216 (1932).
[66] J. G. Oldroyd, *Proc. Cambridge Philos. Soc.* **43**, 100 (1947).
[67] R. von Mises, *Nachr. Ges. Wiss. Goettingen, Math.-Phys. Kl.* p. 532 (1913).
[68] W. Prager and P. G. Hodge, "Theory of Perfectly Plastic Solids." Wiley, New York, 1951.
[69] R. Hill, "The Mathematical Theory of Plasticity." Oxford Univ. Press (Clarendon) London 1950.
[70] K. H. Sweeny and K. D. Geckler, *J. Appl. Phys.* **25**, 1135 (1954).
[71] J. V. Robinson, *Trans. Soc. Rheol.* **1**, 15 (1957).
[72] R. Roscoe, *Br. J. Appl. Phys.* **3**, 267 (1952).
[73] S. G. Ward and R. L. Whitmore, *Br. J. Appl. Phys.* **1**, 286 (1950).
[74] E. J. Hinch and L. G. Leal, *J. Fluid Mech.* **46**, 685 (1971).
[75] H. Brenner, *Int. J. Multiphase Flow* **1**, 195 (1974).
[76] G. Segré and A. Silberberg, *J. Fluid Mech.* **14**, 115 (1962); **14**, 136 (1962).
[77] H. L. Goldsmith and S. G. Mason, *J. Fluid Mech.* **12**, 88 (1962).
[78] A. Karnis, H. L. Goldsmith, and S. G. Mason, *Can. J. Chem. Eng.* **44**, 181 (1966).
[79] A. Karnis, H. L. Goldsmith, and S. G. Mason, *J. Colloid Interface Sci.* **22**, 531 (1966).
[80] F. Gauthier, H. L. Goldsmith, and S. G. Mason, *Trans. Soc. Rheol.* **15**, 297 (1971).
[81] R. G. Cox and H. Brenner, *Chem. Eng. Sci.* **23**, 147 (1968).
[82] B. P. Ho and L. G. Leal, *J. Fluid Mech.* **65**, 365 (1974).
[83] C. D. Han, unpublished research (1978).
[84] K. Weissenberg, *Proc. Int. Congr. Rheol., 1st* p. 29. North-Holland Publ., Amsterdam, 1948.
[85] M. A. Newab and S. G. Mason, *J. Phys. Chem.* **62**, 1248 (1958).
[86] R. O. Maschmeyer and C. T. Hill, *in* "Fillers and Reinforcements for Plastics" (R. D. Deanin and N. R. Schott, eds.), Adv. Chem. Ser. No. 134, p. 95. Am. Chem. Soc., Washington, D.C., 1974.
[87] J. Mewis and A. B. Metzner, *J. Fluid Mech.* **62**, 593 (1974).
[88] G. K. Batchelor, *J. Fluid Mech.* **41**, 545 (1970).
[89] N. S. Clarke, *J. Fluid Mech.* **52**, 781 (1972).
[90] R. G. Cox, *J. Fluid Mech.* **44**, 791 (1970); **45**, 625 (1971).
[91] G. K. Batchelor, *J. Fluid Mech.* **44**, 419 (1970).
[92] G. K. Batchelor, *J. Fluid Mech.* **46**, 813 (1971).
[93] T. E. Kizor and F. A. Seyer, *Trans. Soc. Rheol.* **18**, 271 (1974).
[94] C. B. Weinberg and J. D. Goddard, *Int. J. Multiphase Flow* **1**, 465 (1974).
[95] F. Nazem and C. T. Hill, *Trans. Soc. Rheol.* **18**, 87 (1974).

[96] R. O. Maschmeyer and C. T. Hill, *Trans. Soc. Rheol.* **21**, 195 (1977).
[97] P. K. Agarwal, E. B. Bagley, and C. T. Hill, *Polym. Eng. Sci.* **18**, 282 (1978).
[98] N. J. Mills, *J. Appl. Polym. Sci.* **15**, 2791 (1971).
[99] S. Newman and Q. A. Trementozzi, *J. Appl. Polym. Sci.* **9**, 3071 (1965).
[100] J. M. Charrier and J. M. Rieger, *Fibre Sci. Technol.* **7**, 161 (1974).
[101] S. Wu, *Polym. Eng. Sci.* **19**, 638 (1979).
[102] L. Czarnecki and J. L. White, *J. Appl. Polym. Sci.* **25**, 1217 (1980).
[103] W. H. Herschel and R. Bulkley, *Proc. Am. Soc. Test. Mater.* **26**, 621 (1926).
[104] W. H. Herschel and R. Bulkley, *Kolloid-Z.* **39**, 291 (1926).
[105] P. R. Crowley and A. S. Kitzes, *Ind. Eng. Chem.* **49**, 888 (1957).
[106] H. A. Kearsey and A. G. Cheney, *Trans. Inst. Chem. Eng.* **39**, 91 (1961).
[107] R. Murdoch and H. A. Kearsey, *Trans. Inst. Chem. Eng.* **38**, 165 (1960).
[108] A. H. P. Skelland, "Non-Newtonian Flow and Heat Transfer." Wiley, New York, 1967.
[109] C. D. Han and Y. W. Kim, *J. Appl. Polym. Sci.* **18**, 2589 (1974).
[110] J. D. Goddard, *J. Non-Newtonian Fluid Mech.* **1**, 1 (1976).
[111] J. D. Goddard, *J. Rheol.* **22**, 615 (1978).
[112] J. G. Oldroyd, *Proc. Cambridge Philos. Soc.* **43**, 383 (1947); **44**, 396, 521 (1974); **44**, 200, 214 (1948).
[113] W. P. Graebel, *in* "Second Order Effects in Elasticity, Plasticity, and Fluid Dynamics" (M. Reiner and D. Abir, eds.), p. 636. Macmillan, New York, 1964.
[114] A. Slibar and P. R. Paslay, *in* "Second Order Effects in Elasticity, Plasticity, and Fluid Dynamics" (M. Reiner and D. Abir, eds.), p. 314. Macmillan, New York, 1964.
[115] J. L. White, *J. Non-Newtonian Fluid Mech.* **5**, 177 (1979).
[116] J. L. White and H. Tanaka, *J. Non-Newtonian Fluid Mech.* **8**, 1 (1981).
[117] E. P. Plueddemann and G. L. Stark, *Proc. Annu. Conf.—Reinf. Plast. Compos. Inst., Soc. Plast. Ind.* **32**, 4-C (1977).
[118] R. C. Hartlein, *Ind. Eng. Chem. Prod. Res. Dev.* **10**, 92 (1971).
[119] E. P. Plueddemann, *in* "Interfaces in Polymer Matrix Composites" (E. P. Plueddemann, ed.), Composite Materials, Vol. 6, p. 174. Academic Press, New York, 1974.
[120] M. W. Ranney, S. E. Berger, and J. G. Marsden, *in* "Interfaces in Polymer Matrix Composites" (E. P. Pluedemann, ed.), Composite Materials, Vol. 6, p. 132. Academic Press, New York, 1974.
[121] Union Carbide Corp., "Properties of Commercially Available Union Carbide Silane Coupling Agents," Tech. Bull. F-46110C. New York, 1978.
[122] Dow Corning Corp., "Selection Guide to Dow Corning Organosilane Chemicals," Tech. Bull. 23-181B-77. Midland, Michigan, 1977.
[123] Kenrich Petrochemicals Inc., "Titanate Coupling Agents for Filled Polymers," Tech. Bull. KR-0975-2; Tech. Bull. KR-1076-5. Bayonne, New Jersey, 1975.
[124] E. P. Plueddemann, *Proc. Annu. Conf.—Reinf. Plast./Compos. Inst., Soc. Plast. Ind.* **25**, 13-D, (1970).
[125] E. P. Plueddemann and G. L. Stark, *Mod. Plast.* **54**(9), 102 (1977).
[126] L. R. Daley and F. Rodriguez, *Polym. Eng. Sci.* **9**, 428 (1969).
[127] S. J. Monte and P. F. Bruins, *Mod. Plast.* **51**(12), 68 (1974).
[128] S. J. Monte, *Mod. Plast. Encycl.* **53**(10A), 161 (1976).
[129] E. H. Kerner, *Proc. Phys. Soc., London, Sect. B* **69**, 808 (1956).
[130] C. Van der Poel, *Rheol. Acta* **1**, 198 (1958).
[131] Z. Hashin and S. Shtrikman, *J. Mech. Phys. Solids* **11**, 127 (1963).
[132] Z. Hashin, *J. Appl. Mech.* **29**, 143 (1962); **32**, 630 (1965).
[133] B. J. Budiansky, *J. Mech. Phys. Solids* **13**, 223 (1965).
[134] J. C. Smith, *J. Res. Natl. Bur. Stand., Sect. A* **78**, 355 (1974); **79**, 419 (1975).

[135] J. C. Halpin, *J. Compos. Mater.* **2**, 4366 (1968).
[136] J. C. Halpin and N. J. Pagano, *J. Compos. Mater.* **3**, 720 (1969).
[137] J. M. Charrier, *Polym. Eng. Sci.* **15**, 731 (1975).
[138] L. Nicolais, *Polym. Eng. Sci.* **15**, 137 (1975).
[139] J. C. Halpin and J. L. Kardos, *Polym. Eng. Sci.* **16**, 344 (1976).
[140] J. L. Kardos and J. Raisoni, *Polym. Eng. Sci.* **15**, 183 (1975).
[141] L. E. Nielsen, "Mechanical Properties of Polymers and Composites," Vol. 2. Dekker, New York, 1974.
[142] L. E. Nielsen, *Appl. Polym. Symp.* No. 12, p. 249 (1969).
[143] F. R. Schwarzl, H. W. Bree, and C. J. Nederveen, *Proc. Int. Congr. Rheol., 4th*, Part 3, p. 241. Wiley (Interscience), New York, 1965.
[144] J. C. Smith, *Polym. Eng. Sci.* **16**, 394 (1976).
[145] L. A. Goettler, *Proc. Annu. Conf.—Reinf. Plast./Compos. Inst., Soc. Plast. Ind.* **25**, 14-A (1970).
[146] D. L. Faulkner and L. R. Schmidt, *Polym. Eng. Sci.* **17**, 657 (1977).
[147] J. Leidner and R. T. Woodham, *J. Appl. Polym. Sci.* **18**, 1639 (1974).
[148] R. F. Fedors and R. F. Landel, *J. Polym. Sci., Polym. Phys. Ed.* **13**, 579 (1975).
[149] J. Lusis, R. T. Woodham, and M. Xanthos, *Polym. Eng. Sci.* **13**, 139 (1973).
[150] S. R. Moghe, *Rubber Chem. Technol.* **49**, 1160 (1976).
[151] A. Y. Coran, P. Hamed, and L. A. Goettler, *Rubber Chem. Technol.* **49**, 1167 (1976).
[152] J. E. O'Conner, *Rubber Chem. Technol.* **50**, 945 (1977).
[153] H. F. Mark, S. M. Atlas, and E. Cernia, eds., "Man-Made Fibers." Vols. 1, 2, and 3. Wiley (Interscience), New York, 1967.
[154] A. Ziabicki, "Fundamentals of Fiber Formation." Wiley, New York, 1976.
[155] C. D. Han, "Rheology in Polymer Processing," Chap. 8. Academic Press, New York, 1976.
[156] J. L. White and H. Tanaka, *J. Appl. Polym. Sci.* **26**, 579 (1981).
[157] I. I. Rubin, "Injection Molding." Wiley (Interscience). New York, 1972.
[158] C. D. Han, "Rheology in Polymer Processing." Chap. 11. Academic Press, New York, 1976.
[159] L. R. Schmidt, *Polym. Eng. Sci.* **17**, 666 (1977).
[160] Y. Chan, J. L. White, and Y. Oyanagi, *Polym. Eng. Sci.* **18**, 268 (1978).
[161] J. P. Bell, *J. Compos. Mater.* **3**, 244 (1969).
[162] W. K. Lee and H. H. George, *Polym. Eng. Sci.* **18**, 146 (1978).
[163] K. Okuno and R. T. Woodhams, *Polym. Eng. Sci.* **15**, 308 (1975).
[164] H. Ishida and J. L. Koenig, *Polym. Eng. Sci.* **18**, 128 (1978).
[165] J. G. Mohr, S. S. Oleesky, G. D. Shook, and L. S. Meyer, "SPI Handbook of Technology and Engineering of Reinforced Plastics/Composites," 2nd ed. Van Nostrand-Reinhold, New York, 1973.
[166] P. F. Bruins, ed., "Unsaturated Polyester Technology." Gordon & Breach, New York, 1976.
[167] Owens-Corning Fiberglas Corp., "FRP—An Introduction to Fiberglass-Reinforced Plasticol Composites," Tech. Bull. 1-PL-6305-A. Granville, Ohio, 1976.
[168] Owens-Corning Fiberglas Corp., "Structural SMC: Material, Process, and Performance Review," Tech. Bull. 5-TM-8364-A. Granville, Ohio, 1979.
[169] B. Kleinmeier and G. Menges, *SPE Annu. Tech. Conf., 35th Prepr.* p. 7 (1977).
[170] F. J. Parker, *Proc. Annu. Conf.—Reinf. Plast./Compos. Inst., Soc. Plast. Ind.* **32**, 6-F (1977).
[171] M. Miwa, A. Nakayama, T. Ohsawa, and A. Hasegawa, *J. Appl. Polym. Sci.* **23**, 2957 (1979).

[172] K. Whybrew, Ph.D. Thesis, Univ. of Nottingham, Nottingham, 1972.
[173] D. H. Thomas, Ph.D. Thesis, Univ. of Nottingham, Nottingham, 1975.
[174] M. J. Owens, D. H. Thomas, and M. S. Ford, *Proc. Annu. Conf.—Reinf. Plast./Compos. Inst., Soc. Plast. Ind.* **33**, 20-B (1978).
[175] K. S. Gandhi and R. Burns, *Trans. Soc. Rheol.* **20**, 489 (1976).
[176] A. B. Isham, *Soc. Automot. Eng., Detroit, Mich.* Pap. 760333 (1976).
[177] A. B. Isham, *Soc. Automot. Eng., Detroit, Mich.* Pap. 780354 (1978).
[178] D. L. Simpkins, *Soc. Automot. Eng., Detroit, Mich.* Pap. 790165 (1979).
[179] E. G. Schwarz, F. E. Critchfield, L. P. Tackett, and P. M. Tarin, *S. P. I. 34th Annu. Conf. Reinf. Plast.* 14-C, 1979.
[180] R. M. Gerkin, L. F. Lawler, and E. G. Schwarz, *J. Cell. Plast.* **15**(1), 51 (1979).
[181] W. E. Becker, ed., "Reaction Injection Molding." Van Nostrand-Reinhold, New York, 1979.

4

Dispersed Flow of Heterogeneous Polymeric Systems

4.1 INTRODUCTION

Heterogeneous polymeric systems may include semicrystalline homopolymers, block copolymers, segmented elastomers known as thermoplastic elastomers, and polymer blends. Although there are some copolymers that are homogeneous, most of them show microphase separation. Figure 4.1 shows an electron photomicrograph of a triblock copolymer of styrene–butadiene–styrene (SBS) cast from a mixed solvent of THF/methyl ethyl ketone [1]. In the photograph, the light area represents the discrete polystyrene phase and the dark area represents the continuous polybutadiene phase. It appears that the domain size of the discrete phase is less than 0.1 μm. Various experimental techniques, such as electron microscopy, small-angle x-ray diffraction, light scattering, thermal analysis, and dynamic mechanical analysis, have been used investigating the question of compatibility and microphase morphology of heterogeneous polymeric systems.

The morphologies of such systems are determined by the composition of each component. Furthermore, in the case of block copolymers, the method of sample preparation (i.e., polymerization technique) is important, and in the case of polymer blends, the history of mixing. The resulting microstructure exerts a profound influence on the physical/mechanical properties of these materials. Therefore, in the past, much effort has been spent on investigating the relationships between the morphology and the properties of block copolymers and polymer blends [2–12].

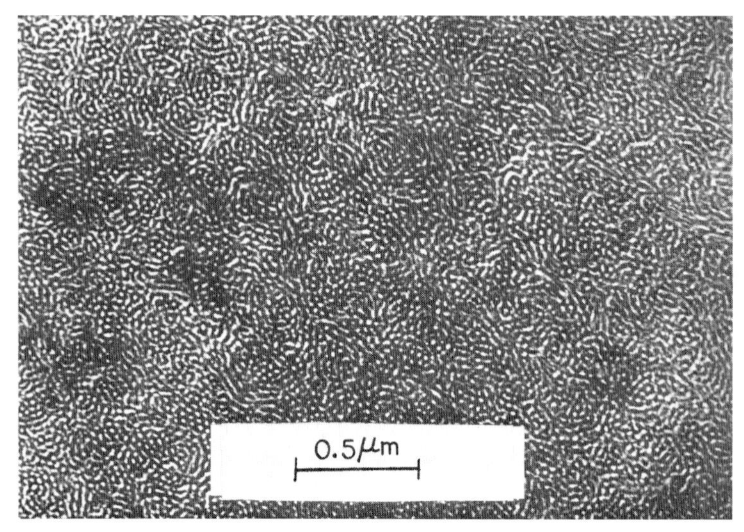

Multiphase Polymers

Fig. 4.1 Photomicrograph of a triblock copolymer of styrene–butadiene–styrene (SBS) [1].

Polymer blends can be divided into two broad categories, namely, compatible (miscible) and incompatible (immiscible). There are various definitions used for these terms among researchers. One frequently used definition is based on dynamic mechanical measurements. An incompatible blend is defined as one which exhibits only transitions that can be assigned to transitions in the constituent components. In other words, according to this definition, incompatible polymer blends are characterized as exhibiting more than one glass transition temperature [13, 14]. This does not mean that the components are incompatible in the thermodynamic sense [15]. It is possible to have a situation where thermodynamically compatible polymers can exist as discrete phases in contact with others [15, 16]. It is of interest to note that blends of two incompatible polymers are usually opaque because of the light scattering from rather large domain sizes, whereas block copolymers themselves are usually transparent because of much smaller domain sizes.

By the use of polymer blends, physical/mechanical properties can be altered to produce useful materials with a wide range of applications. Desirable properties can be achieved more easily, in many instances, by proper blend selection than by polymerization of a new polymer.

Furthermore, the plastics industry uses blends of two rigid polymers, poly(2,6-dimethyl-1,4-phenylene oxide) (PPO) with polystyrene (PS), in order to improve melt flow, and polyvinylchloride (PVC) with acrylonitrile–

butadiene–styrene (ABS) to improve toughness or to reduce mold shrinkage [17]. For another example, the rubber industry often uses blends of different rubbers (e.g., blends of natural rubber with synthetic polybutadiene) in order to achieve certain desired mechanical properties. The natural rubber imparts more resilience to the styrene–butadiene rubber (SBR). It is of interest to note that the addition of an elastomer to a rigid polymer results in an increase of the modulus and is known as rubber reinforcement. For instance, blending nitrile rubber with polyvinylchloride (PVC) gives rise to combined effects of flexibility and hardness, and the blending of SBR in polystyrene (PS) imparts some flexibility and shock resistance. Figure 4.2 gives photomicrographs of blends of SBR in PS containing various amounts of rubber as the discrete phase [18].

Elastomer modified thermoplastics as a class of materials, ranging from simple mechanical blends on the one hand to sophisticated grafted terpolymers on the other, are a proven technological success. Incorporating a dispersed rubber phase can significantly raise the impact toughness of brittle polymers (e.g., high impact polystyrene). This is particularly true of the acrylonitrile–butadiene–styrene (ABS) terpolymers. In the ABS resin,

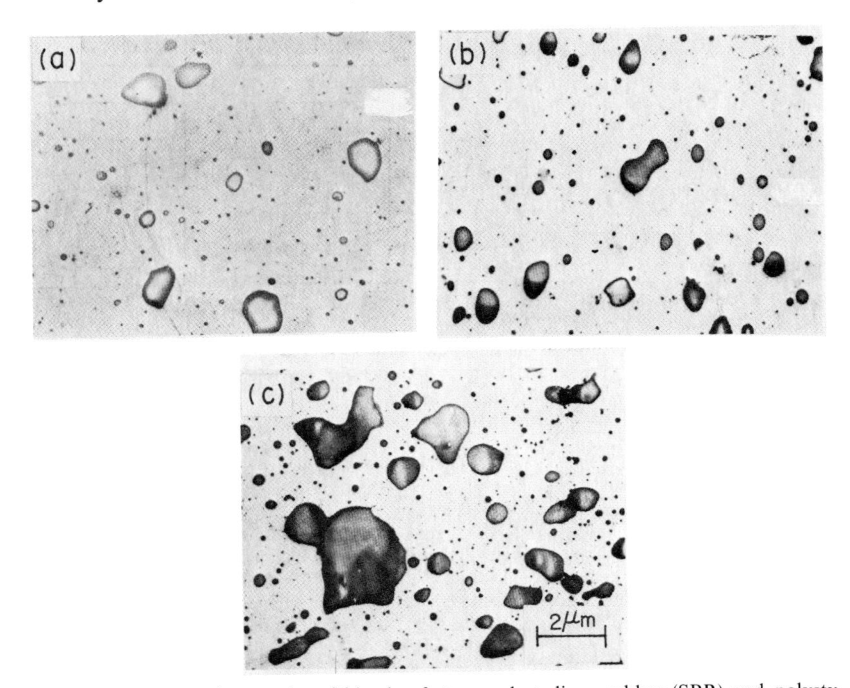

Fig. 4.2 Photomicrographs of blends of styrene–butadiene rubber (SBR) and polystyrene, for various amounts of rubber (wt. %) [18]: (a) 5; (b) 10; (c) 15. From H. Keskkula *et al.*, *J. Appl. Polym. Sci.* **15**, 351. Copyright © 1971. Reprinted by permission of John Wiley & Sons, Inc.

the individual elastomer particles (polybutadience) act as stress concentration sites and, under an applied load, numerous crazes are generated and propagated without the formation of major cracks. Figure 4.3 gives photomicrographs of some commercial ABS resins [18]. These photomicrographs indicate the heterogeneous nature and discreteness of the rubber phase and its detailed morphology. It is of interest to note that the polymer forming the continuous phase (light area) is occluded by the polymer forming the discrete phase (dark area).

Blends of a partially cured monoolefin copolymer rubber (ranging from 40% to 85%) and a polyolefin thermoplastic are commercially available [19]. The monoolefin copolymer rubber is typically a saturated ethylene–propylene copolymer rubber (EPM) or a saturated ethylene–propylene–

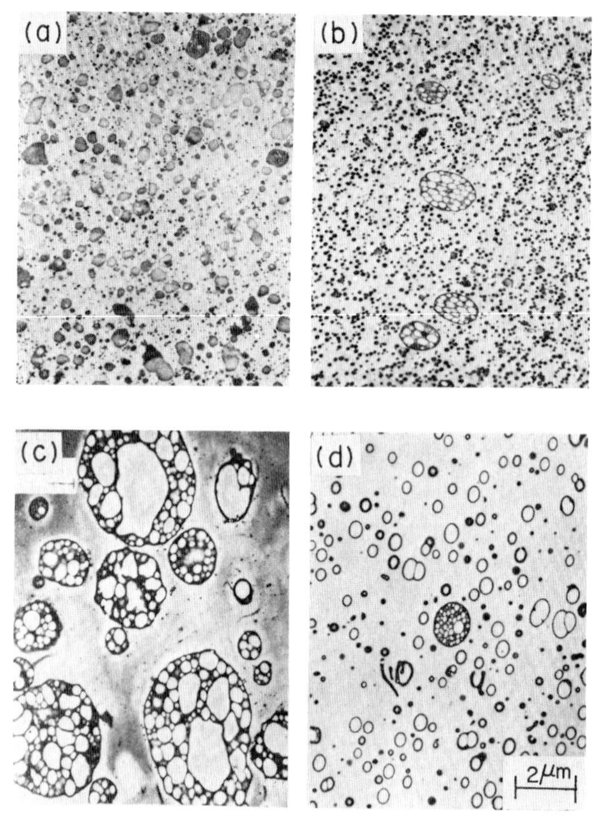

Fig. 4.3 Photomicrographs of acrylonitrile–butandiene–styrene (ABS) terpolymers [18]: (a) Marbon Chemical Co., Cycolac TD-1001; (b) Monsanto Co., Lustron I-440; (c) Dow Chemical Co., Tybrene 213; (d) Toyo Rayon Co., TH Resin. From H. Keskkula *et al.*, *J. Appl. Polym. Sci.* **15**, 351. Copyright © 1971. Reprinted by permission of John Wiley & Sons, Inc.

nonconjugated diene terpolymer rubber (EPDM). It is partially cured by the action of a vulcanizing agent (e.g., a peroxide). The polyolefin resin may be low- or high-density polyethylene or polypropylene. Such thermoplastic elastomers are [illegible] produced for such shaping operations as extrusion, injectio[illegible] or compression molding [20].

There are, of [illegible] compatible blends. Several research groups [21–34], notably [illegible]-workers [22–28], investigated the compatibility of various p[illegible]ds by means of dynamic mechanical analysis and thermal analy[illegible]

In incompatib[illegible]ends, one component forms the discrete phase (as either droplets [illegible]ks) dispersed in the other component forming the continuous ph[illegible]g. 4.2). The question as to which of the two fluids forms the di[illegible]e depends on, among other parameters, the volume ratio of each [illegible]ent, the viscosity ratio and the elasticity ratio of the individual com[illegible]nts, and also on the interfacial tension between the two fluids. Thus, depending on the composition ratio, the less viscous fluid may form droplets and be dispersed into the more viscous fluid, or the more viscous fluid may form droplets and be dispersed into the less viscous fluid. In practice, uniform droplets are not obtainable, and the average size of droplets and their size distribution may depend very much on the blending history (i.e., mixing time, the intensity of mixing, and the type of mixing equipment).

Van Oene [35] advanced a theoretical foundation to interpret the mode of dispersion in terms of the droplet size, interfacial tension, and differences in the viscoelastic properties of the two components, when both components are viscoelastic. According to his theory, the component with larger normal stress functions will form droplets dispersed in the other component, and differences in viscosity, shear rate (or shear stress), extrusion temperature, and residence time spent in mixing (or flow) influence only the homogeneity of the dispersion and not the mode of dispersion.

Today, many polymer processing operations (e.g., fiber spinning, injection molding, film extrusion) make use of additives, sometimes as processing aids (e.g., plasticizers, lubricants), and sometimes to achieve certain desired physical/mechanical properties in the final products (e.g., antistatic fiber, flame-retardant fiber). The additives are used in small quantities and, being generally low molecular weight materials compared to the base polymer, they usually form the discrete phase (either droplets or long streaks) dispersed in the base polymer during the fabrication process. Therefore one may surmise that the method of mixing can strongly influence the state of dispersion, location, and even shape, of the additive in the base polymer, and hence the rheological properties of the mixture. Figure 4.4 gives a photomicrograph of an antistatic nylon fiber in which the antistatic agent, elongated along the fiber axis, is dispersed in the nylon matrix. There are

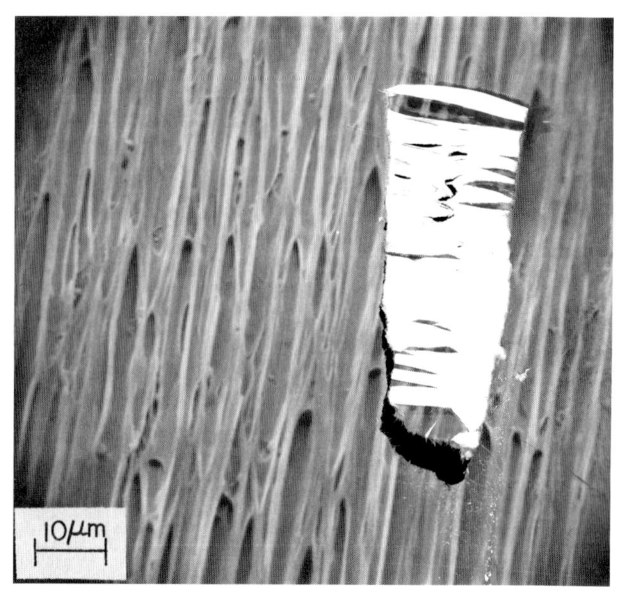

Fig. 4.4 Photomicrograph of a drawn nylon fiber in which the hollow areas represent the antistatic component peeled off from the fiber.

numerous similar examples which can readily be found in the polymer processing industry, but limitations of space here do not permit us to mention all such practices.

In this chapter, we shall first discuss the deformation of single droplets in shearing and elongational flow fields, and then some experimental observations of the bulk rheological properties of concentrated emulsions and two-phase polymer blends. The seemingly peculiar *bulk* rheological properties of such heterogeneous systems will be explained by use of constitutive equations derived on the basis of phenomenological theories. Finally we shall discuss the fiber spinning and injection molding of heterogeneous polymeric systems, emphasizing the close relationships that exist between the rheological properties and processability, and between the morphology and mechanical properties.

4.2 DEFORMATION OF SINGLE DROPLETS IN SHEARING AND EXTENSIONAL FLOWS

Since the *bulk* rheological behavior of concentrated emulsions and many polymer blends giving rise to the heterogeneous phase is intimately related to the deformation of droplets suspended in the continuous phase, it is quite apparent that a better understanding of the phenomenon of the deformation of droplets in another liquid is of fundamental and practical importance to the control of various polymer processing operations which deal with dispersed multiphase polymeric systems. In the past, numerous studies, experimental and theoretical, have been made to gain a better understanding of the phenomenon of droplet deformation.

Taylor [36] is the first who investigated the deformation of a viscous liquid droplet in uniform shearing and plane hyperbolic flow fields. Subsequently, a number of researchers [37–41] carried out theoretical studies on droplet deformation. The majority of these researchers dealt with the deformation of Newtonian droplets suspended in another Newtonian fluid, subjected to either uniform shearing flow or uniaxial elongational flow. Such theoretical studies have been tested by experimental investigations [42–47].

Figure 4.5 gives a schematic describing uniform shearing flow and plane hyperbolic flow fields in rectangular coordinates. Taylor [36] shows that in steady *uniform shearing* flow, the droplet deforms into a spheroid, and the shape of the droplet depends on the viscosity ratio of the droplet phase to the medium k and the ratio of the product of the local shear stress and the droplet radius to the interfacial tension We. In the case when the interfacial tension effect dominates the viscous effect (i.e., $k = O(1)$, $We \ll 1$), the deformation D and the orientation angle α (see Fig. 4.5) of the droplet are expressed as:

$$D = We[(19k + 16)/(16k + 16)], \qquad \alpha = \tfrac{1}{4}\pi \tag{4.1}$$

where

$$k = \bar{\eta}_0/\eta_0 \tag{4.2}$$

$$D = (L - B)/(L + B) \tag{4.3}$$

$$We = \eta_0 \dot{\gamma} a/\sigma \tag{4.4}$$

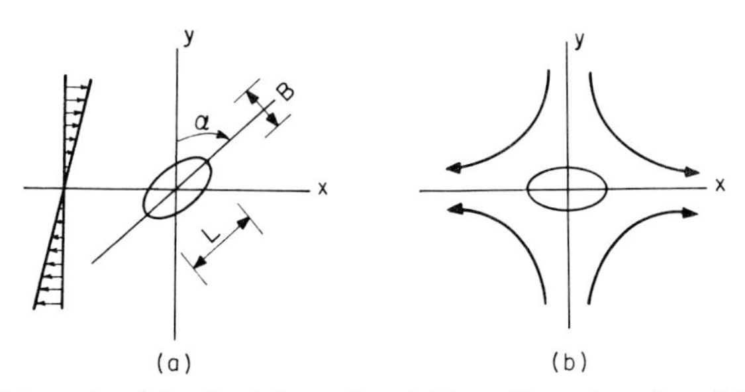

Fig. 4.5 Schematic of droplet deformation: (a) in uniform shear flow field; (b) in plane hyperbolic flow field.

in which L and B are the length and breadth of the deformed droplet, $\bar{\eta}_0$, η_0 are the viscosities of the droplet phase and suspending medium, respectively, a is the droplet radius, σ is the interfacial tension, and $\dot{\gamma}$ is the rate of deformation (i.e., shear rate). It may be noted that over the entire range of k from zero to infinity, $(19k + 16)/(16k + 16)$ in Eq. (4.1) varies only from 1.0 to 1.187, so that D is nearly equal to We. On the other hand, when the interfacial tension effect is negligible compared to the viscosity effect (i.e., $We = O(1)$, $k \gg 1$), Taylor's analysis yields

$$D = 5/4k, \qquad \alpha = \tfrac{1}{2}\pi \tag{4.5}$$

Taylor [36] observed experimentally that at low rates of deformation, the experimental observation of droplet deformation agreed with this theory, in both uniform shear and plane hyperbolic flow. Similar experimental studies on droplet deformation were made by others. Bartok and Mason [42, 43] experimentally observed the deformation and internal circulation in large droplets, which deformed into spheroids in uniform shear flow (e.g., Couette flow), and determined that the deformation of droplets agreed with Taylor's theory. It was also observed, by Rumscheidt and Mason [44], that internal circulation occurred inside droplets suspended in a medium subjected to plane hyperbolic flow.

Of particular importance is the experimental work of Rumscheidt and Mason [44], who studied the deformation of droplets over a wide range of viscosity ratio and interfacial tension. They observed four different types of deformation in *uniform shear* flow and two types of deformation in *plane hyperbolic* flow, depending on the viscosity ratio, as summarized in Fig. 4.6.

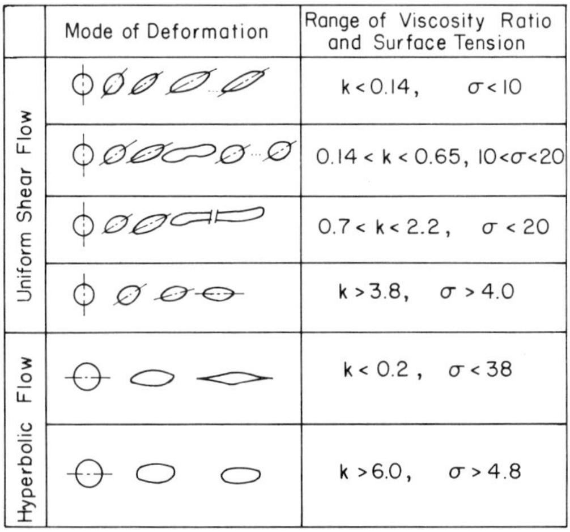

	Mode of Deformation	Range of Viscosity Ratio and Surface Tension
Uniform Shear Flow		$k < 0.14$, $\sigma < 10$
		$0.14 < k < 0.65$, $10 < \sigma < 20$
		$0.7 < k < 2.2$, $\sigma < 20$
		$k > 3.8$, $\sigma > 4.0$
Hyperbolic Flow		$k < 0.2$, $\sigma < 38$
		$k > 6.0$, $\sigma > 4.8$

Fig. 4.6 Schematic showing the mode of droplet deformation in different flow fields [44].

They noted that the experimentally observed deformation agreed well with Taylor's theory [Eq. (4.1)] at low values of k.

The behavior of droplets in *nonuniform* shear flow, for instance in Poiseuille flow, differs from that in *uniform* shear flow. Note that the velocity gradient varies with radial distance in Poiseuille flow, while it is constant in uniform shear flow. It is then not difficult to expect that in nonuniform shear flow the extent of droplet deformation varies depending on the location of the droplet in the plane of shear, whereas in uniform shear flow this is not the case. Therefore any information about droplet deformation under a uniform shear flow field is not readily translatable to that under a nonuniform shear flow. It is worth pointing out that from the point of view of polymer processing operations, droplet deformation in Poiseuille flow (and in extensional flow) fields is of far greater importance than in uniform shear flow fields.

It has been shown by Goldsmith and Mason [48] that in circular Poiseuille flow, Newtonian droplets suspended in a Newtonian medium migrate toward the tube axis. It was also reported by Gauthier *et al.* [49] that for Newtonian droplets suspended in a pseudoplastic liquid subjected to circular Poiseuille flow, a two-way migration was observed, i.e., the droplets initially close to the tube axis migrated toward the tube wall and those close to the tube wall migrated toward the tube axis and attained an equilibrium position between the axis and the wall. However, in a viscoelastic medium, the droplets were observed to always migrate toward the tube axis. This fact indicates that the behavior of droplets suspended in another liquid depends on the type of flow field and the nature of the suspending medium.

Chaffey *et al.* [50] carried out theoretical studies on the migration of deformed droplets in nonuniform shear flow. Their results show that deformed droplets migrate in the direction perpendicular to the tube wall and the migration is always away from the wall. On the other hand, theoretical studies by Hetsroni *et al.* [51] show that a neutrally buoyant droplet suspended in circular Poiseuille flow migrates toward the wall.

Most of the studies referred to above have dealt with Newtonian systems. In their experimental studies, Tavgac [46] and Lee [47] have studied the effect of fluid elasticity on the extent of the deformation of droplets in uniform shear and extensional flow fields. Tavgac [46] reports that he could observe a difference in deformability between Newtonian and viscoelastic droplets in Couette flow, and that the effect of fluid elasticity on the droplet deformation varies with the viscosity ratio. On the other hand, Lee [47] reports that he could *not* observe a discernible difference in deformability between Newtonian and viscoelastic droplets in plane hyperbolic flow. He noted that the deformation of viscoelastic droplets appears to be the same as that of

Newtonian droplets having a viscosity equal to the zero-shear viscosity of the viscoelastic droplet.

Chaffey and Brenner [37] made an effort to improve Taylor's theory [36]. They obtained *second-order* solutions in terms of the deformation parameter E, defined by

$$E = We[(19k + 16)/(16k + 16)] \tag{4.6}$$

Note that Taylor's equation [Eq. (4.1)] is a *first-order* solution in E.

Without putting any restriction on the interfacial tension or viscosity ratio, except for small deformations, Cox [38] developed a *first-order* theory for the deformation of a droplet in a general *time-dependent* shearing flow field. By making a series expansion of the velocity field in terms of a perturbation parameter, he obtained theoretical expressions for the shape and orientation angle of a droplet. For steady uniform shearing flow, Cox's theory gives:

$$D = \frac{5(19k + 16)}{4(k + 1)[(20/We)^2 + (19k)^2]^{1/2}} \tag{4.7}$$

$$\alpha = \tfrac{1}{4}\pi + \tfrac{1}{2}\tan^{-1}(19kWe/20) \tag{4.8}$$

It is seen that Cox's theory reduces to Taylor's in both of the following cases: (i) $k = O(1)$, $We \ll 1$; and (ii) $k \gg 1$, $We = O(1)$.

Using the method of series expansion (perturbation technique) proposed by Cox, Barthès-Biesel and Acrivos [39] obtained a *second-order* solution in terms of a perturbation parameter for the shape of a droplet in a time-dependent shearing flow. Representative droplet shapes predicted by their theory are given in Fig. 4.7.

Choi and Schowalter [41] also extended Cox's theory to describe the deformation of droplets in a *moderately concentrated* emulsion of Newtonian liquids. They obtained the deformation D and the orientation angle α of

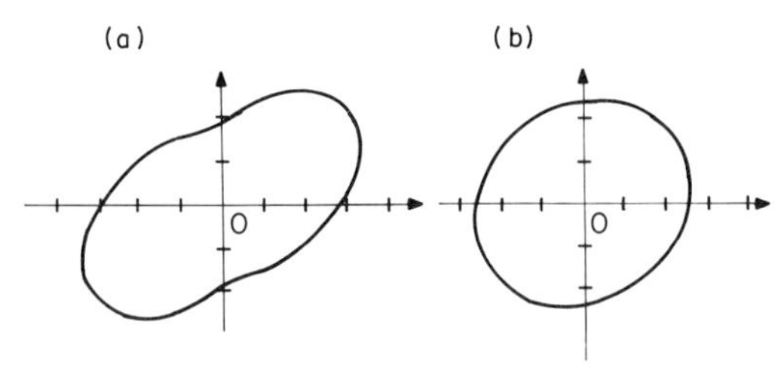

Fig. 4.7 Theoretically predicted shapes of a Newtonian droplet suspended in the Newtonian medium, subjected to a uniform shear flow [39]: (a) $k = 1.0$, $We = 0.13$; (b) $k = 6.4$, $We = 0.05$.

droplets in steady uniform shear flow as follows:

(i) when $We \ll 1, k = O(1)$

$$D = We\left(\frac{19k + 16}{16k + 16}\right)\left[1 + \frac{5(5k + 2)}{4(k + 1)}\phi\right] \tag{4.9}$$

$$\alpha = \frac{\pi}{4} + We\left[\frac{(19k + 16)(2k + 3)}{80(k + 1)}\right]\left[1 + \frac{5(19k + 16)}{4(k + 1)(2k + 3)}\phi\right] \tag{4.10}$$

(ii) when $We \ll 1, k \gg 1$

$$D = \frac{5(19k + 16)}{4(k + 1)[(20/We)^2 + (19k)^2]^{1/2}}\left[1 + \frac{5(5k + 2)}{4(k + 1)}\phi\right] \tag{4.11}$$

$$\alpha = \tfrac{1}{4}\pi + \tfrac{1}{2}\tan^{-1}(19kWe/20) \to 0 \tag{4.12}$$

(iii) when $We = O(1), k \gg 1$

$$D = (5/4k)(1 + \tfrac{25}{4}\phi) \tag{4.13}$$

$$\alpha = \tfrac{1}{4}\pi + \tfrac{1}{2}\tan^{-1}(19kWe/20) \tag{4.14}$$

in which ϕ is the volume fraction of the droplets in the emulsion. It can be seen that this theory reduces to Cox's and Taylor's theories as the value of ϕ approaches zero.

Figure 4.8 gives a comparison of experimental data of droplet deformation with existing theories. It is seen that the Cox theory shows better agreement of the deformation parameter D with the data than the Choi–Schowalter theory, especially when $k = 3.6$. Furthermore, the dependence of D on We is quite nonlinear, in contrast to Taylor's theory predicting a

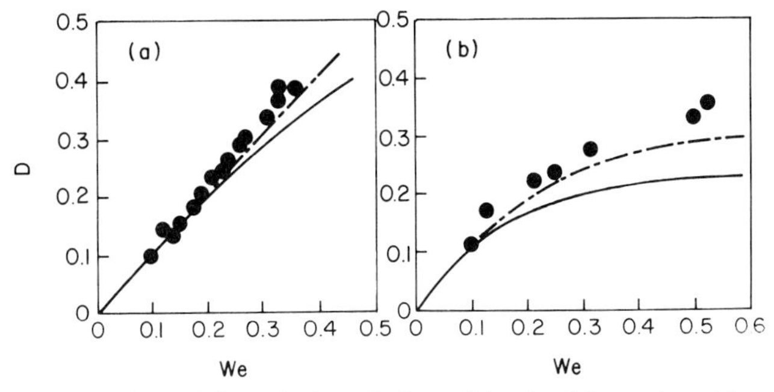

Fig. 4.8 Comparison of theoretical predictions of droplet deformation with experimental results, for different values of viscosity ratio k [41]: (a) 0.08; (b) 3.6. The data are taken from the work of Torza *et al.* [45]; the broken line (—·—) represents the Cox theory [38]: the solid line (——) represents the Choi–Schowalter theory [41].

linear relationship between D and We, which is valid only for slightly deformed droplets.

It should be mentioned that all existing theories are restricted to Newtonian droplets suspended in a Newtonian medium. Hence, such theories should be modified in situations where Newtonian droplets are suspended in a viscoelastic medium, and viscoelastic droplets are suspended in either a Newtonian or viscoelastic medium.

Figure 4.9 gives a photograph displaying the shape of a droplet subjected to plane hyperbolic flow (i.e., in a "four-roll" apparatus), in which the arrows indicate the direction of rotation of the rollers. Turner and Chaffey [40] considered hyperbolic radial flow (a steady elongational flow) to predict theoretically the deformation of a Newtonian droplet suspended in another Newtonian medium.

In order to investigate the effect of fluid elasticity on droplet deformation, Chin and Han [52] used the Coleman–Noll second-order fluid [see Eq. (2.53) in Chapter 2]. They considered the deformation of a viscoelastic droplet suspended in another viscoelastic liquid, subject to a steady elongational flow. Using the perturbation technique, they obtained the first-order solution, and thereby obtained the following expression for predicting the droplet

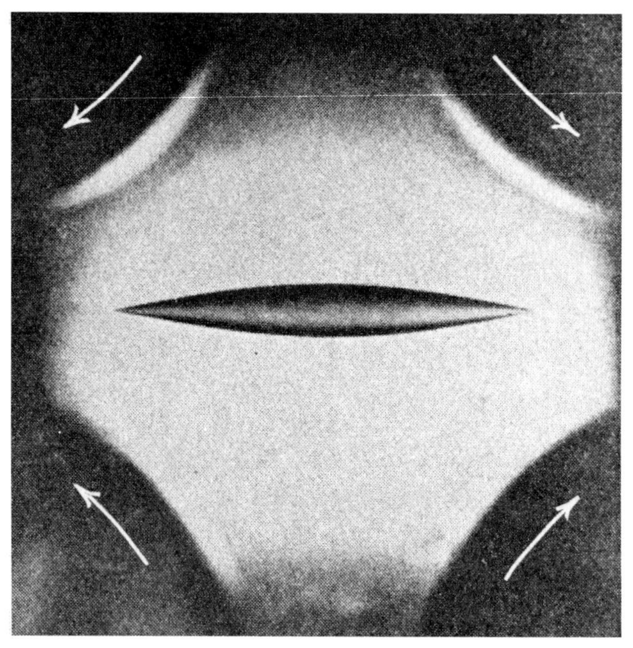

Fig. 4.9 Photograph showing a droplet elongated in a "four-roll" apparatus [36].

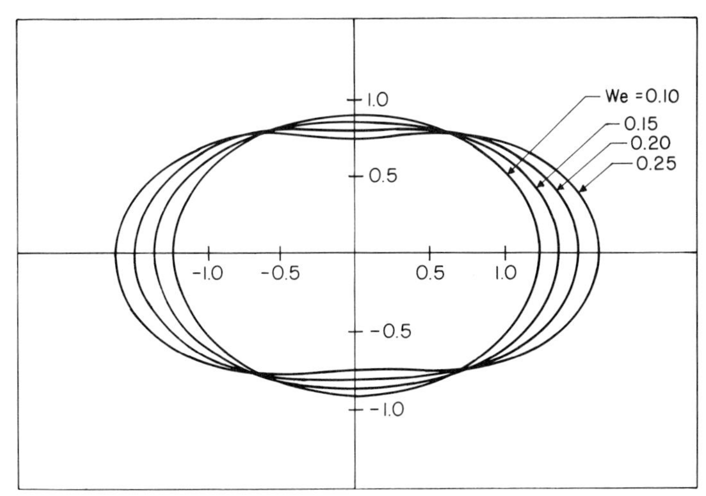

Fig. 4.10 Theoretically predicted shapes of a viscoelastic droplet suspended in a viscoelastic medium, subjected to a steady elongational flow, where $k = 0.1, \varepsilon = 0.1$, and $\bar{\varepsilon}/\varepsilon = 0.1$ [52]. From H. B. Chin and C. D. Han, *J. Rheol.* **23**, 557. Copyright © 1979. Reprinted by permission of John Wiley & Sons, Inc.

shape $r^{\dagger}$:

$$r = 1 + We\,Z_0 P_2(\mu) + We^2 Z_0[Z_{01} P_2(\mu) + Z_{02} P_4(\mu)] + \varepsilon\, We[Z_1 P_2(\mu) + Z_2 P_4(\mu)] \tag{4.15}$$

in which

$$Z_0 = \frac{2(19k + 16)}{(16k + 16)}, \qquad Z_{01} = \frac{9k^2 - 8k + 12}{28(k + 1)^2}, \qquad Z_{02} = \frac{111k + 96}{126(k + 1)} \tag{4.16}$$

$P_n(\mu)^{\ddagger}$ is the Legendre polynomial of degree n, We is defined by Eq. (4.4), Z_1 and Z_2 are complicated functions of system parameters, and ε is the perturbation parameter defined by

$$\varepsilon = \beta \dot{\gamma}_E / \eta_0 \tag{4.17}$$

where β is a material constant (representing fluid elasticity) appearing in the Coleman–Noll second-order model [see Eq. (2.53) in Chapter 2], γ_E is the rate of strain (elongation rate), and η_0 is the zero-shear viscosity of the medium, also appearing in the second-order model.

Figure 4.10 gives droplet shapes, determined by the use of Eq. (4.15), at different values of the parameter We. It is seen that droplet deformation increases with We. Figure 4.11 gives representative streamlines inside and

† $r = 1$ represents a sphere, meaning that r is nondimensionalized.
‡ $\mu = \cos\theta$, in which θ is a component of spherical coordinates.

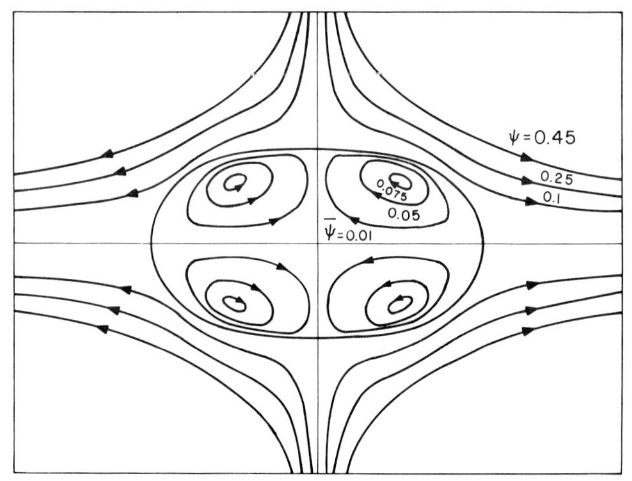

Fig. 4.11 Steamlines outside and inside a droplet placed in an elongational flow field, where $k = 0.1$, $We = 0.2$, $\varepsilon = 0.1$, and $\bar{\varepsilon}/\varepsilon = 0.1$ [52]. From H. B. Chin and C. D. Han, *J. Rheol.* **23**, 557. Copyright © 1979. Reprinted by permission of John Wiley & Sons, Inc.

outside the deformed droplet. Note that an internal circulation inside the droplet is predicted. The predicted circulation results from the use of the boundary condition that the tangential stress is continuous at the surface of the droplet.

Figure 4.12 gives plots of the *apparent* deformation D versus the viscosity ratio k with We as parameter. It is of interest to note that for k values up to approximately 0.1, D tends to decrease as k increases, and that at a k value

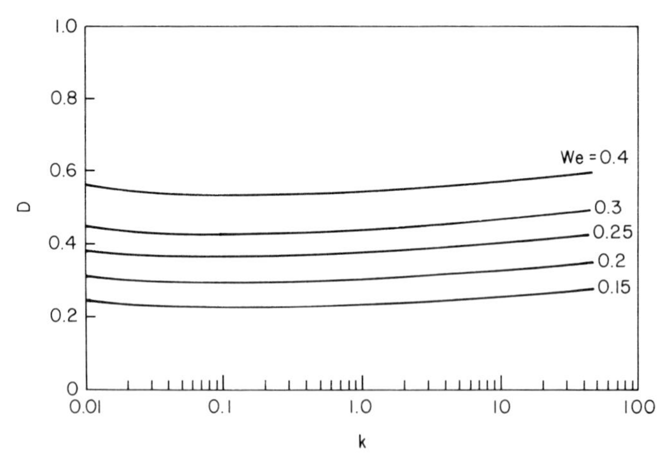

Fig. 4.12 Theoretical prediction of *apparent* deformation of a droplet in a steady elongational flow, where $\varepsilon = 0.1$ and $\bar{\varepsilon}/\varepsilon = 0.1$ [52]. From H. B. Chin and C. D. Han, *J. Rheol.* **23**, 557. Copyright © 1979. Reprinted by permission of John Wiley & Sons, Inc.

of about 0.1, D starts to increase slowly with k. The *apparent* deformation D used in Fig. 4.12 is calculated using the definition given by Eq. (4.3), for which L was evaluated from r at $\theta = 0$ and B from r at $\theta = \pi/2$, using Eq. (4.15).

Table 4.1 gives a summary of the computed results describing the effect of the perturbation parameter ε (representing the fluid elasticity) on the *apparent* deformation D. It is seen that D increases with increasing We, and also increases (slowly) with increasing ε. It should be mentioned that Eq. (4.15) was obtained using the perturbation technique, by assuming that the magnitude of the perturbation parameter ε is very small [52]. Therefore, it is not surprising to see the relatively minor contribution of ε to the *apparent* deformation D.

Figure 4.12 and Table 4.1 clearly show that the parameter We has a much greater influence on the droplet deformation than the viscosity ratio k and elasticity parameter ε have. It can be concluded from the above observation that in steady elongational flow, the medium viscosity plays a much more important role in determining the droplet deformation than the medium elasticity and the droplet phase viscosity.

TABLE 4.1

Effect of Medium Elasticity Parameter on Droplet Deformation[a]

(i) Apparent deformation D for $k = 0.1$, $\bar{\varepsilon}/\varepsilon = 0.1$

We	$\varepsilon = 0.001$	$\varepsilon = 0.01$	$\varepsilon = 0.1$	$\varepsilon = 0.3$
0.05	0.0765	0.0766	0.0771	0.0782
0.10	0.1529	0.1530	0.1538	0.1555
0.15	0.2276	0.2277	0.2286	0.2307
0.20	0.2996	0.2997	0.3006	0.3027
0.25	0.3683	0.3684	0.3692	0.3710
0.30	0.4329	0.4330	0.4336	0.4349
0.40	0.5487	0.5488	0.5488	0.5489

(ii) Apparent deformation D for $k = 10$, $\bar{\varepsilon}/\varepsilon = 0.1$

We	$\varepsilon = 0.001$	$\varepsilon = 0.01$	$\varepsilon = 0.1$	$\varepsilon = 0.3$
0.05	0.0875	0.0876	0.0889	0.0916
0.10	0.1734	0.1737	0.1759	0.1809
0.15	0.2560	0.2564	0.2594	0.2661
0.20	0.3341	0.3345	0.3382	0.3462
0.25	0.4069	0.4073	0.4114	0.4205
0.30	0.4738	0.4743	0.4787	0.4884
0.40	0.5895	0.5900	0.5948	0.6051

[a] From H. B. Chin and C. D. Han, *J. Rheol.* **25**, 557. Copyright © 1979. Reprinted by permission of John Wiley & Sons, Inc.

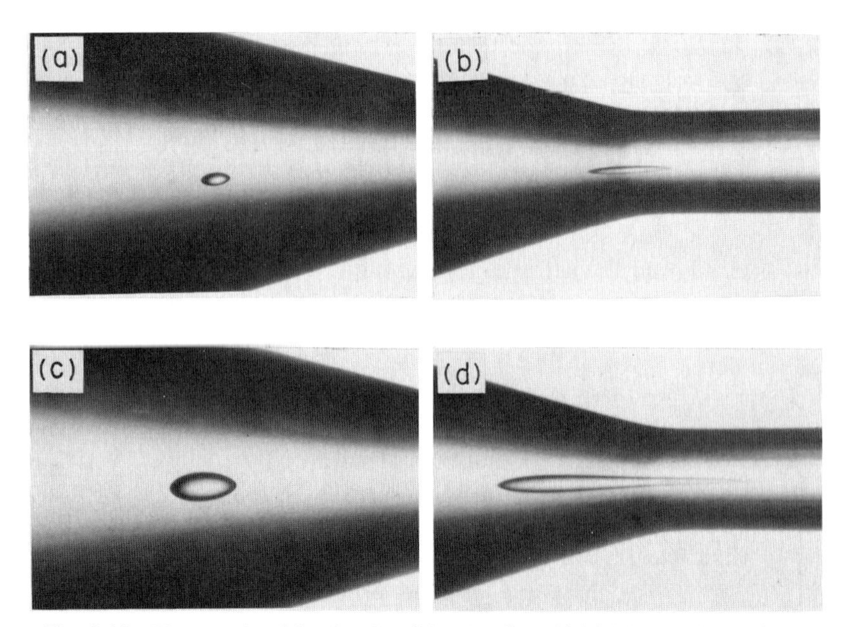

Fig. 4.13 Photographs of droplets describing the effect of initial droplet size on the extent of droplet deformation [52]: (a) and (b) for $a = 0.68$ mm; (c) and (d) for $a = 1.20$ mm. The droplet phase is a 2% PIB solution, the suspending medium is a 2% Separan solution, and the apparent wall shear rate is 75.4 sec^{-1}. From H. B. Chin and C. D. Han, *J. Rheol.* **23**, 557. Copyright © 1979. Reprinted by permission of John Wiley & Sons, Inc.

Chin and Han [52] have performed an experimental study of droplet deformation, using a flow channel constructed of Plexiglas, and consisting of a reservoir section, a conical section, and a straight cylindrical tube section.

Figure 4.13 presents photographs showing the effect of initial droplet size on the extent of droplet deformation. Figure 4.14 shows the effect of shear rate and Fig. 4.15 shows the effect of medium viscosity. It is seen that larger droplets give rise to a greater deformation than smaller droplets; that the higher the shear rate, the greater the droplet deformation; and that the high-viscosity medium (i.e., the 4% Separan solution) yields a greater deformation of the droplet than the low-viscosity medium (i.e., the 2% Separan solution). This experimental observation is in agreement with the theoretical prediction that as the parameter *We* increases, the deformation increases. According to the definition of *We* given by Eq. (4.4), the effect of *We* on the predicted droplet shape can be interpreted as indicating that droplet deformation is increased by any of the following: (a) an increase in the medium viscosity η_0; (b) an increase in the elongation rate $\dot{\gamma}_E$; (c) an increase in droplet size a; (d) a decrease in interfacial tension σ.

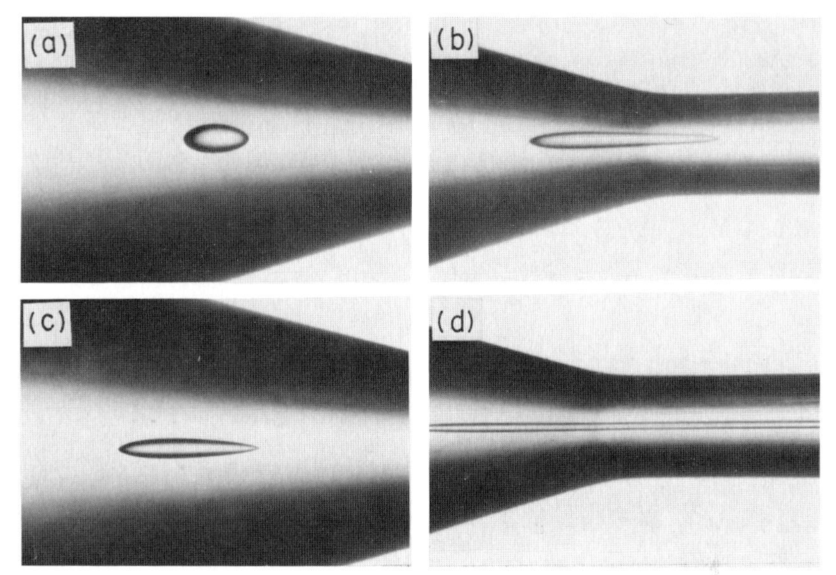

Fig. 4.14 Photographs of droplets describing the effect of shear rate on the extent of droplet deformation [52]: (a) and (b) at $\dot{\gamma} = 13.7\ \text{sec}^{-1}$; (c) and (d) at $\dot{\gamma} = 128.7\ \text{sec}^{-1}$. The droplet phase is a 6% PIB solution, the suspending medium is a 2% Separan solution, and the initial droplet size is 1.20 mm. From H. B. Chin and C. D. Han, *J. Rheol.* **23**, 557. Copyright © 1979. Reprinted by permission of John Wiley & Sons, Inc.

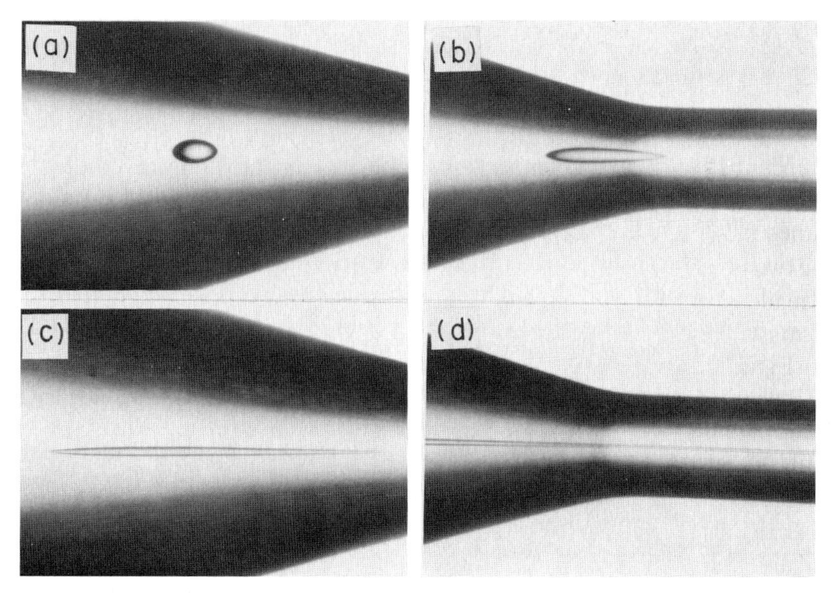

Fig. 4.15 Photographs of droplets describing the effect of suspending medium viscosity on the extent of droplet deformation [52]: (a) and (b) with a 2% Separan solution; (c) and (d) with a 4% Separan solution. The droplet phase is a 6% PIB solution, the initial droplet size is 0.91 mm, and the apparent wall shear rate is 15.0 sec^{-1}. From H. B. Chin and C. D. Han, *J. Rheol.* **23**, 557. Copyright © 1979. Reprinted by permission of John Wiley & Sons, Inc.

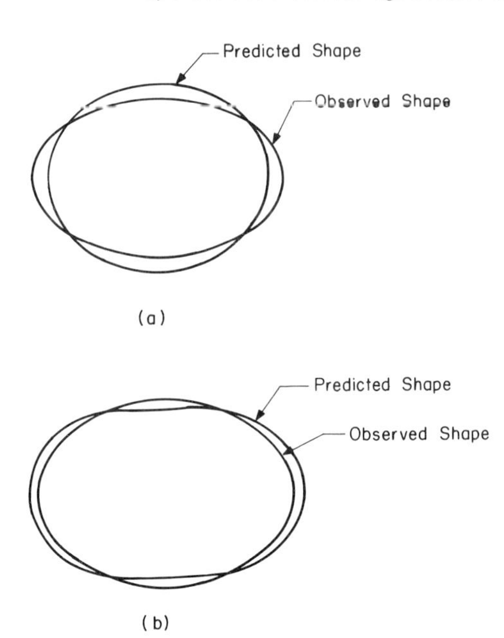

Fig. 4.16 Comparison of the theoretically predicted droplet shape with the experimentally observed one [52]: (a) the droplet of 10% PIB solution suspended in the medium of 4% Separan solution ($k = 0.037$; $We = 0.067$; $\bar{\varepsilon}/\varepsilon = 1.67$; $\dot{\gamma}_E = 0.0042\ \text{sec}^{-1}$); (b) the droplet of 6% PIB solution suspended in the medium of 2% Separan solution ($k = 0.055$; $We = 0.19$; $\bar{\varepsilon}/\varepsilon = 0.58$; $\dot{\gamma}_E = 0.024\ \text{sec}^{-1}$). From H. B. Chin and C. D. Han, *J Rheol.* **23**, 557. Copyright © 1979. Reprinted by permission of John Wiley & Sons, Inc.

Figure 4.16 shows a comparison of the experimentally observed droplet shapes in the region of the converging section of the flow channel where steady extensional flow prevails, with the theoretically predicted ones determined from the use of Eq. (4.15). It can be said that the agreement between the two is reasonable. It should be pointed out, however, that the comparison is made at small values of *We* because Eq. (4.15) is valid only for *small* deformations.

Figure 4.17 gives plots of *apparent* deformation (D) versus *We* for 6% PIB droplets suspended in 2% Separan solution. It is seen that the deformation, theoretically predicted by the use of Eq. (4.15), is in good agreement with the experimentally observed one *only* at low values of *We*, and deviates considerably at large values of *We*. There are two primary reasons for this discrepancy. First, the theoretical prediction is based on the assumption of small deformations. Second, as may be seen in Fig. 4.16, as *We* increases, the theoretically predicted droplet develops a pinch at the center along the major axis, thus resembling a dumbbell. However, no such dumbbell shape was observed during the experiment. Since D was calculated by using the

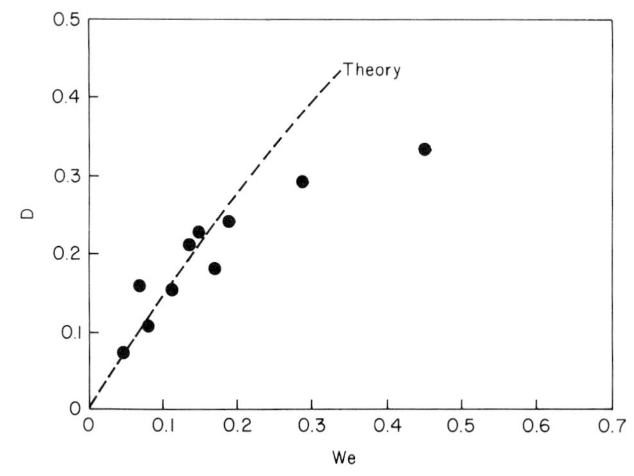

Fig. 4.17 Comparison of the experimentally observed droplet deformation with the theoretically predicted one [52]. The droplet phase is a 6% PIB solution and the suspending medium is a 2% Separan solution ($k = 0.55$; $\bar{\varepsilon}/\varepsilon = 0.58$). From H. B. Chin and C. D. Han, *J. Rheol.* **23**, 557. Copyright © 1979. Reprinted by permission of John Wiley & Sons, Inc.

major and minor axes of the deformed droplet, the theoretically predicted value of D is expected to be larger than the experimentally observed one. The minor axis of the theoretically predicted droplet shape is always smaller than that of the experimentally observed one, particularly for larger values of We.

4.3 RHEOLOGICAL BEHAVIOR OF TWO-PHASE POLYMER BLENDS AND CONCENTRATED EMULSIONS

In processing heterogeneous polymeric systems, one needs information about their bulk rheological properties in a specific flow field, relevant to the particular processing technique chosen. The state of dispersion (i.e., morphology) of a two-phase polymer blend is influenced by processing conditions (e.g., extrusion temperature, pressure drop), which in turn influence the rheological properties [35, 53–64]. It therefore becomes essential to relate the rheological properties of polymer blends to their state of dispersion in flow.

It seems appropriate that before we attempt to theorize, we should first present some representative, experimentally observed examples of the rheological behavior of concentrated emulsions and, then, discuss such behavior with the aid of photomicrographs describing the state of dispersion. In view of the fact that the morphology of polymer blends in the molten state may be considered very similar to that of concentrated emulsions, we shall discuss the rheological behavior of two-phase polymer blends and rubber modified polymers. The existence of a similarity in the rheological behavior between polymer blends and concentrated emulsions will be demonstrated on the basis of experimental investigations published in the literature.

4.3.1 The Rheological Behavior of Concentrated Emulsions

We shall discuss below some experimental studies of the bulk (macroscopic) rheological behavior of concentrated emulsions consisting of two Newtonian liquids and emulsions consisting of two viscoelastic liquids. An interpretation of the experimental results is presented with the aid of photomicrographs describing the state of dispersion of the emulsion systems. Later in this chapter, we shall discuss phenomenological theories that will help us interpret the experimental observations.

Han and King [65] have taken rheogoniometric measurements for various composition ratios of the following emulsion systems: (a) a Newtonian–Newtonian system consisting of low molecular weight polybutene (Indopol L100) and glycerine; (b) a viscoelastic–viscoelastic system consisting of a 2 wt.% aqueous solution of polyacrylamide (Separan AP 30) and a 6 wt.% solution of polyisobutylene (PIB) in decalin.

Viscosity measurements of the Newtonian–Newtonian system consisting of glycerine and Indopol L100 are given in Fig. 4.18. Note that the open and closed symbols represent data taken with the emulsion samples

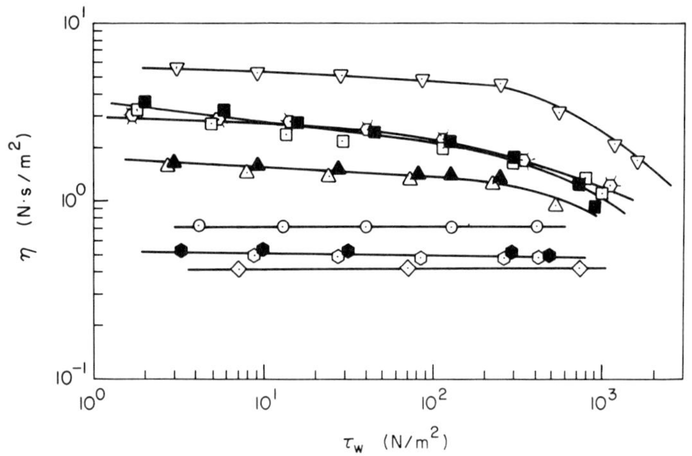

Fig. 4.18 Viscosity versus shear stress for the Indopol L100/glycerine emulsions (by vol.) [65]: (▽) Indopol/glycerine = 50/50; (⊡, ■) Indopol/glycerine = 30/70; (⬡) Indopol/glycerine = 70/30; (△, ▲) Indopol/glycerine = 10/90; (⊙) glycerine; (⬡, ⬢) Indopol/glycerine = 90/10; (◇) Indopol L100. From C. D. Han and R. G. King, *J. Rheol.* **24**, 213. Copyright © 1980. Reprinted by permission of John Wiley & Sons, Inc.

prepared at two different mixing conditions. It can be concluded that to all intents and purposes, the two emulsion samples yielded the same values of bulk viscosity. It is also seen in Fig. 4.18 that certain emulsions exhibit shear-thinning behavior of bulk viscosity as the shear stress is increased. Similar observations were also reported by Suzuki *et al.* [66] and Vadas *et al.* [67].

In their work, Suzuki *et al.* [66] used several organic liquids (benzene, toluene, decalin, n-hexane, etc.) as the continuous phase and deionized water as the droplet phase. They also made visual observations of the dispersion state of *dilute* emulsions while the emulsions were forced to flow through a capillary, and attributed the observed non-Newtonian shear-thinning behavior to a change of the dispersion state of the droplets with increasing rate of shear. They reported further that no deformation of the emulsion droplets under shear (e.g., droplets as large as 100 μm in diameter at rates of shear as high as 2000 $\sec^{-1}$) could be observed with the emulsions studied. It should be mentioned that in their experiments, the range of shear stress varied from 0.01 to 1.7 N/m^2, which is about 3 orders of magnitude smaller than that for the glycerine–Indopol system given in Fig. 4.18.

Vadas *et al.* [67] reported that concentrated emulsions (50 and 70 vol. %) of amyl acetate in aqueous glycerol exhibited non-Newtonian (shear-thinning) behavior, as shown in Fig. 4.19. Note that both amyl acetate and

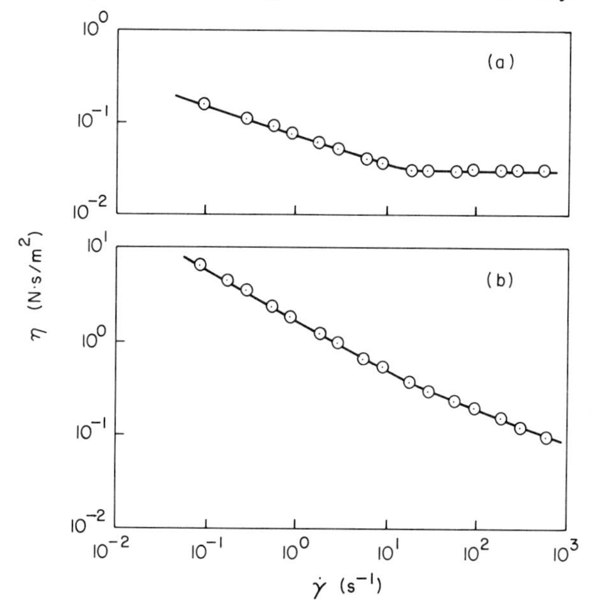

Fig. 4.19 Viscosity versus shear rate for the amyl acetate (AA)/glycerine (G) emulsions (by vol.) [67]: (a) AA/G = 50/50; (b) AA/G = 70/30. From E. B. Vadas *et al.*, *Trans. Soc. Rheol.* **20**, 373. Copyright © 1976. Reprinted by permission of John Wiley & Sons, Inc.

aqueous glycerol are Newtonian liquids. The mean droplet size of the emulsions studied was approximately 2 μm in both concentrations. They note that at very low shear rates, the breakup of interparticle structures may be primarily responsible for the decrease in *apparent* viscosity, whereas at high shear rates, droplets deformation occurring as a result of the combined effects of droplet crowding and shear must have played a role in the decrease in *apparent* viscosity.

Sherman [68] has pointed out the importance of droplet size and its distribution in the non-Newtonian behavior observed of concentrated oil/water emulsions. He considered the effects of the hydrodynamic interactions between the droplets on the bulk flow properties of concentrated emulsions, and pointed out that at very low shear rates, the effect will be very pronounced. It will be particularly pronounced when the droplet size is very small for a given value of volume fraction because the volume of the continuous phase, which is immobilized within the particle aggregates, will be greater. The rate of droplet aggregation and the extent of interaction between the droplets are also affected by droplet size. Interestingly enough, Sherman [68] showed that the relative viscosity of emulsions decreases as the droplet size is increased, as shown in Fig. 4.20. Richardson [69] also noted an effect of droplet size on the bulk viscosity of emulsions, namely, that the more uniform the droplet size, the greater is the viscosity of the emulsion.

Figure 4.21 gives plots of bulk viscosity versus concentration of Indopol L100 for glycerine/Indopol L100 emulsions, prepared by cross plotting the

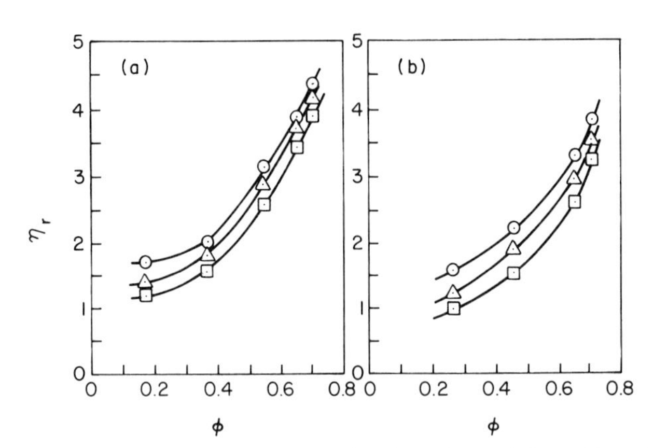

Fig. 4.20 Relative viscosity versus volume fraction of water/oil emulsions [68]: (a) stabilized with sorbitan sesquioleate; (⊙) $d_m = 1.4$ μm. (△) $d_m = 2.0$ μm, (⊡) $d_m = 3.3$ μm; (b) stabilized with sorbitan trioleate: (⊙) $d_m = 1.4$ μm, (△) $d_m = 2.0$ μm, (⊡) $d_m = 3.3$ μm. From *Proc. Int. Cong. Rheol., 4th*, p. 605. Copyright © 1965. Reprinted by permission of John Wiley & Sons, Inc.

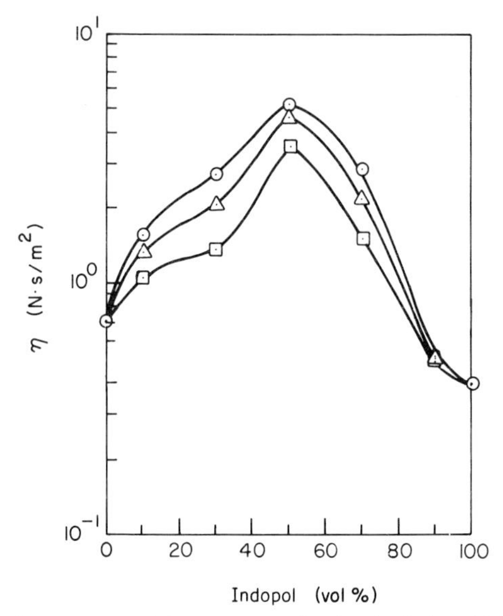

Fig. 4.21 Viscosity versus emulsion concentration for the Indopol L100/glycerine system at various shear stresses (N/m²) [65]: (⊙) 10; (△) 100; (⊡) 500. From C. D. Han and R. G. King, *J. Rheol.* **24**, 213. Copyright © 1980. Reprinted by permission of John Wiley & Sons, Inc.

data given in Fig. 4.18 with shear stress τ_w as parameter. It is seen that the bulk viscosity of the emulsions goes through a maximum as the concentration of Indopol L100 is increased.

Figure 4.22 gives photographs displaying the state of dispersion (i.e., morphology) of Indopol L100/glycerine emulsions of various composition ratios. It is seen that the droplet size and its distribution vary from one volume concentration to another. Of particular note are the morphologies of the emulsions containing 70% and 90% of Indopol L100. These resemble the morphologies of the ABS resins shown in Fig. 4.3. According to Han and King [65], repeated visual observations have confirmed the unique morphology of these emulsions, i.e., droplets inside droplets, leading to the conclusion that the complex morphology is inherent in the particular composition ratios.

Emulsions containing 10%, 30%, and 50% Indopol L100 had small and relatively uniform droplets (somewhere between 1 and 5 μm). In these emulsions, little droplet deformation was observed even when the emulsion samples were subjected to shear rate as high as 1000 sec^{-1} in the cone-and-plate instrument [65].

Figure 4.23 gives plots of first normal stress difference $\tau_{11} - \tau_{22}$ versus shear stress τ_w for four Indopol L100/glycerine emulsions prepared at two

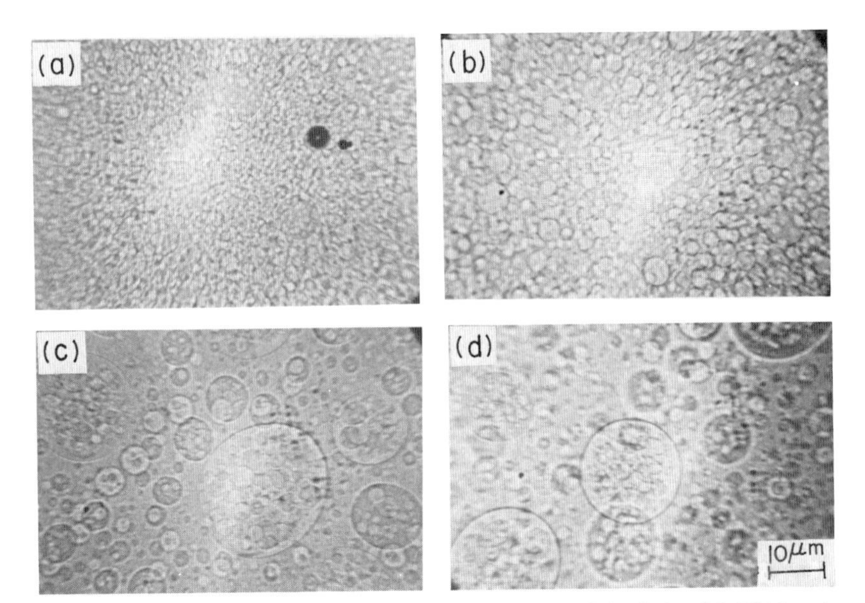

Fig. 4.22 Photographs showing the state of dispersion of the Indopol L100/glycerine emulsions (by vol.) [65]: (a) Indopol/glycerine = 10/90; (b) Indopol/glycerine = 30/70; (c) Indopol/glycerine = 70/30; (d) Indopol/glycerine = 90/10. From C. D. Han and R. G. King, *J. Rheol.* **24**, 213. Copyright © 1980. Reprinted by permission of John Wiley & Sons, Inc.

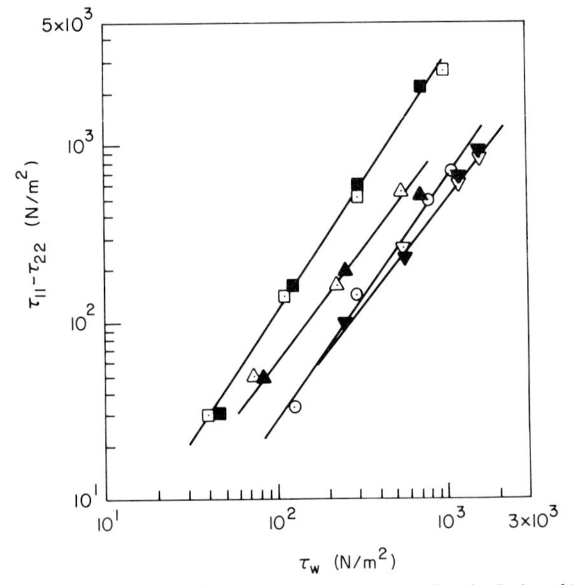

Fig. 4.23 First normal stress difference versus shear stress for the Indopol L100/glycerine emulsions (by vol.) [65]: (⊡, ■) Indopol/glycerine = 30/70; (△, ▲) Indopol/glycerine = 10/90; (⊙) Indopol/glycerine = 70/30; (▽, ▼) Indopol/glycerine = 50/50. From C. D. Han and R. G. King, *J. Rheol.* **24**, 213. Copyright © 1980. Reprinted by permission of John Wiley & Sons, Inc.

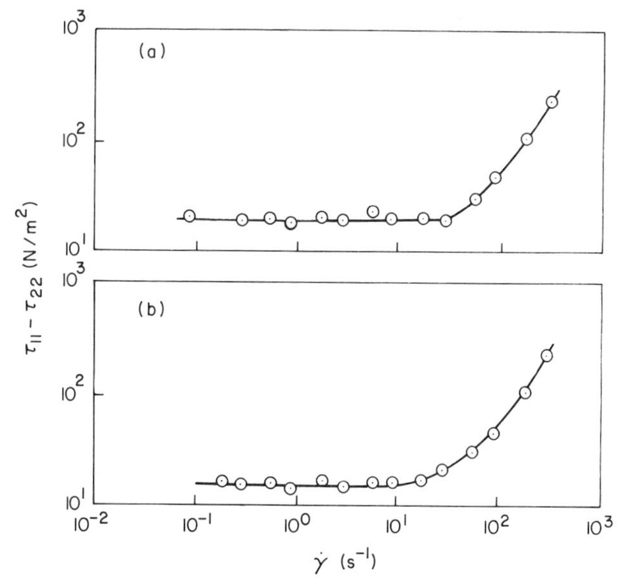

Fig. 4.24 First normal stress difference versus shear rate for the amyl acetate (AA)/glycerine (G) emulsions (by vol.) [67]: (a) AA/G = 50/50; (b) AA/G = 70/30. From E. B. Vadas *et al.*, *Trans. Soc. Rheol.* **20**, 373. Copyright © 1976. Reprinted by permission of John Wiley & Sons, Inc.

different mixing conditions. (The open and closed symbols identify the two sets of data.) Not much difference in the measured values of $\tau_{11} - \tau_{22}$ is shown for the two sets of data. It is of great interest to note in Fig. 4.23 that these emulsions, consisting of two Newtonian liquids, exhibit normal stress effects.

Vadas *et al.* [67] also reported that concentrated emulsions, consisting 50 and 70 vol.% of amyl acetate in aqueous glycerol (both are Newtonian liquids), exhibited large normal stress effects, as shown in Fig. 4.24, when they were subjected to shearing motion in a Weissenberg rheogoniometer. They noted that the first normal stress difference of the emulsions studied is an order of magnitude greater than the shear stress.

Another easy and simple, though qualitative, way of testing normal stress effects is the "rod climb-up" experiment. Figure 4.25 gives photographs showing the difference in the rod climb-up behavior between the two Newtonian liquids (glycerine and Indopol) and an emulsion consisting of glycerine and Indopol L100. It is clearly seen that the emulsion exhibits mild rod climb-up behavior, whereas both glycerine and Indopol show the opposite behavior, i.e., the liquid climbs up the wall of the beaker. This observation gives credence to the experimentally measured values of $\tau_{11} - \tau_{22}$ for the emulsions of glycerine and Indopol L100.

Fig. 4.25 The rod climb-up effect [65]: (a) glycerine; (b) Indopol L100; (c) Indopol/glycerine = 70/30 (by vol.). From C. D. Han and R. G. King, *J. Rheol.* **24**, 213. Copyright © 1980. Reprinted by permission of John Wiley & Sons, Inc.

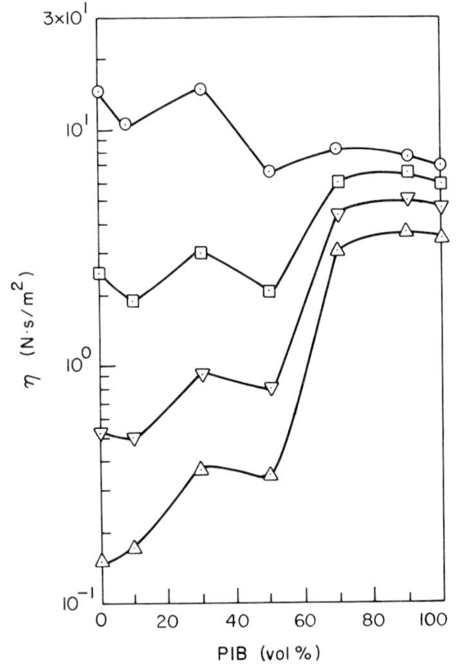

Fig. 4.26 Viscosity versus emulsion concentration for the PIB/Separan system at various shear stresses (N/m^2) [65]: (⊙) 10; (⊡) 30; (▽) 60; (△) 100. From C. D. Han and R. G. King, *J. Rheol.* **24**, 213. Copyright © 1980. Reprinted by permission of John Wiley & Sons, Inc.

Figure 4.26 gives plots of η versus concentration of PIB, and Fig. 4.27 gives plots of $\tau_{11} - \tau_{22}$ versus concentration of PIB for emulsions consisting of a 2 wt.% aqueous solution of polyacrylamide (Separan AP 30) and a 6 wt.% solution of polyisobutylene (PIB) in decalin. It is seen that η goes through a maximum and a minimum, and $\tau_{11} - \tau_{22}$ goes through a minimum, as the concentration of PIB is increased. Figure 4.28 gives photographs of Separan/PIB emulsions of various composition ratios. It is seen that the droplet size and its distribution vary from one volume concentration to another. These photographs show the initial droplet sizes and their distribution before the rheogoniometer was started. One can surmise that under shearing motion, droplets (in particular, large droplets) will deform and hydrodynamic interactions (crowding) will play an important role in determining the extent of droplet deformation and hence the macroscopic rheological properties of the emulsions.

Doppert and Overdiep [70] used emulsions obtained with two mutually incompatible polymer solutions, polyacrylonitrile (PAN) and polyurethane

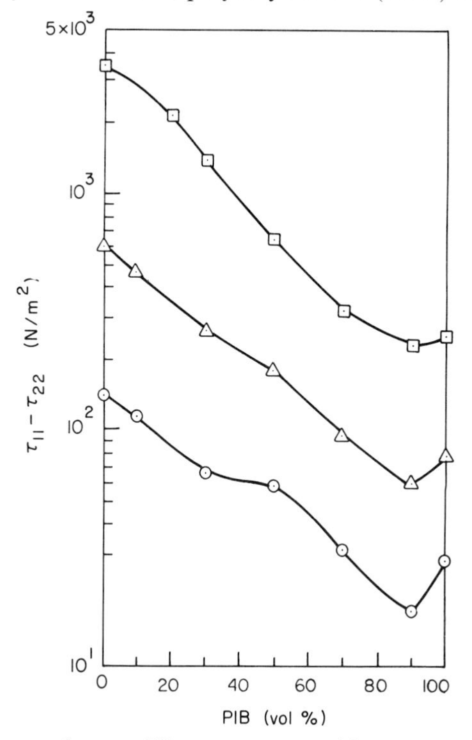

Fig. 4.27 First normal stress difference versus emulsion concentration for the PIB/Separan system at various shear stresses (N/m^2) [65]: (⊙) 50; (△) 100; (⊡) 200. From C. D. Han and R. G. King, *J. Rheol.* **24**, 213. Copyright © 1980. Reprinted by permission of John Wiley & Sons, Inc.

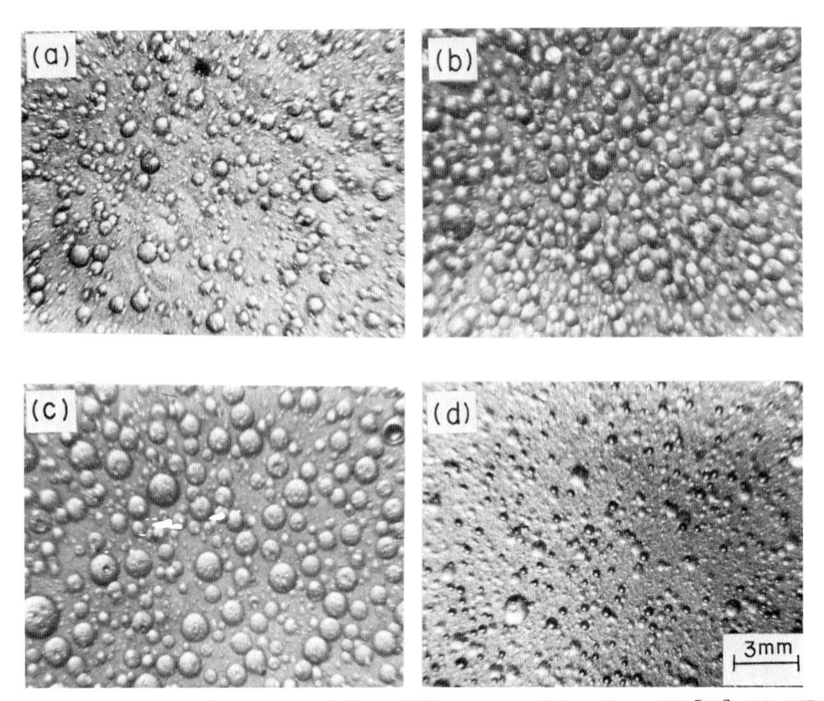

Fig. 4.28 Photomicrographs of the PIB/Separan emulsions (by vol.) [65]: (a) PIB/Separan = 10/90; (b) PIB/Separan = 30/70; (c) PIB/Separan = 50/50; (d) PIB/Separan = 90/10. From C. D. Han and R. G. King, *J. Rheol.* **24**, 213. Copyright © 1980. Reprinted by permission of John Wiley & Sons, Inc.

(PU) dissolved in a common solvent (N-methylpyrrolidone), and determined the apparent bulk viscosity of the emulsion. According to them, *phase inversion* occurred at a volume fraction of about 0.5 and the average size of the dispersed droplets was approximately 5 μm. They reported further that when the PAN droplets were dispersed in a PU solution, the droplets remained *almost spherical*, whereas when the PU droplets were dispersed in a PAN solution the droplets were highly deformed into long *streaks*. Figure 4.29 gives the viscosity–composition curves of the PAN/PU emulsion system. It is seen that the PAN solution is more viscous than the PU solution, and that the viscosity of the emulsions does not follow the simple additivity rule in terms of volume fraction.

Using a capillary viscometer, Meissner *et al.* [71] measured the viscosities of emulsions of polyacrylonitrile (PAN) and cellulose acetate (CA) dissolved in dimethylformamide. They report that the viscosity of some emulsions is even lower than that of the less viscous component, as given in Fig. 4.30. It is seen that the viscosity–composition curves go through a minimum and/or a maximum, depending on the level of shear stress at

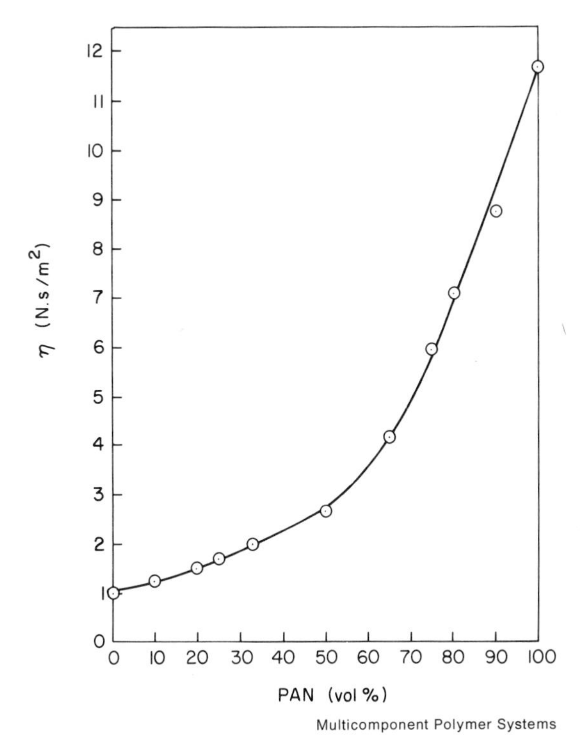

Multicomponent Polymer Systems

Fig. 4.29 Viscosity versus emulsion concentration of polyacrylonitrile (PAN)/polyurethane (PU) system at shear stress $\tau_w = 2 \times 10^3$ N/m^2 [70].

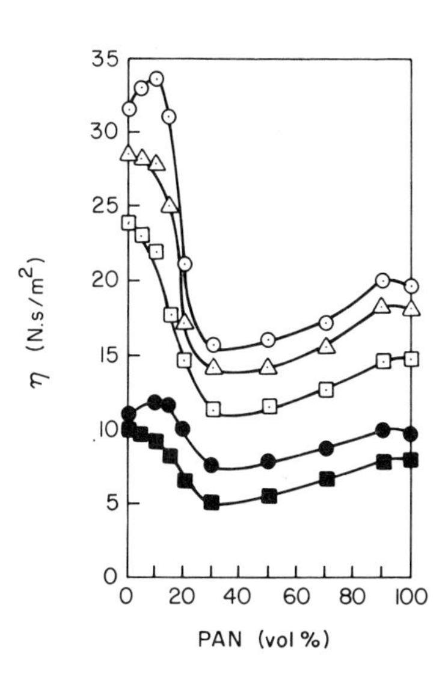

Fig. 4.30 Viscosity versus emulsion concentration for the polyacrylonitrile (PAN)/cellulose acetate (CA) system at various shear stresses (N/m^2) [71]. Open symbols at 20°C: (⊙) 0.918×10^3; (△) 2.754×10^3; (⊡) 9.186×10^3. Closed symbols at 40°C: (●) 0.918×10^3; (■) 9.186×10^3.

which the emulsion samples were tested. No information, however, is reported about the state of dispersion of the PAN/CA emulsion system investigated.

As will be discussed below, there is sufficient theoretical evidence that the non-Newtonian and normal stress effects, observed with Newtonian droplets suspended in a Newtonian medium, are attributable to the *deformability* of the droplets [41, 72, 73]. For instance, Hyman and Skalak [72] analyzed the motion of Newtonian droplets suspended in a Newtonian medium subjected to circular Poiseuille flow, and showed that the deformation of the droplets results in a significant reduction in the pressure gradient compared to that necessary for a suspension of rigid spheres. In other words, the resistance to flow (i.e., pressure drop divided by flow rate) decreases as the flow rate increases, manifesting a mechanism of non-Newtonian behavior.

Therefore it is essential to understand the phenomenon of droplet deformation in terms of the physical/rheological properties of the fluids involved and flow conditions, in order to intelligently discuss the theoretical development which permits us to predict the *bulk* rheological properties of emulsions.

4.3.2 The Rheological Behavior of Two-Phase Polymer Blends

In the past, a number of researchers have reported measurements of the rheological properties of thermoplastic blends [35, 53–64, 74–84], elastomer blends [85–87], and rubber modified thermoplastics [88–95]. These investigators have considered the effects of blending ratio on the viscosity and/or normal stress effects of blends and their variation with mechanical test variables or processing variables (e.g., shear rate, melt temperature). We shall now discuss some representative rheological measurements of two-phase polymer blends and rubber modified polymers.

Figure 4.31 gives the effect of blending ratio on melt viscosity of blends of polyoxymethylene (POM) with copolyamide (CPA), with shear stress as parameter. It is of interest to note that the viscosity goes through a maximum at low stress levels and through a minimum at high stress levels. While investigating this behavior, Tsebrenko *et al.* [96] examined the details of the morphological state of the POM/CPA blends by freezing the extrudate samples in liquid nitrogen. Figure 4.32 gives a photomicrograph of the extrudate collected in the region just past the die entrance, and Fig. 4.33 gives a photomicrograph of the extrudate collected at the down-stream side of the circular die employed. It is seen that fine fibrils of the POM phase (light area) are dispersed in the CPA phase (dark area).

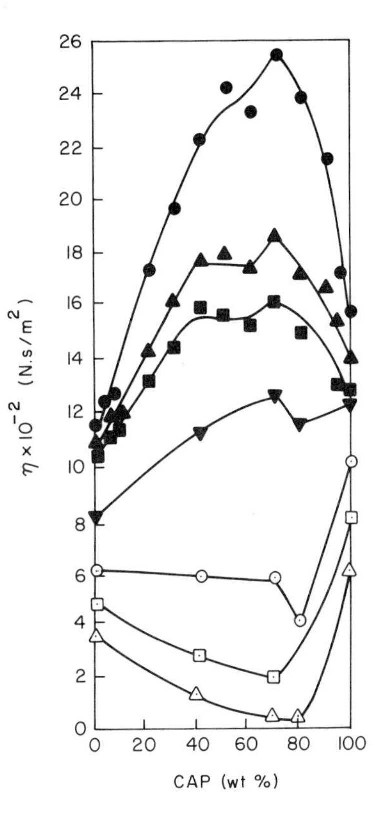

Fig. 4.31 Viscosity versus blend composition for blends of polyoxymethylene (POM) and copolyamide (CPA) at 190°C, for various wall shear stresses (N/m^2) [61]: (●) 1.27×10^4; (▲)3.93×10^4; (■) 5.44×10^4; (▼) 6.30×10^4; (⊙) 1.26×10^5; (⊡) 1.93×10^5; (△) 3.16×10^5. From T. I. Ablazova *et al.*, *J. Appl. Polym. Sci.* **19**, 1781. Copyright © 1975. Reprinted by permission of John Wiley & Sons, Inc.

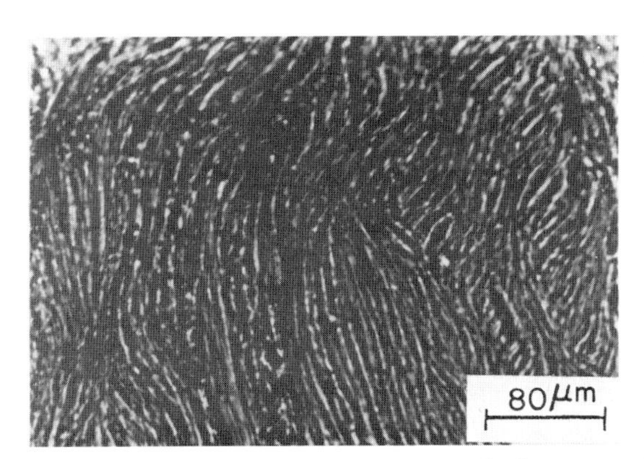

Fig. 4.32 Photomicrograph of the longitudinal section of a frozen extrudate of a blend of POM and CPA in the entrance region of the die (region C in Fig. 4.34) [96]. Reproduced from M. V. Tsebrenko *et al.*, *Polymer* **17**, 831 (1976), by permission of the publishers, IPC Business Press Ltd. ©

Fig. 4.33 Photomicrograph of the longitudinal section of a frozen extrudate of a blend of POM and CPA in the fully developed region of the die (region D in Fig. 4.34) [96]. Reproduced from M. V. Tsebrenko *et al.*, *Polymer* **17**, 831 (1976), by permission of the publishers, IPC Business Press Ltd. ©

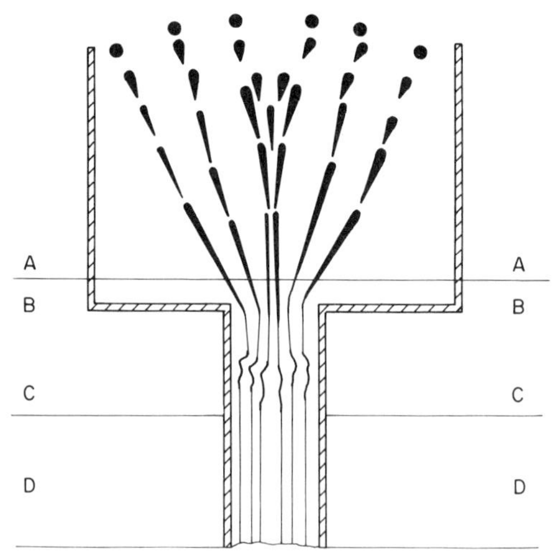

Fig. 4.34 Schematic describing the fibrillation process in the entrance region and in the duct [96]. Reproduced from M. V. Tsebrenko *et al.*, *Polymer* **17**, 831 (1976), by permission of the publishers, IPC Business Press Ltd. ©

In offering an explanation for the occurrence of fibrillation (stratification) observed experimentally, Tsebrenko *et al.* [96] presented the schematic given in Fig. 4.34. The dispersed droplets, formed initially during the mixing operation, are elongated as they enter the die entrance (region A in Fig. 4.34) under the effect of tensile stresses acting in the direction of converging streamlines. Then the elongated droplets recoil as they pass the die entrance (region C in Fig. 4.34). Finally, rearrangement of the elongated droplets (shown as threadlike fibrils in Fig. 4.33) occurs at the downstream side of the capillary (region D in Fig. 4.34), giving rise to fibrils parallel to the capillary axis.

We can now explain, at least in part, why the viscosity–composition curves may go through a maximum and a minimum, depending on the level of shear stresses applied. At low shear stress levels, there would be little deformation of the droplets in the entrance region, and strong interactions among droplets would be expected to occur as the two-phase mixture flows through the capillary. Such a state of dispersion can give rise to a bulk viscosity greater than that of the constituent components. As the level of shear stress is increased, the droplets are elongated and thus the chances of interactions occurring between droplets will be reduced. This would give rise to bulk viscosity lower than that obtained at low stress levels. At sufficiently high levels of shear stress, the majority of droplets may form long threadlike fibrils, well aligned in the direction of flow (i.e., parallel to the capillary wall). Under such a state of dispersion, the mixture may require less pressure drop for flow in the capillary than the individual components would, and hence the bulk viscosity can go through a minimum when plotted against blending ratio. The rationale for this expectation is based on the existence of long, flexible fibrils parallel to the direction of flow. Such well-aligned fibrils will give a minimum contribution to the wall shear stress.

Let us now turn our attention to the experimental observations of both the viscosity and elasticity of two-phase viscoelastic polymer blends. Figure 4.35 gives plots of viscosity η and first normal stress difference $\tau_{11} - \tau_{22}$ versus shear rate for blends of polymethylmethacrylate (PMMA) and polystyrene (PS) at 200°C. The data for the shear rates below about 10 $\sec^{-1}$ were obtained with a Weissenberg rheogoniometer and the data at high shear rates were obtained with a Han slit/capillary rheometer. It is seen that crossovers in the η and $\tau_{11} - \tau_{22}$ curves occur as the shear rate is increased. Figure 4.36 shows the effect of blending ratio on η and $\tau_{11} - \tau_{22}$ for the PMMA/PS blends, with *shear rate* as parameter, and Fig. 4.37 gives similar plots, with *shear stress* as parameter. Some interesting and important observations may be made from Figs. 4.36 and 4.37.

First, at low shear rates or low shear stresses, the viscosity of the blends decreases *monotonically* as the amount of the less viscous component, PS,

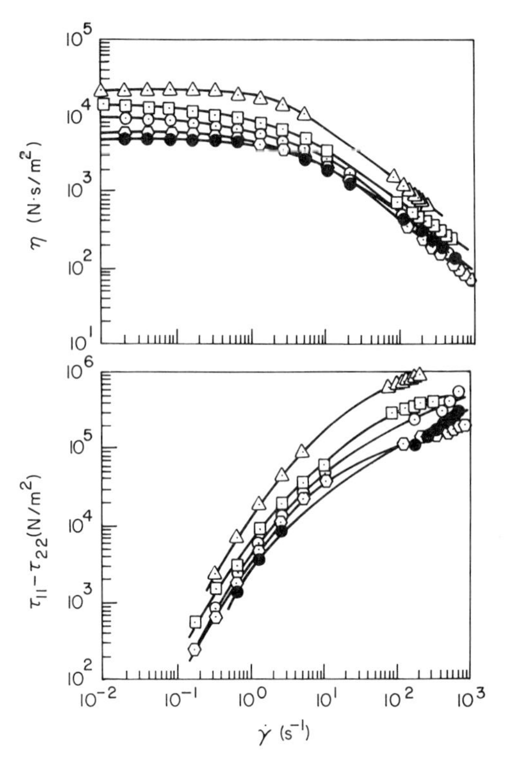

Fig. 4.35 Viscosity and first normal stress difference versus shear rate for blends of polymethylmethacrylate (PMMA) and polystyrene (PS) at 200°C [82]: (●) PS; (△) PMMA; (⊡) PS/PMMA = 25/75 (by wt.); (⊙) PS/PMMA = 50/50 (by wt.); (⬡) PS/PMMA = 75/25 (by wt.).

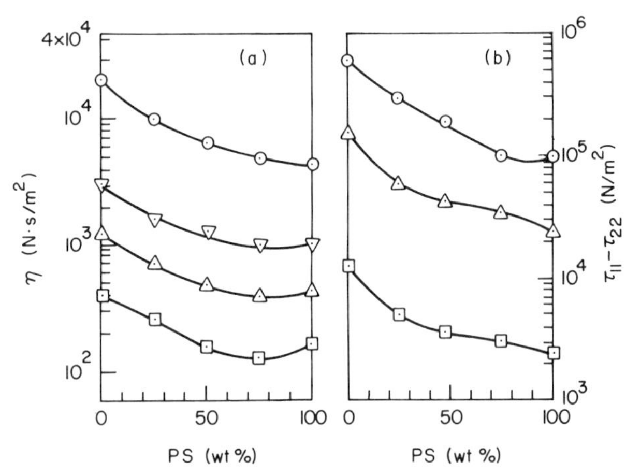

Fig. 4.36 Viscosity and first normal stress difference versus blend composition for blends of PMMA and PS at 200°C, at various shear rates (s^{-1}): (a) (⊙) 0.5; (▽) 30; (△) 100; (⊡) 400. (b) (⊡) 1.0; (△) 10; (⊙) 100.

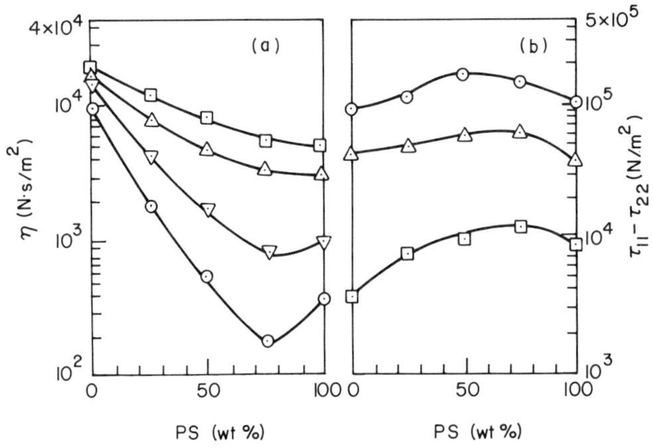

Fig. 4.37 Viscosity and first normal stress difference versus blend composition for blends of PMMA and PS at 200°C, at various wall shear stresses (N/m^2): (a) (⊡) 10^3, (△) 10^4, (▽) 3×10^4, (⊙) 5×10^4; (b) (⊡) 10^4, (△) 3×10^4, (⊙) 5×10^4.

is increased. However, as the level of shear rate (or shear stress) is increased, the viscosity of the blends goes through a *minimum*. Note that the dependence of viscosity on blending ratio shows the same trend whether shear rate or shear stress is used.

However, the dependence of $\tau_{11} - \tau_{22}$ on blending ratio shows different trends, depending on whether shear rate or shear stress is used. Bear in mind that the blend of PMMA and PS consists of two phases, discrete and continuous [35]; hence, the shear rate at the interface between the phases would be discontinuous, because of the difference in their viscosities. Therefore the the use of shear rate in correlating the data may be subject to serious criticism. However, it is reasonable to assume that the shear stress at the interface between the phases would be continuous and therefore the use of wall shear stress in correlating the data is less vulnerable to criticism. It is then interesting to see that when shear stress is used as parameter, the $\tau_{11} - \tau_{22}$ curves go through a *maximum*. In other words, the blend that shows a *minimum* in viscosity tends to show a *maximum* in $\tau_{11} - \tau_{22}$. Before we attempt to theorize about this particular experimental observation, let us examine the rheological behavior of other blend systems.

Figure 4.38 shows the effect of blending ratio on η and $\tau_{11} - \tau_{22}$ for blends of high-density polyethylene (HDPE) and polystyrene (PS) at 200°C, with *shear stress* as parameter. It is seen that the blends show a minimum in viscosity. Also, a strong maximum in viscosity is seen at a blending ratio of about 75 wt.% HDPE, regardless of the values of shear stress applied. On the other hand, the first normal stress difference of the blends show a maximum and also a minimum as the shear stress is increased. The particular

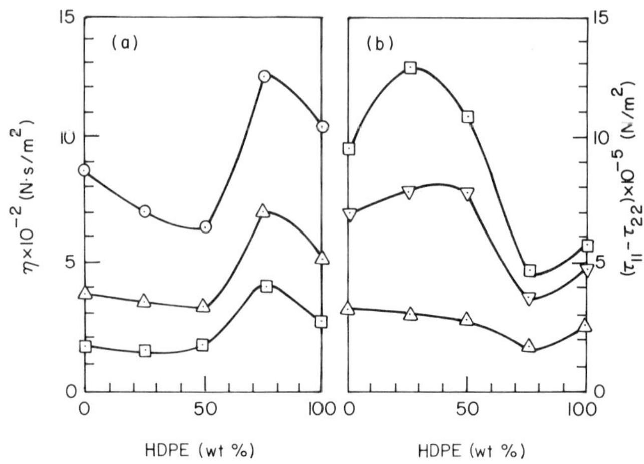

Fig. 4.38 Viscosity and first normal stress difference versus blend composition for blends of high-density polyethylene (HDPE) and polystyrene (PS) at 200°C, at various wall shear stresses (N/m^2) [55]: (a) (⊙) 4×10^4, (△) 6×10^4, (⊡) 9×10^4; (b) (△) 6×10^4, (▽) 9×10^4, (⊡) 10^5. From C. D. Han and Y. W. Kim, *Trans. Soc. Rheol.* **19**, 245. Copyright © 1975. Reprinted by permission of John Wiley & Sons, Inc.

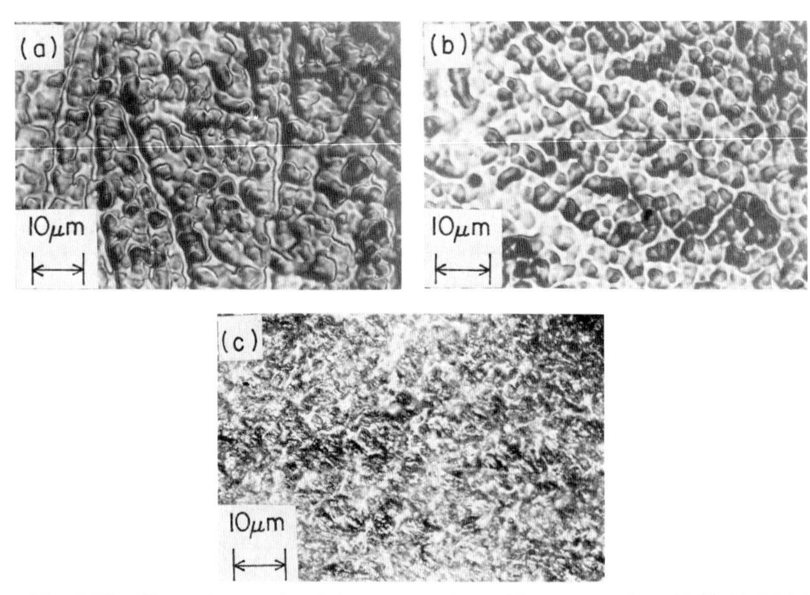

Fig. 4.39 Photomicrographs of the cross sections of frozen extrudates [55]: (a) HDPE/PS = 25/75 (by wt.), $\tau_w = 0.63 \times 10^5$ N/m^2; (b) HDPE/PS = 50/50 (by wt.), $\tau_w = 0.61 \times 10^5$ N/m^2; (c) HDPE/PS = 75/25 (by wt.), $\tau_w = 0.69 \times 10^5$ N/m^2. From C. D. Han and Y. W. Kim, *Trans. Soc. Rheol.* **19**, 245. Copyright © 1975. Reprinted by permission of John Wiley & Sons, Inc.

blend that gives a maximum in η also gives a minimum in $\tau_{11} - \tau_{22}$, and the trend of viscosity behavior of this blend system is almost opposite to the trend of normal stress effect, which is consistent with what has been observed with the PMMA/PS blend system.

Figure 4.39 gives photomicrographs of extrudates of three HDPE/PS blends, in which the dark areas are the PS phase and the white areas are the HDPE phase. It is seen that the morphology of the blend (HDPE/PS = 75/25), which gives a maximum in viscosity and a minimum in normal stress, is quite different from the morphologies of the other two blends. No discrete droplets are seen in the HDPE/PS = 75/25 blend, making it difficult to identify one component as the discrete phase and the other as the continuous phase (forming an interlocked morphology). Such a state of dispersion would make the mixture more resistant to flow because of the strong interactions between the phases. It is as if they pull at each other, and this may be the reason why a maximum in viscosity is observed in Fig. 4.38. It should be emphasized that the maximum in viscosity observed in Fig. 4.38 occurs at very high shear stresses. Therefore the reason for the existence of a maximum in η here is quite different from the situation where a maximum in η was observed in the POM/CPA blend system, shown in Fig. 4.38. It is the intrinsic morphology of the HDPE/PS = 75/25 blend that gives rise to a maximum in η, whereas in the POM/CPA blend system the hydrodynamic effect (i.e., the interactions among underformed droplets) is responsible for the maximum in η observed.

Figure 4.40 gives plots of recoverable shear strain S_R [defined by Eq. (2.74) in Chapter 2] versus shear stress τ_w for HDPE/PS blends. We have already discussed in Chapter 2 that such plots are useful for comparing the elasticity

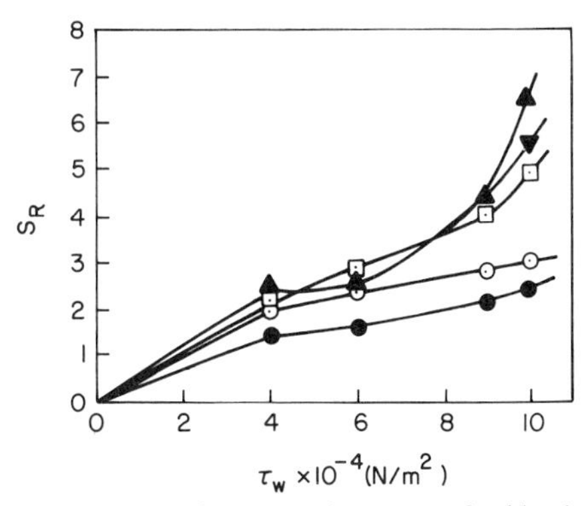

Fig. 4.40 Recoverable shear strain versus shear stress for blends of HDPE and PS, at various blend compositions (by wt.) [64]: (⊙) HDPE; (⊡) PS; (▲) HDPE/PS = 25/75; (▼) HDPE/PS = 50/50; (●) HDPE/PS = 75/25.

of one fluid with that of others. It is seen that the S_R for the blend with an interlocked morphology (HDPE/PS = 75/25) is lower, at all shear stresses, than that of either one of the constituent components. The magnitude of the recoverable shear strain is a measure of the stored elastic energy of deformation in flow. Therefore, a possible inference is that dispersions of viscoelastic droplets have an additional mode for the accumulation of the free energy of deformation, and that dispersions with a large interfacial area, such as those with an "interlocked" morphology, would exhibit a lower recoverable free energy of deformation after the cessation of flow [64].

Figure 4.41 gives the effect of blending ratio on the melt viscosity and normal stress for blends of polypropylene (PP) and polystyrene (PS) with shear stress as parameter. Here again, the η curves show a minimum, and the $\tau_{11} - \tau_{22}$ curves show a maximum, at certain blending ratios.

For a given two-phase blend system, the location of a minimum in η and a maximum in $\tau_{11} - \tau_{22}$ depends on the morphological state of the blends. Note that at a given stress level, small droplets would deform less than large ones. Thus, a blend containing small droplets would be expected to exhibit a viscosity greater than one containing large droplets. In other words, when two blends have the same mode of dispersion, the *bulk* rheological properties of the blends would depend on the droplet size of the discrete phase and its distribution.

It is of interest to note that Van Oene [35] carried out extrusion experiments with blends of HDPE and PS by varying the length of the capillary,

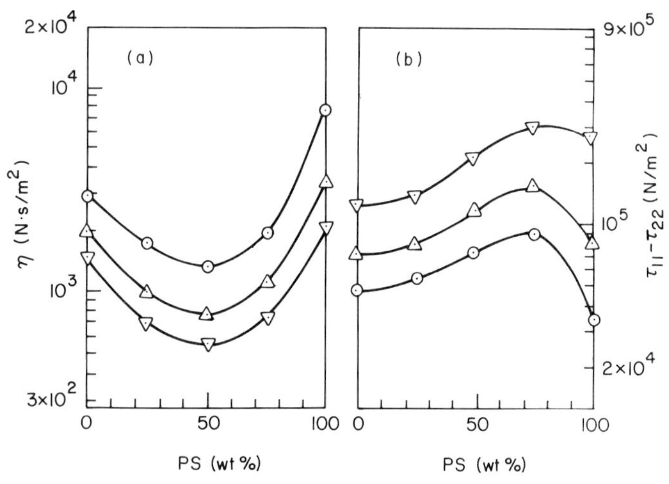

Fig. 4.41 Viscosity and first normal stress difference versus blends of polypropylene (PP) and polystyrene (PS) at 200°C, at various wall shear stresses (N/m²): (a) (⊙) 2×10^4, (△) 3×10^4, (▽) 4×10^4; (b) (⊙) 2×10^4, (△) 3×10^4, (▽) 5×10^4.

which changed the residence time. He found that there was no marked change in extrudate morphology, except that extrusion through a short capillary resulted in a coarser dispersion. He observed further that a once-extruded mixture could be reextruded without change in extrudate structure, although reextrusion made the dispersions more uniform. Similar experiments were also carried out by Tsebrenko *et al.* [60], who used blends of polyoxymethylene and copolyamides. They observed that reextrusion increased the apparent viscosities of the blends and reextruded extrudates had more uniform and finer droplets dispersed in the continuous phase. This, then, confirms that the large decrease in the apparent viscosity of the two-phase polymer blends observed above is due to the deformation of the droplet phase dispersed in the continuous phase.

Figure 4.42 gives the effect of blending ratio on the melt viscosity and steady-state shear compliance [defined by Eq. (2.71) in Chapter 2] for blends of polystyrene (PS) and styrene–methacrylic acid copolymer (S–MAA). Note, in Fig. 4.42, that shear rate is used as parameter, instead of shear stress. We have discussed in Chapter 2 that plots of shear compliance J_e versus shear rate $\dot{\gamma}$ yield the same information as plots of recoverable shear strain S_R versus shear stress τ_w. It is not, therefore, a coincidence that the PS/S–MMA blend system shows maxima in viscosity at those blending ratios which show minima in shear compliance, and that it shows minima in viscosity at those blending ratios which show maxima in shear compliance.

At this point, it is quite appropriate to offer an explanation of why certain blends show a maximum (or maxima) in elastic effect (in terms of

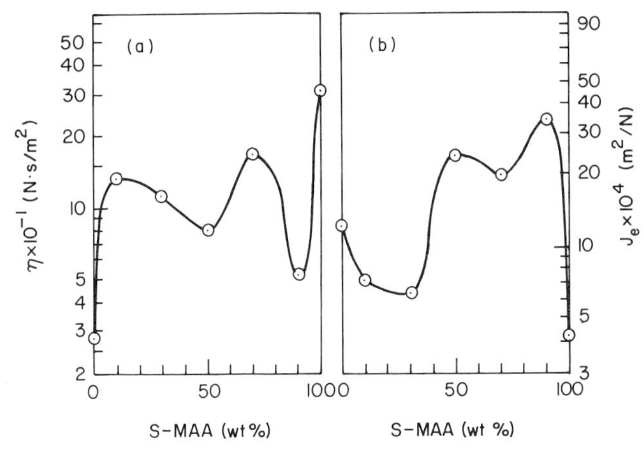

Fig. 4.42 Viscosity and steady-state shear compliance versus blend composition for blends of polystyrene (PS) and styrene–methacrylic acid copolymer (S–MAA) at shear rate of 9 sec^{-1} [74]. From K. Iwakura and T. Fujimura, *J. Appl. Polym. Sci.* **19**, 1427. Copyright © 1975. Reprinted by permission of John Wiley & Sons, Inc.

$\tau_{11} - \tau_{22}$, S_R, or J_e). There are at least two reasons for this. One is the presence of interfacial tension between the phases. As will be discussed below, concentrated emulsions consisting of two *Newtonian* liquids can, at least theoretically, exhibit normal stress effects, primarily due to the presence of interfacial tension. The second reason, at present lacking a rigorous theoretical analysis, is that the long, flexible threadlike droplets suspended in the continuous medium (see Fig. 4.33) may be considered similar to long flexible macromolecules in a polymer solution. It should be remembered that, in Chapter 3, we have shown that a Newtonian liquid containing glass fibers can exhibit normal stress effects (in terms of $\tau_{11} - \tau_{22}$ or "rod climbup" phenomenon). In the flow of two-phase polymer blends, and when the morphology of the system permits it, the observed increase in normal stress above the level of the constituent components might come from a combination of the two effects mentioned above, namely, part from the presence of interfacial tension and part from the hydrodynamic effect.

Let us turn our attention back to the viscosity–composition curve of two-phase blends. Figure 4.43 gives plots of *zero-shear* viscosity versus

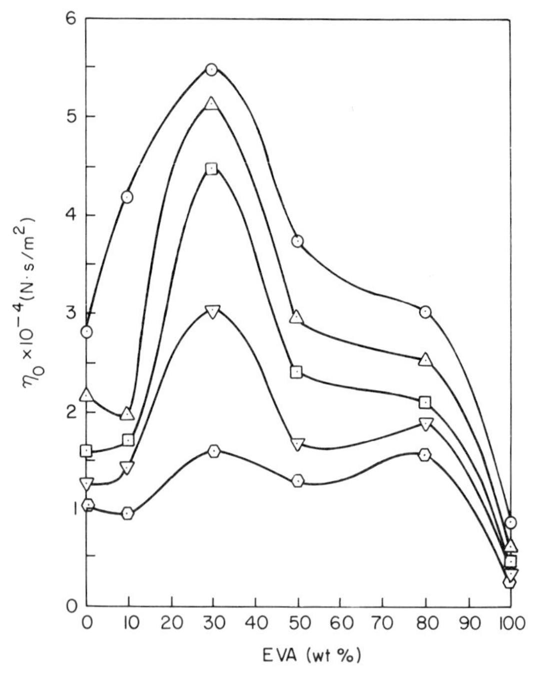

Fig. 4.43 Zero-shear viscosity versus blend composition for blends of high-density polyethylene (HDPE) and ethylene–vinyl acetate (EVA), at various temperatures (°C) [75]: (⊙) 160; (△) 170; (□) 180; (▽) 190; (⬡) 200.

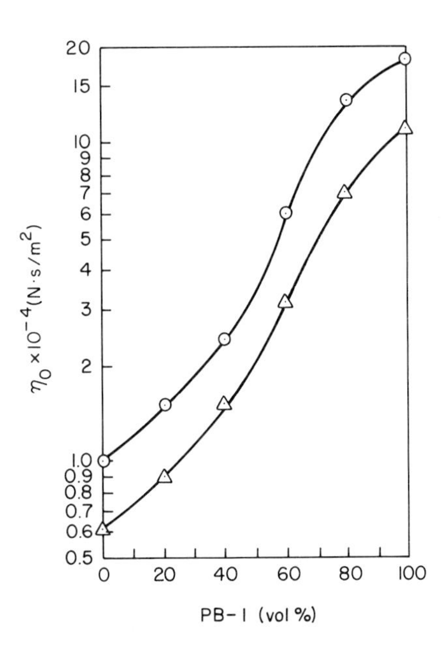

Fig. 4.44 Zero-shear viscosity versus blend composition for blends of high-density polyethylene (HDPE) and polybutene-1 (PB-1), at various temperatures (°C) [77]: (○) 161; (△) 182. From S. Vemura and M. Takayanagi, *J. Appl. Polym. Sci.* **10**, 113. Copyright © 1966. Reprinted by permission of John Wiley & Sons, Inc.

blending ratio for blends of ethylene–vinyl acetate copolymer (EVA) and high-density polyethylene (HDPE) at various temperatures. It is seen that in most cases the viscosities of the blends are greater than those of the constituent components. Since we are concerned here with zero-shear viscosity, we can safely assume that there was hardly any deformation of droplets at the condition under which the data were obtained. When the deformation of droplets is absent, the mixture is expected to give viscosities greater than the viscosity of the less viscous of the two polymers. However the exceedingly large viscosities observed in Fig. 4.43 might have come either from strong interactions between droplets or from a particular mode of dispersion unknown to us.†

Figure 4.44 gives plots of *zero-shear* viscosity versus blending ratio for blends of polybutene-1 (PB-1) and high-density polyethylene (HDPE) at two different temperatures. Interestingly enough, the viscosity–composition curves show an S shape, not exhibiting a maximum. It is reported that a phase inversion occurs at the blending ratio identified by the point of inflection of the curves [77]. It would seem that, in this particular blend system, the morphological state was such that there was little interaction among droplets at the conditions under which the data were obtained.

† The authors [75] did not report the morphology of the blend system investigated.

The diversity of the viscosity–composition curves observed above may be summarized as given in Fig. 4.45: Type I, the viscosity decreases (or increases) monotonically with blend composition, Type II, the viscosity decreases (or increases), exhibiting an S shape; Type III, the viscosity–composition curve goes through a minimum (or minima); Type IV, the viscosity–composition curve goes through a maximum (or maxima). All of these viscosity–composition curves depend on the mode of dispersion (i.e., the morphological state of a blend), which in turn depends on the blending ratio, and the deformation and thermal histories of the mixing operations. Therefore, a diversity of viscosity–composition curves is to be expected when one relates the bulk (macroscopic) rheological behavior of two-phase polymer blends to their morphologies. That is, Type I occurs when the droplets have relatively little interaction and the viscosity of the droplet phase is much greater than that of the suspending medium; Type II occurs when there is a phase inversion at a certain blending ratio and there is little interaction among droplets; Type III occurs when the droplets get sufficiently elongated, giving rise to threadlike fibrils that are aligned in the flow direction; Type IV occurs either when there are strong interactions among droplets at low shear rates (or low shear stresses) or when the blend has an "interlocked" morphology.

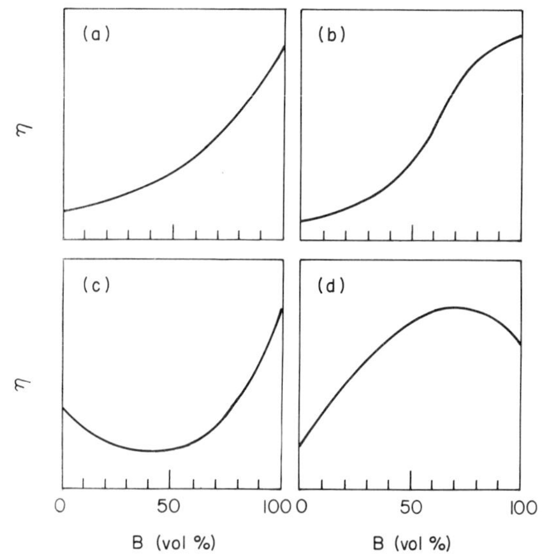

Fig. 4.45 Schematic summarizing various shapes of blend viscosity curve: (a) viscosity increases monotonically with blend composition; (b) viscosity increases with blend composition, exhibiting an S-shape curve; (c) viscosity goes through a minimum; (d) viscosity goes through a maximum.

When dealing with elastomer modified thermoplastics, it has been observed in general that the level of shear viscosity increases with rubber loading and normal stresses, when plotted against shear stress, decrease with rubber loading [88–95]. Some investigators [88, 90–92] have noted that certain elastomer modified thermoplastics (e.g., ABS resins) exhibit yield values in steady and/or oscillatory shear flow, and/or in steady elongational flow. Note that, as discussed in Chapter 3, yield values are commonly observed in polymer melts filled with high volume fraction loadings of small rigid particles [97–101]. It can be said therefore that the rheological behavior of elastomer modified thermoplastics has a resemblance to that of highly filled thermoplastics. Lee [88] and Münstedt [94] have argued that agglomeration and network formation by the rubber globules is the cause for the existence of the yield values they observed. Note that much of the elastomer modified thermoplastics (e.g., HIPS, ABS resins) has considerable graft or crosslinked rubber globules.

At this juncture, it should be mentioned that measurement of the rheological properties of a two-phase polymer blend is meaningful only when one controls the morphology (i.e., phase distribution) of the material during flow. In the situation where phase distribution evolves during flow from the reservoir section to the downstream end of a capillary die, the use of the conventional (i.e., plunger-type) capillary viscometer is unsuitable for the determination of the rheological properties of two-phase polymer blends. This is because in the plunger-type viscometer the converging flow from the barrel into the capillary entrance tends to elongate the dispersed droplets along the flow direction and, then, the elongated droplets recoil downstream in the flow channel to yield an equilibrium shape [102]. (See also Figs. 4.13–4.15.) The details of such phase distribution evolution can differ with instrument. This would be the case with the capillary and cone-and-plate geometries.

4.4 THEORETICAL CONSIDERATION OF THE RHEOLOGICAL BEHAVIOR OF EMULSIONS

We shall now turn to the theoretical aspects of predicting the *bulk* rheological properties of emulsions. More specifically, we will show from the phenomenological point of view that the deformation of droplets, in an emulsion consisting of two Newtonian fluids, gives rise to non-Newtonian and normal stress effects. In order to determine the effects of shear-induced droplet deformation and rotation on the macroscopic (bulk) rheological properties of emulsions, one needs to consider the relation between the macroscopic and microscopic motions of a flowing emulsion of *deformable* particles of any shape.

In the past, a number of investigators [41, 73, 103–110] have made attempts at predicting theoretically the bulk rheological properties of emulsions. The first attempt was made by Taylor [103], who extended Einstein's work [111] which dealt with suspensions of *rigid* spherical particles. By assuming that droplets remain spherical during flow, Taylor derived a theoretical expression for the bulk (or effective) viscosity η of a dilute emulsion consisting of two Newtonian fluids, yielding

$$\eta = \eta_0\{1 + [(5k + 2)/(2k + 2)]\phi\} \tag{4.18}$$

where η_0 is the viscosity of the suspending medium, k is the ratio of the droplet phase viscosity to the suspending medium viscosity and ϕ is the volume fraction of the droplet phase. It is seen that as k approaches infinity (i.e., for solid particles), Eq. (4.18) reduces to the expression for the bulk viscosity of a dilute suspension derived first by Einstein [111]. (See Eq. (3.1) in Chapter 3.]

Considering a time-dependent flow, Fröhlich and Sack [104] have shown that a dilute emulsion of *elastic spheres* suspended in a Newtonian liquid may exhibit elastic behavior. Assuming that the emulsion may be considered to be an incompressible and isotropic medium, and that inertia effects may be neglected, they obtained a rheological equation of state for a dilute emulsion, yielding

$$(1 + \lambda_1\, d/dt)\boldsymbol{\tau} = 2\eta(1 + \lambda_2 d/dt)\mathbf{d} \tag{4.19}$$

in which η, λ_1, λ_2 are given

$$\begin{aligned} \eta &= \eta_0(1 + \tfrac{5}{2}\phi) \\ \lambda_1 &= (3\eta_0/2G)(1 + \tfrac{5}{3}\phi) \\ \lambda_2 &= (3\eta_0/2G)(1 - \tfrac{5}{2}\phi) \end{aligned} \tag{4.20}$$

where G is the modulus of *spherical elastic* particles obeying Hooke's law. Note that Eq. (4.19) predicts the bulk viscosity of an emulsion at steady shear flow, and the elastic behavior of an emulsion in a *time-dependent* flow.

Following the approach of Fröhlich and Sack [104], Oldroyd [105] considered a dilute emulsion of one Newtonian liquid in another. He derived a rheological equation of state essentially the same as Eq. (4.19), except that the time derivative appearing in Eq. (4.19) has the right invariant properties for general validity (i.e., convected derivative or Jaumann derivative). According to Oldroyd [105], the bulk rheological properties of an emulsion may be represented by the following expressions for the appropriate parameters appearing in Eq. (4.19),

$$\eta = \eta_0\left[1 + \left(\frac{5k+2}{2k+2}\right)\phi + \frac{(5k+2)^2}{10(k+1)^2}\phi^2\right] \tag{4.21}$$

$$\lambda_1 = \frac{(19k+16)(2k+3)}{40(k+1)}\left(\frac{\eta_0 a}{\sigma}\right)\left[1 + \frac{(19k+16)}{5(k+1)(2k+3)}\phi\right] \tag{4.22}$$

$$\lambda_2 = \frac{(19k+16)(2k+3)}{40(k+1)}\left(\frac{\eta_0 a}{\sigma}\right)\left[1 - \frac{3(19k+16)}{10(k+1)(2k+3)}\phi\right] \tag{4.23}$$

in which a is the radius of droplets, $k = \overline{\eta}_0/\eta_0$ (the viscosity ratio of the droplet phase to the continuous phase), and σ is the interfacial tension.

It should be noted that in the studies of both Fröhlich and Sack [104] and Oldroyd [105], both inertia effects and hydrodynamic interactions between droplets are neglected, and the rate of particle deformation is accounted for by a matching of velocities at a phase interface, assuming that the deviation of the particle from a spherical shape is sufficiently small so that interfacial matching conditions can be imposed at the surface of the original sphere. There is, however, a subtle and important difference between the approach of Fröhlich and Sack [104] and that of Oldroyd [105]. In the former, the bulk *elastic* property of an emulsion comes from the elastic nature (i.e., elastic modulus G) of the suspended spheres, whereas in the latter it comes from the presence of the interfacial tension [see Eqs. (4.22) and (4.23)] between the two *Newtonian* liquids, where the interfacial tension provides a force tending to restore the deformed drop to its equilibrium (spherical) shape. It is of interest to note, in Eqs. (4.21)–(4.23), that the Oldroyd theory predicts the bulk viscosity of an emulsion, dependent only on the volume fraction (ϕ) and the viscosities of the individual phases ($\overline{\eta}_0$ and η_0), but independent of the size of the droplet. On the other hand, the elastic parameters (λ_1 and λ_2) vary directly with the droplet radius a, and inversely with the interfacial tension between the two liquids, in addition to depending on $\overline{\eta}_0$, η_0, and ϕ.

By considering a suspension of Hookean elastic spheres dispersed in a Newtonian fluid undergoing an arbitrary time-dependent homogeneous deformation far from the particle, Goddard and Miller [106] developed a rheological equation of state for a dilute emulsion. Essentially they extended the Jeffery results [112], which considered *rigid* ellipsoidal particles, to the case of *deformable* elastic particles (initially of spherical shape), and determined the effects of shear-induced particle deformation and rotation on suspension behavior where Brownian effects are absent. Using an averaging technique (i.e., taking the volume averages of the stress and deformation rate over the entire sample), Goddard and Miller [106] derived the following explicit form of rheological equation of state,

$$\boldsymbol{\tau} + \lambda_1\, \mathscr{D}\boldsymbol{\tau}/\mathscr{D}t = 2\eta_0\big[(1 + \tfrac{5}{2}\phi)\mathbf{d} + (1 - \tfrac{5}{3}\phi)\lambda_1\, \mathscr{D}\mathbf{d}/\mathscr{D}t + \tfrac{25}{7}\phi\lambda_1(\mathbf{d}^2 - \tfrac{1}{3}\mathrm{tr}(\mathbf{d}^2)\mathbf{I})\big] \tag{4.24}$$

where

$$\lambda_1 = 3\eta_0/2G \tag{4.25}$$

in which η_0 is the viscosity of the suspending medium and G is the modulus of the Hookean elastic particles. It is of interest to note that Eq. (4.24) may be considered a special case of the Oldroyd eight-constant model. (See Eq. (2.43) in Chapter 2.)

It is of further interest to note that the nonlinear terms appearing in Eq. (4.24) have evidently resulted from the effect of the shear-induced ellipticity of the particle on the flow fields of the suspending medium, i.e., the ellipticity of the deformed droplet gives rise to *additional* stresses at the droplet surface. It should be mentioned, however, that, in the derivation of Eq. (4.24), the motion of the initially spherical particles is assumed to be one of purely *homogeneous* deformation, in which case the velocity gradient field is *homogeneous inside* the particle. Therefore, under a *homogeneous* deformation, a spherical particle will be transformed at any instant into an ellipsoid, and there exists a homogeneous stress field inside the deforming ellipsoid which matches with the fluid stresses on its surface. In other words, the effect of interfacial tension is neglected.

It can be shown that, in steady shearing flow, Eq. (4.24) predicts "shear-thinning" viscosity and normal stress effects,

$$\eta = \frac{\eta_0[(1 + \frac{5}{2}\phi) + \lambda_1^2\dot{\gamma}^2(1 - \frac{5}{3}\phi)]}{1 + \lambda_1^2\dot{\gamma}^2} \tag{4.26}$$

$$\tau_{11} - \tau_{22} = \frac{\frac{25}{3}\eta_0\lambda_1\dot{\gamma}^2\phi}{1 + \lambda_1^2\dot{\gamma}^2} \tag{4.27}$$

$$\tau_{22} - \tau_{33} = -\frac{\eta_0\lambda_1\dot{\gamma}^2(\frac{50}{21} - \frac{25}{14}\lambda_1^2\dot{\gamma}^2)\phi}{1 + \lambda_1^2\dot{\gamma}^2} \tag{4.28}$$

Note that the non-Newtonian behavior of the dispersion results from the nonlinear terms in Eq. (4.24), which come physically from the deformation of the particles in the emulsion. As λ_1 approaches zero (i.e., for rigid particles), Eq. (4.24) reduces to Einstein's equation.

Assuming that both the droplet phase and the suspending medium are incompressible Newtonian fluids and the emulsion is sufficiently dilute for interactions between droplets to be negligible, Frankel and Acrivos [73] developed a rheological equation of state in a time-dependent shearing flow field. In their theoretical development, they applied the averaging technique of Batchelor [113] to formulate a continuum theory, by considering the deformation of single droplets. Their analysis therefore yields two different sets of equations: one expression relating the stress to the rate of strain and to a measure of the local anisotropy, and a set of differential equations describing the variation of the *anisotropy* as a function of time and of the rate of strain. Their equations are difficult to handle in general, except for some special cases.

For steady (or weakly time-dependent) flows, Frankel and Acrivos [73] reduced their general equations to the following expressions:

$$\tau = 2\eta_0\left\{1 + \left(\frac{5k+2}{2k+2}\right)\phi\right\}\mathbf{d} + \left(\frac{\eta_0^2 a}{\sigma}\right)\left\{-\frac{1}{40}\left(\frac{19k+16}{k+1}\right)^2\frac{\mathscr{D}\mathbf{d}}{\mathscr{D}t} + \frac{3(19k+16)(25k^2+41k+4)}{140(k+1)^3}S_d(\mathbf{d}:\mathbf{d})\right\} \tag{4.29}$$

where

$$S_d(\mathbf{d}:\mathbf{d}) = \tfrac{1}{2}[\mathbf{d}^2 + (\mathbf{d}^T)^2 - \tfrac{2}{3}\mathrm{tr}(\mathbf{d}^2)\mathbf{I}] \tag{4.30}$$

Note that Eq. (4.29) is identical to the rheological equation of state derived earlier by Schowalter *et al.* [109], who used steady-state hydrodynamic analysis.

It is worth mentioning that the nonlinear terms in kinematic variables in Eq. (4.29) give rise to "fluid memory" effects attributable to the droplet surface dynamics (i.e., the time-dependent droplet deformation). For a steady shearing flow, it can be shown that Eq. (4.29) yields,

$$\eta = \eta_0\left\{1 + \left(\frac{5k+2}{2k+2}\right)\phi\right\} \tag{4.31}$$

$$\tau_{11} - \tau_{22} = \frac{\eta_0^2\dot{\gamma}^2 a}{40\sigma}\left(\frac{19k+16}{k+1}\right)^2\phi \tag{4.32}$$

$$\tau_{22} - \tau_{33} = -\frac{\eta_0^2\dot{\gamma}^2 a\phi}{280\sigma}\left[\frac{(19k+16)(29k^2+61k+50)}{(k+1)^3}\right] \tag{4.33}$$

It is seen that Eq. (4.29) predicts the bulk elastic property of a dilute emulsion, resulting from the presence of a finite interfacial tension σ, which always acts so as to oppose any deformation of a droplet from the spherical shape. In other words, the deformability of a droplet is responsible for the normal stress effects of an emulsion consisting of two Newtonian liquids, subjected to steady shearing flow.

According to Frankel and Acrivos [73], Eq. (4.29) can be recast into the following form,

$$\begin{aligned} \tau + \Lambda \frac{\mathscr{D}\tau}{\mathscr{D}t} &= 2\eta_0 \left\{ 1 + \left(\frac{5k+2}{2k+2} \right) \phi \right\} \left\{ \mathbf{d} + \Lambda \frac{\mathscr{D}\mathbf{d}}{\mathscr{D}t} \right\} \\ &+ \left(\frac{\eta_0^2 a}{\sigma} \right) \phi \left\{ -\frac{1}{40} \left(\frac{19k+16}{k+1} \right)^2 \frac{\mathscr{D}\mathbf{d}}{\mathscr{D}t} \right. \\ &+ \left. \frac{3(19k+16)(25k^2+41k+4)}{140(k+1)^3} S_d(\mathbf{d}:\mathbf{d}) \right\} \end{aligned} \tag{4.34}$$

where

$$\Lambda = \frac{(2k+3)(19k+16)}{40(k+1)} \left(\frac{\eta_0 a}{\sigma} \right) \tag{4.35}$$

Note that Eq. (4.34) may be considered to be a special case of the Oldroyd eight-constant model [see Eq. (2.43) in Chapter 2].

For a steady shearing flow, Eq. (4.34) yields,

$$\begin{aligned} \eta = \frac{\eta_0}{1+\Lambda^2\dot{\gamma}^2} &\left\{ 1 + \left(\frac{5k+2}{2k+2} \right) \phi \right. \\ &\left. + \Lambda^2\dot{\gamma}^2 \left[1 + \left(\frac{5k+2}{2k+2} \right) \phi - \frac{(19k+16)}{(2k+2)(2k+3)} \phi \right] \right\} \end{aligned} \tag{4.36}$$

$$\tau_{11} - \tau_{22} = \frac{\eta_0^2 \dot{\gamma}^2 a \phi}{40\sigma(1+\Lambda^2\dot{\gamma}^2)} \left(\frac{19k+16}{k+1} \right)^2 \tag{4.37}$$

$$\begin{aligned} \tau_{22} - \tau_{33} = -\frac{\eta_0^2 \dot{\gamma}^2 a \phi}{40\sigma(1+\Lambda^2\dot{\gamma}^2)} &\left\{ \frac{1}{2} \left(\frac{19k+16}{k+1} \right)^2 \right. \\ &\left. - \left[\frac{3(19k+16)(25k^2+41k+4)}{14(k+1)^3} \right] (1+\Lambda^2\dot{\gamma}^2) \right\} \end{aligned} \tag{4.38}$$

It is seen that Eq. (4.34) predicts non-Newtonian viscosity and normal stress effects in steady shearing flow. Note, however, that Eqs. (4.32) and (4.33), and Eqs. (4.37) and (4.38), cease to apply when the interfacial tension becomes very small.

Barthès-Biesel and Acrivos [110] extended the Frankel–Acrivos analysis, which is a first-order approximation, by including the second-order terms of the regular perturbation expansion, and obtained a rheological equation of state for a dilute emulsion in time-dependent flow. Their equations are very lengthy and space limitations do not permit us to reproduce their results here.

The analyses presented above deal with dilute emulsions. Choi and Schowalter [41], however, have considered moderately concentrated emulsions of Newtonian droplets dispersed in a Newtonian medium, and derived a rheological equation of state. Following Simha's approach of the cell model [114], they have taken into account interactions between neighboring droplets and considered the effects of concentration, and of deformation, on the *bulk* rheological properties of moderately concentrated emulsions of deformable particles. Otherwise, the mathematical technique employed by them is very close to that presented earlier by Cox [38] and Frankel and Acrivos [73].

After some approximations Choi and Schowalter [41] derived a rheological equation of state which has the following form,

$$(1 + h_1 \mathscr{D}/\mathscr{D}t)\boldsymbol{\tau} = 2\eta_a(1 + h_2 \mathscr{D}/\mathscr{D}t)\mathbf{d} \tag{4.39}$$

where

$$\eta_a = \eta_0\left[1 + \left(\frac{5k+2}{2k+2}\right)\phi + \frac{5(5k+2)^2}{8(k+1)^2}\phi^2\right] \tag{4.40}$$

$$h_1 = \frac{(19k+16)(2k+3)}{40(k+1)}\left(\frac{\eta_0 a}{\sigma}\right)\left[1 + \frac{5(19k+16)}{4(k+1)(2k+3)}\phi\right] \tag{4.41}$$

$$h_2 = \frac{(19k+16)(2k+3)}{40(k+1)}\left(\frac{\eta_0 a}{\sigma}\right)\left[1 + \frac{3(19k+16)}{4(k+1)(2k+3)}\phi\right] \tag{4.42}$$

Note that the forms of Eqs. (4.40)–(4.42) are very similar to those of Eqs. (4.21)–(4.23). There is, however, an important difference between the Oldroyd analysis and the Choi–Schowalter analysis; namely, the former assumes a spherical droplet shape, whereas the latter takes account of the effect of droplet deformation and interactions between neighboring droplets.

For a steady shearing flow, Eq. (4.39) predicts both the non-Newtonian and normal stress effects, yielding

$$\eta = \eta_a(1 + h_1 h_2 \dot{\gamma}^2)/(1 + h_1^2\dot{\gamma}^2) \tag{4.43}$$

$$\tau_{11} - \tau_{22} = 2\eta_a(h_1 - h_2)\dot{\gamma}^2/(1 + h_1^2\dot{\gamma}^2) \tag{4.44}$$

$$\tau_{22} - \tau_{33} = -\eta_a(h_1 - h_2)\dot{\gamma}^2/(1 + h_1^2\dot{\gamma}^2) \tag{4.45}$$

What has been shown above is that the phenomenological theory predicts the existence of shear-thinning behavior and normal stress effects in an emulsion consisting of two Newtonian liquids. Such a prediction has become possible by the consideration of the deformability of droplets and the presence of a finite value of interfacial tension. We are now assured that the experimental observations of non-Newtonian and normal stress effects in Newtonian emulsions (see Figs. 4.18, 4.19, 4.23, and 4.24) are real physical phenomena, not experimental artifacts.

Figures 4.46 and 4.47 give plots of viscosity versus shear rate and plots of first normal stress difference versus shear rate, respectively, predicted by the Choi–Schowalter model, Eq. (4.39), and the Frankel–Acrivos model, Eq. (4.34). Note that in preparing Figs. 4.46 and 4.47, the following numerical values of the physical parameters were used: $\eta_0 = 7.6$ poises (the viscosity of glycerine), $k = 0.552$, $\sigma = 10^{-2}$ N/m (10 dyn/cm), $a = 5$ μm. These numerical values were to approximately simulate the Indopol L100/glycerine emulsion system investigated experimentally, as given in Figs. 4.18 and 4.23.

For comparison purposes, the experimental measurements of bulk viscosity and first normal stress difference for the Indopol L100/glycerine emulsion system are given in Figs. 4.48 and 4.49. It should be remembered that, as shown in Fig. 4.22, the emulsions used has nonuniform droplet sizes

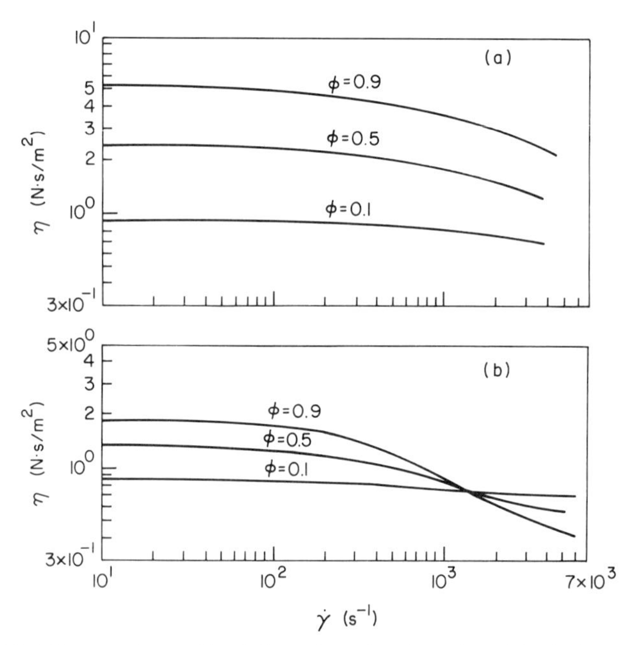

Fig. 4.46 Theoretically predicted shear-dependent behavior of the bulk viscosity of the concentrated emulsions consisting of two Newtonian liquids [52]: (a) Choi–Schowalter model; (b) Frankel–Acrivos model. From C. D. Han and R. G. King, *J. Rheol.* **24**, 213, Copyright © 1980. Reprinted by permission of John Wiley & Sons, Inc.

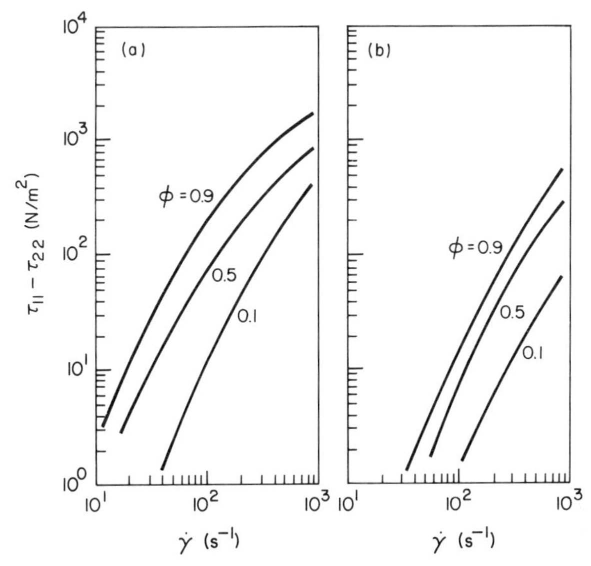

Fig. 4.47 Theoretically predicted first normal stress difference of the concentrated emulsions consisting of two Newtonian liquids [52]: (a) Choi–Schowalter model; (b) Frankel–Acrivos model. From C. D. Han and R. G. King, *J. Rheol.* **24**, 213. Copyright © 1980. Reprinted by permission of John Wiley & Sons, Inc.

and complex morphological states at certain composition ratios. Nevertheless, it is encouraging to see that the phenomenological theories available predict, at least qualitatively, the rather complex rheological behavior of concentrated emulsions.

It is worth mentioning that Eqs. (4.36) and (4.43) indicate that the bulk

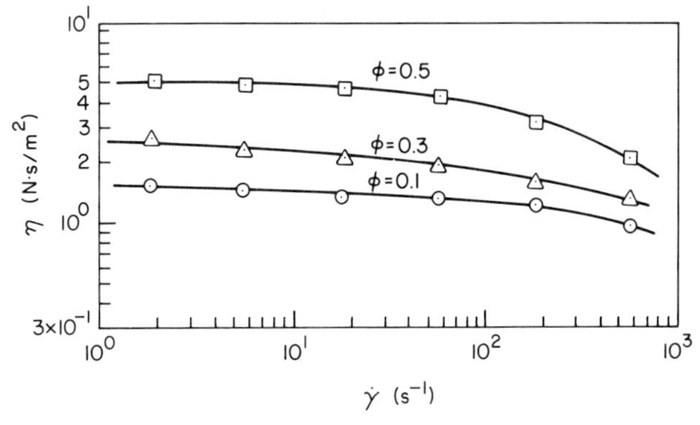

Fig. 4.48 Experimentally observed shear-dependent behavior of the bulk viscosity of the Indopol L100/glycerine emulsions [52]. From C. D. Han and R. G. King, *J. Rheol.* **24**, 213. Copyright © 1980. Reprinted by permission of John Wiley & Sons, Inc.

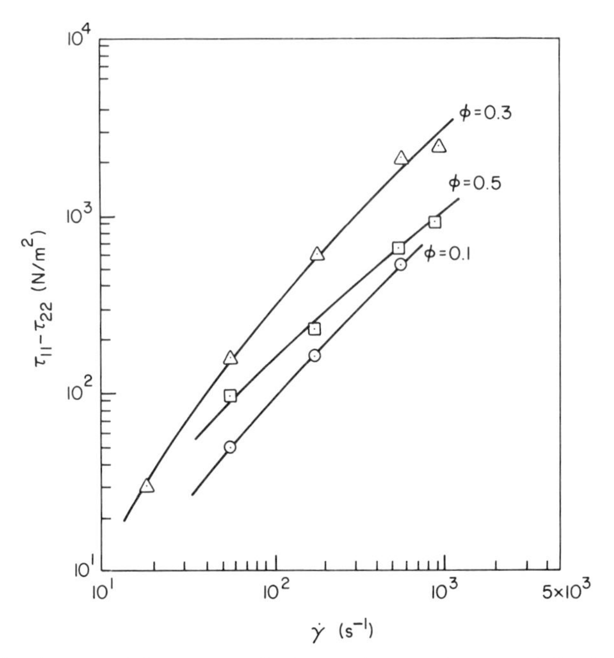

Fig. 4.49 Experimentally observed first normal stress difference of the Indopol L100/glycerine emulsions [52]. From C. D. Han and R. G. King, *J. Rheol.* **24**, 213. Copyright © 1980. Reprinted by permission of John Wiley & Sons, Inc.

viscosity of an emulsion decreases as the droplet size a, the viscosity of the suspending medium η_0, the shear rate $\dot{\gamma}$, and the interfacial tension σ are increased. Note that an increase in shear rate will bring an increase in droplet deformation, and that large droplets can be deformed more readily than small droplets at otherwise identical flow conditions. It can be concluded therefore that the shear-dependent behavior of the emulsions experimentally observed may be attributable, at least in part, to the deformability of the droplets.

The extensional flow of droplets in an emulsion is of practical interest in those polymer processing operations which make use of liquid additives. For instance, the fiber industry uses liquid additives (e.g., antistatic agents), which are elongated in the fiber-forming material under stretching. The elongated droplets, when solidified, sometimes reinforce the strength of the finished fiber. Some specific commercial applications of this concept will be discussed later in this chapter.

Using the various constitutive equations suggested above for dilute emulsions, we can derive some theoretical expressions for *bulk* elongational viscosity in a uniaxial stretching. They are:

For the Goddard–Miller model [Eq. (4.24)]:

$$\eta_{\mathrm{E}} = 3\eta_0\left[1 + \tfrac{5}{2}\phi + \tfrac{25}{14}\lambda_1\dot{\gamma}_{\mathrm{E}}\phi\right] \tag{4.46}$$

For the Frankel–Acrivos model [Eq. (4.34)]:

$$\eta_{\mathrm{E}} = 3\eta_0\left\{1 + \left(\frac{5k+2}{2k+2}\right)\phi + \frac{\eta_0 a\dot{\gamma}_{\mathrm{E}}}{4\sigma}\left[\frac{3(19k+16)(25k^2+41k+4)}{140(k+1)^3}\right]\phi\right\} \tag{4.47}$$

For the Choi–Schowalter model [Eq. (4.39)]:

$$\eta_{\mathrm{E}} = 3\eta_0\left\{1 + \left(\frac{5k+2}{2k+2}\right)\phi + \left[\frac{5(5k+2)^2}{8(k+1)^2}\right]\phi^2\right\} \tag{4.48}$$

It is seen that the *bulk* (effective) elongational viscosity increases with emulsion concentration, and also with extension rate, for the first two models, but not for the Choi–Schowalter model. Little experimental study is available to test these theoretical predictions.

4.5 PROCESSING OF HETEROGENEOUS POLYMERIC SYSTEMS

We have discussed above how the bulk rheological properties of heterogeneous polymeric systems (and concentrated emulsions) are affected by their morphology, which in turn is influenced by flow conditions. It may be readily surmised that the mechanical/physical properties of the final product of heterogeneous polymeric systems, similar to the situation with particulate-filled polymeric systems discussed in Chapter 3, are also influenced by their morphology. Therefore a better understanding of processing–morphology–property relationships in heterogeneous polymeric systems is very important.

In processing polymer blends, for instance, the method of blending, and the deformation and thermal histories control the morphology and, consequently, the mechanical/physical properties of the finished product. In the past, a number of researchers [115–126] have formulated polymer blends and have identified the variables necessary for correlating the morphology to the mechanical/physical properties of the blends in the solid state.

Also, some efforts were spent on developing theories [127–130] in order to predict the mechanical (static or dynamic) properties of polymer–polymer composites, and experimental studies [131–140] were carried out to test the theories.

We shall now discuss the fiber spinning and injection molding of heterogeneous polymeric systems in order to illustrate how the mechanical properties of the final product are influenced by the morphology, which in turn is influenced by processing conditions.

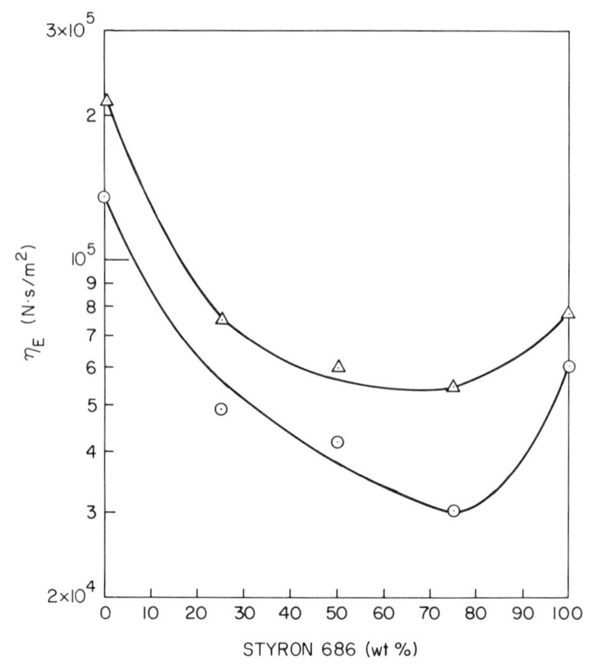

Fig. 4.50 Apparent elongational viscosity versus blend composition for blends of HDPE and STYRON 686, at various elongation rates (s^{-1}) [141]: (△) 0.5; (⊙) 1.0. From C. D. Han and Y. W. Kim, *J. Appl. Polym. Sci.* **18**, 2589. Copyright © 1974. Reprinted by permission of John Wiley & Sons, Inc.

4.5.1 Melt Spinning of Two-Phase Polymer Blends

The fiber industry makes use of blends of two or more polymers, or one polymer with an additive or additives, in order to obtain certain desired mechanical and/or physical properties of the finished fiber. When two fiber-forming materials are spun together, the fundamental question arises as to the role that each plays in the process of fiber formation. Undoubtedly, there must be some interaction between the components in the process. The problem becomes complicated when the two materials form separate phases.

Using polymer blends of polystyrene (STYRON 686 and STYRON 678) and high-density polyethylene (HDPE), Han and Kim [141] have made an experimental study of melt spinning in order to investigate the question of spinnability and fiber morphology when two incompatible polymers are spun together.

Figure 4.50 gives plots of *apparent* elongational viscosity† versus blending ratio for blends of STYRON 686 and HDPE, and Fig. 4.51 gives similar

† Apparent elongation viscosity as determined with a fiber spinning apparatus.

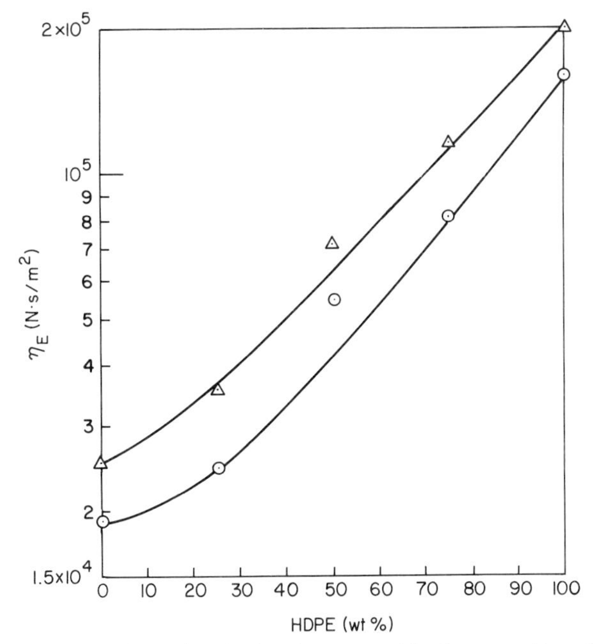

Fig. 4.51 Apparent elongational viscosity versus blend composition for blends of STYRON 678 and HDPE, at various elongation rates (s^{-1}) [141]: (△) 0.5; (⊙) 1.0. From C. D. Han and Y. W. Kim, *J. Appl. Polym. Sci.* **18**, 2589. Copyright © 1974. Reprinted by permission of John Wiley & Sons, Inc.

plots for blends of STYRON 678 and HDPE. It is seen in Fig. 4.50 that the *apparent* elongational viscosity goes through a minimum at a blending ratio of STYRON 686/HDPE = 75/25. On the other hand, Fig. 4.51 shows that, as the amount of HDPE is increased in the blends of STYRON 678/HDPE, the *apparent* elongational viscosity increases correspondingly, and does not go through a minimum. In other words, the two blend systems respond quite differently in the fiber-forming operation. To explain the observed elongational flow behavior in the melt spinning of the two-phase polymer systems investigated, one needs information on the morphological state of blends subjected to a uniaxial stretching.

Figure 4.52 gives plots of shear viscosity versus shear rate for the three homopolymers, STYRON 678, STYRON 686, and HDPE. It is seen that the HDPE is more viscous than the STYRON 678, and that the STYRON 686 is more viscous than the HDPE. Since the shear rate in the spinnerette hole was in the neighborhood of 600 sec^{-1} [141], we would expect that, in the blends of HDPE/STYRON 678, the HDPE forms the discrete phase and the STYRON 678 forms the continuous phase, and that in the blends of STYRON 686/HDPE, the STYRON 686 forms the discrete phase and the HDPE forms the continuous phase. This expectation is based on a general

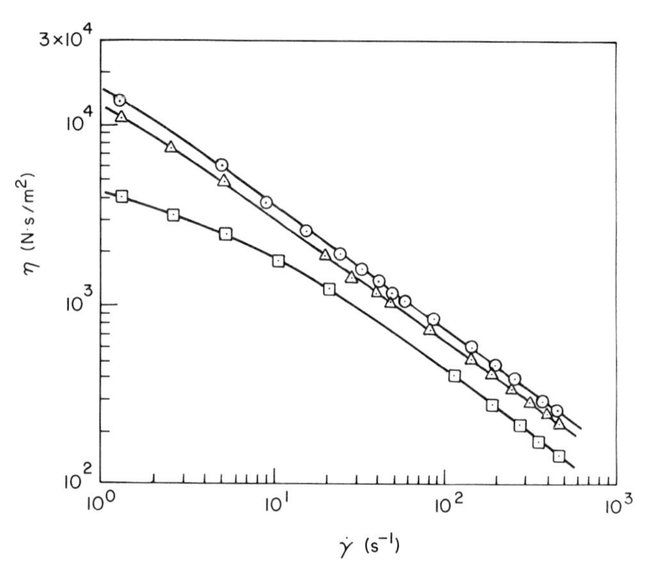

Fig. 4.52 Viscosity versus shear rate for three homopolymers at 200°C: (△) HDPE; (⊙) STYRON 686; (⊡) STYRON 678.

tendency that the more viscous component of the two forms the discrete phase and the less viscous component forms the continuous phase.

In order to explain the observed elongational flow behavior of the two blend systems shown in Figs. 4.50 and 4.51, let us examine the elongational flow behavior of the individual components, HDPE, STYRON 678, and STYRON 686, as shown in Fig. 4.53. It is seen that the *apparent* elongational viscosity of STYRON 686 is lower than that of HDPE, implying that in the blends of STYRON 686/HDPE, at a given tensile stress the droplets (STYRON 686) would be deformed more easily than the continuous phase (HDPE). On the other hand, it is seen in Fig. 4.53 that the *apparent* elongational viscosity of HDPE is higher than that of STYRON 678, implying that in the blends of HDPE/STYRON 678, at a given tensile stress the droplet phase (HDPE) would not deform as readily as the continuous phase (STYRON 678). This observation is described schematically in Fig. 4.54. In other words, the blend system of STYRON 686/HDPE shows a minimum in *apparent* elongational viscosity, as given in Fig. 4.50, due to the presence of elongated droplets of STYRON 686 suspended in the molten threadline. On the other hand, the blend system of HDPE/STYRON 678 shows a monotonically increasing trend of *apparent* elongational viscosity as the amount of HDPE is increased, as given in Fig. 4.51, due to the presence of barely deformed droplets of HDPE suspended in the molten threadline.

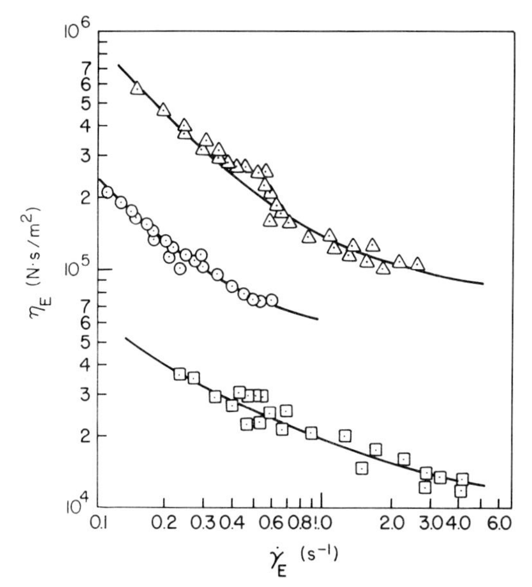

Fig. 4.53 Apparent elongational viscosity versus elongation rate for three homopolymers 200°C [141]: (△) high-density polyethylene (HDPE); (⊙) polystyrene (STYRON 686); (⊡) polystyrene (STYRON 678) From C. D. Han and Y. W. Kim, *J. Appl. Polym. Sci.* **18**, 2589. Copyright © 1974. Reprinted by permission of John Wiley & Sons, Inc.

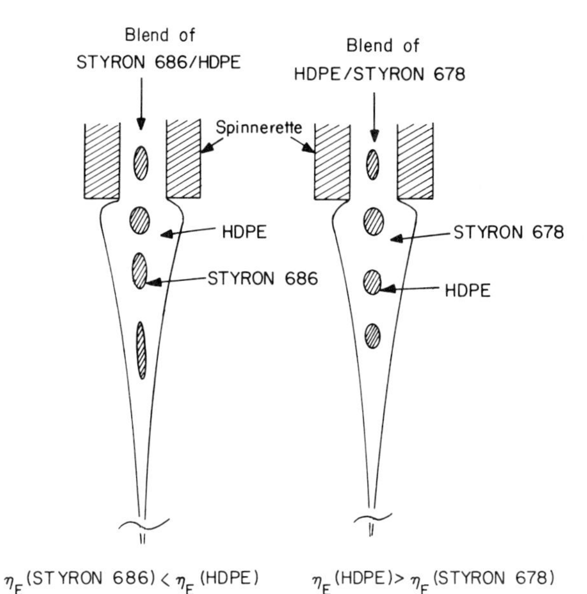

Fig. 4.54 Schematic illustrating the state of dispersion and droplet deformation in the two blend systems [141]: STYRON 686/HDPE and HDPE/STYRON 678 blends. From C. D. Han and Y. W. Kim, *J. Appl. Polym. Sci.* **18**, 2589. Copyright © 1974. Reprinted by permission of John Wiley & Sons, Inc.

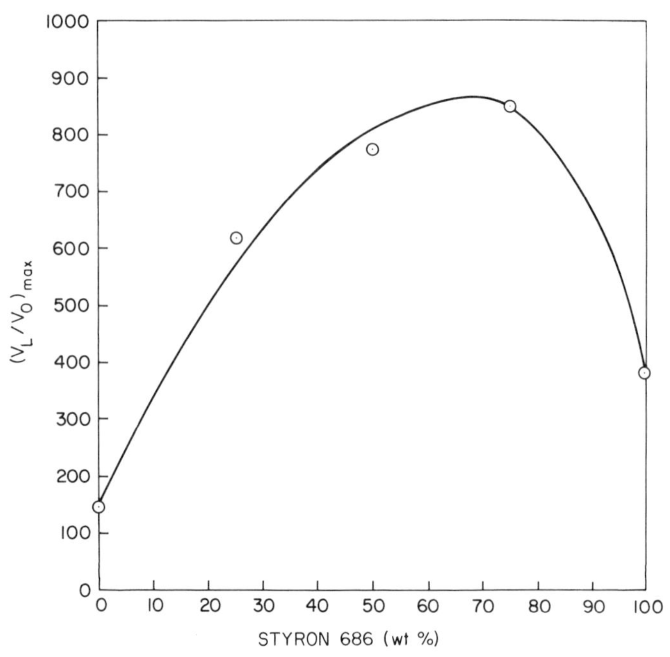

Fig. 4.55 Maximum draw-down ratio versus blend composition for blends of HDPE and STYRON 686 [141]. From C. D. Han and Y. W. Kim, *J. Appl. Polym. Sci.* **18**, 2589. Copyright © 1974. Reprinted by permission of John Wiley & Sons, Inc.

Let us now use the maximum draw-down ratio $(V_L/V_0)_{max}$ for comparing the spinnability of the two blend systems. Here V_L denotes the take-up velocity of the threadline and V_0 the linear velocity of the melt in the spinnerette. It can be said that the larger the value of $(V_L/V_0)_{max}$, the more spinnable a material is.

Figure 4.55 gives plots of $(V_L/V_0)_{max}$ versus blending ratio for STYRON 686/HDPE blend systems, and Fig. 4.56 for HDPE/STYRON 678 blend systems. A close look at the elongational behavior given in Figs. 4.50 and 4.51, and the spinnability characteristics given in Figs. 4.55 and 4.56 would seem to indicate that the material having the lower elongational viscosity gives rise to a better spinnability than the material having the higher elongational viscosity. In other words, there is a clear indication that elongational viscosity may be correlated to spinnability. However, further study is needed to establish a firm correlation between elongational viscosity and spinnability.

The tensile properties of the two-phase fibers obtained at various draw-down ratios are given in Figs. 4.57 and 4.58 for blends of HDPE with STYRON 686, and in Figs. 4.59 and 4.60 for blends of HDPE with STYRON

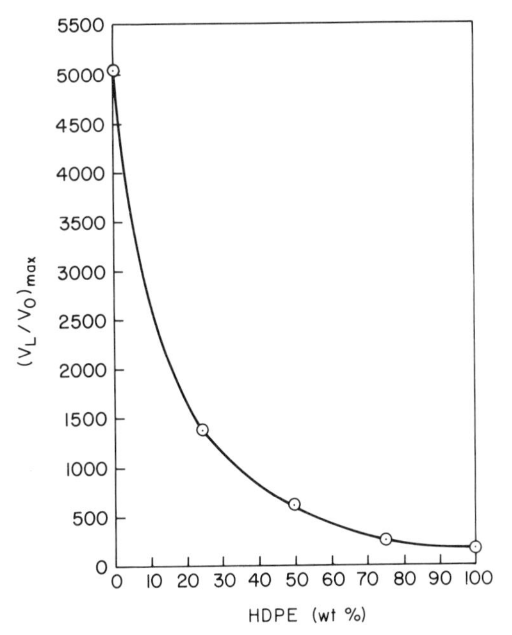

Fig. 4.56 Maximum draw-down ratio versus blend composition for blends of STYRON 678 and HDPE [141]. From C. D. Han and Y. W. Kim, *J. Appl. Polym. Sci.* **18**, 2589. Copyright © 1974. Reprinted by permission of John Wiley & Sons, Inc.

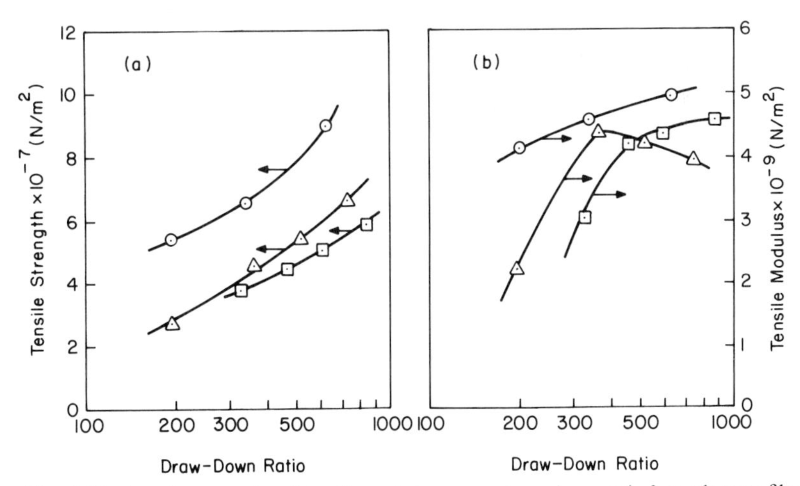

Fig. 4.57 Tensile strength and tensile modulus versus draw-down ratio for melt-spun fibers of STYRON 686/HDPE blends: (⊙) STYRON 686/HDPE = 25/75 (by wt.); (△) STYRON 686/HDPE = 50/50; (⊡) STYRON 686/HDPE = 75/25.

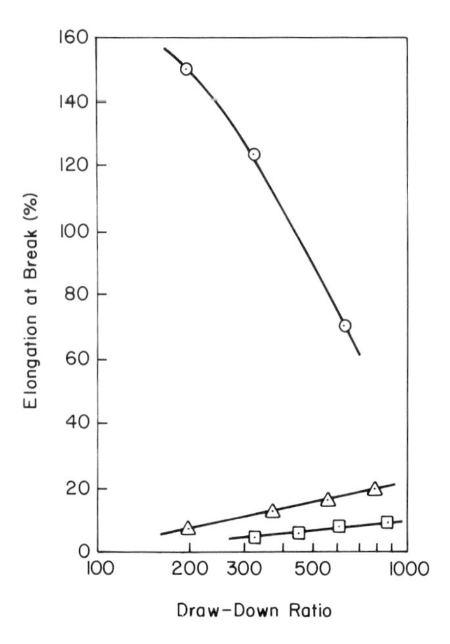

Fig. 4.58 Elongation at break versus draw-down ratio for melt-spun fibers of STYRON 686/HDPE blends. Symbols are the same as in Fig. 4.57.

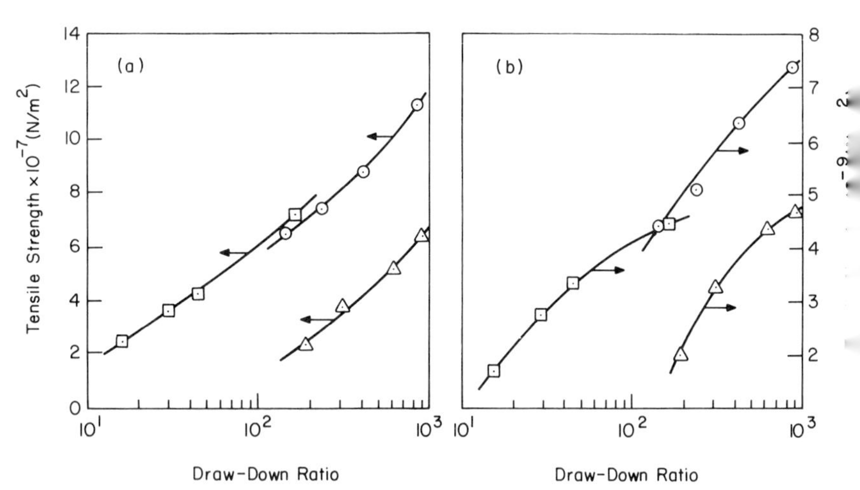

Fig. 4.59 Tensile strength and tensile modulus versus draw-down ratio for melt-spun fibers of HDPE/STYRON 678 blends: (⊙) HDPE/STYRON 678 = 25/75 (by wt.); (△) HDPE/STYRON 678 = 50/50; (⊡) HDPE/STYRON 678 = 75/25.

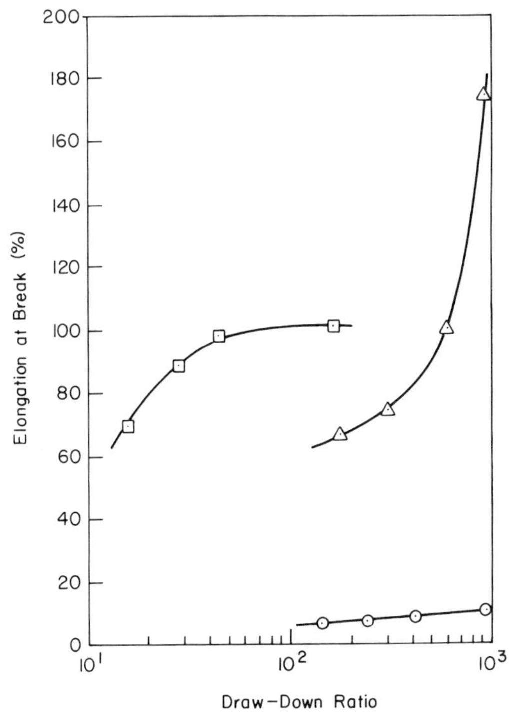

Fig. 4.60 Elongation at break versus draw-down ratio for melt-spun fibers of HDPE/STYRON 678 blends. Symbols are the same as in Fig. 4.59.

678. In order to facilitate our discussion of the rather complex tensile behavior of the two-phase blend fibers, given in Figs. 4.57–4.60, the tensile properties of two HDPE homopolymers are given in Figs. 4.61 and 4.62.

Note that many partially crystalline polymers (e.g., HDPE) exhibit the process of cold drawing at room temperature, and deformations of several hundred percent or more are possible without fracture, and that the elongation at break generally follows behavior opposite to that of the tensile modulus (see Figs. 4.61 and 4.62).

It is seen in Figs. 4.57–4.60 that except the blend of STYRON 686/HDPE = 25/75, the elongation at break for the two-blend system does not follow behavior opposite to that of the tensile modulus. This may be explained by examining the morphological state of each blend. Referring to Fig. 4.58, in the blend of STYRON 686/HDPE = 25/75, the decrease in elongation at break, as the draw-down ratio is increased, may be attributable primarily to the continuous HDPE phase, which might have been stretched with relatively little interference from the discrete PS phase during

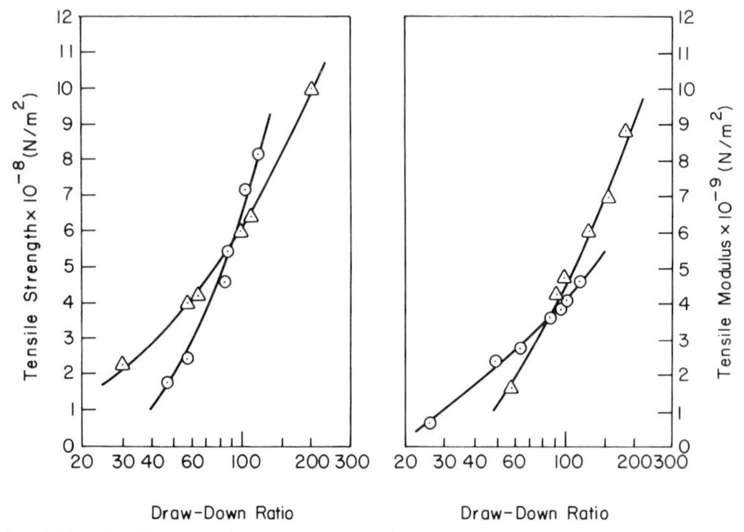

Fig. 4.61 Tensile strength and tensile modulus versus draw-down ratio for melt-spun fibers of two high-density polyethylenes having different molecular weight distributions: (⊙) $\bar{M}_n = 0.48 \times 10^4$, $\bar{M}_w = 2.26 \times 10^5$; (△) $\bar{M}_n = 1.33 \times 10^4$, $\bar{M}_w = 1.75 \times 10^5$.

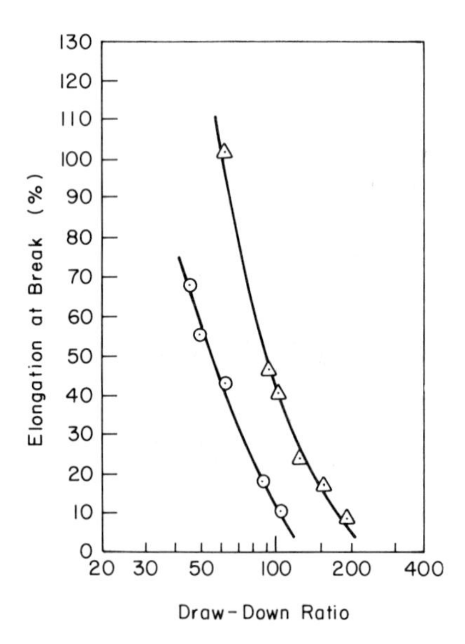

Fig. 4.62 Elongation at break versus draw-down ratio for melt-spun fibers of two high-density polyethylenes. Symbols are the same as in Fig. 4.61.

melt spinning. However, as the amount of the PS phase is increased (i.e., in the blends of STYRON 686/HDPE = 50/50 and 75/25), the deformation of the HDPE phase during melt spinning might have been difficult due to the interference of the PS phase. Note in Fig. 4.58 that the elongations at break of the two blends (STYRON 686/HDPE = 50/50 and 75/25) are rather small. This may be attributable to the large amount of the brittle PS phase present in these blends. The reason why the elongations at break of these blends show an increasing trend may be due to the fact that the deformation of the continuous HDPE phase might have taken place at room temperature during tensile testing.

Referring to Fig. 4.60, in the blend of HDPE/STYRON 678 = 25/75, the exceedingly small values of elongation at break may be due to the large amount of the PS phase that forms the continuous phase (see Fig. 4.54), and an increasing trend of elongation at break may be attributable to the deformation of the discrete HDPE phase during tensile testing at room temperature. Note in Fig. 4.60 that the HDPE/STYRON 678 = 50/50 blend shows a rapidly increasing trend of elongation as the draw-down ratio is increased. Recall that the HDPE phase is the discrete phase, suspended in the continuous STYRON 678 phase, and that the HDPE phase exhibits the process of cold drawing at room temperature. Therefore, at some optimal blending ratios, the HDPE phase may be elongated very much with relatively little interference from the STYRON 678 phase. However, as the amount of the HDPE phase is increased, the domain size of the HDPE phase may become bigger, possibly bringing about some interactions among different domains. This would make the elongation of the HDPE phase somewhat difficult when subjected to cold drawing at room temperature. This might have been the case with the elongational behavior of the blend of HDPE/STYRON 678 = 75/25, whose elongation shows, initially, an increasing trend, and then a leveling off as the draw-down ratio is increased, as given in Fig. 4.60.

In the STYRON 686/HDPE blend system (see Fig. 4.57), both tensile strength and modulus decrease as the amount of the PS phase is increased. In view of the fact that the STYRON 686 phase forms the discrete phase (see Fig. 4.54) and it is a brittle material, little reinforcement is to be expected from an increase in the amount of the STYRON 686 in this blend system.

On the other hand, in the HDPE/STYRON 678 blend system (see Fig. 4.59), the HDPE forms a discrete phase that is cold drawable at room temperature (see Fig. 4.54). Therefore the HDPE phase can act as reinforcement, and hence can improve the strength properties of the blend. The effectiveness of reinforcement would depend on the morphological state of the blend. For instance, a greatly elongated discrete phase resembling glass fibers would be expected to considerably improve the strength properties of two-phase polymers. It should be remembered that, as discussed in the early part of this chapter, the deformability of the discrete phase in a two-phase liquid system depends, among many other factors, on both the rheological properties of the individual phases and the processing conditions (e.g., shear stress, temperature). It is clear, then, that a good understanding of the processing–structure–property relationships in heterogeneous polymeric systems is very important for obtaining improvements in the mechanical/physical properties of their final products.

4.5.2 Injection Molding of Two-Phase Polymer Blends

In recent years there has been a growing interest in the injection molding of blends of two or more thermoplastic polymers, or a polymer containing an additive (or additives) as a processing aid [142, 143]. Due to the heterogeneous nature of polymer blends in general, the ultimate mechanical/physical properties of molded specimens depend very much on the state of dispersion of one component in the other.

Figure 4.63 gives plots of tensile strength versus blending ratio and Fig. 4.64 gives plots of percent elongation at break versus blending ratio, of

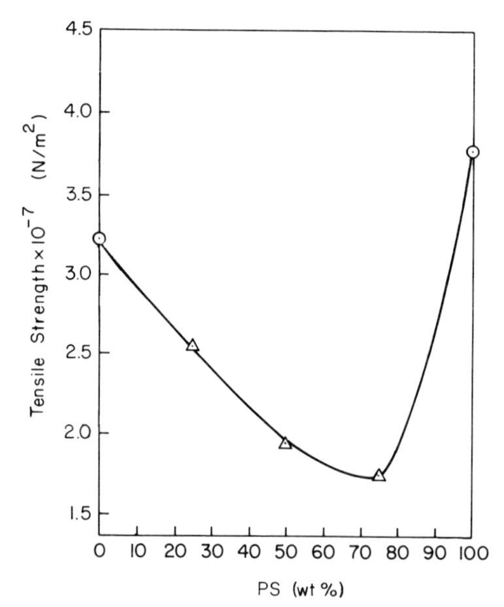

Fig. 4.63 Tensile strength versus blending ratio for injection-molded specimens of blends of polypropylene (PP) and polystyrene (PS).

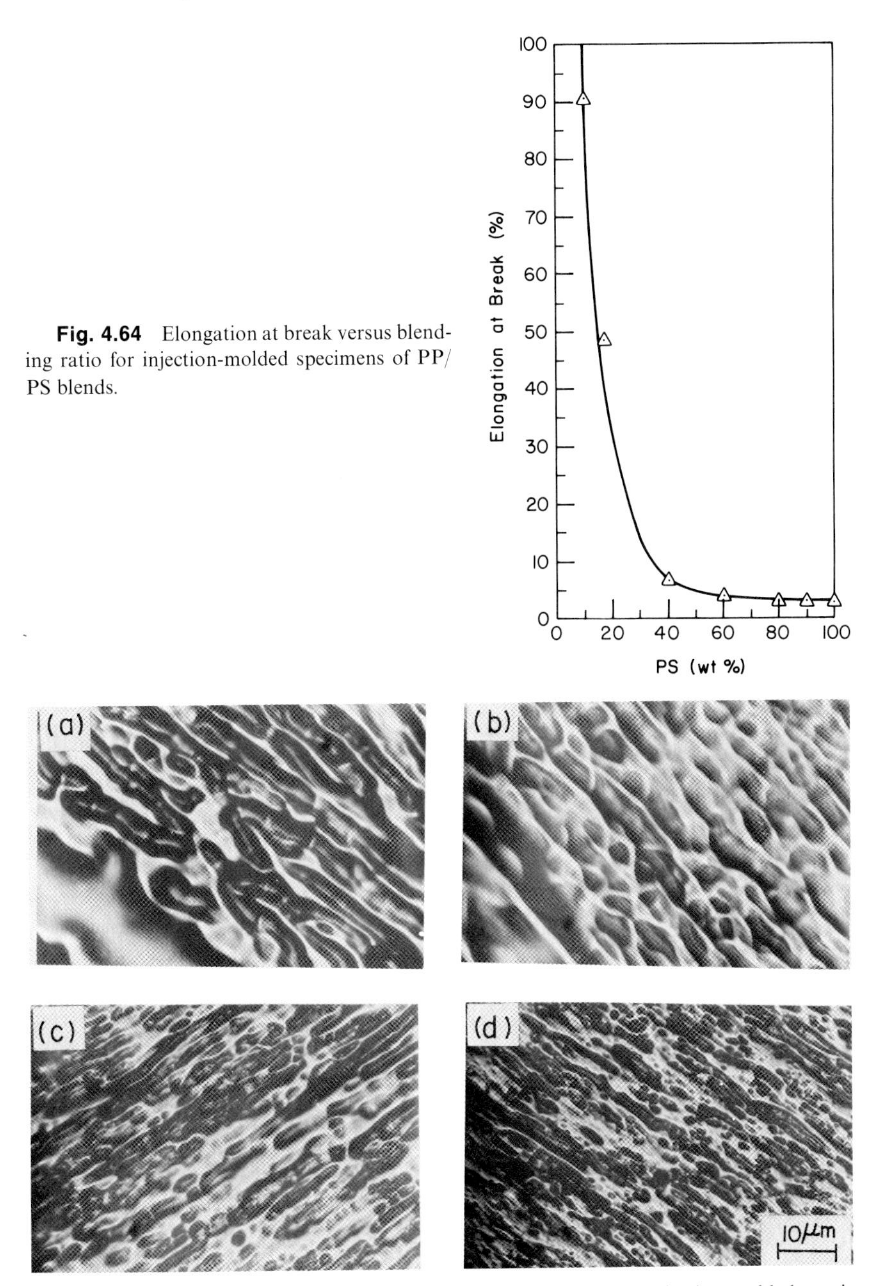

Fig. 4.64 Elongation at break versus blending ratio for injection-molded specimens of PP/PS blends.

Fig. 4.65 Photomicrographs of the longitudinal sections of the injection-molded specimens of PP/PS blends at different blending ratios (by wt.): (a) PP/PS = 10/90; (b) PP/PS = 40/60; (c) PP/PS = 60/40; (d) PP/PS = 80/20.

injection-molded specimens of polystyrene (PS)/polypropylene (PP) blends. It is seen that the tensile strength goes through a minimum at a blending ratio of 75 wt.% PS and 25 wt.% PP. This may be explained by the photomicrographs given in Fig. 4.65, showing that the blends have a microstructure which contains a discrete phase (PS) dispersed in a continuous phase (PP). Note that PS and PP are recognized as incompatible polymers and, therefore, when a specimen is under tension, the fracture path may preferentially follow the weak interface between the polymer phases, or the fracture may initiate at the interface.

The tensile modulus, tensile strength, and elongation at break of injection-molded specimens of blends of high-impact polystyrene (HIPS) and polypropylene (PP) are plotted against blending ratio in Figs. 4.66–4.68. Note that the HIPS is itself a two-phase polymer, containing deformable rubbery particles dispersed in the PS (see Fig. 1.4 in Chapter 1 for photomicrographs of typical HIPS's). It is seen in Figs. 4.66–4.68 that the blend containing 10% HIPS behaves most unusually in elongation, undergoing a 960% elongation at break (more than three times the value for PP). Furthermore, this particular blend gives rise to a hardness (modulus) superior to that of both HIPS and PP, without a sacrifice of tensile strength. Figure 4.69 gives photomicrographs of HIPS/PP blends of various blending ratios. It is seen that the blend containing 10% HIPS has a state of dispersion, which is more uniform than other blends. Thus the discrete phase (HIPS) appears to reinforce the continuous phase (PP).

Asar *et al.* [143] report that the impact properties of an injection-molded blend of polypropylene (PP) and ethylene–propylene–diene terpolymer (EPDM) depend on blend composition, processing variables, and testing

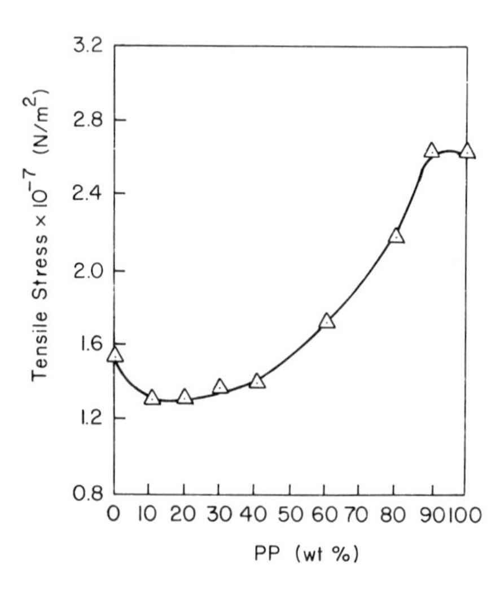

Fig. 4.66 Tensile strength versus blending ratio for injection-molded specimens of blends of high-impact polystyrene (HIPS) and polypropylene (PP).

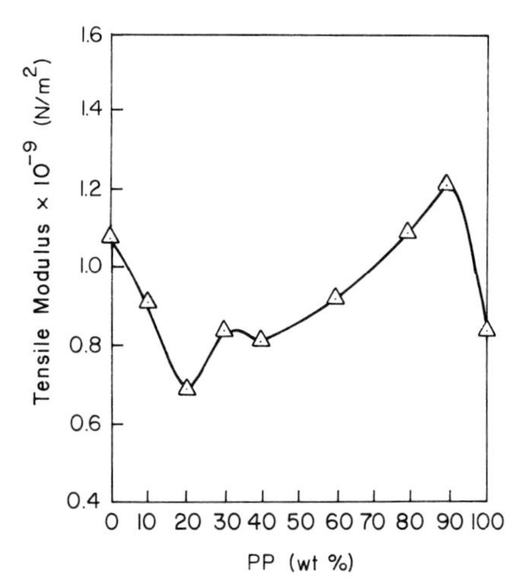

Fig. 4.67 Tensile modulus versus blending ratio for injection-molded specimens of HIPS/PP blends.

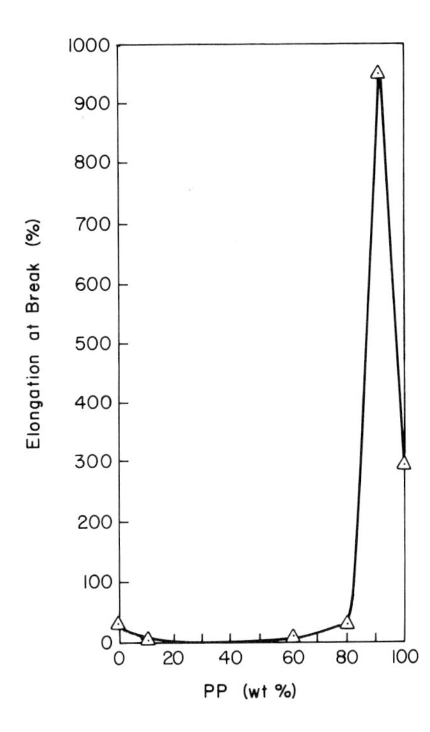

Fig. 4.68 Percent elongation at break versus blending ratio for injection-molded specimens of HIPS/PP blends.

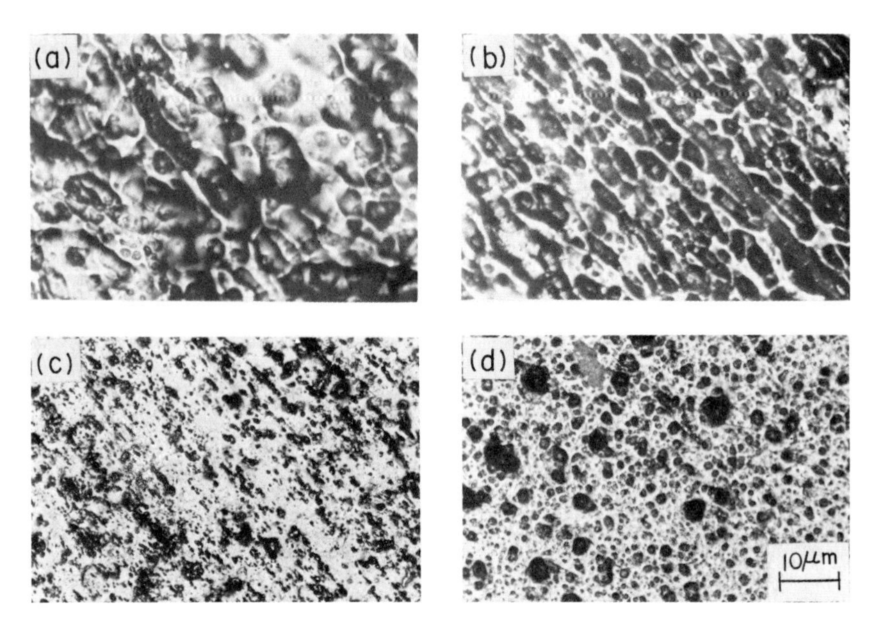

Fig. 4.69 Photomicrographs of the longitudinal sections of injection-molded specimens of HIPS/PP blends at different blending ratio (by wt.): (a) HIPS/PP = 80/20; (b) HIPS/PP = 60/40; (c) HIPS/PP = 20/80; (d) HIPS/PP = 10/90.

conditions. They observed that local variations in EPDM concentration and domain sizes can result in a twofold difference in the total energy absorbed during impact. This clearly indicates that the processing conditions employed (e.g., method of mixing, injection molding conditions), which in turn control the morphology (i.e., the distribution of the EPDM phase and domain sizes), have a profound influence on the mechanical properties of polymer blends.

The poor mechanical properties of incompatible polymer blends are believed to be due to poor interfacial bonding between the components involved.[†] It should therefore be possible to improve the mechanical properties by improving the interfacial bonding (or interfacial adhesion) by using an emulsifier or additive, which may strengthen the interfacial bonding by some mechanism. Some attempts have indeed been made with moderate success [144–148].

One important class of such additives is the block and graft copolymers, with blocks or segments having the same chemical compositions as those of the two polymers to be blended [144, 145]. An A–B block or graft copolymer will tend to accumulate at, orient, and bridge the interface between polymers

[†] This is the case particularly for blends of a crystalline polymer with an amorphous glassy polymer, whereas for two amorphous polymers this may not necessarily be the case.

A and B, and thus reduce the interfacial tension and improve the compatibility and adhesion. Block copolymers are usually believed to be more effective than graft copolymers, because the former can orient more readily than the latter. Wu [149] has pointed out that the dispersion, morphology, and adhesion of the component phases are greatly affected by the internal energies, which thereby play an important role in determining the mechanical properties of a heterogeneous polymer blend.

Heikens and co-workers [147, 148] report that a graft copolymer of polystyrene (PS) and low-density polyethylene (LDPE), when added to mechanical blends of PS and LDPE, improved certain mechanical properties (e.g., yield strength and tensile strength), as shown in Fig. 4.70. They attribute the improvement observed to the role of adhesive that the graft copolymer plays, strengthening the interfacial bonding between the PS and the LDPE and thus decreasing the stress concentrations around the dispersed polymer particles at yield. Figure 4.71 gives electron scanning micrographs of fracture surfaces of PS/LDPE blends with and without a graft copolymer. It is clearly seen that the fracture surface of the ordinary PS/LDPE has no indication of adhesion between the two polymers, whereas the fracture surface of the blend with a graft copolymer shows that particles (discrete phase) are locally connected to the surrounding medium (continuous phase). This supports the contention that, when graft copolymer is added to the LDPE/PS blend, it adheres to both phases, reinforcing the interface.

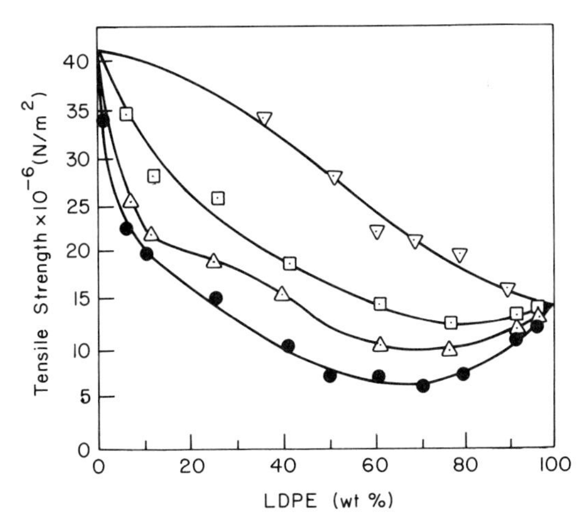

Fig. 4.70 Effect of graft copolymer on the tensile strength of blends of polystyrene (PS) and low-density polyethylene (LDPE) at various amounts of graft copolymer added in the dispersed phase (wt. %) [147]: (●) 0.0; (△) 5; (□) 30; (▽) 100. Reproduced from W. M. Barentsen and D. Heikens, *Polymer* **14**, 579 (1973), by permission of the publishers, IPC Business Press Ltd. ©

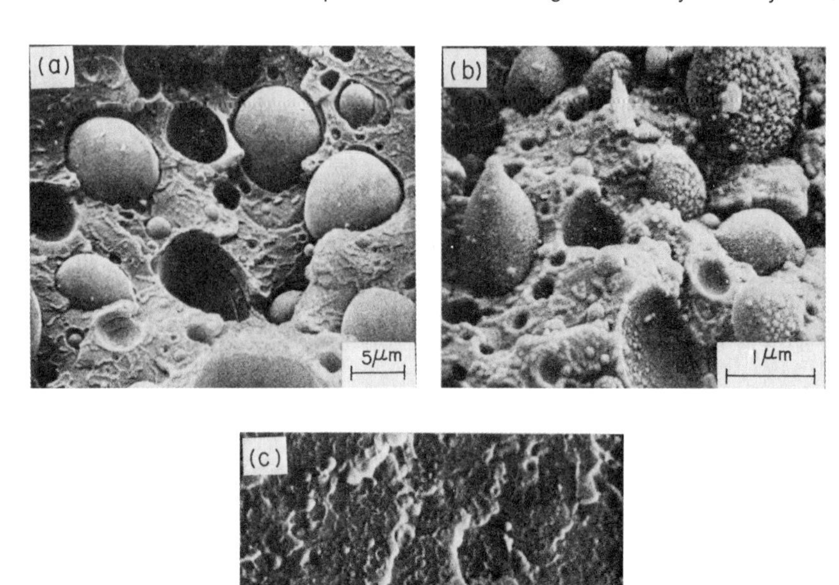

(c)
2μm

Fig. 4.71 Electron scanning photomicrographs of fracture surfaces for PS/LDPE blends with and without a graft copolymer [148]: (a) PS/LDPE = 75/25 (by wt.); (b) PS/LDPE/Copolymer = 75/25/1.25; (c) PS/LDPE/Copolymer = 75/25/7.5. Reproduced from W. M. Barentsen *et al.*, *Polymer* **15**, 119 (1974), by permission of the publishers, IPC Business Press Ltd. ©

An improvement of the mechanical properties of two-phase polymer blends is also reported by Locke and Paul [150, 151], who used blends of polyethylene and polyvinylchloride with chlorinated polyethylene, and by Ide and Hasgawa [152], who used blends of polypropylene and nylon 6 with a graft copolymer.

Problems

4.1 Consider the situation where a spherical droplet, suspended in another liquid, flows along the centerline of a conical die. Assuming that the initial droplet diameter is d_p, which is quite small compared to the opening of the conical die whose diameter is D (i.e., $d_p \ll D$), you are asked to derive expressions describing the droplet shapes in the conical die for the following different circumstances: (a) both the droplet and the suspending medium are Newtonian fluids; (b) the droplet is a Newtonian fluid and the suspending medium is a viscoelastic fluid represented by the Zaremba–DeWitt model, Eq. (2.42); (c) the droplet is a viscoelastic fluid represented by the Zaremba–DeWitt model and the suspending medium also is represented by the Zaremba–DeWitt model. You may assume that the velocity field under consideration is given by Eq. (2.24), that is, there is no circulatory pattern in the conical section of the die. State clearly any further assumptions that you wish to make in answering the question.

4.2 Figure 4.72 gives plots of elastic compliance J_e versus blend composition for blends of ethylene–vinyl acetate (EVA) and high-density polyethylene (HDPE) at various temperatures, at a shear stress of 5×10^4 N/m^2. It is seen that, at a blend ratio of about 10 wt.% of EVA, J_e increases with melt temperature. Note that, over the range

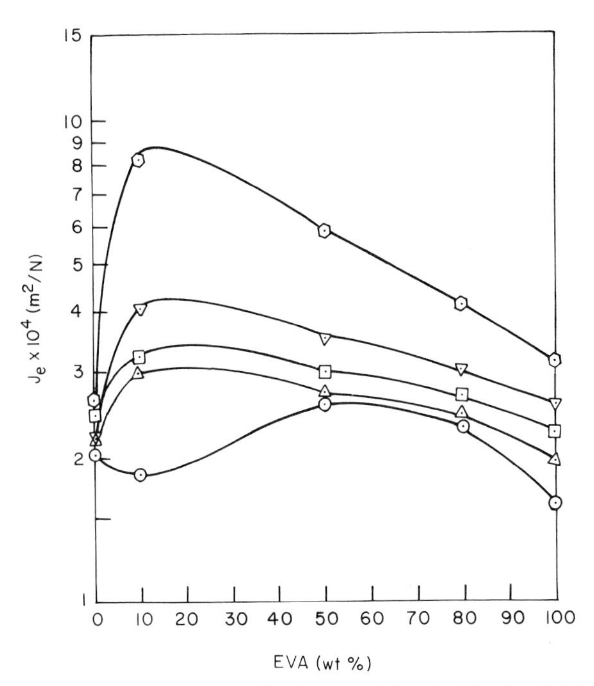

Fig. 4.72 Elastic compliance versus blend composition for blends of high-density polyethylene (HDPE) and ethylene–vinyl acetate (EVA), at various temperatures (°C): (⊙) 160; (△) 170; (□) 180; (▽) 190; (⬡) 200.

of temperature investigated, the HDPE forms the discrete phase suspended in the EVA forming the continuous phase. Give a phenomenological explanation as to why J_e increases with melt temperature.

4.3 You are given two different polymer blend systems whose elongational viscosities have the following relationships:

(a) Blend I: $(\eta_E)_A > (\eta_E)_B$
(b) Blend II: $(\eta_E)_A < (\eta_E)_B$.

In both cases, polymer A forms the discrete phase suspended in polymer B, the continuous phase. Which of the blends, at an equal weight fraction (i.e., 50 wt.% of A and 50 wt.% of B), will have the better spinning characteristics, and why?

4.4 Derive Eqs. (4.26)–(4.28).
4.5 Derive Eqs. (4.31)–(4.33).
4.6 Derive Eqs. (4.36)–(4.38).
4.7 Derive Eqs. (4.43)–(4.45).
4.8 Derive Eqs. (4.46)–(4.48).
4.9 Derive the expressions for steady, *equal* biaxial elongational viscosity for:

(i) the Goddard–Miller model defined by Eq. (4.24);
(ii) the Frankel–Acrivos model defined by Eq. (4.34);
(iii) the Choi–Schowalter model defined by Eq. (4.39).

REFERENCES

[1] M. Shen, *in* "Multiphase Polymers" (S. L. Cooper and G. M. Estes, eds.), Adv. Chem. Ser. No. 176, p. 181. Am. Chem. Soc., Washington, D.C., 1979.
[2] P. F. Bruins, ed., "Polyblends and Composites," Appl. Polym. Symp. No. 15. Wiley (Interscience), New York, 1970.
[3] S. L. Aggarwal, ed., "Block Copolymers." Plenum, New York, 1970.
[4] G. E. Molau, ed., "Colloidal and Morphological Behavior of Block and Graft Copolymers." Plenum, New York, 1971.
[5] N. A. J. Platzer, ed., "Multicomponent Polymer Systems," Adv. Chem. Ser. No. 99. Am. Chem. Soc., Washington, D.C., 1971.
[6] H. H. Keskkula, ed., "Polymer Modification of Rubbers and Plastics," Appl. Polym. Symp. No.7. Wiley (Interscience), New York, 1969.
[7] L. H. Sperling, ed., "Recent Advances in Polymer Blends, Grafts, and Blocks." Plenum, New York, 1974.
[8] N. A. J. Platzer, ed., "Copolymers, Polyblends, and Composites," Adv. Chem. Ser. No. 142. Am. Chem. Soc., Washington, D.C., 1975.
[9] D. R. Paul and S. Newman, eds., "Polymer Blends." Academic Press, New York, 1978.
[10] J. J. Burke and V. Weiss, eds., "Block and Graft Copolymers." Syracuse Univ. Press, Syracuse, New York, 1973.

[11] A. Noshay and J. E. McGrath, "Block Copolymers: Overview and Critical Survey." Academic Press, New York, 1977.
[12] S. L. Cooper and G. M. Estes, eds., "Multiphase Polymers," Adv. Chem. Ser. No. 176. Am. Chem. Soc., Washington, D.C., 1979.
[13] C. D. Armeniades and E. Baer, *in* "Introduction to Polymer Science and Technology" (H. S. Kaufman and J. J. Falcetta, eds.), Chap. 6. Wiley, New York, 1977.
[14] W. J. MacKnight, F. E. Karasz, and J. R. Fried, *in* "Polymer Blends" (D. R. Paul and S. Newman, eds.), Vol. 1, Chap. 5. Academic Press, New York, 1978.
[15] S. Krause, *in* "Polymer Blends" (D. R. Paul and S. Newman, eds.), Vol. 1, Chap. 2. Academic Press, New York, 1978.
[16] L. J. Hughes and G. L. Brown, *J. Appl. Polym. Sci.* **5**, 580 (1961).
[17] S. Newman, *in* "Polymer Blends" (D. R. Paul and S. Newman, eds.), Vol. 2, Chap. 13. Academic Press, New York, 1978.
[18] H. Keskkula, S. G. Turley, and R. F. Boyer, *J. Appl. Polym. Sci.* **15**, 351 (1971).
[19] W. K. Fischer, U.S. Patents 3,758,643 (1973); 3,806,558 (1974); 3,835,201 (1974); 3,862,106 (1975).
[20] E. N. Kresge, *in* "Polymer Blends" (D. R. Paul and S. Newman, eds.), Vol. 2, Chap. 20. Academic Press, New York, 1978.
[21] L. M. Robeson and A. B. Furtek, *J. Appl. Polym. Sci.* **23**, 645 (1979).
[22] D. C. Wahrmund, D. R. Paul, and J. W. Barlow, *J. Appl. Polym. Sci.* **22**, 2155 (1978).
[23] T. R. Nassar, D. R. Paul, and J. W. Barlow, *J. Appl. Polym. Sci.* **23**, 85 (1979).
[24] R. N. Mohn, D. R. Paul, J. W. Barlow, and C. A. Cruz, *J. Appl. Polym. Sci.* **23**, 575 (1979).
[25] C. A. Cruz, D. R. Paul, and J. W. Barlow, *J. Appl. Polym. Sci.* **23**, 589 (1979).
[26] D. C. Wahrmund, R. E. Bernstein, J. W. Barlow, and D. R. Paul, *Polym. Eng. Sci.* **18**, 677 (1978).
[27] R. E. Bernstein, D. C. Wahrmund, J. W. Barlow, and D. R. Paul, *Polym. Eng. Sci.* **18**, 683 (1978).
[28] R. E. Bernstein, D. C. Wahrmund, J. W. Barlow, and D. R. Paul, *Polym. Eng. Sci.* **18**, 1220 (1978).
[29] O. Olabisi, L. M. Robeson, and M. T. Shaw, "Polymer-Polymer Miscibility." Academic Press, New York, 1979.
[30] D. R. Paul and J. O. Altamirano, *in* "Copolymers, Polyblends, and Composites" (N. A. J. Platzer, ed.), Adv. Chem. Ser. No. 142, p. 371. Am. Chem. Soc., Washington, D.C., 1975.
[31] T. Nishi and T. T. Wang, *Macromolecules* **8**, 909 (1975).
[32] R. L. Imken, D. R. Paul, and J. W. Barlow, *Polym. Eng. Sci.* **16**, 593 (1976).
[33] T. K. Kwei, G. D. Patterson, and T. T. Wang, *Macromolecules* **9**, 780 (1976).
[34] J. V. Koleske, *in* "Polymer Blends" (D. R. Paul and S. Newman, eds.), Vol. 2, Chap. 22. Academic Press, New York, 1978.
[35] H. Van Oene, *J. Colloid Interface Sci.* **40**, 448 (1972).
[36] G. I. Taylor, *Proc. R. Soc. London, Ser. A* **146**, 501 (1934).
[37] C. E. Chaffey and H. Brenner, *J. Colloid Interface Sci.* **24**, 258 (1967).
[38] R. G. Cox, *J. Fluid Mech.* **37**, 601 (1969).
[39] D. Barthès-Biesel and A. Acrivos, *J. Fluid Mech.* **61**, 1 (1973).
[40] B. M. Turner and C. E. Chaffey, *Trans. Soc. Rheol.* **13**, 411 (1969).
[41] S. J. Choi and W. R. Schowalter, *Phys. Fluids* **18**, 420 (1975).
[42] W. Bartok and S. G. Mason, *J. Colloid Sci.* **13**, 293 (1958).
[43] W. Bartok and S. G. Mason, *J. Colloid Sci.* **14**, 13 (1959).
[44] F. D. Rumscheidt and S. G. Mason, *J. Colloid Sci.* **16**, 238 (1961).
[45] S. Torza, R. C. Cox, and S. G. Mason, *J. Colloid Interface Sci.* **38**, 395 (1972).
[46] T. Tavgac, Ph.D. Thesis (Chem. Eng.), Univ. of Houston, Houston, Texas, 1972.

[47] W. K. Lee, Ph.D. Thesis (Chem. Eng.), Univ. of Houston, Houston, Texas, 1972.
[48] H. L. Goldsmith and S. G. Mason, *J. Colloid Sci.* **17**, 448 (1962).
[49] F. Gauthier, H. L. Goldsmith, and S. G. Mason, *Trans. Soc. Rheol.* **15**, 297 (1971).
[50] C. E. Chaffey, H. Brenner, and S. G. Mason, *Rheol. Acta* **4**, 56 (1965); **4**, 64 (1965).
[51] G. Hetsroni, S. Haber, H. Brenner, and T. Greenstein, "Progress in Heat and Mass Transfer." (G. Hestroni, S. Sideman, and J. P. Hartnett, eds.), Vol. 6, p. 591, Pergamon, London, 1972.
[52] H. B. Chin and C. D. Han, *J. Rheol.* **23**, 557 (1979).
[53] C. D. Han and T. C. Yu, *J. Appl. Polym. Sci.* **15**, 1163 (1971).
[54] C. D. Han and T. C. Yu, *Polym. Eng. Sci.* **12**, 81 (1972).
[55] C. D. Han and Y. W. Kim, *Trans. Soc. Rheol.* **19**, 245 (1975).
[56] C. D. Han, Y. W. Kim, and S. J. Chen, *J. Appl. Polym. Sci.* **19**, 2831 (1975).
[57] Y. W. Kim and C. D. Han, *J. Appl. Polym. Sci.* **20**, 2905 (1976).
[58] C. D. Han, *Proc. Int. Congr. Rheol., 7th* p. 66. Chalmers Univ. Technol., Gothenburg, Sweden, 1976.
[59] C. D. Han, *J. Appl. Polym. Sci.* **18**, 481 (1974).
[60] M. B. Tsebrenko, M. Jakob, M. Y. Kuchinka, A. V. Yudin, and G. V. Vinogradov, *Int. J. Polym. Mater.* **3**, 99 (1974).
[61] T. I. Ablazova, M. B. Tsebrenko, A. B. V. Yudin, G. V. Vinogradov, and B. V. Yarlykov, *J. Appl. Polym. Sci.* **19**, 1781 (1975).
[62] G. V. Vinogradov, B. V. Yarlykov, M. B. Tsebrenko, A. V. Yudin, and T. I. Ablazova, *Polymer* **16**, 609 (1975).
[63] C. D. Han, "Rheology in Polymer Processing," Chap. 7. Academic Press, New York, 1976.
[64] H. Van Oene, *in* "Polymer Blends" (D. R. Paul and S. Newman, eds.), Vol. 1, Chap. 7. Academic Press, New York, 1978.
[65] C. D. Han and R. G. King, *J. Rheol.* **24**, 213 (1980).
[66] K. Suzuki, T. Watanabe, and S. Ono, *Proc. Int. Congr. Rheol., 5th* **2**, 339. Univ. Park Press, Baltimore, Maryland, 1970.
[67] E. B. Vadas, H. L. Goldsmith, and S. G. Mason, *Trans. Soc. Rheol.* **20**, 373 (1976).
[68] P. Sherman, *Proc. Int. Congr. Rheol., 4th* Part 3, p. 605. Wiley (Interscience), New York, 1965.
[69] E. G. Richardson, *in* "Flow Properties of Disperse Systems" (J. J. Hermans, ed.), p. 39. North-Holland Publ., Amsterdam, 1953.
[70] H. L. Doppert and W. S. Overdiep, *in* "Multicomponent Polymer Systems" (N. A. J, Platzer, ed.), Adv. Chem. Ser. No. 99, p. 53. Am. Chem. Soc., Washington, D.C., 1971.
[71] W. Meissner, W. Berger, and H. Hoffman, *Faserforsch. Textiltech.* **19**, 407 (1968).
[72] W. A. Hyman and R. Skalak, *AIChE J.* **18**, 149 (1972).
[73] N. A. Frankel and A. Acrivos, *J. Fluid Mech.* **44**, 65 (1970).
[74] K. Iwakura and T. Fujimura, *J. Appl. Polym. Sci.* **19**, 1427 (1975).
[75] T. Fujimura and K. Iwakura, *Kobunshi Ronbunshu* **31**, 617 (1974).
[76] T. Fujimura and K. Iwakura, *Kobunshi Kogaku* **27**(301), 323 (1970).
[77] S. Uemura and M. Takayangi, *J. Appl. Polym. Sci.* **10**, 113 (1966).
[78] M. Natov, L. Peeva, and E. Djagarova, *Polym. Sci., Part C* No. 16, p. 4197 (1968).
[79] R. F. Heitmiller, R. Z. Naar, and H. H. Zabusky, *J. Appl. Polym. Sci.* **8**, 873 (1964).
[80] W. F. Busse and R. Longworth, *J. Polym. Sci.* **58**, 49 (1962).
[81] A. S. Hill and B. Maxwell, *Polym. Eng. Sci.* **10**, 289 (1970).
[82] C. D. Han, K. U. Kim, J. Parker, N. Siskovic, and C. R. Huang, *Appl. Polym. Symp.* No. 21, p. 191 (1973).
[83] B. L. Lee and J. L. White, *Trans. Soc. Rheol.* **19**, 481 (1975).
[84] A. P. Plochocki, *Trans. Soc. Rheol.* **20**, 287 (1976).

[85] V. L. Folt and R. W. Smith, *Rubber Chem. Technol.* **46**, 1193 (1973).
[86] C. K. Shih, *Polym. Eng. Sci.* **16**, 742 (1976).
[87] C. J. Nelson, G. N. Avgeropoulos, F. C. Weissert, and G. G. A. Böhm, *Angew. Makromo. Chem.* **60/61**, 49 (1977).
[88] T. S. Lee, *Proc. Int. Congr. Rheol., 5th* **4**, 421. Univ. Park Press, Baltimore, Maryland, 1970.
[89] C. D. Han, *J. Appl. Polym. Sci.* **15**, 2591 (1971).
[90] A. Zosel, *Rheol. Acta* **11**, 229 (1972).
[91] H. Münstedt, *Proc. Int. Congr. Rheol., 7th* p. 496. Chalmers Univ. Technol., Gothenburg, Sweden, 1976.
[92] Y. Aoki, *Nippon Reoroji Gakkaishi* **7**, 20 (1979).
[93] K. Itoyama and A. Soda, *J. Appl. Polym. Sci.* **23**, 1723 (1979).
[94] H. Münstedt, *Eng. Found. Conf., Asilomar, Calif., 1980.*
[95] H. Tanaka and J. L. White, *Polym. Eng. Review* **1**, 89 (1981).
[96] M. V. Tsebrenko, A. V. Yudin, T. I. Ablazova, and G. Vinogradov, *Polymer* **17**, 831 (1976).
[97] F. M. Chapman and T. S. Lee, *SPE J.* **26**(1), 37 (1970).
[98] G. V. Vinogradov, A. Y. Malkin, E. P. Plotnikova, O. Y. Sabsai, and N. E. Nikolayeva, *Int. J. Polym. Mater.* **2**, 1 (1972).
[99] N. Minagawa and J. L. White, *J. Appl. Polym. Sci.* **20**, 501 (1976).
[100] V. M. Lobe and J. L. White, *Polym. Eng. Sci.* **19**, 617 (1979).
[101] H. Tanaka and J. L. White, *Polym. Eng. Sci.* **20**, 949 (1980).
[102] C. D. Han and K. Funatsu, *J. Rheol.* **22**, 113 (1978).
[103] G. I. Taylor, *Proc. R. Soc. London, Ser. A* **138**, 41 (1932).
[104] H. Fröhlich and R. Sack, *Proc. R. Soc. London, Ser. A* **185**, 415 (1946).
[105] J. Oldroyd, *Proc. R. Soc. London, Ser. A* **218**, 122 (1953).
[106] J. D. Goddard and C. Miller, *J. Fluid Mech.* **28**, 657 (1967).
[107] R. Roscoe, *J. Fluid Mech.* **28**, 273 (1967).
[108] S. J. Choi, Ph.D. Thesis (Chem. Eng.), Princeton Univ., Princeton, New Jersey, 1972.
[109] W. R. Schowalter, C. E. Chaffey, and H. Brenner, *J. Colloid Interface Sci.* **26**, 152 (1968).
[110] D. Barthès-Biesel and A. Acrivos, *Int. J. Multiphase Flow* **1**, 1 (1973).
[111] A. Einstein, *Ann. Phys.* **19**, 289 (1906); **34**, 591 (1911).
[112] G. B. Jeffery, *Proc. R. Soc. London, Ser. A* **102**, 161 (1922).
[113] G. K. Batchelor, *J. Fluid Mech.* **41**, 545 (1970); **44**, 419 (1970).
[114] R. Simha, *J. Appl. Phys.* **23**, 1020 (1952).
[115] H. Iino, M. Tsukasa, T. Nagafune, K. K. Sekaicho, and Y. Minoura, *Nippon Gomu Kyokaishi* **40**, 726 (1967).
[116] Y. Minoura, H. Iono, and M. Tsukasa, *J. Appl. Polym. Sci.* **9**, 1299 (1965).
[117] C. C. Lee, W. Rovatti, S. M. Skinner, and E. G. Bobalek, *J. Appl. Polym. Sci.* **9**, 2047 (1965).
[118] W. Rovatti and E. G. Bobalek, *J. Appl. Polym. Sci.* **7**, 2269 (1963).
[119] R. E. Robertson and D. R. Paul, *J. Appl. Polym. Sci.* **17**, 2579 (1973).
[120] J. R. Stell, D. R. Paul, and J. W. Barlow, *Polym. Eng. Sci.* **16**, 496 (1976).
[121] E. Cuddihy, J. Moacanin, and A. Rembaum, *J. Appl. Polym. Sci.* **9**, 1385 (1965).
[122] S. Newman and S. Strella, *J. Appl. Polym. Sci.* **9**, 2297 (1965).
[123] C. D. Han, C. A. Villamizar, Y. W. Kim, and S. J. Chen, *J. Appl. Polym. Sci.* **21**, 353 (1977).
[124] W. J. MacKnight, J. Stoelting, and F. E. Karasz, *in* "Multicomponent Polymer Systems" (N. A. J. Platzer, ed.), Adv. Chem. Ser. No. 99, p. 29. Amer. Chem. Soc., Washington, D.C., 1971.

[125] G. Ajroldi, G. Gatta, P. D. Gugelmetto, R. Rettore, and G. P. Talamini, *in* "Multicomponent Polymer Systems" (N. A. J. Platzer, ed.), Adv. Chem. Ser. No. 99, p. 119. Am. Chem. Soc., Washington, D.C., 1971.
[126] G. Kraus and K. W. Rollman, *in* "Multicomponent Polymer Systems" (N. A. J. Platzer, ed.), Adv. Chem. Ser. No. 99, p. 189. Am. Chem. Soc., Washington, D.C., 1971.
[127] K. Fujino, I. Owaga, and H. Kawai, *J. Appl. Polym. Sci.* **8**, 2147 (1964).
[128] M. Takayanagi, S. Eumura, and S. Minami, *J. Polym. Sci., Part C* **5**, 113 (1964).
[129] E. H. Kerner, *Proc. Phys. Soc. Sect. B* **69**, 808 (1956).
[130] Z. Hashin and S. Shtrikman, *J. Mech. Phys. Solids* **11**, 127 (1963).
[131] T. Horino, Y. Oyawa, J. Soen, and H. Kawai, *J. Appl. Polym. Sci.* **9**, 2261 (1965).
[132] T. Okamoto and M. Takayanagi, *J. Polym. Sci. Part C* **23**, 597 (1968).
[133] M. Matsuo, T. K. Kwei, D. Klempner, and H. L. Frisch, *Polym. Eng. Sci.* **10**, 327 (1970).
[134] G. Kraus, K. W. Rollman, and J. T. Gruver, *Macromolecules* **3**, 92 (1970).
[135] K. Marcinin, A. Romanov, and V. Pollak, *J. Appl. Polym. Sci.* **16**, 2239 (1972).
[136] N. K. Kalfoglou and H. L. Williams, *J. Appl. Polym. Sci.* **17**, 1377 (1973).
[137] J. L. Work, *Polym. Eng. Sci.* **13**, 46 (1973).
[138] R. A. Dickie, *J. Appl. Polym. Sci.* **17**, 45, 65, 79 (1973).
[139] D. Kaplan and N. W. Tschoegl, *Polym. Eng. Sci.* **14**, 43 (1974).
[140] R. A. Dickie, *in* "Polymer Blends" (D. R. Paul and S. Newman, eds.), Vol. 1, Chap. 8. Academic Press, New York, 1978.
[141] C. D. Han and Y. W. Kim, *J. Appl. Polym. Sci.* **18**, 2589 (1974).
[142] C. D. Han, C. A. Villamizar, Y. W. Kim, and S. J. Chen, *J. Appl. Polym. Sci.* **21**, 353 (1977).
[143] H. K. Asar, M. B. Rhodes, and R. Salovey, *in* "Multiphase Polymers" (S. L. Cooper and G. M. Ester, eds.), Adv. Chem. Ser. No. 176, p. 489. Am. Chem. Soc., Washington, D.C., 1979.
[144] N. G. Gaylord, *in* "Copolymers, Polyblends, and Composites" (N. A. J. Platzer, ed.), Adv. Chem. Ser. No. 142, p. 76. Am. Chem. Soc., Washington, D.C., 1975.
[145] D. R. Paul, *in* "Polymer Blends" (D. R. Paul and S. Newman, eds.), Vol. 2, Chap. 12. Academic Press, New York, 1978.
[146] C. E. Locke and D. R. Paul, *J. Appl. Polym. Sci.* **17**, 2791 (1973).
[147] W. M. Barentsen and D. Heikens, *Polymer* **14**, 579 (1973).
[148] W. M. Barentsen, D. Heikens, and P. Piet, *Polymer* **15**, 119 (1974).
[149] S. Wu, *in* "Polymer Blends" (D. R. Paul and S. Newman, eds.), Vol. 1, Chap. 6. Academic Press, New York, 1978.
[150] D. R. Paul, C. E. Locke, and C. E. Vinson, *Polym. Eng. Sci.* **13**, 202 (1973).
[151] C. E. Locke and D. R. Paul, *Polym. Eng. Sci.* **13**, 308 (1973).
[152] F. Ide and A. Hasegawa, *J. Appl. Polym. Sci.* **18**, 963 (1974).

5

Droplet Breakup in Dispersed Two-Phase Flow

5.1 INTRODUCTION

The dispersion of one liquid in another is of practical importance in chemical processes that involve the preparation of emulsions and in polymer processing operations that involve the dispersion of a liquid additive or the preparation of polymer blends consisting of two or more polymers. Dispersion of one liquid in another (and hence mixing) involves the *breakup* of droplets in the continuous medium under appropriate flow fields.

In the past, several research groups [1–11] have reported droplet breakup experiments. Their experiments can be classified into two types: (1) droplet breakup during steady flow, and (2) droplet breakup during transient flow. For instance, Taylor's experiment was performed at steady flow conditions [1]. The apparatus was set in motion at a slow speed and adjusted until the flow was steady and stationary. The speed was then increased until droplet breakup was observed. Taylor reports that, under steady extensional flow (using the "four-roll" apparatus), the droplet broke up at low speed, whereas under steady shear flow (using the "parallel band" apparatus), the droplet did not break up until a very high speed was attained. One can then surmise that the type of flow field (i.e., a shearing flow field or an extensional flow field) and the type of flow (i.e., steady flow or transient flow) play important roles in determining the effectiveness of dispersion.

In this chapter, we shall discuss breakup phenomena of single droplets under different flow fields. From the practical point of view, one wishes to control the droplet size and its distribution in an emulsion (or in a blend) so as to minimize the pressure drop required during the fabrication operation and to maximize the mechanical properties of the finished product. From the theoretical point of view, the phenomenon of droplet breakup may be considered as a problem of flow instability, and some serious attempts [11–18] have been made at predicting the condition (or conditions) at which droplet breakup occurs.

The primary objective of this chapter is to put our current understanding of both the experimental and theoretical aspects of droplet breakup into a proper perspective, so that the problems involved with the processing of *dispersed* liquid–liquid polymeric systems can be better understood.

5.2 EXPERIMENTAL OBSERVATIONS OF DROPLET BREAKUP

We have seen in Chapter 4 that the dynamical and viscous forces acting on the surface of a small spherical droplet will tend to distort it, whereas the interfacial tension will tend to resist these forces and keep the drop spherical. However, should the dynamical and viscous forces become larger than the interfacial forces, the droplet will deform and *eventually* break into small droplets. It is, then, of practical and fundamental importance to understand the critical conditions for the occurrence of droplet breakup in terms of the flow conditions and the physical/rheological properties of the liquid systems concerned.

5.2.1 Droplet Breakup in a Shearing Flow Field

In a Couette flow experiment (with uniform shearing flow), an initially spherical droplet will be deformed to an ellipsoidal shape at relatively low shear rates, and, as the shear rate is increased further, the droplet will be deformed to a long threadlike cylinder and eventually will be broken into small droplets.

Taylor [1] first postulated that droplet breakup would occur when the surface tension forces, tending to keep a droplet in one piece, could no longer balance the viscous forces, tending to burst it. Taylor showed theoretically that, at breakup, the *apparent* deformation D of a droplet has the value 0.5. The apparent deformation D is given by

$$D = (L - B)/(L + B) \tag{5.1}$$

where L is the major axis (i.e., length) and B is the minor axis (i.e., breadth) of a deformed droplet. Taylor defined a dimensionless group E, given by

$$E = We[(19k + 16)/(16k + 16)] \tag{5.2}$$

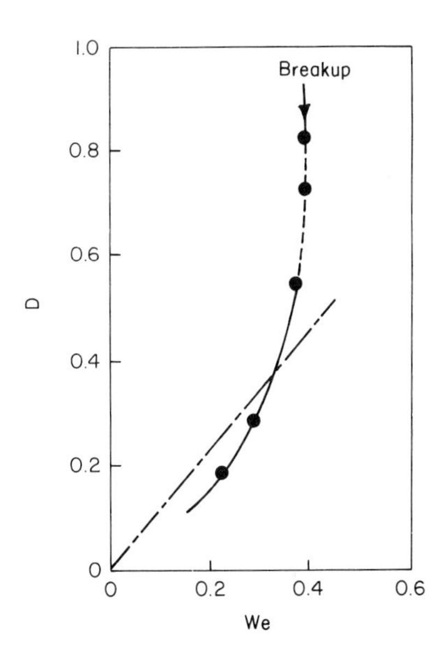

Fig. 5.1 Droplet deformation curve in a four-roll apparatus [1], in which a black lubricating oil is the droplet phase and a syrup is the suspending medium. The broken line (-··-) represents the theoretical prediction due to Taylor, $D = 1.09\ We$.

where k is the viscosity ratio of droplet phase to suspending medium ($\bar{\eta}_0/\eta_0$) and We is a ratio of viscous to interfacial tension forces defined as

$$We = \eta_0 a\dot{\gamma}/\sigma \tag{5.3}$$

in which a is the droplet radius, $\dot{\gamma}$ is the shear rate, and σ is the interfacial tension. At droplet breakup, E theoretically also has the value 0.5. Taylor then demonstrated experimentally that, for k values of 0.1 to 1.0, droplet breakup occurs at

$$D_c = E_c = 0.5 \sim 0.6 \tag{5.4}$$

in which the subscript c refers to the condition of droplet breakup. This was later confirmed by Rumscheidt and Mason [3]. Figure 5.1 gives plots of D versus We, including the critical condition at which breakup occurs. Note, however, that at very high values of k, Taylor predicted a limiting deformation without breakup in uniform shearing flow.

Taylor [1] postulated further that when inertia effects on both phases are negligible in comparison to the viscous effects, the *critical* condition for breakup of Newtonian droplets suspended in a Newtonian medium may be correlatable by the following relationship:

$$(We)_c = f(k) \tag{5.5}$$

in which $(We)_c$ denotes the critical value of We. Figure 5.2 gives plots of E_c versus k, displaying the effect of viscosity ratio on the ratio of viscous to interfacial forces required for droplet breakup in uniform shearing flow. It is

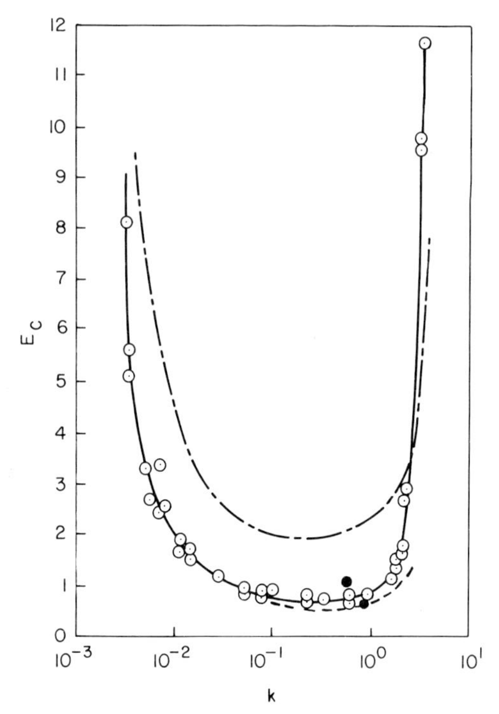

Fig. 5.2 Critical value of droplet deformation versus viscosity ratio for droplet breakup in uniform shearing flow [8]: (⊙) data due to Tavgac [8]; (●) data due to Taylor [1]; (–·–) curve drawn through the data obtained by Karam and Bellinger [6]; (----) curve drawn through the data obtained by Torza *et al.* [5].

seen that, at a given value of $k = \bar{\eta}_0/\eta_0$, if the droplet is exposed to a uniform shearing flow field which produces a value of *We* greater than the critical value $(We)_c$, that droplet will deform and break. That is, when values of *We* fall above and inside the U-shaped contour the droplet is unstable, whereas outside the contour it is stable. It is of interest to note that there appear to exist limits at both ends of the U-shaped contour, below or above which breakup cannot be obtained.

Karam and Bellinger [6] report that, in uniform shearing flow, droplet breakup would occur within the range of viscosity ratios $0.005 \leq k \leq 3.0$, and Tavgac [8] reports the range $0.0033 \leq k \leq 3.7$. Grace [4], however, has performed breakup experiments with the viscosity ratio ranging from 10^{-6} to 10^{3} and reports that breakup occurs at k values as low as 10^{-6}, and does *not* occur at k values above 3.5, as shown in Fig. 5.3. The practical significance of the general correlation given in Figs. 5.2 and 5.3 lies in that it allows us to predict the critical condition for droplet breakup. For example, we can deter-

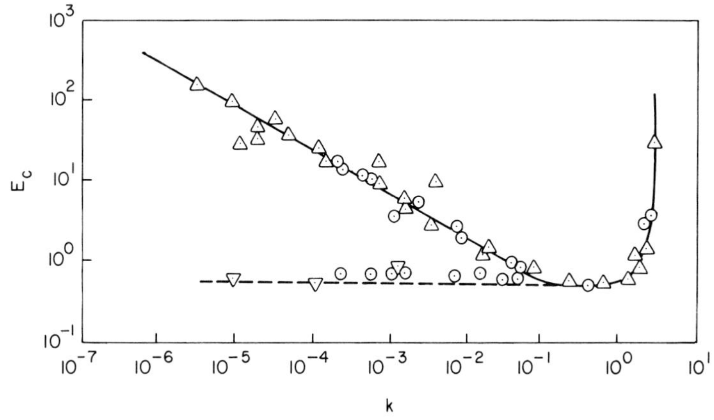

Fig. 5.3 Effect of viscosity ratio on the critical value of droplet deformation in Couette flow, with various values of medium viscosity (N sec/m^2) [4]: (⊙) 4.55; (△) 50.2; (▽) 281.5. The solid line represents the breakup by fracture of main droplet and the broken line represents the breakup by tip streaming.

mine the critical shear rate $\dot{\gamma}_c$ required to break a droplet of radius a at a suspending medium viscosity η_0 and an interfacial tension σ for any viscosity ratio ranging from, say, 10^{-6} to 3.5 in uniform shear flow.

Torza *et al.* [5] report, on the other hand, that the mechanism of droplet breakup depends more on the rate of increase in shear rate ($d\dot{\gamma}/dt$) than on the viscosity ratio. They have found, from their experimental study in uniform shear flow, that the product of $\dot{\gamma}_c$ and droplet radius a remains nearly constant for a given viscosity ratio. Figure 5.4 gives photographs describing the

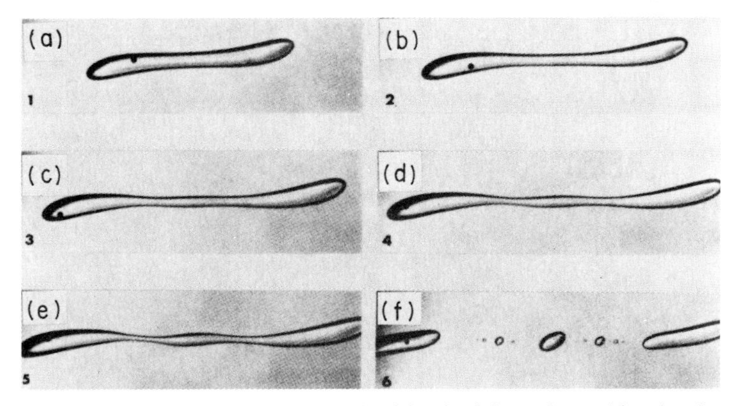

Fig. 5.4 Photographs describing the mode of droplet deformation and breakup in a shear flow field as the shear rate is increased to a critical value [5].

mode of droplet deformation and breakup in a shear flow field as the $\dot{\gamma}$ is increased to $\dot{\gamma}_c$. It shows that the droplet was pulled to a greater length before necking caused it to break up and giving rise to relatively bigger satellite droplets. Another interesting observation by Torza *et al.* is that at high shear rates, the droplet is pulled out into a long threadlike form which eventually becomes varicose and breaks up into a series of smaller droplets. It may be worth noting that the dependence of droplet breakup on $d\dot{\gamma}/dt$ and the fact that a long threadlike droplet becomes varicose before it breaks up suggest that the breakup of a droplet is influenced by hydrodynamic disturbances, which may arise from the change of shear rate.

What is common, however, in the different shapes of the contour given in Figs. 5.2 and 5.3 is that for the breakup of droplets to occur, We (or E) must pass through a minimum within the range of viscosity ratios, say, $k = 0.1 \sim 1.0$, whereas, outside this range, the value of the viscous force required to break the droplet increases very rapidly. This implies that at very low or very high viscosity ratios, it becomes very difficult to break a drop in uniform shear flow. Note, however, that according to Karam and Bellinger [6], the minimum value of E_c is about 2, whereas according to Torza *et al.* [5] and Grace [4], it is about 0.6. Grace [4] explains that this difference is attributable to the different values of interfacial tension used. According to Grace, the static equilibrium value of interfacial tension is expected to be one-half to one-third of the dynamic value. Grace and Torza *et al.* used the dynamic value, whereas Karam and Bellinger used the static equilibrium one.

Figure 5.5 gives plots of critical draw ratio of a droplet versus viscosity ratio, where the critical draw ratio is the length L_c required for breakup divided by the original droplet diameter $2a_0$. The general correlation in Fig. 5.5 is very similar to the correlation between E_c and k given in Fig. 5.3, with a minimum occurring in a viscosity ratio range of 0.2–1.0, and a rapid

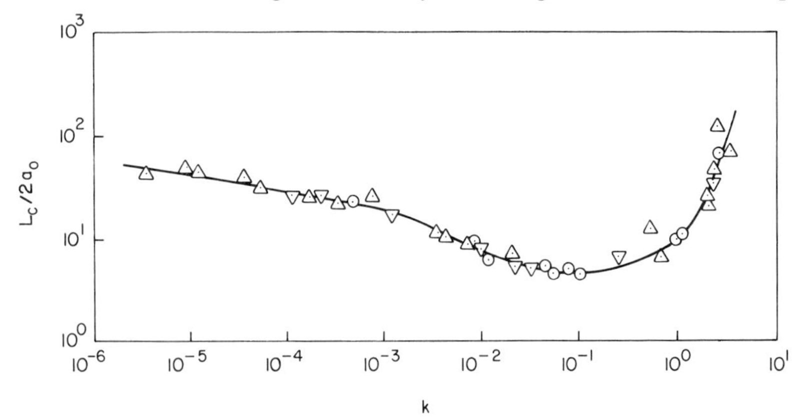

Fig. 5.5 Effect of viscosity ratio on the critical draw ratio for breakup in Couette flow with various values of medium viscosity (N sec/m^2) [4]: (⊙) 4.55; (△) 50.2; (▽) 281.5.

rise in either direction from this minimum. The similarity between the two correlations suggests that for breakup to occur at a critical value of We (or E), a critical droplet draw ratio must be attained.

As one may suspect, the region of droplet breakup (i.e., the contour constructed for Newtonian droplets suspended in another Newtonian medium) may be different in situations when at least one of the two phases (i.e., either the droplet phase or the suspending medium or both) is viscoelastic. For such situations, using the dimensional analysis approach, Flumerfelt [7] has proposed the use of the following dimensionless groups

$$(We)_c = f(k, \bar{\lambda}_1 \dot{\gamma}_c) \tag{5.6}$$

for viscoelastic droplets suspended in a Newtonian medium, and

$$(We)_c = f(k, \lambda_1 \dot{\gamma}_c) \tag{5.7}$$

for Newtonian droplets suspended in a viscoelastic medium. In Eqs. (5.6) and (5.7), $\bar{\lambda}_1$ and λ_1 are the relaxation times of the droplet phase and suspending medium, respectively. Flumerfelt [7] suggests determining the values of $\bar{\lambda}_1$ and λ_1 from a chosen constitutive equation with the aid of rheological data available for evaluating the material constants in the rheological model. When both the droplet phase and suspending medium are viscoelastic, another dimensionless parameter, either $\bar{\lambda}_1/\lambda_1$ or $\lambda_1/\bar{\lambda}_1$ is needed in order to take into account the relative elasticity of the two fluids concerned.

At present, the effect of fluid elasticity on the deformation and breakup of liquid droplets is *not* understood completely. From the experimental point of view, some confusion appears to exist in the literature. Flumerfelt [7] reports that breakup of a given droplet (Newtonian) would occur at lower shear rates in a Newtonian medium than in a viscoelastic medium, all other parameters being equal. He further observes that the shear stress requirements, however, are greater for the breakup of droplets in a Newtonian medium than in a viscoelastic medium under similar conditions. Tavgac [8] observes that, depending on the fluid systems involved, the elastic forces of the viscoelastic systems may have a stabilizing or destabilizing effect in the breakup process. More specifically, he observes that an elastic medium contributes to the breakup of the Newtonian droplets at high values of the viscosity ratio k and tends to stabilize the droplet at low values of k.

Let us now discuss the phenomenon of droplet breakup when single droplets pass through a cylindrical tube. Using a cylindrical tube, Chin and Han [11] performed experiments in order to determine the critical conditions for droplet breakup in terms of the rheological properties of both the droplet phase and the suspending medium, initial droplet size, and apparent wall shear rate. The rheological properties of the liquids used in their experiments are those given in Figs. 2.2 and 2.3 in Chapter 2.

Figure 5.6 gives photographs showing the effect of initial droplet size on

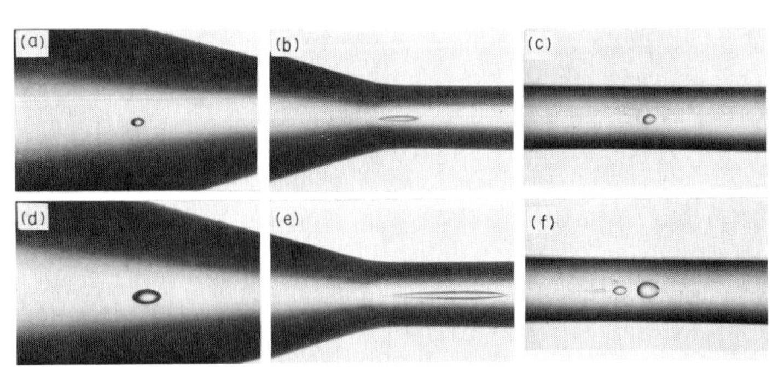

Fig. 5.6 Photographs of droplets describing the effect of initial droplet size on breakup patterns [11]: (a), (b), and (c) for $a = 0.53$ mm; (d), (e), and (f) for $a = 0.91$ mm. The droplet phase is a 2% PIB solution, the suspending medium is a 2% Separan solution, and the apparent wall shear rate is 39.5 sec^{-1}. The rheological properties of the liquids are given in Figs. 2.2 and 2.3 (see Chapter 2). From H. B. Chin and C. D. Han, *J. Rheol.* **24**, 1. Copyright © 1980. Reprinted by permission of John Wiley & Sons, Inc.

the breakup of a 2% PIB droplet suspended in a 2% Separan solution. It is seen that breakup occurs with a droplet radius of 0.091 cm, but not with a droplet radius of 0.053 cm. Figure 5.7 gives photographs displaying the effect of shear rate on the breakup of a 2% PIB droplet suspended in a 2% Separan solution. It is seen that breakup occurs at high shear rate, but not a low shear rate. Figure 5.8 gives photographs displaying the effect of medium viscosity on the breakup of a 10% PIB droplet suspended in 2% and 4% Separan solutions. It is seen that breakup does not occur in the 2% Separan

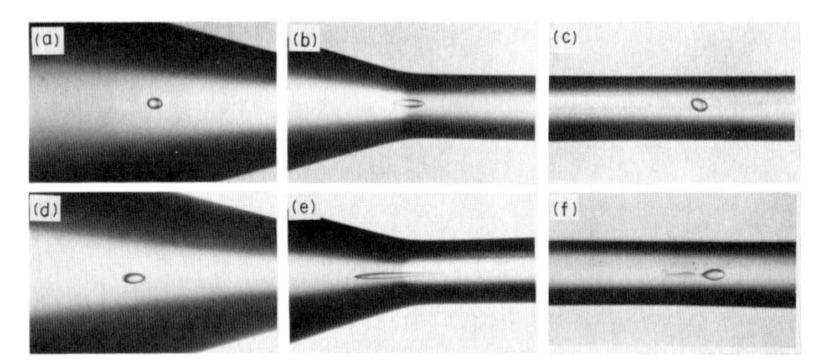

Fig. 5.7 Photographs of droplets describing the effect of shear rate on breakup patterns [11]: (a), (b), and (c) at $\dot{\gamma} = 13.7\ \text{sec}^{-1}$; (d), (e), and (f) at $\dot{\gamma} = 102.4\ \text{sec}^{-1}$. The droplet phase is a 2% PIB solution, the suspending medium is a 2% Separan solution, and the initial droplet size is 0.68 mm. The rheological properties of the liquids are given in Figs. 2.2 and 2.3 (see Chapter 2). From H. B. Chin and C. D. Han, *J. Rheol.* **24**, 1. Copyright © 1980. Reprinted by permission of John Wiley & Sons, Inc.

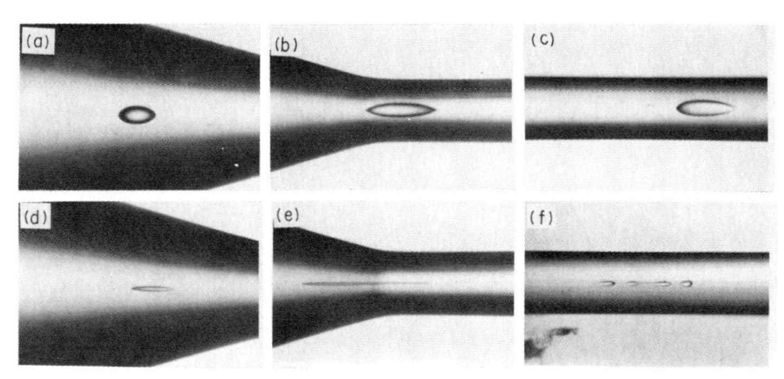

Fig. 5.8 Photographs of droplets describing the effect of suspending medium viscosity on breakup patterns [11]: (a), (b) and (c) with a 2% Separan solution, $\dot{\gamma} = 102.4\ \text{sec}^{-1}$, $a = 1.20$ mm; (d), (e), and (f) with a 4% Separan solution, $\dot{\gamma} = 6.6\ \text{sec}^{-1}$, $a = 0.53$ mm. The droplet phase is a 10% PIB solution. The rheological properties of the liquids are given in Figs. 2.2 and 2.3 (see Chapter 2). From H. B. Chin and C. D. Han, *J. Rheol.* **24**, 1. Copyright © 1980. Reprinted by permission of John Wiley & Sons, Inc.

solution, even at very high shear rate and with a large droplet size. However, breakup occurs in the 4% Separan solution at very low shear rate with a small droplet size. From the observations made above, it can be concluded that the dimensionless parameter *We* has a strong influence on the critical conditions for droplet breakup. It is worth noting in the photographs described above, that breakup occurs always after the long threadlike liquid cylinder passes through the entrance section of the cylindrical tube. Chin and Han [11] noted that breakup always occurred in the region where the elongated droplet started to *recoil*, and they attributed the breakup of the extended long liquid cylinder to the change of fluid velocity in the entrance region.

Figure 5.9 gives plots of $(We)_c$ versus k for various droplets suspended in a 2% Separan solution. The Separan solution being a non-Newtonian and viscoelastic fluid, the *apparent* shear viscosity corresponding to the particular shear rate being considered was used in calculating $(We)_c$. It is seen in Fig. 5.9 that there is no clear trend as to whether or not the breakup of Newtonian droplets requires lower values of *We* than viscoelastic droplets. This may be attributable in part to the practical difficulty encountered during the experiment in accurately determining the breakup conditions.

Note that Fig. 5.9 shows some important parameters which control the breakup of droplets in nonuniform shearing flow. Since the dimensionless parameter *We* represents the ratio of *apparent* shear stress of the suspending medium (tending to break the droplet) and the interfacial force trying to resist the breakage, one can conclude from Fig. 5.9 that the greater the interfacial tension (or the less the *apparent* shear stress of the suspending medium),

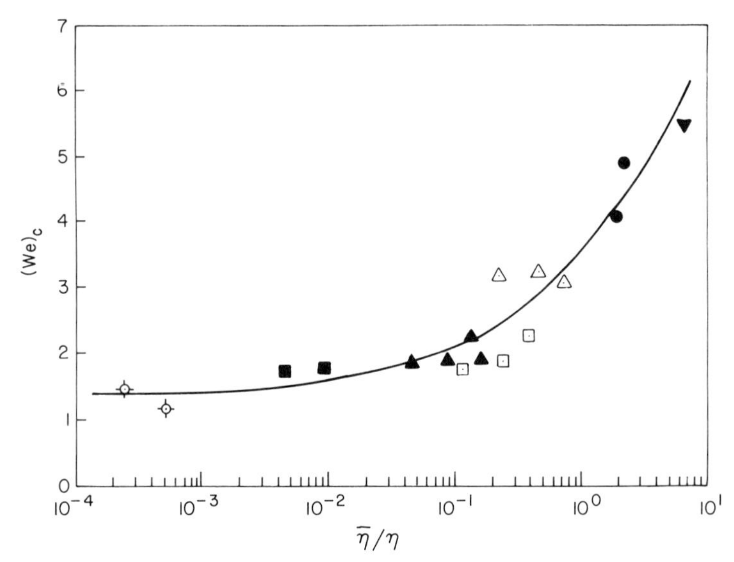

Fig. 5.9 $(We)_c$ versus *apparent* viscosity ratio for various droplets suspended in a 2% Separan solution [11]. Droplets are: (■) 0.6% PIB; (▲) 2% PIB; (●) 6% PIB, (-◇-) benzene; (□) Indopol L50, (△) Indopol L100. From H. B. Chin and C. D. Han, *J. Rheol.* **24**, 1. Copyright © 1980. Reprinted by permission of John Wiley & Sons, Inc.

the less likely will breakup occur. Since the breakup will occur only when a spherical droplet is elongated sufficiently to yield a *long threadlike* liquid cylinder, the variables controlling the extent of deformation (see Chapter 4) should also control the critical conditions for breakup. We can therefore speculate that the greater the elasticity of the droplet phase, the less likely will breakup occur, because the droplet phase elasticity is expected to have the same effect as the interfacial force, i.e., it will try to resist breakup. In other words, viscoelastic droplets will be more *stable* than Newtonian droplets.

5.2.2 Droplet Breakup in an Extensional Flow Field

In a plane hyperbolic flow experiment, which generates essentially extensional flow, an initially spherical droplet will be deformed to an ellipsoidal shape at relatively low extension rates, and to a long threadlike cylinder as the extension rate is increased further, eventually breaking up into small droplets. Figure 5.10 shows photographs displaying breakup patterns in a plane hyperbolic flow field.

Figure 5.11 gives plots of E_c versus k, obtained by the use of the Taylor "four-roll" apparatus. Note that E_c in Fig. 5.11 contains the elongation rate

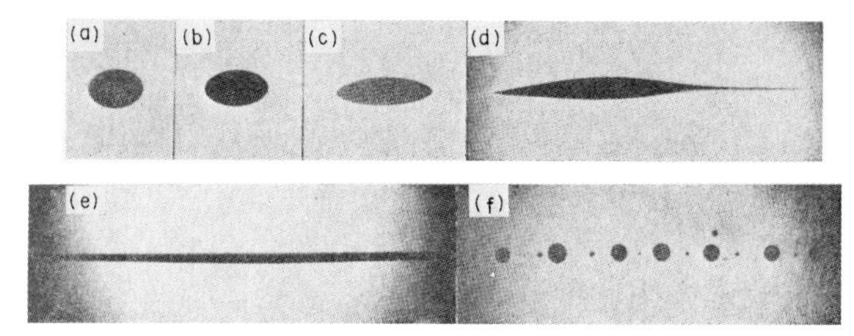

Fig. 5.10 Photographs describing the mode of droplet deformation and breakup in a four-roll apparatus, at various values of viscosity ratio k [1]: (a) $k = 0.0003$, $We = 0.18$, $D = 0.19$; (b) $k = 0.0003$, $We = 0.28$, $D = 0.29$; (c) $k = 0.0003$, $We = 0.41$, $D = 0.54$; (d) $k = 0.9$, $We = 0.39$, (e) $k = 0.9$, $We = 0.39$; (f) $k = 0.9$.

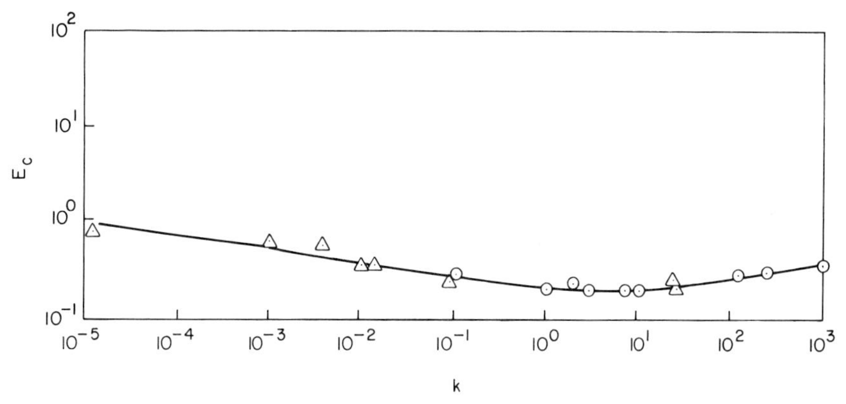

Fig. 5.11 Effect of viscosity ratio on the critical value of droplet deformation for breakup in hyperbolic flow field (four-roll apparatus), with various values of medium viscosity (N sec/m^2) [4]: (⊙) 1.1; (△) 4.55.

$\dot{\gamma}_E$, instead of shear rate $\dot{\gamma}$. It is seen that a minimum in E_c occurs at viscosity ratios between 1 and 5, with a slow increase in the value of E required for breakup at high viscosity ratios up to 10^3 and at low viscosity ratios down to values as small as 10^{-5}. It is also seen that the minimum value of E_c is about 0.2, this value being only one-third of the value of 0.6 observed in similar plots under uniform shearing flow. Figure 5.12 gives a comparison of the critical value of E required for droplet breakup in *uniform* steady shearing flow (realized by the use of a Couette apparatus) with that in plane hyperbolic flow (realized by the use of a four-roll apparatus), over a wide range of viscosity ratios. Of particular interest in Fig. 5.12 is that droplets of viscosity ratio greater than, say, 3.5 can be broken up into smaller droplets in a steady *extensional* flow field, whereas this is *not* possible in a *uniform* steady shearing flow field. It is seen that, over the entire range of viscosity ratios tested, the four-roll apparatus requires much lower values of E_c than the Couette

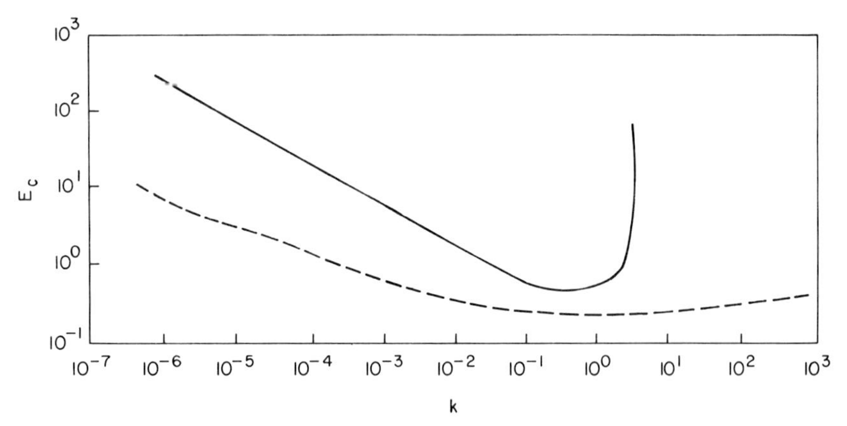

Fig. 5.12 Comparison of critical value of deformation for droplet breakup in shearing and hyperbolic flow fields [4]. The solid line (–) represents the shearing flow field, and the dotted line (---) represents the hyperbolic flow field.

apparatus, implying that the elongational flow field is much more *effective* for breaking up droplets than the uniform shearing flow field. This indicates further that the total energy expenditure required for the breakup of droplets will be lower in an elongational flow field than in a uniform shearing flow field. This then suggests how we may design an effective apparatus for breaking droplets (i.e., the design of an emulsification device). Later in this chapter we will discuss the design of equipment for droplet breakup further, in conjunction with polymer processing operations.

5.3 THEORETICAL CONSIDERATIONS OF DROPLET BREAKUP

Droplet breakup is a transient phenomenon in nature. Therefore any theoretical attempt to understand this phenomenon better requires the solution of unsteady-state equations of motion. The instability of a droplet is governed by the local flow conditions and the rheological and interfacial properties of the fluids involved. Tomotika [12, 13] was the first to carry out a hydrodynamic stability analysis of the problem, explaining Taylor's experimental observation [1]. Tomotika [13] assumed a uniform rate of strain imposed on a liquid cylinder, having a uniform cross section initially, and showed from his analysis that if the ratio of the viscosities of the two fluids is neither zero nor infinity, the maximum instability (i.e., droplet breakup) always occurs at a certain definite value of the wavelength of the assumed initial disturbance. Mikami *et al.* [14] extended Tomotika's analysis to predict the time for breakup and the size of broken droplets in steady extensional flow. Lee [9] also carried out a Tomotika-type analysis for Newtonian

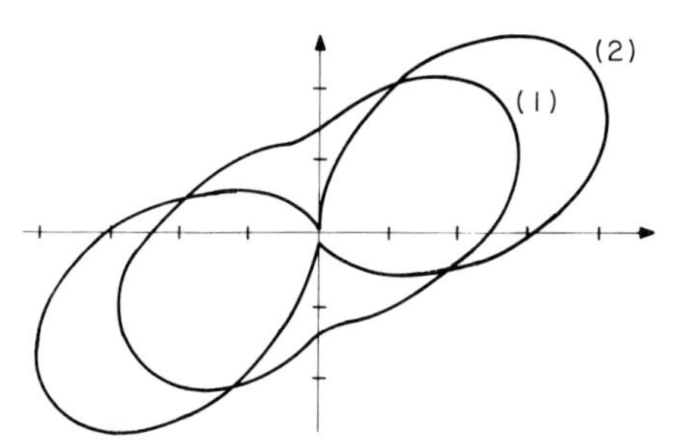

Fig. 5.13 Theoretically predicted mode of deformation and breakup of a droplet in a simple shearing flow [15]: (1) $t = 0.2$ sec; (2) $t = 2.2$ sec. Note that breakup is preceded by a pinching of the middle of the droplet.

droplets, and then extended it to a class of viscoelastic fluids. Using a perturbation technique, Chin and Han [11] carried out a hydrodynamic stability analysis for the breakup of viscoelastic droplets suspended in a viscoelastic medium.

Using a perturbation technique, Barthès-Biesel and Acrivos [15] investigated the existence of steady-state solutions for the droplet shape at large deformations, and applied the theory of linear stability analysis to the steady-state solutions in order to determine the critical conditions for droplet breakup. By carrying out a very elaborate numerical computation, these authors predicted the shape of a droplet during the process of breakup, as shown in Fig. 5.13. They predicted also the critical condition for breakup of Newtonian droplets suspended in a Newtonian medium, as given in Fig. 5.14.

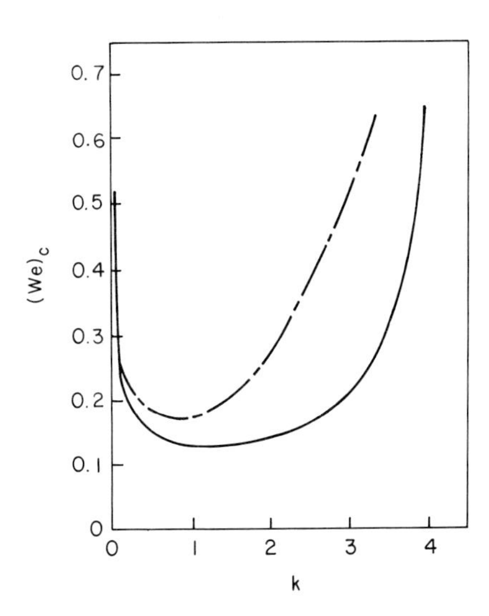

Fig. 5.14 Comparison of the theoretical prediction of droplet breakup curve with experimental results [15]. The solid curve (–) represents the theoretical prediction by Barthès-Biesel and Acrivos [15], and the broken curve (-··-) represents the curve drawn over the data obtained by Torza *et al.* [5].

Acrivos and Lo [16] developed a theoretical criterion for the breakup of a droplet pulled out into a long threadlike form by employing the "slender-body" theory [19, 20]. It should be stated, however, that these studies were concerned only with Newtonian fluid systems.

5.3.1 Hydrodynamic Stability Analysis of Droplet Breakup

We shall consider the stability of a long liquid cylinder suspended in another liquid flowing through a cylindrical tube. Our primary interest is in finding the critical conditions for breakup of the liquid cylinder into smaller droplets, in terms of the rheological properties of both the droplet phase and the suspending medium, the interfacial tension, and flow conditions. Emphasis will be placed on finding the role, if any, that the fluid *elasticity* may play in the occurrence of breakup in a nonuniform shearing flow field, i.e., in cylindrical Poiseuille flow.

In the analysis, we will assume that an axisymmetric disturbance, which is periodic with the direction of flow and exponential in time, is introduced into the primary flow, and that the disturbance either grows or decays, depending on the rheological/physical properties of the fluids concerned and the flow condition. Figure 5.15 gives a photograph displaying the breakup patterns of a long liquid cylinder suspended in another liquid, subjected to an external disturbance [21].

We shall consider two situations: (1) the stability of the system when it is disturbed by long waves, and (2) the stability of the system when it is disturbed by waves of all lengths, under the condition of inertialess flow.

Consider a long neutrally buoyant liquid cylinder suspended in another liquid, subject to cylindrical Poiseuille flow, as schematically shown in Fig. 5.16. Assume that the liquid cylinder is located on the central axis of the tube, that both phases are mutually immiscible, and that there are no external body forces.

In view of the axial symmetry of the flow, we may now assume an axisymmetric disturbance which is periodic with z and exponential in time t, so that an arbitrary perturbation variable q may be represented in the general form:

$$q(r, z, t) = \hat{q}(r) \exp[i\alpha(z - ct)] \tag{5.8}$$

in which $\hat{q}(r)$ is the amplitude of a periodic function, α is a real positive number (commonly referred to as the wave number, defined by $2\pi/L$, L being the wavelength) and $c = c_r + ic_i$ is a complex number. It is apparent from Eq. (5.8) that stability is determined by the sign of the imaginary part of the complex number c. That is, positive values of c_i imply that the amplitude of the disturbance increases with time t and thus the flow is *unstable*, and negative values of c_i indicate that the flow is *stable* [22, 23].

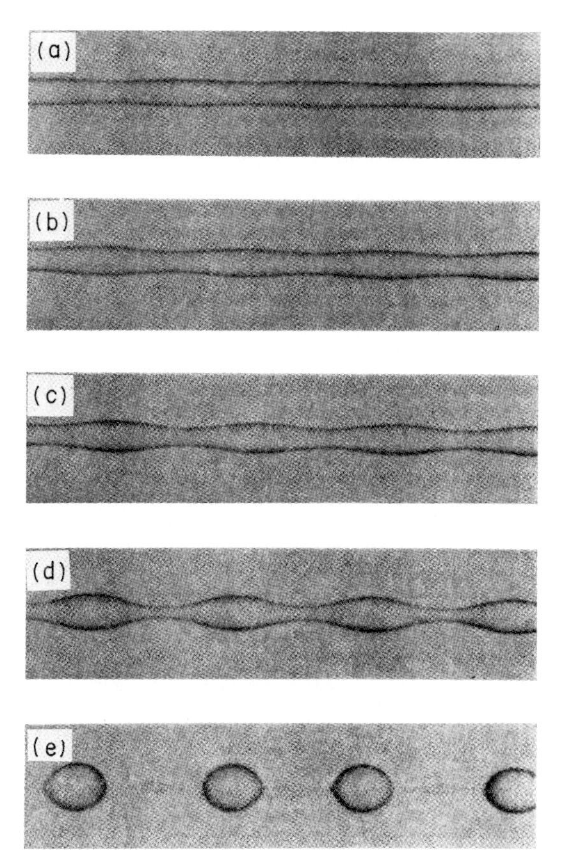

Fig. 5.15 Photographs describing the growth of the disturbance leading to breakup of a stationary liquid thread ($r_0 = 0.005$ cm) [21]. Pale 4 oil is the droplet phase and silicone oil is the suspending medium.

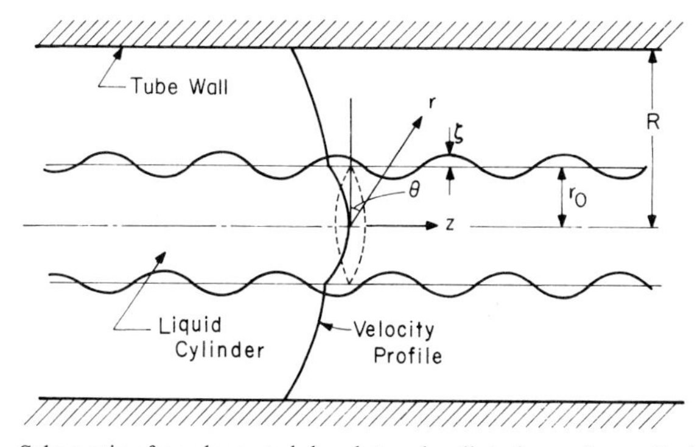

Fig. 5.16 Schematic of an elongated droplet under disturbance in a cylindrical tube [11]. From H. B. Chin and C. D. Han, *J. Rheol.* **24**, 1. Copyright © 1980. Reprinted by permission of John Wiley & Sons, Inc.

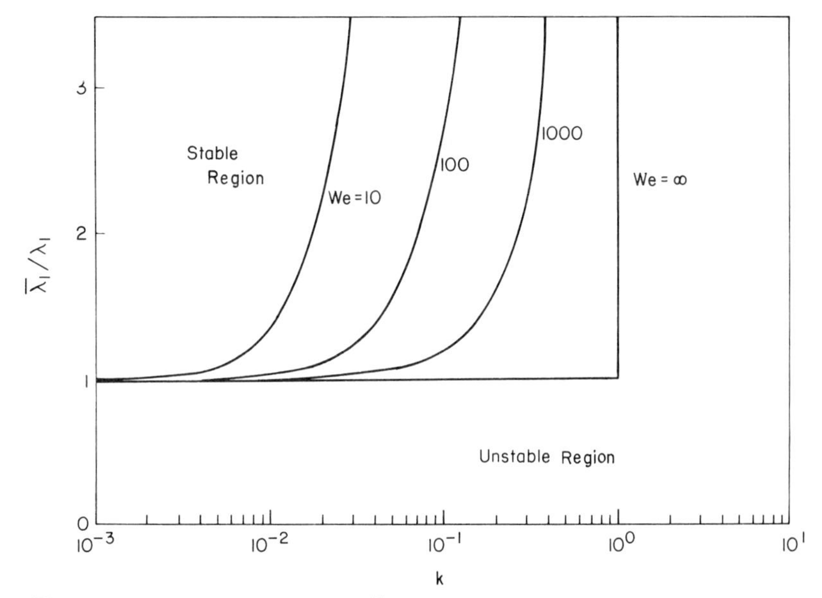

Fig. 5.17 Elasticity parameter ratio $\overline{\lambda}_1/\lambda_1$ versus viscosity ratio k for an $r_0/R = 0.1$, with We as parameter ($Re = 0.01$; $\lambda_1 = 20$) [11]. From H. B. Chin and C. D. Han, *J. Rheol.* **24**, 1. Copyright © 1980. Reprinted by permission of John Wiley & Sons, Inc.

For small values of α, which correspond to long wavelengths, we can apply the method of regular perturbation to obtain approximate solutions. Using the classical Maxwell model [see Eq. (2.37) in Chapter 2], Chin and Han [11] carried out a linear hydrodynamic stability analysis.

Figure 5.17 gives plots showing the regions of stability in term of the viscosity ratio k and elasticity parameter ratio, $\overline{\lambda}_1/\lambda_1$ with dimensionless number We as parameter, for an r_0/R value of 0.1, in which $\overline{\lambda}_1$ and λ_1 are the relaxation time constants, appearing in the linear Maxwell model, for the dispersed phase (i.e., droplet phase) and the continuous phase, respectively. It is seen that, as We increases, the region of stability increases. The effect of We can be interpreted as either the effect of medium viscosity η_0 or the effect of interfacial tension σ. In reference to Fig. 5.17, as either the medium viscosity η_0 increases or the interfacial tension σ decreases, the region of stability increases. Therefore it can be concluded that an increase in medium viscosity stabilizes the liquid cylinder and an increase in interfacial tension destabilizes it.

Figure 5.18 gives plots showing the regions of stability in terms of the viscosity ratio k and the elasticity parameter ratio $\overline{\lambda}_1/\lambda_1$ with λ_1 as parameter for an r_0/R value of 0.1. It is seen that as the medium elasticity parameter λ_1 increases, the stability region also increases. Comparison of Fig. 5.17 with

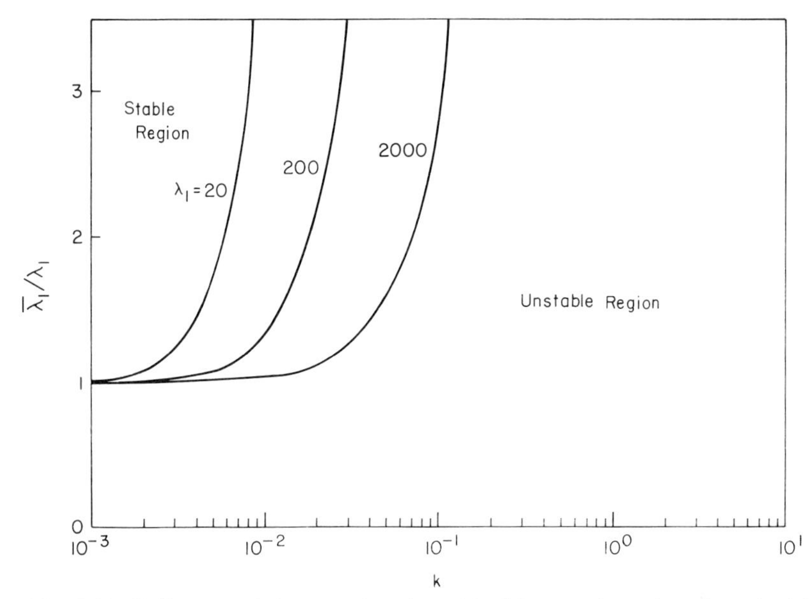

Fig. 5.18 $\overline{\lambda}_1/\lambda_1$ versus k for an $r_0/R = 0.1$, with different values of medium elasticity parameter λ_1 ($We = 10$; $Re = 0.01$) [11]. From H. B. Chin and C. D. Han, *J. Rheol.* **24**, 1. Copyright © 1980. Reprinted by permission of John Wiley & Sons, Inc.

Fig. 5.18 shows that the medium elasticity has an effect very similar to that of *We*.

In reality, it is not likely that a two-phase liquid system will have viscosity and elasticity parameter ratios identically equal to 1.0, and also have a finite value of interfacial tension. Therefore, when *We* approaches infinity (or the interfacial tension σ approaches zero), and at the same time, the viscosity and elasticity parameter ratios become identically 1.0, such a system may be considered, to all intents and purposes, a single-phase one. Thus the system is stable, as shown in Fig. 5.17.

Let us now consider the situation where inertia effects are negligible, and assume that the radius R of the tube is sufficiently large compared to the radius r_0 of the liquid cylinder (i.e., $R/r_0 \to \infty$) so that the wall effect is negligible. Under these conditions, Chin and Han [11] obtained the following relation:

$$\begin{aligned} -\frac{(1-\alpha^2)}{2i\alpha c We}\{I_1(\alpha)\Delta_1 - [\alpha I_0(\alpha) - I_1(\alpha)]\Delta_2\} \\ = \overline{\lambda}_p k[\alpha I_0(\alpha) - I_1(\alpha)]\Delta_1 - \overline{\lambda}_p k[(1+\alpha^2)I_1(\alpha) - \alpha I_0(\alpha)]\Delta_2 \\ + \lambda_p[\alpha K_0(\alpha) + K_1(\alpha)]\Delta_3 + \lambda_p[(1+\alpha^2)K_1(\alpha) + \alpha K_0(\alpha)]\Delta_4 \end{aligned} \tag{5.9}$$

where

$$\Delta_1 = \begin{vmatrix} \alpha I'_1(\alpha) & -K_1(\alpha) & -\alpha K'_1(\alpha) \\ I_0 + 2\alpha I_1(\alpha) & K_0(\alpha) & K_0 - \alpha K_1(\alpha) \\ \bar{\lambda}_p k \alpha I_0(\alpha) & -\lambda_p K_1(\alpha) & \bar{\lambda}_p \alpha K_0(\alpha) \end{vmatrix} \tag{5.10}$$

$$\Delta_2 = \begin{vmatrix} I_1(\alpha) & -K_1(\alpha) & -\alpha K'_1(\alpha) \\ I_0(\alpha) & K_0(\alpha) & K_0 - \alpha K_1(\alpha) \\ \bar{\lambda}_p k I_1(\alpha) & -\bar{\lambda}_p K_1(\alpha) & \bar{\lambda}_p \alpha K_0(\alpha) \end{vmatrix} \tag{5.11}$$

$$\Delta_3 = \begin{vmatrix} I_1(\alpha) & \alpha I'_1(\alpha) & -\alpha K'_1(\alpha) \\ I_0(\alpha) & I_0(\alpha) + \alpha I_1(\alpha) & K_0(\alpha) - \alpha K_1(\alpha) \\ \bar{\lambda}_p k I_1(\alpha) & \bar{\lambda}_p k \alpha I_0(\alpha) & \bar{\lambda}_p \alpha K_0(\alpha) \end{vmatrix} \tag{5.12}$$

$$\Delta_4 = \begin{vmatrix} I_1(\alpha) & \alpha I'_1(\alpha) & -K_1(\alpha) \\ I_0(\alpha) & I_0(\alpha) + \alpha I_1(\alpha) & K_0(\alpha) \\ \bar{\lambda}_p k I_1(\alpha) & \bar{\lambda}_p k \alpha I_0(\alpha) & -\lambda_p K_1(\alpha) \end{vmatrix} \tag{5.13}$$

in which $I_n(x)$ and $K_n(x)$ are modified Bessel functions of order n, the prime on the Bessel functions denotes a differentiation with respect to their argument, and λ_p and $\bar{\lambda}_p$ are defined by

$$\lambda_p = (1 - i\alpha c\lambda_1)^{-1} \qquad \bar{\lambda}_p = (1 - i\alpha c\bar{\lambda}_1)^{-1} \tag{5.14}$$

It is worth mentioning that the above analysis reduces to that of Tomotika [13] when the fluid elasticity is neglected.

Note that Eq. (5.9) is an eigenvalue equation, whose solution gives the growth rate of unstable wavelike disturbances as functions of their wavelengths, and the physical/rheological properties of the fluids concerned. For convenience, let us define the stability parameter S_p by

$$S_p = -i\alpha c \tag{5.15}$$

in which c is a complex number. Equation (5.15) allows us to rewrite Eq. (5.14) as

$$\lambda_p = (1 + S_p\lambda_1)^{-1}, \qquad \bar{\lambda}_p = (1 + S_p\bar{\lambda}_1)^{-1} \tag{5.16}$$

Our problem is now to substitute Eqs. (5.10)–(5.13) and Eq. (5.16) into Eq. (5.9), and then find the values of S_p in terms of the system parameters, i.e.,

$$F(S_p, \alpha, k, \lambda_1, \bar{\lambda}_1, We) = 0 \tag{5.17}$$

We shall now consider some representative results obtained from the analysis of short wave disturbances at low Reynolds numbers [11]. Note that this analysis has no restrictions on the wave number α of the disturbances. Using Eq. (5.17), the value of the stability parameter S_p, defined by Eq. (5.15),

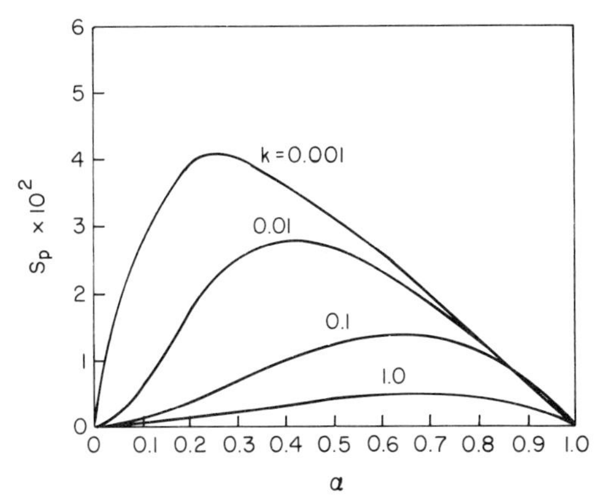

Fig. 5.19 Stability parameter S_p versus wave number α for a Newtonian system (i.e., $\lambda_1 = \bar{\lambda}_1 = 0$; $We = 10$), with k as parameter [11]. From H. B. Chin and C. D. Han, *J. Rheol.* **24**, 1. Copyright © 1980. Reprinted by permission of John Wiley & Sons, Inc.

may be calculated for values of α ranging from 0 to 1.0 for a number of values of system parameters.

Figure 5.19 gives plots of stability parameter S_p versus wave number α for Newtonian systems, with k as parameter. It should be noted that positive values of S_p indicate instability of the system, and that the magnitude of S_p may be considered as a measure of the rate of growth of the disturbances. In other words, the greater the value of S_p, the faster the disturbance will grow, and at a maximum value of S_p, the disturbance grows at the *maximum* rate. This interpretation of S_p will become clear when Eq. (5.15) is substituted in Eq. (5.8). Note that the value of α corresponding to maximum instability varies with the value of k. The practical implication of Fig. 5.19 is that the larger the value of k, the smaller the value of S_p and thus the slower the rate of growth of the disturbance. In other words, as the droplet phase viscosity becomes greater than the medium viscosity, the growth rate of disturbance (hence the rate of droplet breakup) becomes slower. Figure 5.20 shows plots of S_p versus α with k as parameter, for viscoelastic systems. Comparison of Fig. 5.20 with Fig. 5.19 shows that the rate of growth of the disturbances is much greater in viscoelastic systems than in Newtonian systems.

Figure 5.21 shows the effect of the medium elasticity parameter λ_1 on the rate of growth of instability for viscoelastic systems, and Fig. 5.22 shows the effect of the elasticity parameter ratio $\bar{\lambda}_1/\lambda_1$. It is seen that the larger λ_1 is, the greater is the value of S_p, and thus the faster the disturbance will grow, and that the larger $\bar{\lambda}_1/\lambda_1$ is, the greater is the value of S_p again. It can be concluded therefore that once a disturbance is introduced to the system, both the medium elasticity parameter λ_1 and the droplet phase elasticity parameter $\bar{\lambda}_1$ enhance the rate of growth of instability.

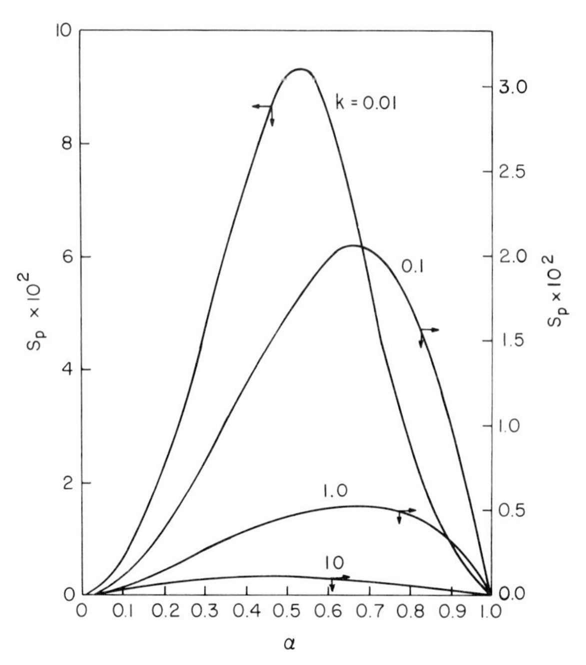

Fig. 5.20 S_p versus α for a viscoelastic system, with k as parameter ($\lambda_1 = 40$; $\bar{\lambda}_1/\lambda_1 = 0.1$; $We = 10$) [11]. From H. B. Chin and C. D. Han, *J. Rheol.* **24**, 1. Copyright © 1980. Reprinted by permission of John Wiley & Sons, Inc.

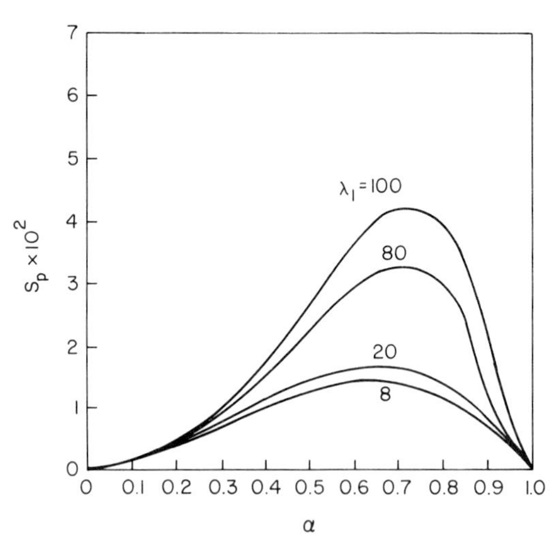

Fig. 5.21 S_p versus α for a viscoelastic system, with λ_1 as parameter ($k = 0.1$; $\bar{\lambda}_1/\lambda_1 = 0.1$; $We = 10$) [11]. From H. B. Chin and C. D. Han, *J. Rheol.* **24**, 1. Copyright © 1980. Reprinted by permission of John Wiley & Sons, Inc.

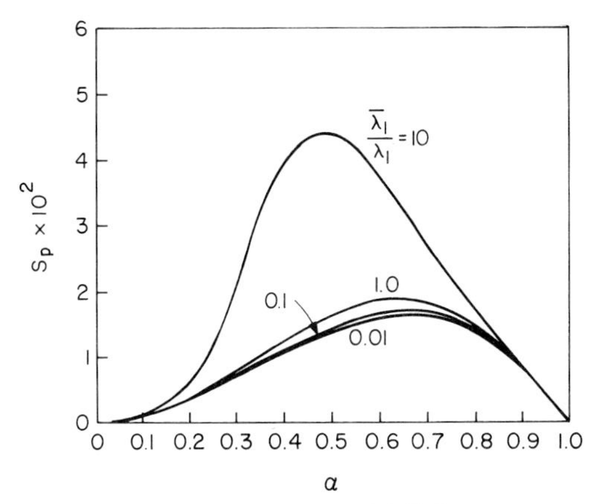

Fig. 5.22 S_p versus α for a viscoelastic system, with $\bar{\lambda}_1/\lambda_1$ as parameter ($k = 0.1$; $\lambda_1 = 20$; $We = 10$) [11]. From H. B. Chin and C. D. Han, *J. Rheol.* **24**, 1. Copyright © 1980. Reprinted by permission of John Wiley & Sons, Inc.

Figure 5.23 shows the effect of We on the rate of growth of instability. It is seen that as We decreases (for instance, as either the interfacial tension increases or the medium viscosity decreases), the rate of growth of instability increases. It can be concluded, therefore, that the medium viscosity stabilizes the system, whereas the interfacial tension destabilizes the system.

In reference to Fig. 5.24, the wavelength Λ of the disturbance may be defined by

$$\Lambda = 2\pi r_0/\alpha \tag{5.18}$$

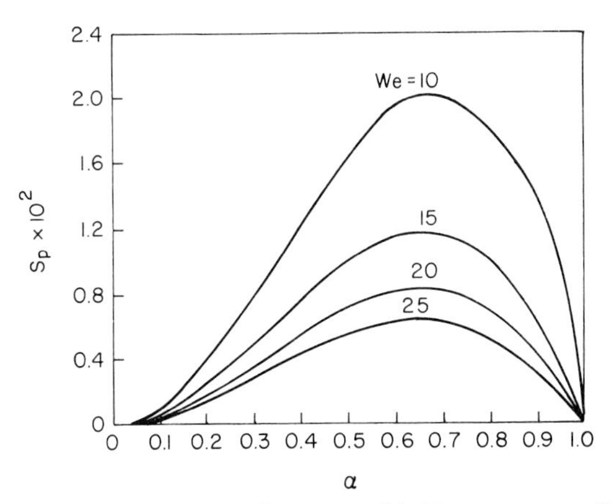

Fig. 5.23 S_p versus α for a viscoelastic system, with We as parameter ($k = 0.1$; $\lambda_1 = 40$; $\bar{\lambda}_1/\lambda_1 = 0.1$) [11]. From H. B. Chin and C. D. Han, *J. Rheol.* **24**, 1. Copyright © 1980. Reprinted by permission of John Wiley & Sons, Inc.

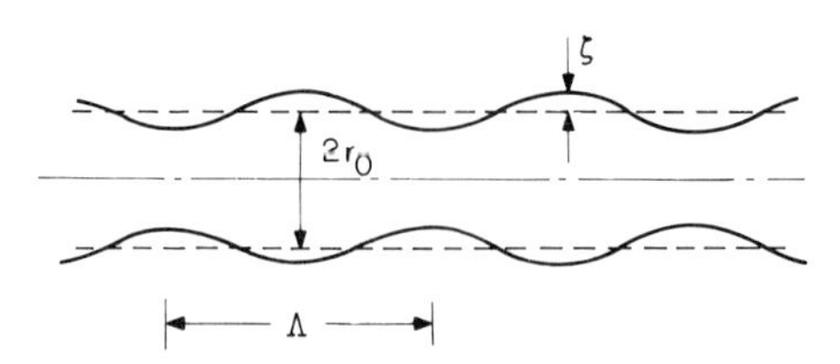

Fig. 5.24 Schematic describing a wavy surface of liquid cylinder under disturbance [11]. From H. B. Chin and C. D. Han, *J. Rheol.* **24**, 1. Copyright © 1980. Reprinted by permission of John Wiley & Sons, Inc.

where r_0 is the radius of the liquid cylinder and α is the wave number of the infinitely small sinusoidal disturbance. It can be seen from Eq. (5.18) that at $\alpha = 1$, the wavelength of the disturbance becomes $2\pi r_0$, which is the circumference of the liquid cylinder. Therefore, disturbances for α greater than 1 are not of practical interest. We can then define the deformation D_c for droplet breakup at $\alpha = 1$ by

$$D_c = (L_c - B)/(L_c + B) = 0.517 \tag{5.19}$$

in which $L_c = 2\pi r_0$ (the value of the wavelength Λ at $\alpha = 1$) and $B = 2r_0$. It is of interest to note that the value of $D_c = 0.517$ given above happens to agree approximately with the critical value 0.5 for the breakup of droplets, as calculated by Rumscheidt and Mason [21] who used Taylor's theory [1]. Rumscheidt and Mason, however, observed experimentally that the breakup of droplets occurred at critical values of deformation greater than 0.5.

In the analysis given above, we have no way of knowing at what value of α the breakup of a liquid cylinder occurs. According to Rayleigh's hypothesis [24], if disturbances of all wave numbers are initially present at infinitely small amplitude, then the disturbance which will dominate the breakup will be the one with the *largest* growth rate. From the results shown in Figs. 5.19–5.23, we can see that there is one wave number α_m for which the disturbance grows more rapidly than for any other wave number. That is, α_m is the wave number which has the *largest* disturbance growth rate (i.e., the largest value of S_p in Figs. 5.19–5.23). If we now assume that breakup occurs at $\alpha = \alpha_m$, we can then calculate the deformation D_B for droplet breakup using the following expression:

$$D_B = (L_B - B)/(L_B + B) \tag{5.20}$$

where L_B is the wavelength of disturbance at $\alpha = \alpha_m$ [see Eq. (5.18)]. Since α_m is always less than 1.0, L_B is always greater than L_c and thus D_B is greater than D_c.

Figure 5.25 shows the effect of k on α_m for viscoelastic systems. This figure is prepared using Fig. 5.20. Note in Fig. 5.25 that α_m initially increases with k, and then decreases as k increases further, going through a maximum. Note

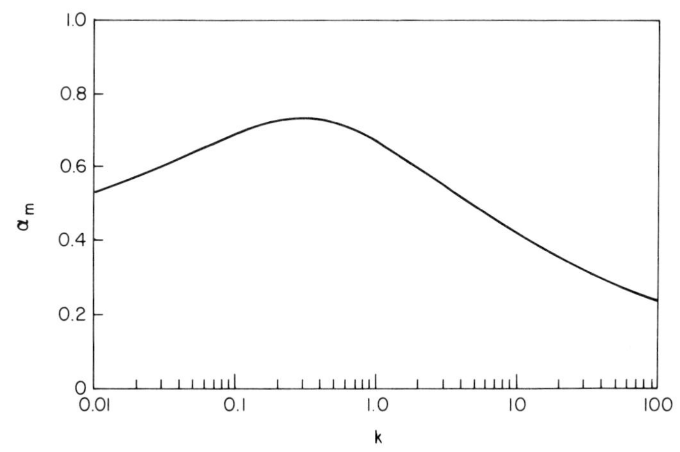

Fig. 5.25 α_m versus k for a viscoelastic system ($\lambda_1 = 40$; $\bar{\lambda}_1/\lambda_1 = 0.1$; $We = 10$) [11]. From H. B. Chin and C. D. Han, *J. Rheol.* **24**, 1. Copyright © 1980. Reprinted by permission of John Wiley & Sons, Inc.

from Eq. (5.18) that the greater the value of α_m, the smaller the value of L_B, and hence the smaller the size of the broken droplets. We can now make some interesting observations about the effect of the rheological properties of the liquid systems under consideration on the size of the broken droplets.

Figure 5.19 shows that for Newtonian systems, the value of α_m increases as k increases. Therefore one can conclude that for a given medium, the larger the droplet phase viscosity, the larger the value of α_m and thus the smaller the broken droplets. Note in Fig. 5.19, however, that the larger the value of k, the smaller the value of S_p, meaning that it takes a longer time for the more viscous liquid cylinder to breakup than for the less viscous liquid cylinder. This means that if the liquid cylinder is very viscous, it will elongate to a cylinder of very small radius and remain stable for a long time before finally breaking up into droplets of very small size.

Figure 5.21 shows that the larger the medium elasticity parameter λ_1, the greater the value of α_m, and thus the smaller the broken droplets will be. Figure 5.22 shows that for given values of λ_1 and k, the larger the droplet phase elasticity parameter $\bar{\lambda}_1$, the smaller the value of α_m, and thus the larger the broken droplets will be. Figure 5.23 shows that for viscoelastic systems, as We increases (i.e., as either the interfacial tension σ decreases or the medium viscosity η_0 increases), the value of α_m tends to decreases and thus larger broken droplets are to be expected. Of particular interest is the role that both the fluid elasticity and the interfacial tension play in the rate of growth of the disturbances. It is seen in Figs. 5.21–5.23 that the growth rate of disturbances (i.e., the value of S_p) increases with an increase in any one of the three parameters: medium elasticity, droplet phase elasticity, interfacial tension. In other words, once disturbances are introduced to the system and instability occurs, both the fluid elasticity and interfacial tension cause an increase in the rate of growth of the disturbances.

To summarize, both the viscosity ratio and the elasticity parameter ratio, as well as interfacial tension, play important roles in determining the rate of growth of disturbances and the size of the resulting broken droplets.

It should be pointed out that the details of a stability analysis of a rheologically complex flow problem, such as the one discussed above, often depends on the choice of a rheological model. Therefore the stability analysis presented above, which is based on the linear Maxwell model, must be interpreted with caution because it is possible for one to reach different conclusions by using other rheological models.

Mikami *et al.* [14] assumed that the extending liquid thread breaks up when the amplitude of the disturbance, which has the largest magnification at a given time t, becomes equal to the thread radius (see Fig. 5.24). They then calculated the time t_B for breakup. Note that they considered the breakup of Newtonian droplets in a Newtonian medium, subjected to a steady uniaxial extensional flow field. Figure 5.26 gives plots of a_B/r_0 versus $\dot{\gamma}_E t_B$ with *We* as parameter for different values of k, in which a_B is the radius of broken droplets, r_0 is the radius of the liquid thread before breakup, and $\dot{\gamma}_E$ is the elongation rate. It is seen that a_B/r_0 decreases with increasing $\dot{\gamma}_E t_B$,

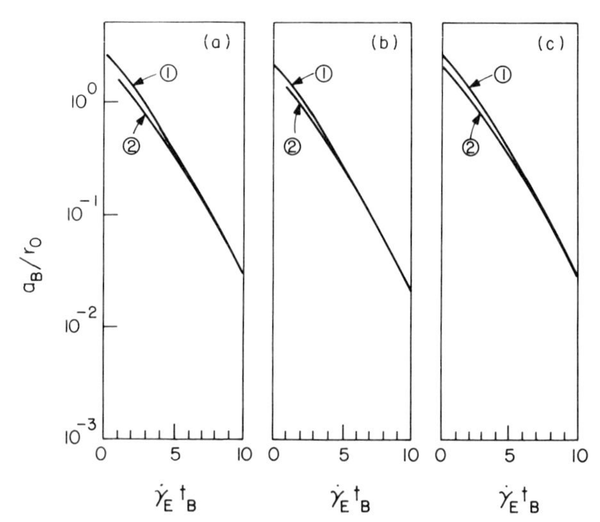

Fig. 5.26 Theoretically predicted results of a_B/r_0 versus $\dot{\gamma}_E t_B$ in an axisymmetric elongational flow field, at various values of viscosity ratio [14]: (a) For $k = 10^{-3}$, ① $We = 10^{-3}$, ② $We = 1.0$; (b) For $k = 1.0$, ① $We = 10^{-3}$, ② $We = 10^{-1}$; (c) For $k = 10^2$, ① $We = 10^{-3}$, ② $We = 10^3$. Reprinted with permission from *International Journal of Multiphase Flow* **2**, T. Mikami *et al.*, Copyright 1975, Pergamon Press, Ltd.

becoming independent of *We* at sufficiently large $\dot{\gamma}_E t_B$. Mikami *et al.* [14] also noted that, except for very large disturbances, the larger k is, the smaller a_B/r_0 will be for a given *We*. This means that if the liquid thread is very viscous, it will elongate to a thread of very small radius and remain stable for a long time before finally breaking up into droplets of very small size.

Using silicone oil as the droplet phase and castor oil as the suspending medium, Mikami *et al.* [14] also carried out breakup experiments. Figure 5.27 gives plots of the experimentally obtained breakup time t_B and breakup thread radius r_B versus $\dot{\gamma}_E$ for different values of k. It is seen that both t_B and r_B decrease with increasing $\dot{\gamma}_E$, and that the larger the viscosity ratio k, the more stable the liquid cylinder and the smaller the values of r_B. This experimental observation is in agreement with the theoretical predictions discussed above.

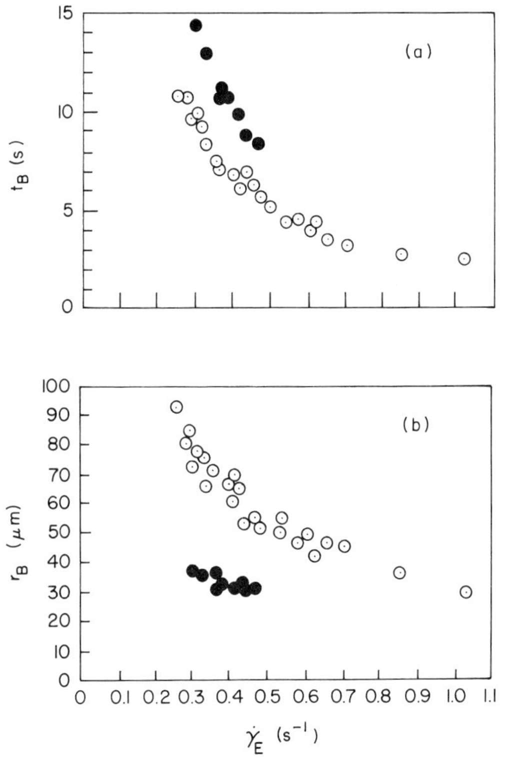

Fig. 5.27 Breakup time and breakup radius versus elongation rate for droplet breakup in a four-roll apparatus [14]: (a) breakup time versus elongation rate; (b) breakup radius versus elongation rate. Symbols represent: (●) $k = 1.46$, $\sigma = 5.2 \times 10^{-2}$ N/m; (⊙) $k = 0.148$, $\sigma = 4.6 \times 10^{-2}$ N/m. Silicone oil is the droplet phase and caster oil is the suspending medium. Reprinted with permission from *International Journal of Multiphase Flow* **2**, T. Mikami *et al.*, Copyright 1975, Pergamon Press, Ltd.

5.3.2 Slender-Body Analysis of Droplet Breakup

As discussed above in reference to the experimental observations the droplets are observed to be long and slender prior to breakup, especially when the droplet viscosity is much smaller than the medium viscosity (i.e., when $k \ll 1$). Hence, any theoretical study, which expands the solution about the spherical shape and retains only two (or three) terms in the series, becomes correspondingly less accurate. Therefore, attempts have been made to develop theories [16, 18, 19] of droplet breakup, using the technique of slender-body theory [19, 20].

Especially worth noting in such attempts is the study of Acrivos and Lo [16], who employed the method of singular perturbation to obtain the solutions they sought. They obtained a deformation curve given by:

$$\left(\frac{\dot{\gamma}_E \eta_0 a}{\sigma}\right)\left(\frac{\overline{\eta}_0}{\eta_0}\right)^{1/6} = \frac{1}{\sqrt{20}}\left[\frac{l}{a}\left(\frac{\overline{\eta}_0}{\eta_0}\right)^{1/3}\right]^{1/2} \Bigg/ \left\{1 + \frac{4}{5}\left[\frac{l}{a}\left(\frac{\overline{\eta}_0}{\eta_0}\right)^{1/3}\right]^3\right\} \tag{5.21}$$

which is plotted in Fig. 5.28.

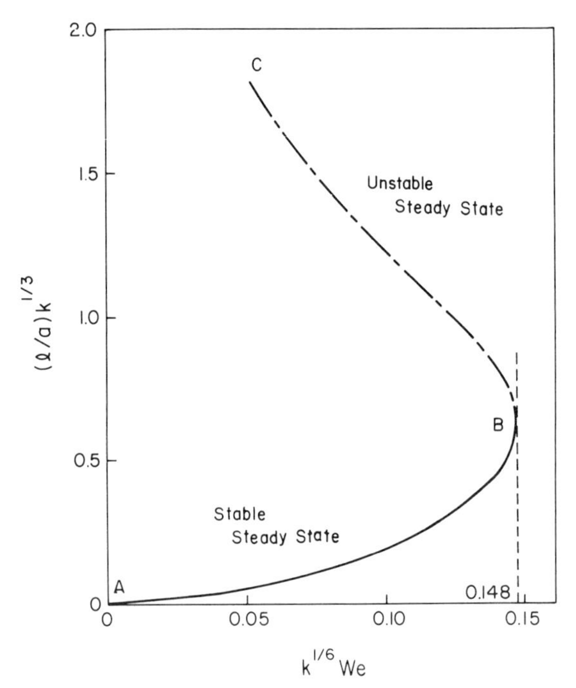

Fig. 5.28 Theoretically predicted curve of droplet breakup [16].

In Eq. (5.21), a denotes the radius of the sphere with the same volume as the droplet and l the half-length of the elongated droplet. Note, in Fig. 5.28, that the broken curve BC represents unstable steady-state solutions which are of no physical significance, and that the solid curve AB represents stable steady-state solutions. Thus, the point B in Fig. 5.28 represents the point of droplet breakup, i.e., according to the Acrivos–Lo analysis, a steady shape of droplet cannot exist for dimensionless shear rates exceeding the critical value:

$$(\dot{\gamma}_c \eta_0 a/\sigma)(\bar{\eta}_0/\eta_0)^{1/6} = 0.148 \tag{5.22}$$

They noted, however, that the criterion developed above for droplet breakup should apply only if $k = \bar{\eta}_0/\eta_0 \ll O(10^{-2})$.

Acrivos and Lo [16] have shown further that in the plane hyperbolic flow field, whose components with respect to Cartesian axes (x, y, z) are

$$v_z = z, \qquad v_y = -y, \qquad v_x = 0 \tag{5.23}$$

the condition for droplet breakup is

$$(We)_c \propto 1/k^{1/6} \tag{5.24}$$

This prediction is in very good agreement with Grace's experimental observations [4], according to which, at the point of breakup,

$$(We)_c = 0.1/k^{0.16} \qquad \text{for} \quad k < 0.1 \tag{5.25}$$

In simple shearing flow

$$v_z = y, \qquad v_y = 0, \qquad v_x = 0 \tag{5.26}$$

and for $k \ll 1$, Acrivos and Lo [16] have shown that

$$(We)_c \propto 1/k^{2/3} \tag{5.27}$$

which is in fair agreement with the corresponding expression

$$(We)_c = 0.17/k^{0.55} \tag{5.28}$$

obtained by fitting Grace's data for $k < 0.1$.

5.4 ENGINEERING IMPLICATIONS OF DROPLET BREAKUP IN POLYMER PROCESSING

In practical situations of polymer processing operations, the phenomenon of droplet breakup is far more complex than that described above in well-controlled shearing or extensional flow experiments. The complexity

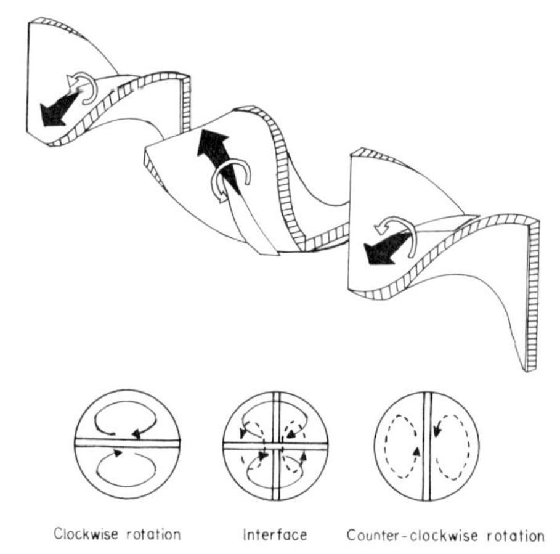

Fig. 5.29 Schematic describing the radial mixing in the Kenics static mixer unit [25].

comes primarily from the flow geometries often encountered in practice. More specifically, seldom does one find either a uniform shearing flow field or a steady extensional flow field in polymer processing equipment (e.g., screw extruder, static mixer).

Figure 5.29 shows a schematic describing the rotation of flow in each mixing element of the Kenics static mixer [25]. It is seen that fluid elements (say, droplets) entering at the center of the stream are forced to the outer wall and back again on a continuous basis, which is attributed to the overall effect of radial mixing. Note that the cross-sectional area, which the fluid elements pass through, is not constant.

In order to facilitate our discussion, let us consider a flow geometry, which consists of a number of short converging–diverging elements, as schematically shown in Fig. 5.30. In reference to Figs. 5.6–5.8, one may conjecture that droplets will get elongated in the converging section, and when the elongation is sufficient, breakup will occur as the elongated droplets

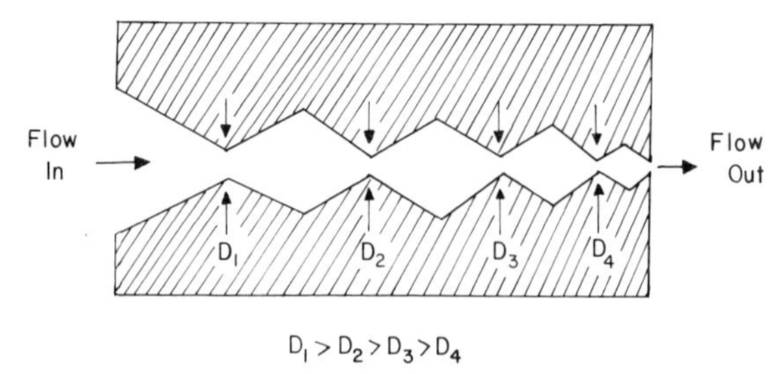

Fig. 5.30 Schematic of a flow channel consisting of a series of diverging and converging sections.

recoil in the diverging section of the mixing element. Converging flow can be described by a combination of shear and extensional flows. Thus the kinematics involved pose considerable mathematical difficulties in analyzing the deformation of many droplets, occurring invariably in real situations in the mixing of two immiscible liquids.

On the basis of the experimental and theoretical evidence discussed above, we can, however, speculate on what might be an optimal design of a mixing device consisting of short converging–diverging elements. It should be remembered that the extent of elongation of a spherical droplet depends, among other things, on the rate of deformation in the converging section, which in turn depends strongly on the opening of the end of the diverging section. In other words, for a given fluid system, the smaller the opening of the diverging section, the larger the rate of shear (or rate of elongation), and therefore the better chance there will be for the droplet to break up.

In reference to Fig. 5.30, let us assume that a droplet breaks up as it passes element I. The size of broken droplets (assuming all have the same size) entering element II must be smaller than the original size in element I. Therefore, if one wishes to have finer dispersion, the opening of the converging section at element II, D_2, must be smaller than the opening D_1, because small droplets require greater rates of deformation than large droplets in order to have the same extent of elongation. This observation then leads us to suggest that the openings of the successive converging sections should be progressively smaller, i.e.,

$$D_1 > D_2 > D_3 > \cdots > D_n \tag{5.29}$$

The rheological properties (viscosity and elasticity), the interfacial tension of the fluid system, and the rate of deformation will determine the relative size of D_i and the number of elements (n) required to achieve a specific state of dispersion desired.

Converging flow may be considered to be a transient flow in the Lagrangian sense. (Hence any analysis dealing with converging flow may be handled using the material coordinates.) For transient flows, however, a dimensionless time should also enter into the analysis. At least for obtaining experimental correlations, one may introduce the following dimensionless variable:

$$\tau = \sigma t / \bar{\eta}_0 a \tag{5.30}$$

Grace [4] reports that for Newtonian fluids, a value of $\tau \cong 50$–100 is necessary for breakup at the critical rate of strain for both shear and hyperbolic flows, with $k = \bar{\eta}_0/\eta_0 \cong 1.0$. Grace's experiments for $k = 1.0$ required total strains

$$(\dot{\gamma} a \eta_0/\sigma)(\sigma t/\bar{\eta}_0 a) = \dot{\gamma} t \eta_0/\bar{\eta}_0 \cong 10^2\text{–}10^3 \tag{5.31}$$

for breakup. Equation (5.31), though empirical, may be considered as a useful guide determining the flow conditions (i.e., residence time t or rate of strain $\dot{\gamma}$) for droplet breakup.

For viscoelastic systems, the relaxation times of the two liquids to be mixed pose an additional complication. When the relaxation times are much larger than the residence time (the time required to pass through the converging section), then the liquids may be considered as elastic solids. In such situations, the breakup in the diverging section of the mixing device, as suggested in Fig. 5.30, may be influenced predominantly by the fluid elasticity, rather than by the interfacial tension. This speculation is based on the experimental observation that the breakup of droplets occurs in the diverging section during the *recoil* of the elongated droplets. The recoil phenomenon of an elastic solid must be governed predominantly by the elasticity of the material, rather than by the interfacial tension. Theoretical analysis of this problem has yet to be tackled, and this is one of the many challenging problems that await solution in enhancing our understanding of the dynamics of mixing two viscoelastic liquids.

Problems

5.1 Tavgac [8] reports that when a Newtonian droplet is suspended in a viscoelastic medium, breakup is obtained at lower shear rates under *transient* conditions than under *steady* conditions. (a) Assuming that Tavgac's experimental observation is correct, give theoretical reasons for it. (b) Will Tavgac's observation also be true for a viscoelastic droplet suspended in a viscoelastic medium? How about for a viscoelastic droplet suspended in a Newtonian medium? Give reasons for your answers.

5.2 Tavgac [8] reports that the shearing stresses required for droplet breakup in a viscoelastic medium are generally much smaller than those required in a Newtonian medium. Assuming that Tavgac's experimental observation is correct, give theoretical reasons for it.

5.3 Lee [9] reports that more regular breakup patterns were obtained in cases where the liquid threads were of small diameter. This would suggest that under the influence of strong interfacial forces, breakup is more regular. Give theoretical reasons why this should be so.

5.4 Lee [9] reports that in the breakup of viscoelastic droplets, the fluid elasticity increases the critical value of $We = (\eta_0 a\dot{\gamma}/\sigma)$ above that found for a Newtonian fluid having the same viscosity. Give theoretical reasons for this.

5.5 Consider a droplet that is placed in a four-roll apparatus and is stretched into a threadlike liquid cylinder. It is reported that the liquid cylinder breaks up into small droplets when the motion of the rollers abruptly stops. The mass balance satisfies the relationship:

$$4\pi a_0^3/3 = \pi r_0^2 l_0 \tag{5.32}$$

in which a_0 is the radius of the initial droplet, and l_0 and r_0 are the length and radius of the elongated liquid cylinder at the roller speed Ω_0 (see Fig. 5.31).

Suppose that when l_0/r_0 is less than 20, the elongated droplet returns to the original (spherical) shape after the rollers stop. Thus $l_0/r_0 = 20$ is the critical value at and beyond which the elongated droplet will break up into smaller droplets *upon cessation* of the motion of the rollers. Experimental evidence in the literature indicates that breakup would not occur under *steady* roller motion (cohesive breakup), even when the roller speed is so high that the length-to-radius ratio of the liquid cylinder reaches about 100. Suppose that $l_c/r_c = 100$ is the value at the roller speed Ω_c, at which breakup of the liquid cylinder occurs due to the cohesive mechanism. Note that the values of l_0/r_0 and l_c/r_c may depend on the rheological properties of the liquid systems.

As the roller speed increases from $\Omega_0(l_0/r_0 = 20)$ to $\Omega_c(l_c/r_c = 100)$, the radius of the liquid cylinder decreases accordingly from r_0 to r_c. The smaller the radius of the liquid cylinder, the larger the number of broken droplets there will be, and the smaller their size. Therefore the roller speeds between Ω_0 and Ω_c may be used to control the size and the number of broken droplets.

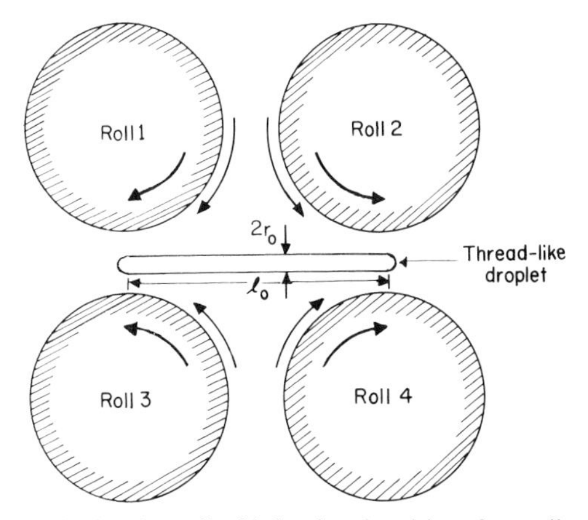

Fig. 5.31 Schematic showing a liquid droplet placed in a four-roll apparatus, in which the droplet is stretched into a threadlike liquid cylinder.

Consider the following combinations of liquid systems;

(i) Newtonian droplet/Newtonian suspending medium;
(ii) Newtonian droplet/viscoelastic suspending medium;
(iii) viscoelastic droplet/Newtonian suspending medium; and
(iv) viscoelastic droplet/viscoelastic suspending medium.

(a) Derive an expression describing the recoil phenomenon of an elongated liquid cylinder, whose length-to-radius ratio is less than l_0/r_0, upon cessation of the motion of the rollers.

(b) Derive the system equations describing the phenomenon of breakup of an elongated liquid cylinder, whose length-to-radius ratio lies between l_0/r_0 and l_c/r_c, upon cessation of the motion of rollers. Assume that a generalized Maxwell fluid with the Jaumann derivative [see Eq. (2.42) in Chapter 2] may represent the rheological behavior of the viscoelastic fluid for either the droplet phase or the suspending medium.

REFERENCES

[1] G. I. Taylor, *Proc. R. Soc. London, Ser. A* **146**, 501 (1934).
[2] W. Bartok and S. G. Mason, *J. Colloid Sci.* **13**, 293 (1958); **14**, 13 (1959).
[3] F. D. Rumscheidt and S. G. Mason, *J. Colloid Sci.* **16**, 238 (1961).
[4] H. P. Grace, *Eng. Found. Res. Conf. Mixing, 3rd, Andover, N. H., 1974.*
[5] S. Torza, R. C. Cox, and S. G. Mason, *J. Colloid Interface Sci.* **38**, 395 (1972).
[6] H. Karam and J. C. Bellinger, *Ind. Eng. Chem., Fundam.* **7**, 576 (1968).
[7] R. W. Flumerfelt, *Ind. Eng. Chem., Fundam.* **11**, 312 (1972).
[8] T. Tavgac, Ph.D. Thesis (Ch. E.), Univ. of Houston, Houston, Texas, 1972.
[9] W.-K. Lee, Ph.D. Thesis (Ch. E.), Univ. of Houston, Houston, Texas, 1972.
[10] C. D. Han and K. Funatsu, *J. Rheol.* **22**, 113 (1978).
[11] H. B. Chin and C. D. Han, *J. Rheol.* **24**, 1 (1980).
[12] S. Tomotika, *Proc. R. Soc. London, Ser. A* **150**, 322 (1935).
[13] S. Tomotika, *Proc. R. Soc. London, Ser. A* **153**, 302 (1936).
[14] T. Mikami, R. G. Cox, and S. G. Mason, *Int. J. Multiphase Flow* **2**, 113 (1975).
[15] D. Barthès-Biesel and A. Acrivos, *J. Fluid Mech.* **61**, 1 (1973).
[16] A. Acrivos and T. S. Lo, *J. Fluid Mech.* **86**, 641 (1978).
[17] G. I. Taylor, *in* "Proceedings of the 11th International Congress on Applied Mechanics" (H. Gortler, ed.), p. 790. Springer-Verlag, Berlin and New York, 1966.
[18] J. M. Rallison and A. Acrivos, *J. Fluid Mech.* **89**, 191 (1978).
[19] G. K. Batchelor, *J. Fluid Mech.* **44**, 419 (1970).
[20] J. B. Keller and S. I. Rubinow, *J. Fluid Mech.* **75**, 705 (1976).
[21] F. D. Rumscheidt and S. G. Mason, *J. Colloid Sci.* **17**, 260 (1962).
[22] C. C. Lin, "The Theory of Hydrodynamic Stability." Cambridge Univ. Press, London, 1955.
[23] C. S. Yih, *J. Fluid Mech.* **27**, 337 (1967).
[24] Lord Rayleigh, *Philos. Mag.* **34**, 145 (1892).
[25] "The Static Mixer Unit and Principles of Operation," Tech. Bull. KTEK-1. Kenics Corp., North Andover, Massachusetts, 1972.

6

Dispersed Flow of Gas-Charged Polymeric Systems

6.1 INTRODUCTION

The use of gases, solutes, or chemicals for generating gases in polymers has long been practiced to produce polymeric foams. Uses of polymeric foams are, to name only a few, in acoustic insulation, thermal insulation, packaging, cushioning, construction of low cost housing, etc.

There are four types of polymeric foam systems [1]: (a) A thermoplastic foam, which is produced by dispersing a mass of gas bubbles in a molten plastic (e.g., polystyrene, polyethylene, polyvinylchloride, cellulose acetate, nylon, ABS) with subsequent solidification; (b) A thermoset foam (e.g., polyurethane, urea formaldehyde, rubber foams, epoxy foams), in which the reactants are foamed before they reach their final state of polymer cure, often reacting as they are foamed, the mixture being finally cured to a thermoset state as part of the stabilization process; (c) A latex which is foamed and stabilized by phase inversion; (d) A system in which cellular structures are produced by the removal of a fugitive phase from a polymer matrix. When the polymer matrix is solidified or cured, the resulting foam has a glassy or elastomeric character (i.e., rigid or flexible), depending on the inherent properties of the polymer matrix and, also, on the extent of foaming (i.e., the foam density). Open, closed, or reticulated cell structure of varying size can be produced with the help of different foaming techniques, as shown in Fig. 6.1 [1]. The closed-cell foams are obtained primarily from thermoplastic resins, and the open-cell foams from thermoset resins [2]. A variety of

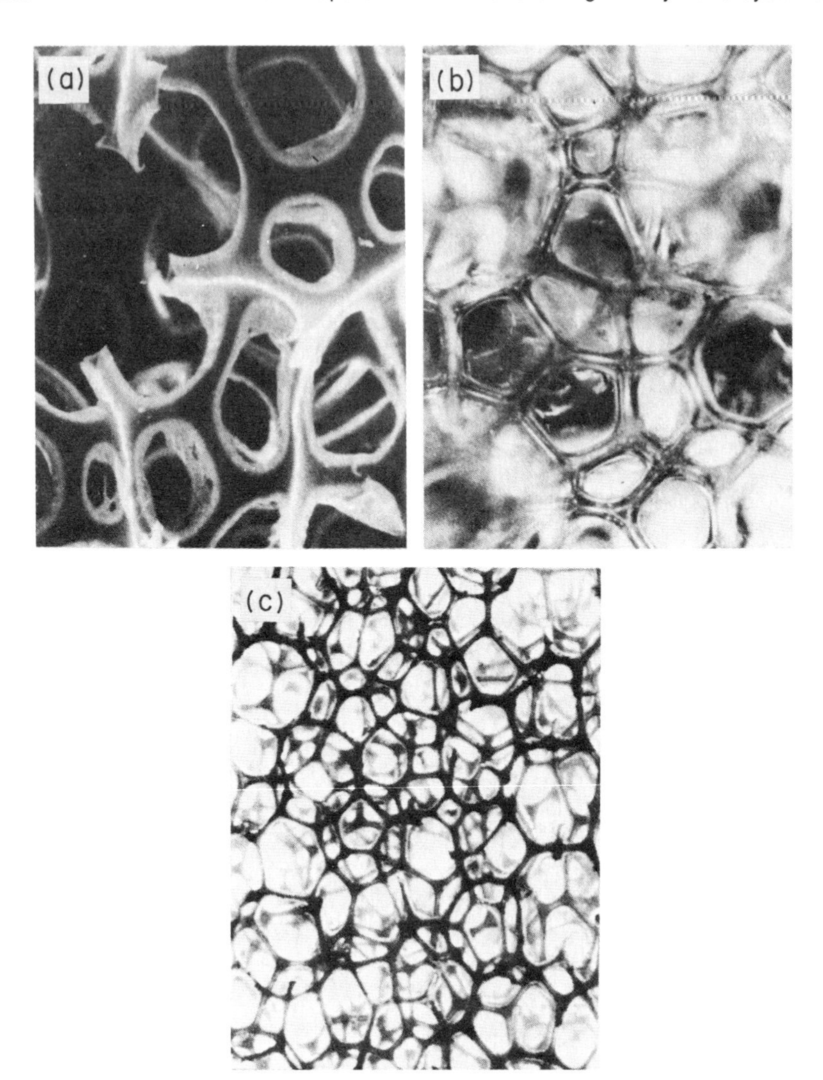

Fig. 6.1 Photographs describing various cell structures of plastic foam [1]: (a) open cell structure; (b) closed cell structure; (c) reticulated cell structure. From "Mechanical Properties of Polymeric Foams," E. A. Meinecke and R. C. Clark, 1973. Courtesy of Technomic Publishing Co., Inc., Westport, CT 06880.

foaming techniques have been developed by the plastics industry, and the use of various polymeric materials in making foam products and the description of various foaming techniques, are well presented in the monographs by Benning [3] and Frisch and Saunders [4].

Polymeric foams are often grouped into two types (low-density foams and high-density foams), according to the relative amount of gas, which is dispersed in the polymer matrix. Low-density foams contain more than nine volumes of gas for each volume of polymer matrix, and the density of this type of foam is less than 0.1 g/cm^3. The density range for medium-density foam is normally considered to be 0.1–0.6 g/cm^3. A high-density foam contains less than 1.5 volumes of gas for each volume of polymer matrix, and its density is greater than 0.6 g/cm^3.

In this chapter, we shall only be concerned with the processing of structural polymeric foams (i.e., high-density foams with a closed-cell structure). With structural foam processes (either injection molding or extrusion), almost all thermoplastic resins can be used together with blowing agents. Blowing agents may be dispersed in a polymer melt, either by injecting gas (e.g., nitrogen, Freon) directly into the melt or by preblending the resin with a chemical blowing agent. Chemical blowing agents are generally available in the form of powder, and are decomposed by heat in the extruder, generating gases (e.g., nitrogen, carbon dioxide). The gas generated will first dissolve, under high pressure, in the polymer melt, and later will be released from the melt as the pressure is decreased, yielding a cellular structure in the final product. Several comprehensive reviews of blowing agents are available in the literature [5–7]. These reviews include discussions of the decomposition of blowing agents and the product of their decomposition.

In line with the scope of this monograph, our primary objective in this chapter is to discuss the fundamental aspects of bubble dynamics and rheology associated with the processing of gas-charged molten polymers via either extrusion or injection molding operations. For this, we shall first review very briefly the structural foam processes of industrial importance, the phenomena of bubble formation and growth in molten polymer, and the diffusion and solution of gases in molten polymer.

6.1.1 Structural Foam Processes of Industrial Importance

Basically, there are two ways of producing a structural foam part by extrusion. One is a modification of the conventional extrusion process, and the other is the Celuka process [8]. In the use of the conventional extrusion process, a preblend of chemical blowing agent and polymer is fed into the extruder where plastication and decomposition of the blowing agent take place under high pressure to prevent any premature expansion before the polymer-foaming gas mixture leaves the die. A temperature gradient is preferably established to initiate decomposition in the last zone of the extruder. The foaming extrudate expands as the mixture emerges from the die in the approximate shape of the desired part. Once the appropriate amount of foaming has been obtained, the expanded extrudate is pulled through the shape cooler, by conventional takeup device, for final shaping of the surface.

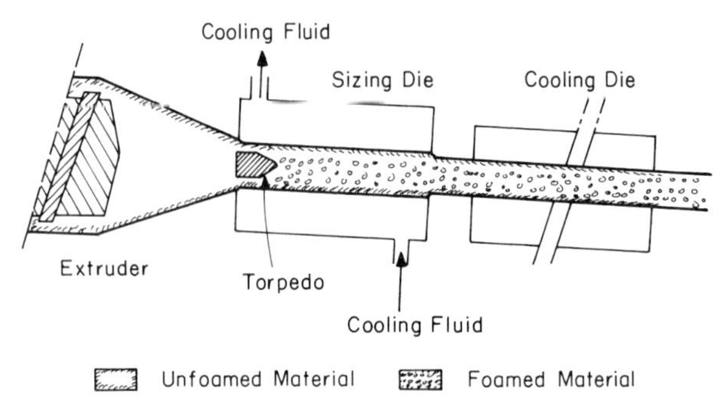

Fig. 6.2 Schematic of the Celuka foam extrusion process.

The Celuka Process [8] consists basically of a conventional extruder, a unique die design, cooling, sizing, and takeup devices, as schematically shown in Fig. 6.2. In this process, resin blended with blowing agent is fed to an extruder and is plasticated in the usual way. The foamable melt is forced through die jaws having the same configuration and dimensions as those desired in the final product. The extrusion die is designed with a central torpedo which directs the flow of the extrudate to the peripheries of the jaws of the die, forming a void in the extrudate into which the foaming occurs by inward expansion toward the center. At the same time, the surface of the extrudate is quickly cooled as it comes in contact with the cooler to form a solid integral skin.

There have been a number of new methods for molding a foamed material introduced in recent years. Nonetheless, any one of them can be considered as a modification of the four main processes for structural foam molding: the Union Carbide low-pressure process [9], the USM high-pressure process [10], the ICI Sandwich process [11], and conventional injection molding. Either inert gas or chemical blowing agents can be used to produce the foam in the first three processes, but chemicals are used almost exclusively in the conventional molding process.

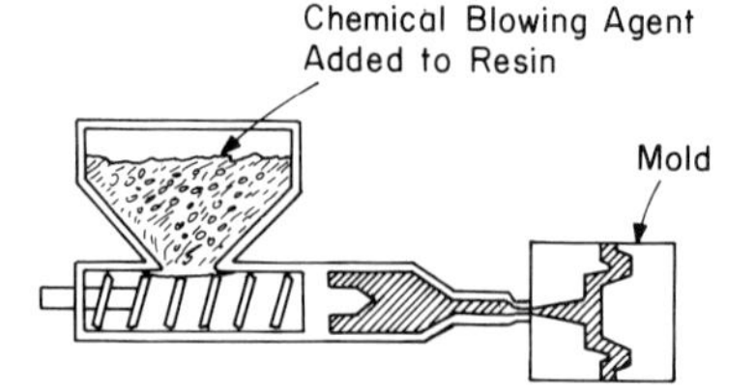

Fig. 6.3 Schematic of a conventional screw injection molding machine.

One of the key points assessing the adaptability of a conventional machine (see Fig. 6.3) for foam molding is fast injection, since foam density, uniformity of cell structure, and optimum surface finish depend upon the speed of injection. In cases where the number or type of foam parts does not warrant capital expenditure on a single-purpose machine, many machine manufacturers modify their conventional units with accumulator systems to increase the speed of injection.

The Union Carbide Process [9] is a low-pressure variation of the basic injection molding process. Basic equipment consists of an extruder, a valve, a hydraulic accumulator, and a mold. In the operation, the resin is melted and mixed in the extruder. This material is pumped by the extruder into the accumulator and held there above the foaming temperature but at a sufficient pressure to prevent foaming. When the accumulator has enough material to fill the mold cavity, a valve is opened and a charge is injected by a hydraulic ram into the mold cavity via multiple nozzles to assure complete filling. The valve is then closed and the accumulator is refilled. Since the mold is under low pressure, the melt immediately expands to fill it. Molding pressures range from 1.4×10^4 to 2.8×10^4 N/m^2 (200–400 psi). Only enough plastic to fill about one half of the mold cavity is delivered by the accumulator; the rest of the cavity being filled by the expansion of the plastic foams. As the foamable mixture flows through the mold cavity, its surface cells collapse when forced against the mold surface, as more material enters the mold from the accumulator, thus forming the characteristic solid skin.

The USM process [10], which is a high-pressure process, provides a more distinct skin and a surface which may be essentially free of splay marks if the correct combination of blowing agent and processing conditions are used. The USM process differs from low-pressure processes in that the mold is first filled with a foamable melt, and then foaming is allowed to occur as a mold member expands. A reciprocating-screw machine can be used for plastication and injection. A typical molding cycle for the USM process is as follows: (1) The plasticator is ready to inject the molten polymer containing the blowing agent after closing of the mold; (2) The nozzle is opened and a screw pushes forward, filling the mold under pressure; (3) When the injection is completed, the nozzle shuts and foaming starts as the mold cavity volume expands at a controlled rate while the plasticator begins to refill the accumulator. The design features of the USM foam process include controlled density by a suitable design of mold expansion motion that results in greater strength in critically stressed sections, combined with weight savings in the remainder of the part. However, it is limited in part geometry. Parts must be simple and/or flat so that proper foaming takes place. Higher mold costs are inevitable because designers must provide retractable inserts to increase the volume of the mold.

A combination of dispersed and stratified multiphase flow has also been used to produce commerical products. For instance, in producing a sandwiched foam product, the gas-charged core component is either coextruded with the skin component through a sheet-forming die, or is coinjection molded in a mold cavity, giving rise to the product, as schematically shown in Fig. 6.4. The sandwich foam coextrusion process is a polymer processing technique that combines the coextrusion process and the foam extrusion process. In producing sandwiched foam products, the core-forming polymer B, containing a blowing agent, is coextruded with the skin-forming polymer A. A large number of combinations of polymer systems may be used for the skin and core components of a sandwiched foam. In the selection of materials, both the core-forming polymer B and the skin-forming polymer A can be the same (except that B contains a blowing agent), or they can be different polymers.

A sandwiched foam product may also be obtained by means of the sandwich injection molding process, known as the ICI sandwich process [11, 12]. In this process, two polymers are injected into the mold cavity through a common sprue with the following sequence: (a) the skin-forming polymer A is injected to partially fill the mold cavity; (b) the core-forming polymer B containing a blowing agent is injected into the mold cavity, forcing the skin-forming polymer A outwards. In this step, the core-forming polymer B fills the mold cavity completely and ensures a good surface finish with a uniform skin layer of polymer A; (c) for the second time, the skin-forming polymer A is injected into the mold cavity to clear the foamable polymer B in the sprue. This step is to prevent the residual foamable polymer B remaining in the sprue from contaminating the skin-forming polymer A of the next molding; (d) the mold is opened a small distance so that the core-forming polymer B can now foam to give a uniform foam structure within a thin skin of polymer A that contains no foam residue.

As is the case with standard injection molding, the success of the structural foam injection molding process depends on choosing the right combination of temperature, pressure, injection speed, mold design, etc. In addition, it is important to keep in mind that the blowing agent must be dispersed uniformly in the melt. In structural foam injection molding, temperatures

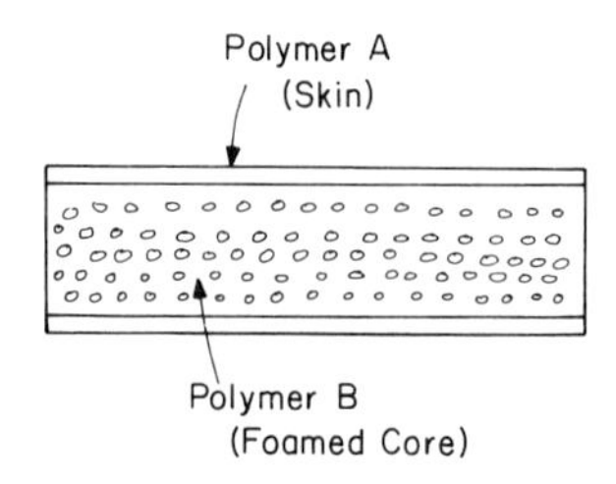

Fig. 6.4 Schematic of the cross section of a sandwiched foam product.

in the feed zone should be kept low and then be increased towards the nozzle in order to minimize the premature decomposition of the blowing agent. It is essential, however, that the melt temperature prior to injection into the mold is high enough for optimum melt viscosity and for full decomposition of the blowing agent [13, 14].

A rapid injection rate is essential for obtaining a molded part of uniform cell size and cell size distribution. Fast fill is desirable for assuring that all the foaming occurs in the mold, and not in the machine. A slow injection rate allows gas bubbles to elongate as they pass through the nozzle or gate, producing a less uniform cell structure, usually with poor physical properties. With slow injection rates, more material will solidify at the mold surface and consequently the resistance to flow will increase. Hence, more pressure is required to fill the mold and this partially counteracts the expansion power of the blowing agent. Complete mold filling can then only be achieved by increasing the quantity of material injected into the mold, giving rise to higher density.

As one may surmise, the gas will expand as the molten polymer containing the blowing agent passes into the mold cavity. Therefore the rate of bubble growth is a controlling factor in obtaining the ultimate physical/mechanical properties desired of the molded article. Some important processing variables which contribute to establishing the final cell size and its distribution in the molded article are: (a) the injection speed; (b) the injection melt temperature, which largely determines the melt viscosity and elasticity; (c) the pressure and temperature gradients set up in the mold. Because of the temperature gradient in the mold cavity, bubble growth is slower near the cold mold wall than in the interior of the mold cavity, developing a skin-core morphology in the molded articles [15–17].

6.1.2 The Phenomenon of Bubble Formation and Growth in Molten Polymer

There are two fundamental steps involved in the formation of a bubble of visible size in a molten polymer saturated with gas. They are nucleation and growth. In structural foam processes, a blowing agent is dissolved in a molten polymer. Bubbles nucleate from the molten polymer and grow by diffusion of blowing agent from the molten polymer to the gas bubbles. In order to obtain a desirable cell structure of a foam product, an understanding of the mechanism of nucleation and growth of a bubble in a foaming process is essential.

Nucleation of many bubbles from supersaturated solutions of gas in polymers is the key to the success of achieving uniform cell structure. When a small number of bubbles are nucleated, they can grow to large size in an almost explosive fashion and often can result in poor cell structure. Highly supersaturated solutions of gases in polymers are extremely unstable and relatively easy to nucleate. Hansen and Martin [18, 19] have reported that well-dispersed particulate materials (e.g., metal particles, sodium aluminum silicate, SiO_2, Fe_2O_3) act as nucleating sites for the formation of bubbles from the dissolved gas and help to create a uniform cell structure in the extruded foam. Figure 6.5 gives photographs displaying the effect of nucleating particulates on the number and size of bubbles present in a polyethylene extrudate [2].

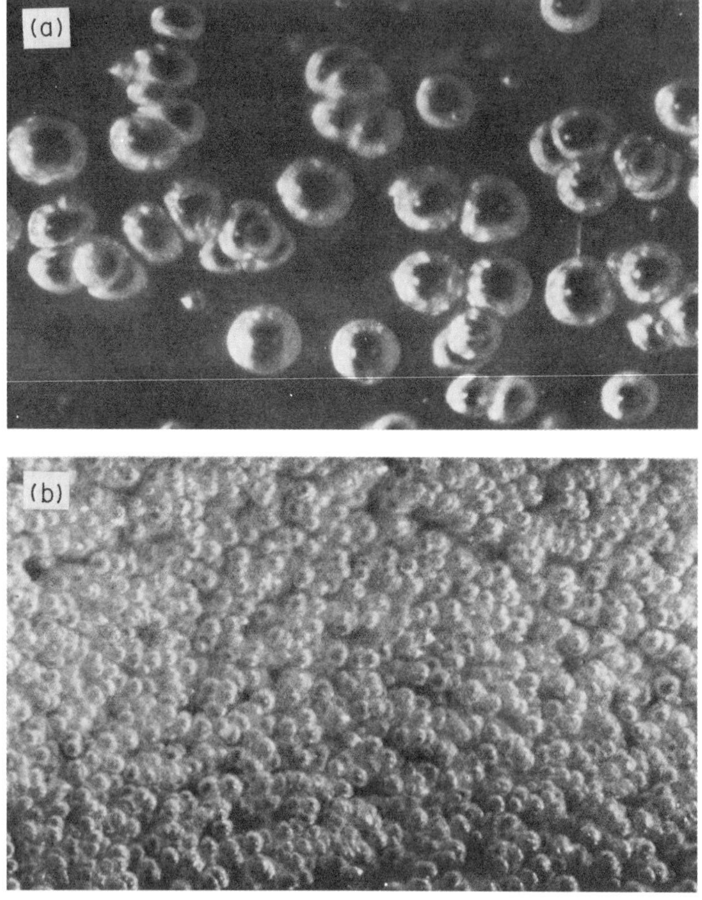

Fig. 6.5 Photographs displaying the effect of nucleating particulates on the number and the size of bubbles in a low-density polyethylene extrudate: (a) without any nucleating agent; (b) with 0.01 vol.% of metal particles. Reprinted from [2] by Courtesy of Marcel Dekker, Inc.

Self-nucleation may occur at high supersaturation. As soon as nucleation relieves the gas concentration sufficiently, bubbles are no longer formed, and the concentration of gas in the polymer melt is further reduced by diffusion into the bubbles, which already exist. When no additional gas is available, the equilibrium (saturation) concentration of gas in the polymer melt is reached. From this time on, bubbles can grow only by diffusion of gas from small bubbles into larger bubbles, by coalescence. The rate of these processes depends on a number of variables, the most important being the temperature and the viscosity of the polymer melt. Stewart [20] has obtained approximate expressions for calculating the number of bubbles formed under a high degree of supersaturation.

In the presence of nucleating agents, one would expect a behavior similar to that described above, except that bubble formation would occur at lower gas concentrations than in the absence of nucleating agents. In other words, heterogeneous nucleation, providing a great abundance of nucleating sites within the supersaturated melt, is much more effective than homogeneous nucleation in facilitating bubble formation and producing uniform cell structure. Hansen and Martin [19] report that a tenfold increase in nucleator particle concentration caused a thousandfold increase in the number of bubbles, and a fiftyfold decrease in the volume of each bubble, at approximately the same level of supersaturation, as shown in Fig. 6.6. It is then clear

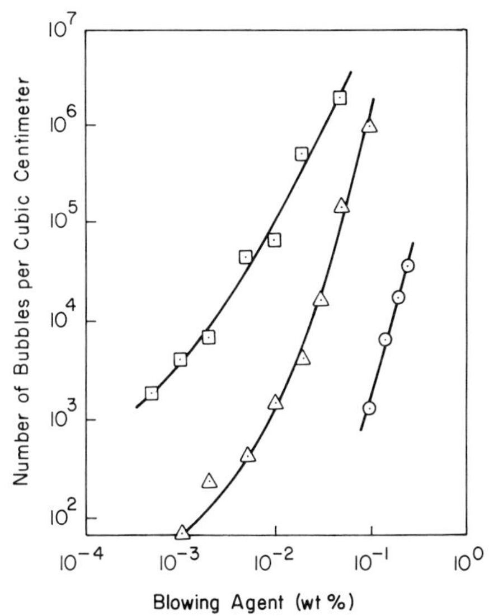

Fig. 6.6 Effect of blowing agent concentration on the number of bubbles present in a low-density polyethylene extrudate, with various blowing agents. (⊙) Azodicarbonamide only; (△) Azodicarbonamide with 0.01 wt.% nucleator particles; (⊡) Azodicarbonamide with 0.1 wt.% nucleator particles. Reprinted from [2] by Courtesy of Marcel Dekker, Inc.

that the degree of supersaturation, and both the number and kind of nucleator particles, determine both cell size and the efficiency of a foaming operation.

After bubbles have been formed, the next stage of the foaming process involves the growth of those bubbles. When nucleated bubbles have attained a certain minimum size, they will continue to grow until the pressure inside them, p_g, is in equilibrium with the external pressure in the polymer melt, p_L. That is, at equilibrium the following force balance should hold:

$$\Delta p = 2\sigma/R \tag{6.1}$$

where $\Delta p = p_g - p_L$, σ is the interfacial tension, and R is the bubble radius. Note that the value of σ is approximately $3 \times 10^{-2} \sim 5 \times 10^{-2}$ N/m ($30 \sim 50$ dyn/cm) for most gas/polymer melt systems [21]. Equation (6.1) indicates that the smaller the bubble radius R, the higher the pressure of gas in the bubble. In other words, the application of external pressures in excess of the pressure inside the bubble will suppress bubble growth.

It is of interest to point out at this juncture that there exists a theory [22] which relates the pressure p_g in a spherical bubble in an elastomer to the shear modulus G of the medium by

$$p_g/G = \tfrac{5}{2} - 2(R_0/R) - \tfrac{1}{2}(R_0/R)^4 + 2\sigma/GR \tag{6.2}$$

where R is the bubble radius at the pressure p_g, R_0 is the initial bubble radius, and σ is the interfacial tension. Equation (6.2) indicates that the critical pressure p_c can be obtained by differentiating Eq. (6.2) with respect to R and setting $d(p_g/G)/dR = 0$. For small values of (R_0/R_c), in which R_c is the critical bubble size at the pressure p_c, and when σ is small compared to GR_c, Eq. (6.2) reduces to

$$p_c = 5G/2 \tag{6.3}$$

as a limiting case.

The above theory [Eq. (6.3)] was tested against experimental results and was found to describe the physical phenomenon rather well [22]. It should be remembered, however, that Eqs. (6.2) and (6.3) are applicable strictly to elastomers, and they are not expected to describe the bubble growth in a viscoelastic polymer melt because the melt viscosity should play an important role in determining the critical pressure in viscoelastic fluids. Unfortunately, at present there is no theory available which permits us to relate the critical pressure to the rheological properties of viscoelastic fluids.

Since the pioneering work of Rayleigh [23] in 1917, the growth/collapse of a stationary, spherical bubble suspended in a stationary infinite liquid medium has been the subject of considerable interest to many researchers [24–38].

Neglecting the hydrodynamic resistance of the liquid to bubble growth, Epstein and Plesset [24] developed a theory of diffusion-controlled growth (and collapse) of a stationary gas bubble suspended in a Newtonian medium. They obtained an approximate solution for bubble growth as

$$R(t) = k't^{1/2} \tag{6.4}$$

in which R is the bubble radius, t is the time, and k' is the rate constant which is a function of diffusion coefficient, solubility of the gas in the liquid, and the initial concentration of the gas in the liquid.

Gent and Tompkins [22] reported that Eq. (6.4) adequately describes the kinetics of bubble growth in purely elastic media, and Stewart [20] also reported that his experimental results of bubble growth in thermoplastic elastomers were in reasonable agreement with the prediction by Eq. (6.4).

Villamizar and Han [39] have measured the growth rate of a single gas bubble (carbon dioxide) in a stationary molten polystyrene. Figure 6.7 gives plots of bubble radius R versus time t for bubbles of different initial sizes. It is seen that the rate of bubble growth (i.e., the slope of the curve) is almost independent of the initial size of the bubbles and it may be represented as

$$R(t) = k't^n \tag{6.5}$$

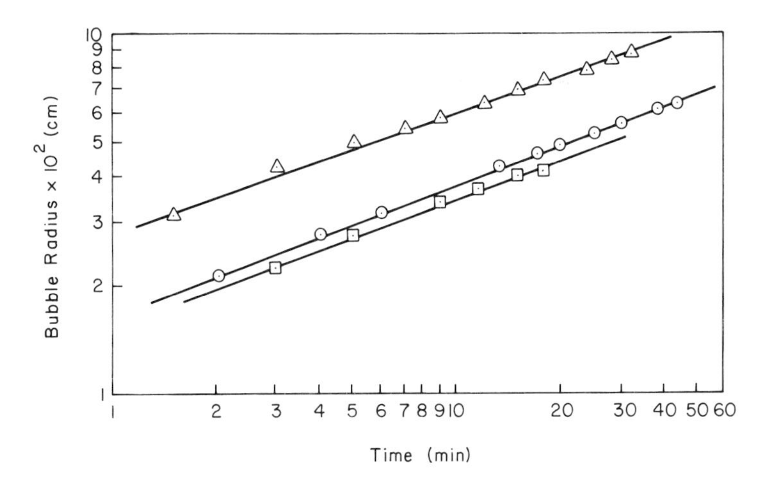

Fig. 6.7 Bubble radius versus time during the period of diffusion-controlled bubble growth under the isothermal condition ($T = 200°C$), with different initial bubble sizes [39]: (△) Bubble No. 1 ($n = 0.315$); (⊙) Bubble No. 2 ($n = 0.348$); (⊡) Bubble No. 3 ($n = 0.328$). The medium is polystyrene and the gas is CO_2.

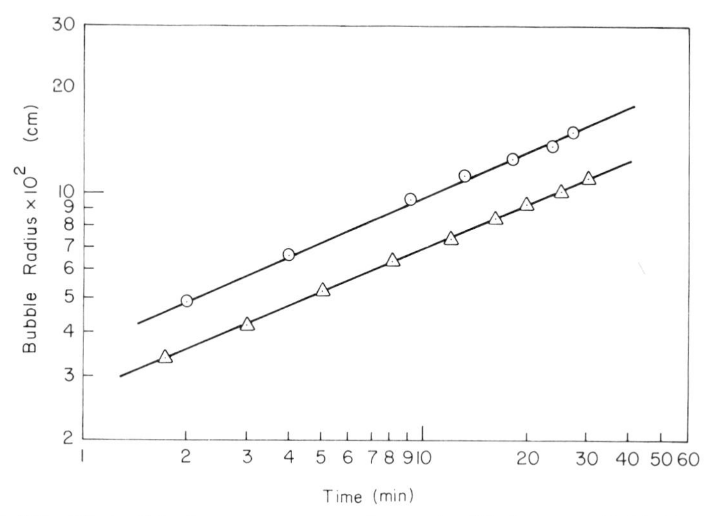

Fig. 6.8 Effect of blowing agent concentration on the diffusion-controlled bubble growth under isothermal condition ($T = 220°C$), with different blowing agent concentrations [39]: (⊙) 0.5 wt. % $NaHCO_3$ ($n = 0.430$); (△) 0.1 wt. % $NaHCO_3$ ($n = 0.431$). The medium is polystyrene and the gas is CO_2.

where n is approximately 0.33 for the particular polymer–gas system investigated. Figure 6.8 gives plots of bubble radius R versus time t at two different levels of initial blowing agent concentration in polystyrene melt at otherwise identical conditions. It is seen that the growth rate is almost independent of the initial concentration of blowing agent, following the power-law relation, Eq. (6.5), with the value of n being approximately equal to 0.43. The effect of melt temperature on the bubble growth rate is given in Fig. 6.9. It is seen that the bubble growth rate increases with melt temperature, yielding the power-law constant $n = 0.31$ at 180°C and $n = 0.47$ at 240°C. It should be noted that an increase in melt temperature implies a decrease in melt viscosity. It is then clear that the melt viscosity plays a significant role in controlling diffusion-controlled bubble growth in a very viscous molten polymer.

Villamizar and Han [39] attributed the observed discrepancy between their experimental results given above (i.e., varying n values) and the Epstein–Plesset theory ($n = 0.5$) to the following possibilities. First, the theory of Epstein and Plesset [Eq. (6.4)] assumes the liquid to be inviscid, which is certainly not true in an experimental system employing a very viscous molten polystyrene. Second, the polystyrene melt possesses an elastic property, and thus the melt elasticity might have played some role in the bubble

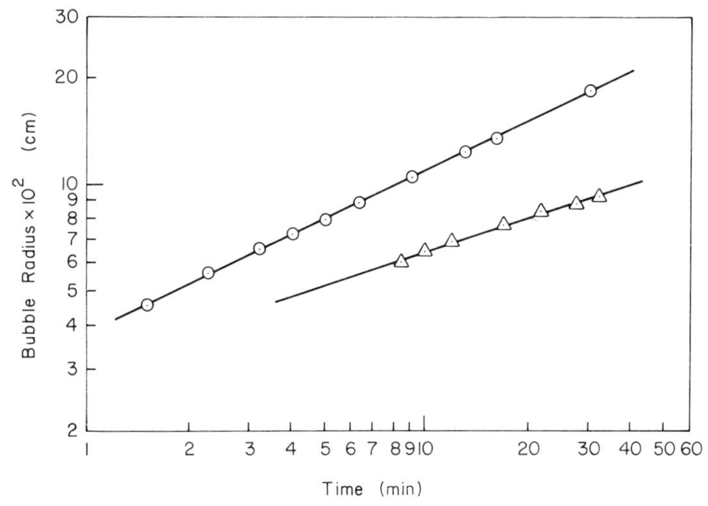

Fig. 6.9 Effect of temperature on the diffusion-controlled bubble growth under the isothermal condition [39]: (⊙) At 240°C ($n = 0.474$); (△) at 180°C ($n = 0.306$). The medium is polystyrene and the gas is CO_2.

growth. Third, the decomposition of the sodium bicarbonate ($NaHCO_3$), used as blowing agent, gives rise not only to CO_2, but also to H_2O in the gas phase. Therefore the presence of H_2O in the gas-charged molten polystyrene might have played a role, however small it may have been, in the evolution of CO_2 gas bubbles from the melt.

Hobbs [40] extended the Epstein–Plesset theory by including the coalescence of small bubbles into bubble growth kinetics and obtained the following expression:

$$R(t) = R_0 + ABt^{1/2} + (1 - A)B^2t/(2R_0 + Bt^{1/2}) \tag{6.6}$$

where R_0 is the initial bubble radius, and A and B are constants which depend on diffusion coefficient, gas solubilities, temperature, and pressure differential in bubbles of different sizes. Note that, for large t, Eq. (6.6) reduces to Eq. (6.4).

In structural foam processes, which deal with viscoelastic polymer melts, the roles that both melt viscosity and melt elasticity play in bubble growth must be included in any theoretical analysis. Barlow and Langlois [27] were the first to include in their analysis both the mass transfer effect of a solute and the hydrodynamic effect of the suspending liquid. They considered a single, spherical gas bubble growing in an infinite stationary, Newtonian liquid at a constant temperature. By assuming that the concentration of the solute in the liquid phase is significantly disturbed by the growing bubble only in a thin shell surrounding the bubble (referred to as the thin shell approximation), they obtained analytical solutions for two limiting cases; namely, the initial and the asymptotic stages of bubble growth, and numerical solution for the intermediate stage. Two important results deduced from their analysis are: (1) initial bubble growth is directly proportional to time, whereas bubble growth at the asymptotic stage is proportional to the square root of time; (2) neither the initial stage solution nor the asymptotic solution can describe the growth rate of the intermediate stage.

Yang and Yeh [29] investigated the effect of shear-dependent viscosity on the growth/collapse of a gas bubble using a power-law model and the Bingham-plastic model. Using a power-law model, Street *et al.* [35] extended the work done by Yang and Yeh by including the effects of mass, momentum, and heat transfer in their analysis. They also used the thin shell approximation in simplifying the mass and heat transfer equations and solved the resulting equations numerically. The important results of their analysis are: (1) the bubble grows faster under isothermal conditions than in nonisothermal cases; (2) the more non-Newtonian a fluid is, the greater is the initial growth rate; (3) the initial growth rate is governed by both the mass and momentum processes; (4) the most significant parameters controlling the growth rate are the diffusivity and concentration of the blowing agent, the viscosity of the medium, and the extent of non-Newtonian behavior of the medium (i.e., the value of the power-law index n).

An attempt to include the elastic effect of a medium in the analysis of bubble growth was first made by Street [30], who, using the Oldroyd three-constant model, investigated the stress field surrounding a spherical bubble. A particularly important outcome of his analysis was that the initial bubble growth rate in a viscoelastic medium was always greater than that in a Newtonian fluid of the same viscosity. He attributed this seemingly unexpected result to the fact that the total stress opposing the bubble growth in a viscoelastic medium, at some distance from the bubble surface, would be lower than that in the corresponding Newtonian fluid, whereas the opposite phenomenon would occur near the bubble surface. Interestingly enough, his analysis showed that medium elasticity had a strong effect on the bubble growth.

Zana and Leal [38], using a modified Oldroyd model, considered the diffusion-induced collapse of a spherical bubble in an infinite viscoelastic medium. They considered the dissolution of a spherical gas bubble which rises by buoyancy through a viscoelastic fluid. The results of their study may be summarized as follows: (i) The rate of bubble collapse decreases rather sharply with an increase in medium viscosity; (ii) During the initial stage, a decrease in retardation time contributes to an increase in bubble collapse rate, but it tends to retard bubble motion for the remainder of the bubble's lifetime. A decrease in relaxation time, however, has the opposite effect; (iii) An increase in surface tension causes an increase in bubble collapse rate, particularly for very small bubbles; (iv) For a gas which has low solubility, the rate of bubble collapse is completely controlled by the mass transfer process.

6.1.3 Solubility and Diffusivity of Gases in Molten Polymers

By this time it should be clear that both solubility and diffusivity of gases in molten polymers play important roles in controlling the bubble formation and growth in structural foam processes. In the past, studies [41–53] were made to experimentally determine the solubility and diffusivity of gases in molten polymers and to correlate these physical/transport properties to molecular parameters and/or thermodynamic variables (i.e., temperature and pressure).

For a given amount of blowing agent used, the amount of gas which dissolves in the molten polymer is a function of the pressure and temperature that are applied to the gas-charged molten polymer. While a gas bubble grows, pressure within the bubble will decrease as its radius becomes larger, so it will grow rapidly until much of the excess dissolved gas in the contiguous area has been utilized in expanding the polymer. Therefore, bubble growth will be governed by the rate of diffusion of dissolved gas to the polymer–gas interface as well as by the degree of supersaturation and viscosity of the melt. Later in this chapter, we shall discuss this problem in great detail on a firm theoretical basis.

The experimental data of solubility reported in the literature appears to be correlatable by Henry's law:

$$X = K_p p_g \tag{6.7}$$

where X is the concentration of gas in the melt [cm^3(STP)/g], in which STP refers to standard temperature and pressure, p_g is the partial pressure of the dissolved gas (atm), and K_p is Henry's law constant [cm^3(STP)/(g)(atm)]. Table 6.1 gives experimental values of Henry's law constant for gases in some molten polymers. Durrill and Griskey [47] report that for certain gas/polymer systems, Henry's law constant K_p follows the Arrhenius behavior with temperature:

$$K_p = K_{po} \exp(-E_s/R_g T) \tag{6.8}$$

in which K_{po} is Henry's law constant at the reference temperature T_0 (°K), E_s is the heat of solution of the gas in the melt (Kcal/mole), R_g is the universal

TABLE 6.1

Some Experimental Values of Henry's Law Constant of Gases in Molten Polymers at 188°C[a]

Polymer	Gas	K_p [cm^3(STP)/(g)(atm)]
Polyethylene	Nitrogen	0.111
	Carbon dioxide	0.275
	Monochlorofluoromethane	0.435
Polypropylene	Nitrogen	0.133
	Carbon dioxide	0.228
	Monochlorofluoromethane	0.499
Polyisobutylene	Nitrogen	0.057
	Carbon dioxide	0.210
Polystyrene	Nitrogen	0.049
	Carbon dioxide	0.220
	Monochlorofluoromethane	0.388
Polymethylmethacrylate	Nitrogen	0.045
	Carbon dioxide	0.260

[a] From Durrill and Griskey [46].

TABLE 6.2

Heats of Solution and Energies of Activation of Gases in Molten Polymers[a]

Polymer	Gas	Temperature range (°C)	E_s (Kcal/mole)	E_d (Kcal/mole)
Polyethylene	Nitrogen	125–188	0.95	2.0
	Carbon dioxide	188–224	−0.80	4.4
Polypropylene	Carbon dioxide	188–224	−1.70	3.0
Polystyrene	Nitrogen	120–188	−1.70	10.1

[a] From Durrill and Griskey [47].

gas constant, and T is the melt temperature (°K). Table 6.2 gives experimentally determined values of E_s for some gases in molten polymers. Equations (6.7) and (6.8) permit us to calculate the maximum amount of gas that can be dissolved in a polymer melt at any given temperature and pressure.

Of particular interest is the solubility of nitrogen (N_2) in polyethylene and polystyrene melts, as given in Fig. 6.10. Note that N_2 is used very

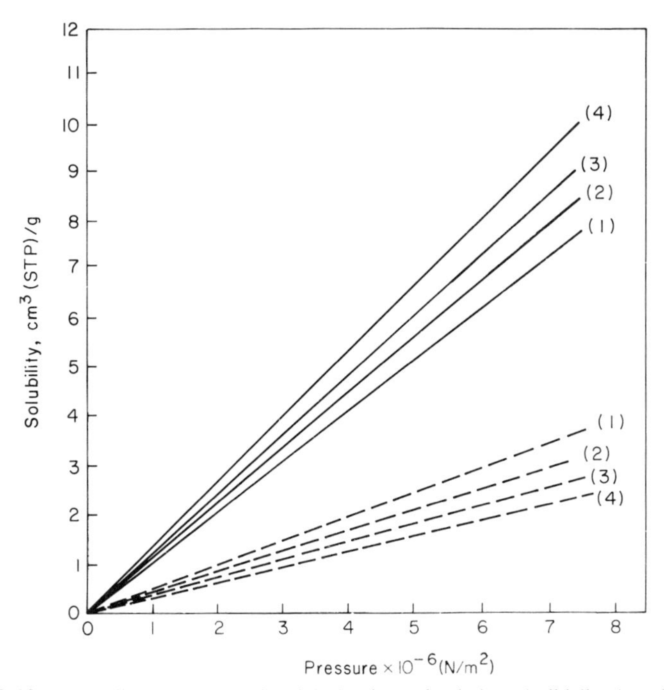

Fig. 6.10 Solubility of nitrogen in high-density polyethylene (solid lines) and in polystyrene (dotted lines) at various temperatures (°C) [56]: (1) 180; (2) 220; (3) 260; (4) 320. The plots are constructed with the assumption that Henry's law is valid over the range of pressure indicated.

frequently in structural foam processes and Fig. 6.10 is constructed using Eqs. (6.7) and (6.8), and the values of E_s given in Table 6.2. The practical implication of Fig. 6.10 is that an increase in melt temperature will make more N_2 dissolve in the polyethylene, whereas it will have an opposite effect on the nitrogen/polystyrene system.

Stiel and Harnish [50] showed that the logarithm of the experimentally determined Henry's law constant K_p is linearly correlatable to $(T_c/T)^2$, in which T_c is the critical temperature (°K) of the gas, for a number of gases in polystyrene. Using the published data in the literature, they obtained the following empirical relationship:

$$\ln K_p = -2.338 + 2.706(T_c/T)^2 \tag{6.9}$$

It would be interesting to observe if a similar correlation may be obtained for gases in other polymer systems.

There is experimental evidence that the form of Henry's law given by Eq. (6.7) does not hold for certain polymer/gas systems, as shown in Figs. 6.11

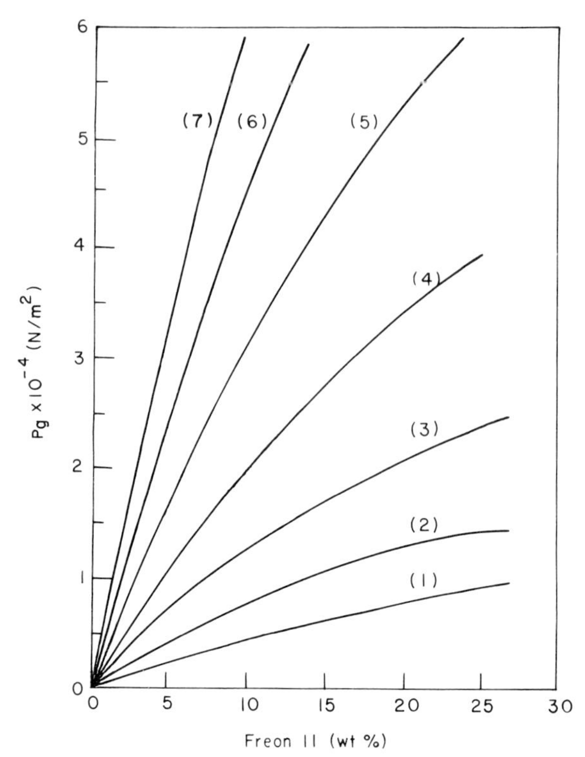

Fig. 6.11 Solubility of Freon 11 in polystyrene at various temperatures (°C) [54]: (1) 25; (2) 40; (3) 60; (4) 80; (5) 100; (6) 120; (7) 140.

and 6.12 [54]. Note that Freon has been widely used as a physical blowing agent in producing low-density foams of polystyrene and low-density polyethylene. It is seen in Figs. 6.11 and 6.12 that the solubility of Freon in the polystyrene melt increases with temperature, and that Henry's law holds for very low concentrations of Freon in polystyrene melt.

Table 6.3 gives experimentally determined diffusion coefficients for gases in some molten polymers. Durrill and Griskey [47] report that for the gas/polymer systems investigated, the dependence of diffusion coefficient D on temperature T is correlatable by an Arrhenius expression:

$$D = D_0 \exp(-E_d/R_g T) \tag{6.10}$$

in which D_0 is the diffusion coefficient at the reference temperature T_0 (°K), E_d is the energy of activation (Kcal/mole), R_g is the universal gas constant, and T is the melt temperature (°K). Values of E_d for some gas/polymer systems are given in Table 6.2. Durrill and Griskey [47] noted that the

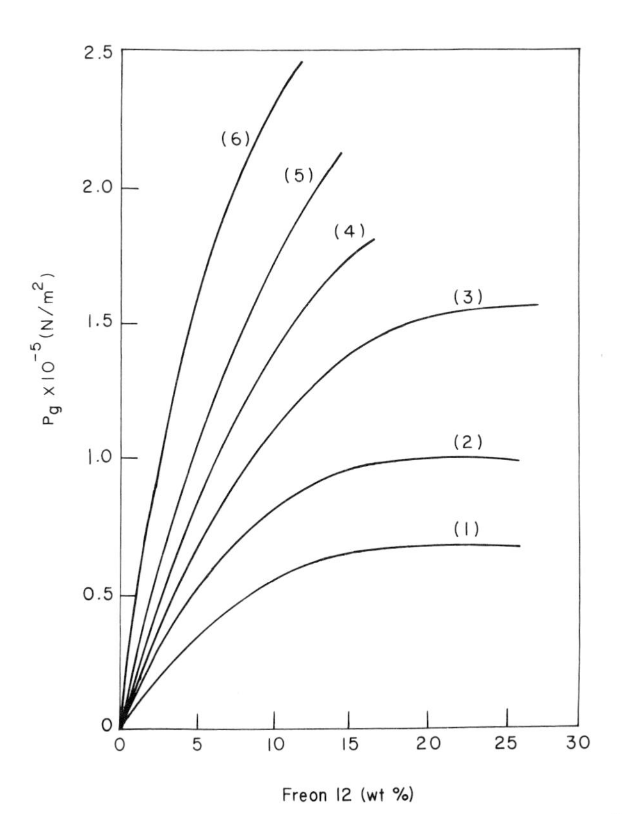

Fig. 6.12 Solubility of Freon 12 in polystyrene at various temperatures (°C) [54]: (1) 25; (2) 40; (3) 80; (4) 100; (5) 120; (6) 140.

TABLE 6.3

Some Experimental Values of Diffusion Coefficient of Gases in Molten Polymers at 188°C[a]

Polymer	Gas	$D \times 10^5$ (cm^2/sec)
Polyethylene	Nitrogen	6.04
	Carbon dioxide	5.69
	Monochlorofluoromethane	4.16
Polypropylene	Nitrogen	3.51
	Carbon dioxide	4.25
	Monochlorofluoromethane	4.02
Polyisobutylene	Nitrogen	2.04
	Carbon dioxide	3.37

[a] From Durrill and Griskey [46].

energy of activation E_d increases with increasing values of the Lennard-Jones force constant of the gas.

In the rest of this chapter, we shall discuss bubble dynamics in the structural foam processing of polymeric liquids, namely, in foam extrusion and foam injection molding. Emphasis will be placed on relating the growth rate of gas bubbles to the rheological properties of the molten polymer, to the processing conditions (e.g., melt temperature, extrusion rate, injection speed), and to the geometry of the die or mold cavity. It is hoped that the topics discussed in this chapter will help achieve a better control of the cell structure and hence the mechanical/physical properties of the foamed products, and develop a criterion (or criteria) for the engineering design of foam extrusion dies and mold cavities.

6.2 BUBBLE DYNAMICS IN FOAM EXTRUSION

6.2.1 Experimental Observation of Bubble Dynamics in Foam Extrusion

A better understanding of bubble dynamics (i.e., either the growth or collapse of bubbles) in a molten polymer is of fundamental and practical importance for controlling the quality of extruded foams. In foam extrusion, the growth of gas bubbles starts when the pressure within the die falls below a critical pressure for bubble inflation, which in turn depends on the concentration of blowing agent, the solubility of the gas in the melt, extrusion temperature, etc.

In order to investigate the bubble growth phenomenon when a molten polymer containing a blowing agent is extruded, Han and co-workers [55, 56] constructed various flow channels with glass windows, permitting them to make visual observations of the dynamic behavior of gas bubbles in the die. Figure 6.13 gives a schematic of the side view of one such flow channel, a rectangular channel. Note that according to Han and Villamizar [56], the melt pressure transducers shown in Fig. 6.13 were used to measure the distributions of wall normal stress (commonly referred to as wall pressure) along the axis of the flow channel as a gas-charged molten polymer was extruded through the die.

Figure 6.14 gives axial pressure profiles of polystyrene containing 0.4 wt.% of Celogen CB (generating N_2). It is seen that the pressure profile starts to deviate from a straight line somewhere in the flow channel, as the melt approaches the die exit. Furthermore, the position at which the pressure profile starts to deviate from a straight line appears to move toward the die exit as the flow rate is increased. On the other hand, axial pressure

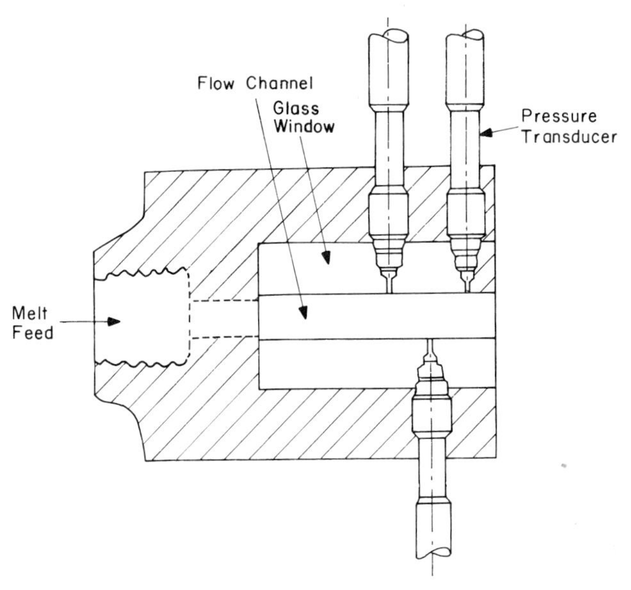

Fig. 6.13 Schematic of a rectangular channel with glass windows, allowing one to measure pressure profile and to observe bubble growth in foam extrusion [56].

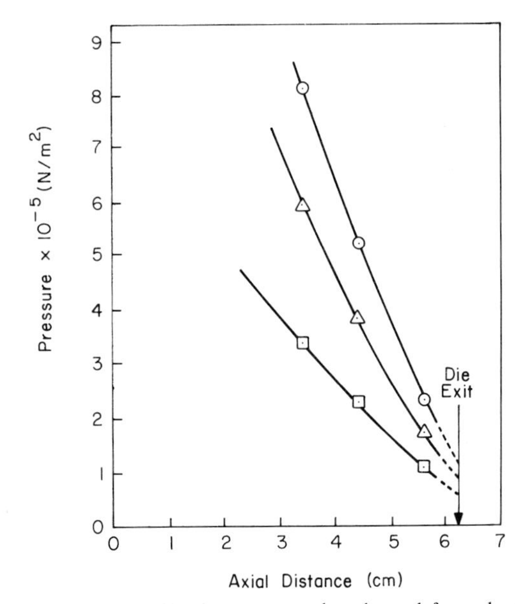

Fig. 6.14 Axial pressure profiles in a rectangular channel for polystyrene ($T = 200°C$) containing 0.4 wt.% Celogen CB (generating N_2) at various volumetric flow rates (cm^3/min) [56]: (⊙) 69.0; (△) 33.2; (⊡) 15.2.

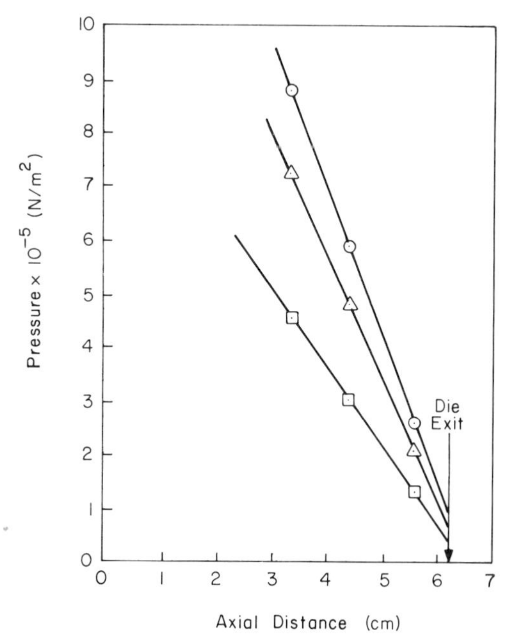

Fig. 6.15 Axial pressure profiles in a slit die for *pure* polystyrene ($T = 200°C$) at various volumetric flow rates (cm^3/min) [56]: (⊙) 51.2: (△) 33.0; (⊡) 16.1.

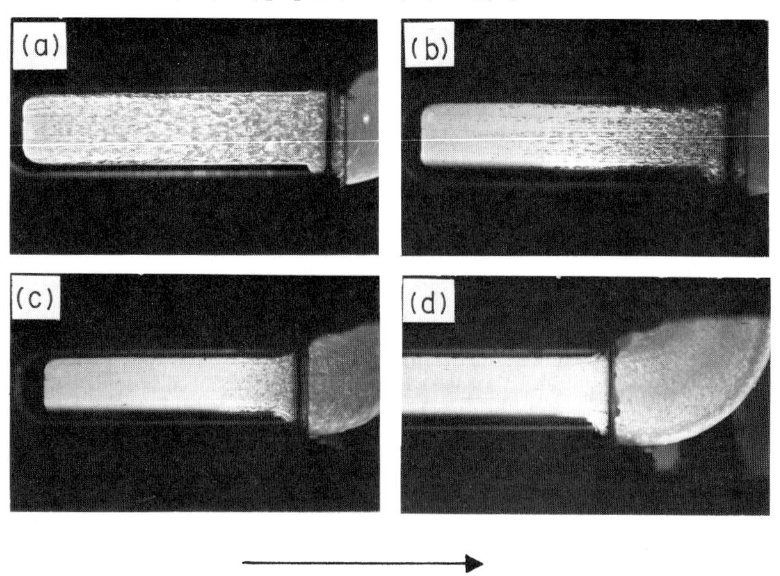

Fig. 6.16 Photographs describing bubble growth in a rectangular channel die for high-density polyethylene ($T = 180°C$) containing 0.4 wt.% Celogen CB (generating N_2) at various volumetric flow rates (cm^3/min) [56]: (a) 3.89; (b) 11.26; (c) 28.3; (d) 39.9.

profiles of pure polystyrene show constant pressure gradients, as given in Fig. 6.15. It can therefore be concluded that the *curved* pressure profiles in Fig. 6.14 are attributable to the two-phase nature of gas-charged molten polymers.

Figure 6.16 shows photographs illustrating the bubble growth phenomenon when a high-density polyethylene containing Celogen CB is extruded through a transparent rectangular channel (see Fig. 6.13) at various flow rates. It is seen that at low flow rates, bubbles start to grow near the die entrance and, as the flow rate is increased, the foaming point is pushed toward the die exit. Figure 6.17 gives representative plots of volumetric flow rate versus bubble inflation position (or foaming point). In their study, Han and Villamizar [56] used a magnifying lens attached to the camera, allowing them to detect bubbles as small as 0.1 mm diameter. It is seen that as the blowing agent concentration is increased, it requires a higher flow rate (hence a higher external pressure) in order to keep the bubble inflation point at the same location in the die.

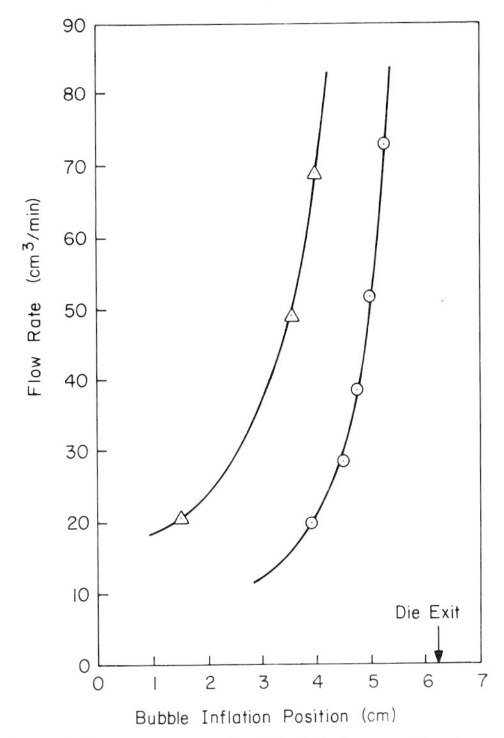

Fig. 6.17 Volumetric flow rate versus bubble inflation position in a rectangular channel for polystyrene ($T = 200°C$) containing Celogen CB of two different concentrations (wt.%) [56]: (⊙) 0.2; (△) 0.4.

TABLE 6.4

Critical Pressure for Bubble Inflation in Gas-Charged Molten Polymers[a]

Blowing agent	Extrusion temperature (°C)	Volumetric flow rate (cm^3/min)	Location[†] of bubble inflation (cm)	Critical pressure $\times 10^{-5}$ (N/m^2)
	(a) Nitrogen in High-Density Polyethylene			
0.2 wt.% Celogen	220	7.4	5.08	1.123
0.2 wt.% Celogen	220	13.0	5.64	1.116
0.2 wt.% Celogen	220	21.8	5.87	1.116
0.2 wt.% Celogen	220	29.6	5.99	1.137
0.2 wt.% Celogen	260	12.0	5.08	0.847
0.2 wt.% Celogen	260	22.4	5.71	0.861
0.2 wt.% Celogen	260	29.3	5.86	0.861
0.2 wt.% Celogen	260	42.1	6.04	0.868
0.4 wt.% Celogen	220	19.9	3.33	3.514
0.4 wt.% Celogen	220	30.5	4.04	3.479
0.4 wt.% Celogen	220	46.5	4.77	3.479
0.4 wt.% Celogen	220	62.6	5.43	3.479
0.4 wt.% Celogen	260	36.3	3.66	3.086
0.4 wt.% Celogen	260	53.3	4.29	3.204
0.4 wt.% Celogen	260	72.1	4.77	3.204
0.8 wt.% Celogen	180	22.4	1.60	7.579
0.8 wt.% Celogen	180	36.3	2.23	7.579
0.8 wt.% Celogen	180	44.6	3.02	7.579
0.8 wt.% Celogen	220	38.8	1.91	5.856
0.8 wt.% Celogen	220	48.6	2.72	5.856
0.8 wt.% Celogen	260	57.0	2.23	5.236
0.8 wt.% Celogen	260	69.4	3.02	5.098
	(b) Carbon Dioxide in Polystyrene			
0.1 wt.% $NaHCO_3$	200	7.3	5.71	0.627
0.1 wt.% $NaHCO_3$	200	14.2	6.15	0.620
0.2 wt.% $NaHCO_3$	200	9.8	5.41	1.398
0.2 wt.% $NaHCO_3$	200	19.6	5.84	1.461
0.2 wt.% $NaHCO_3$	200	27.4	5.99	1.447
0.2 wt.% $NaHCO_3$	200	40.4	6.12	1.392
0.8 wt.% $NaHCO_3$	200	40.1	3.02	6.614
	(c) Nitrogen in Polystyrene			
0.2 wt.% Celogen	200	19.9	3.96	3.899
0.2 wt.% Celogen	200	28.5	4.95	3.789
0.2 wt.% Celogen	200	38.8	4.85	3.962
0.2 wt.% Celogen	200	51.8	5.08	3.893
0.2 wt.% Celogen	200	73.4	5.41	3.652
0.4 wt.% Celogen	200	49.5	3.66	6.545
0.4 wt.% Celogen	200	69.0	3.96	6.477

[a] From Han and Villamizar [56].

[†] The location is measured from the die entrance and the length of the flow channel is 6.35 cm.

Table 6.4 gives the experimentally determined bubble inflation locations for different combinations of polymer and blowing agent under a variety of extrusion conditions (i.e., at different flow rates and melt temperatures). Knowing these locations and having measured pressures along the die axis, one can determine the pressure at which bubble inflation (the so-called critical pressure for bubble growth) starts. The last column of Table 6.4 presents experimentally determined critical pressures for bubble growth under a variety of extrusion conditions. It is worth noting in Table 6.4 that for a given combination of polymer and blowing agent, the critical pressure for bubble inflation is independent of flow rate, but dependent on melt extrusion temperature and blowing agent concentration.

Figure 6.18 gives plots of critical pressure for bubble inflation versus melt extrusion temperature for high-density polyethylene containing various

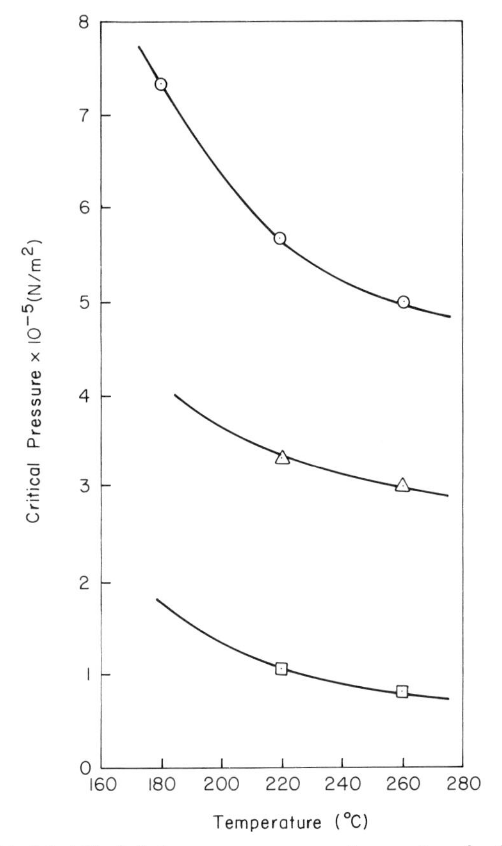

Fig. 6.18 Critical bubble inflation pressure versus temperature for high-density polyethylene containing various amounts of Celogen CB (wt. %) [56]: (⊙) 0.8; (△) 0.4; (⊡) 0.2.

amounts of blowing agent. It is seen that the critical inflation pressure decreases with increasing melt temperature and increases with increasing blowing agent concentration.

At first glance, the decrease in critical inflation pressure with increasing melt temperature may appear to contradict one's intuitive expectation. As a way of explaining this result, plots of volumetric flow rate versus bubble inflation position are given in Fig. 6.19, for high-density polyethylene having various concentrations of blowing agent at different melt temperatures. It is seen that at all levels of blowing agent concentration tested, the flow rate required for causing bubble inflation at the same position in the die increases with melt temperature, which is as expected. In other words, an

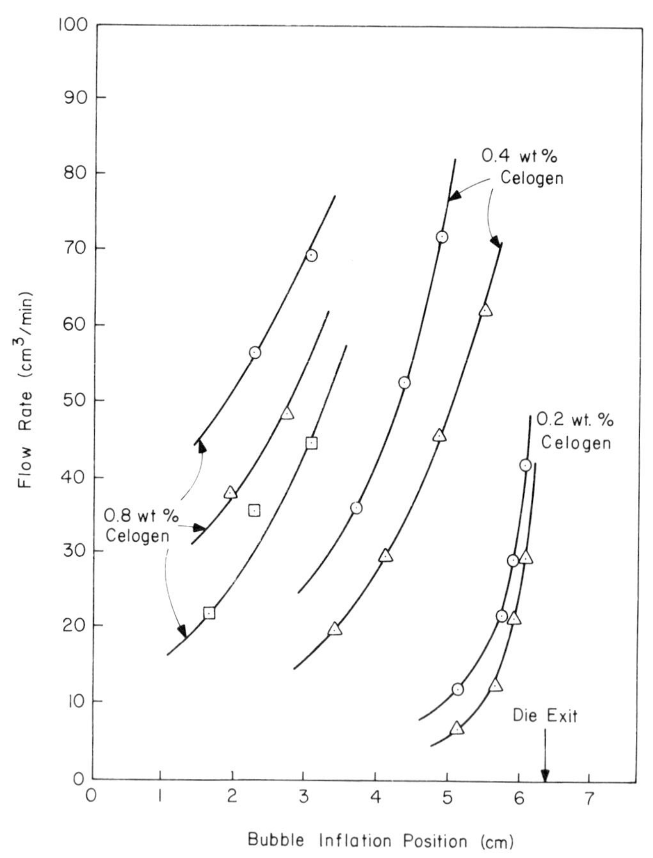

Fig. 6.19 Volumetric flow rate versus bubble inflation position in a rectangular channel for high-density polyethylene containing Celogen CB at various extrusion temperatures (°C) [56]: (⊙) 260; (△) 220; (⊡) 180.

increase in melt temperature will move the bubble inflation position toward the die entrance unless the flow rate is increased to compensate.

It should be noted that an increase in melt temperature causes a decrease in melt viscosity, which, in turn, requires a lower pressure for the material to flow through the die. Note further that an increase in melt temperature causes an increase in the solubility of nitrogen in high-density polyethylene (see Fig. 6.10). Therefore, a smaller amount of gas will be available as microbubbles in the melt as the temperature is increased. It turns out that for the situation under consideration, the extent of the pressure decrease due to the increase in melt temperature is far greater than the extent of the pressure increase expected from the increase in flow rate. Table 6.5 shows representative data supporting this observation.

TABLE 6.5

Effect of Temperature on the Wall Pressure in the Extrusion of High-Density Polyethylene Containing 0.8 wt. % of Celogen CB[a]

Position from the die entrance (cm)	Melt temperature (°C)	Flow rate (cm^3/min)	Wall pressure $\times 10^{-5}$ (N/m^2)
2.54	180	35	7.165
	220	46	6.339
	260	61	5.305
3.17	180	45	7.028
	220	55	6.132
	260	70	4.961

[a] From Han and Villamizar [56].

It should be noted that the critical pressure referred to is the pressure at which a small gas bubble starts to inflate in the melt. This raises the question as to whether a chemical blowing agent, upon decomposition, is completely dissolved in the polymer melt or present as microbubbles. This, in turn, raises the question of the solubility of a gas in the polymer melt.

There are two hypotheses for the state of gas present in a polymer melt [14]. One hypothesis is that at elevated pressures a molten polymer can be supersaturated with a gas-forming component. Then, as pressure is reduced, the gas that is present in excess of the solubility limit comes off from the homogeneous melt, forming microbubbles. Thereafter, these microbubbles grow as pressure is reduced further, and also as more gas diffuses into them. Another hypothesis is that the gas in excess of the solubility limit is never dissolved and remains as microbubbles. Therefore there is a finite volume occupied by free gas (i.e., microbubbles) in the polymer melt containing a gas-forming component. Both hypotheses may be correct for certain polymer/gas systems. To date, however, no clearcut theory or experimental technique appears to have been suggested that may distinguish the two different situations.

Figure 6.20 gives plots of critical inflation pressure versus blowing agent concentration for different combinations of polymer and blowing agent. It is seen that at the same concentration of blowing agent, the polystyrene/Celogen CB system has a critical bubble inflation pressure higher than the polystyrene/$NaHCO_3$ system. It should be remembered that upon decomposition, Celogen CB releases nitrogen, and $NaHCO_3$ releases carbon dioxide. It can be said, therefore, that the polystyrene melt containing nitrogen requires a higher external pressure than the polystyrene containing carbon dioxide in order to suppress the growth of gas bubbles. This may be attributable to the fact that the solubility of N_2 in polystyrene is lower than that of CO_2 in polystyrene (see Table 6.1).

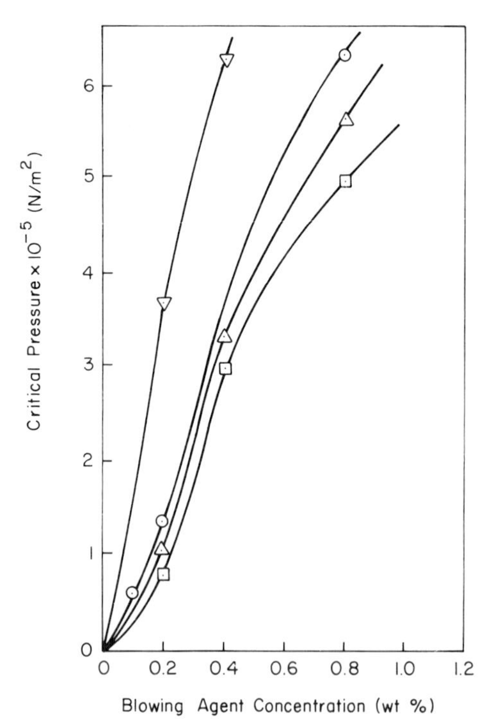

Fig. 6.20 Critical bubble inflation pressure versus blowing agent concentration [56]: (▽) PS with Celogen CB at 200°C; (⊙) PS with $NaHCO_3$ at 200°C; (△) HDPE with Celogen CB at 220°C; (⊡) HDPE with Celogen at 260°C.

On the basis of the above observations, the critical pressure for bubble inflation may be a complicated function of several factors: (a) melt temperature, (b) blowing agent concentration, (c) solubility of the gas in the melt, and (d) the initial bubble size. At present there is no theory published in the literature suggesting how these factors may be correlatable.

The shape of a single gas bubble at various positions along the die axis is of both theoretical and practical interest. Figure 6.21 gives photographs

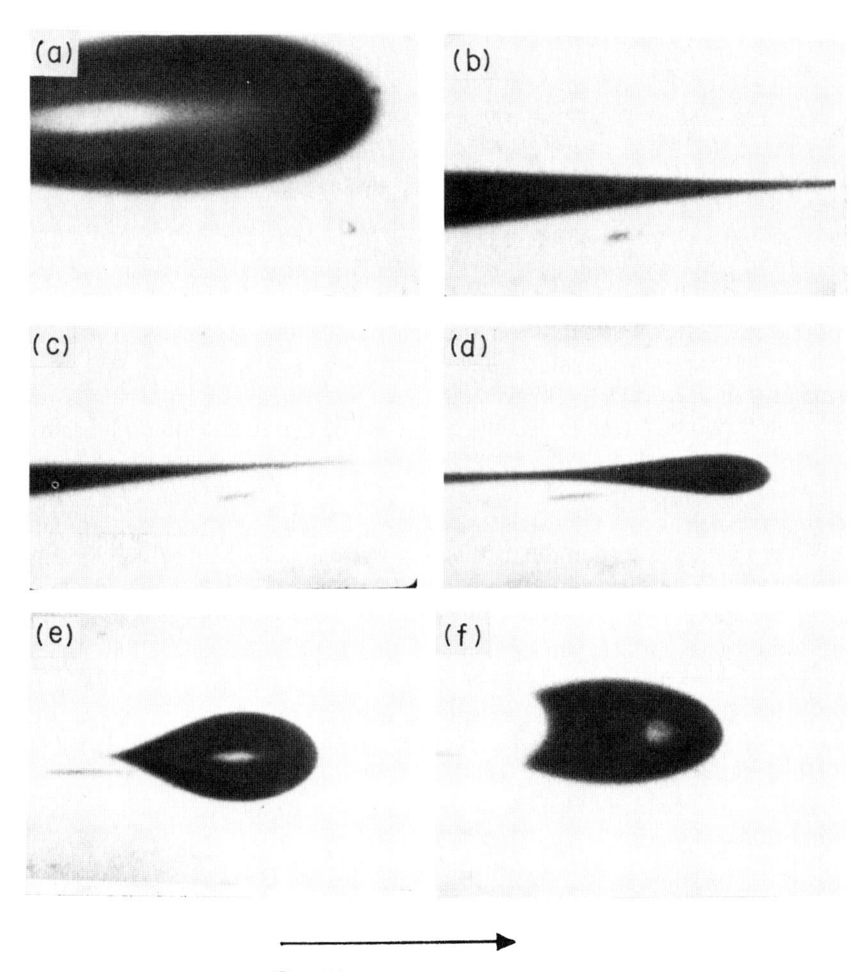

Fig. 6.21 Photographs displaying the bubble dynamics (the shape and size) in the polystyrene melt ($T = 200^\circ$C), at various positions in a flow channel. (a) represents the upstream end of the converging section and (f) represents the downstream end of the fully developed region. See Fig. 6.22 for the locations of the bubble in the flow channel.

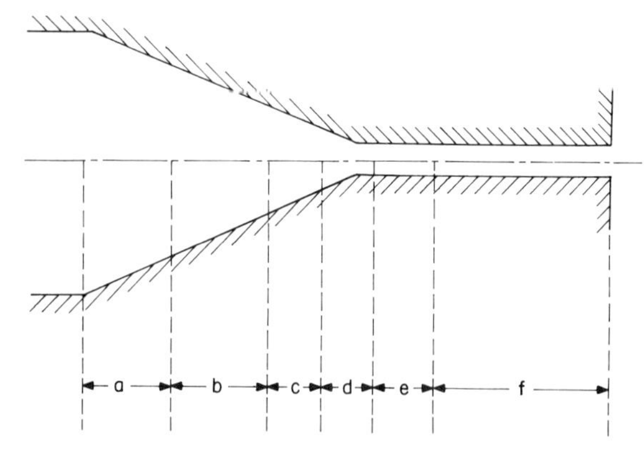

Fig. 6.22 Schematic of the flow channel employed for photographing the bubble dynamics shown in Fig. 6.21.

describing the shape of a gas bubble, initially spherical at the upstream end of the reservoir, that travels along the centerline of the rectangular channel. Figure 6.22 gives a schematic indicating the positions in the flow channel where the photographs were taken of the bubble, given in Fig. 6.21. It is seen in Fig. 6.21 that the bubble gets very elongated as it approaches the die entrance, and then recoils after it passes the entrance region. The shape of gas bubbles very much resembles that of the liquid droplets discussed in Chapter 4.

Figure 6.23 gives photographs showing how a few small gas bubbles, observed just upstream in the rectangular channel, grow as they flow toward the die exit. Note that as a polymer melt containing a blowing agent flows from the region of high pressure to the region of low pressure (see Fig. 6.14), not all bubbles form at the same position in the die. This is attributable in part to the nonuniform dispersion of blowing agent when first mixed with polymer, giving rise to a nonuniform nucleation of gas bubbles, and in part to the shear flow field having a parabolic velocity profile across the flow channel. Due to the nonuniform shear rate across the flow channel, the shape of bubbles varies from position to position, in the direction perpendicular to flow.

At this juncture, it is worth mentioning that using the slender-body theory, Acrivos and Lo [57] predicted the shape of a gas bubble subjected to the steady elongational flow of a viscous Newtonian liquid. To a first approximation, their analysis shows that

$$l/a = \beta(v)[\dot{\gamma}_E \eta_0 a/\sigma]^2 \tag{6.11}$$

where

$$\beta(v) = \tfrac{4}{3}(v+1)(2v+1), \qquad v = 2, 4, 6, \ldots, \tag{6.12}$$

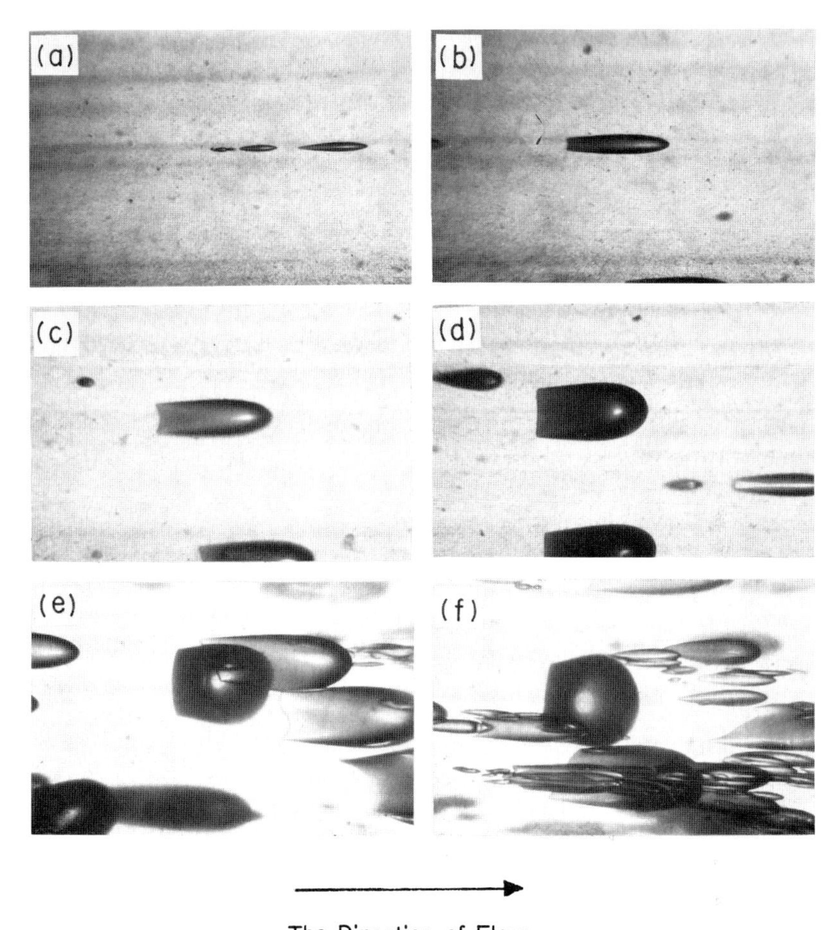

Fig. 6.23 Photographs describing bubble growth at various positions along the die axis for polystyrene ($T = 200°C$) containing 0.4 wt.% $NaHCO_3$, in which the volumetric flow rate is 3.5 cm^3/min [56]. Photograph (a) represents the upstream end and photograph (f) represents the downstream end of the rectangular channel.

in which a is the radius of the sphere with the same volume as the bubble, l is the half-length of the elongated bubble, η_0 is the medium viscosity, $\dot{\gamma}_E$ is the elongation rate, and σ is the interfacial tension. Acrivos and Lo noted that all shapes of bubbles (i.e., for positive even integers of v) described by Eq. (6.11) are unstable except for that corresponding to $v = 2$.

To date, however, there is no theory developed that can predict the shape of a gas bubble subjected to unsteady (in the Lagrangian sense) elongational flow in viscoelastic liquids.

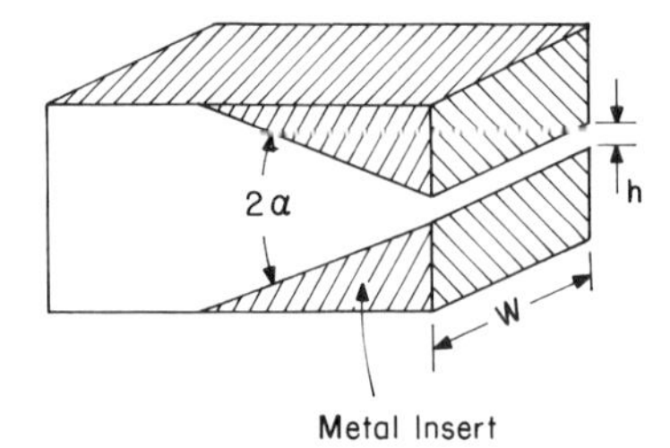

Fig. 6.24 Schematic of a converging flow channel.

Today it is a well-established fact that the entrance angle of an extrusion die has a profound influence on the stress distribution of the polymer in the entrance region [58]. It is therefore reasonable to expect that the die entrance angle will also have a strong influence on bubble dynamics in a gas-charged molten polymer. We shall now discuss bubble dynamics in foam extrusion through a coverging die.

Using a converging flow channel consisting of two nonparallel planes and having glass windows, Yoo and Han [59] investigated bubble dynamics in foam extrusion. Figure 6.24 gives a schematic of the flow channel used by them, and Fig. 6.25 gives photographs of a single gas bubble at different positions along the centerline of the channel having a converging angle of 45°. It is seen in Fig. 6.25 that starting at the upstream end of the die, the size of the gas bubble first appears to increase and then to decrease as the bubble approaches the die exit. Figures 6.26–6.28 give the size change of a gas bubble traveling along the centerline of the converging flow channel. In Figs. 6.26–6.28, A_0 refers to the projected area of the gas bubble at the upstream end of the die, where the bubble is nearly spherical and has a diameter of about 0.1 mm. According to Yoo and Han [59], since there was a slight variation in the bubble size at the reference position, depending on the extrusion conditions (i.e., melt temperature, concentration of blowing agent), the projected area A of a gas bubble was normalized to its initial value A_0.

It is seen in Figs. 6.26–6.28 that the size of gas bubble first increases and then decreases as the bubble approaches the exit of the die. Let us now examine closely the results given in Figs. 6.26–6.28 in order to find how the extrusion conditions affect the dynamic behavior of the gas bubbles. Figure 6.26 shows that as the melt flow rate is increased, the size of the gas bubble decreases. This is understandable because the higher the melt flow rate, the greater will be the normal stresses in the polymer melt, and hence the more difficult it will be for the bubble to grow. Figure 6.27 also shows the effect of melt flow rate, whereas Fig. 6.28 shows the effect of blowing agent concentration on the bubble dynamics. Note in Fig. 6.28 that the bubble produced at the high flow rate has a blowing agent concentration much greater

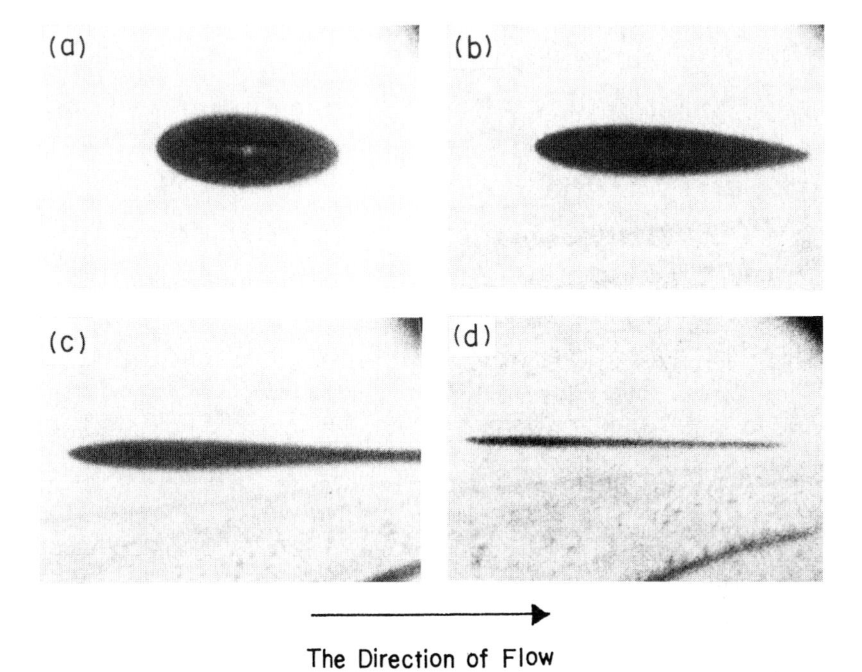

Fig. 6.25 Photographs describing the bubble dynamics (the shape and size) in polystyrene melt ($T = 200°C$), at various positions along the centerline of a converging channel with the converging angle of 45° [59]. Photograph (a) represents the bubble shape at the upstream end, and photograph (d) represents the bubble shape at the exit region of the converging channel.

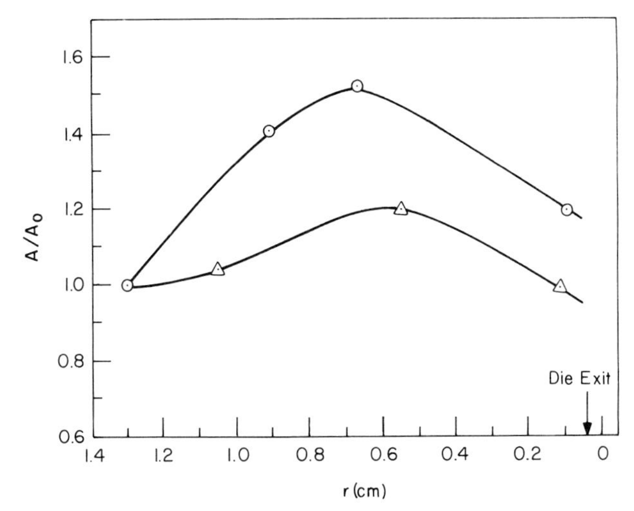

Fig. 6.26 Reduced bubble size versus position along the centerline of a converging flow channel (see Fig. 6.24) at various volumetric flow rates (cm^3/min) [59]: (⊙) 5.5; (△) 22.0. The medium is polystyrene ($T = 200°C$), the gas phase is CO_2 (generated from 0.03 wt.% of $NaHCO_3$), and the die converging angle is 90°.

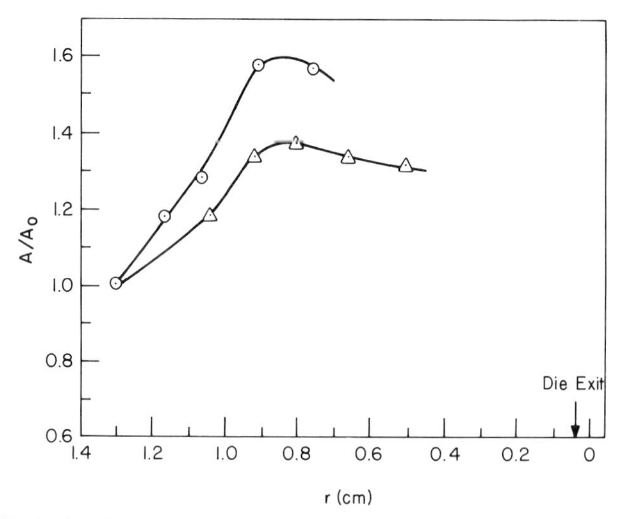

Fig. 6.27 Reduced bubble size versus position along the centerline of a converging flow channel (see Fig. 6.24) at various volumetric flow rates (cm^3/min) [59]: (⊙) 5.7; (△) 20.2. The medium is polystyrene ($T = 200°C$), the gas phase is CO_2 (generated from 0.08 wt.% of $NaHCO_3$), and the die converging angle is 45°.

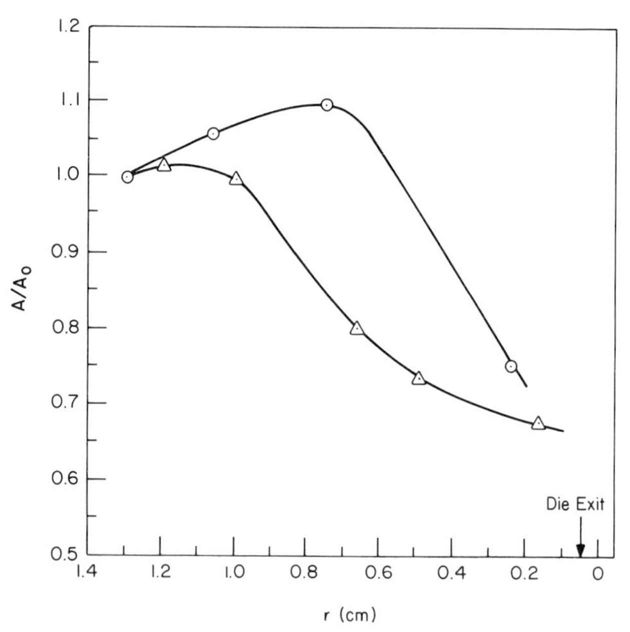

Fig. 6.28 Effect of blowing agent concentration on the reduced bubble size traveling along the centerline of a converging flow channel (see Fig. 6.24) [59]: (⊙) 0.30 wt.% of $NaHCO_3$ and $Q = 3.55$ cm^3/min; (△) 0.05 wt.% of $NaHCO_3$ and $Q = 0.23$ cm^3/min. The medium is polystyrene ($T = 200°C$), the gas phase is CO_2, and the die converging angle is 30°.

than that produced at the low flow rate. Therefore it can be concluded that in this particular situation, the effect of blowing agent concentration plays a far greater role than melt flow rate in influencing the dynamic behavior of gas bubbles.

Although qualitative explanations are given for the experimental results presented above, we need a theoretical foundation to explain why the bubble first grows and then starts to collapse as it approaches the die exit. For this, we shall now present a theoretical analysis of the problem.

6.2.2 Theoretical Consideration of Bubble Dynamics in Foam Extrusion

Let us now consider that a threadlike thin gas bubble, suspended in a viscoelastic medium, moves along the centerline of a converging channel bounded by two nonparallel planes, as schematically shown in Fig. 6.29. A rigorous theoretical analysis of the bubble dynamics observed experimentally (see Fig. 6.25) is a formidable task. We shall therefore choose a crude approach in order to obtain a relatively simple expression, *albeit* approximate, that will help us to interpret, at least *qualitatively*, the experimental results presented in Figs. 6.26–6.28. In order to make our analysis tractable, we assume that the presence of the gas bubble does not disturb the flow in the neighborhood of the bubble. We realize that such an assumption is not valid from a rigorous theoretical point of view. But our objective here is to offer a *qualitative* explanation to the experimental results presented in Figs. 6.26–6.28.

Let us assume that there is *no* secondary flow within the converging flow channel. Using the cylindrical coordinate system (see Fig. 6.29), the velocity

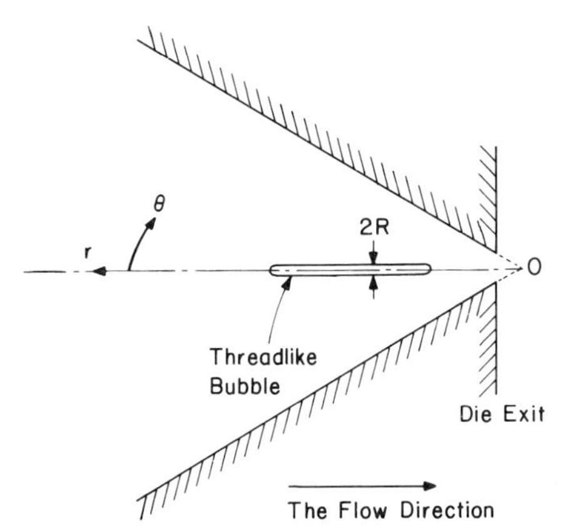

Fig. 6.29 Schematic showing a threadlike thin gas bubble traveling along the centerline of a converging channel [59].

is given by

$$v_r = v_r(r, \theta), \qquad v_\theta = v_z = 0 \tag{6.13}$$

which upon substitution into the continuity equation,

$$\partial(rv_r)/\partial r = 0 \tag{6.14}$$

gives

$$v_r(r, \theta) = f(\theta)/r, \qquad f(\theta) < 0 \tag{6.15}$$

where $f(\theta)$ is an as yet undetermined function depending on θ only. The function $f(\theta)$ in Eq. (6.15) will be determined, however, when the equations of motion are solved for $v_r(r, \theta)$, subject to boundary conditions specified as

$$\text{(i)} \quad v_r(r, \theta = \pm\alpha) = 0 \tag{6.16}$$

$$\text{(ii)} \quad (\partial v_r/\partial\theta)_{\theta=0} = 0 \tag{6.17}$$

The equations of motion in cylindrical coordinates can be written as

$$-\frac{\partial p}{\partial r} + \frac{1}{r}\frac{\partial}{\partial r}(r\tau_{rr}) + \frac{1}{r}\frac{\partial \tau_{r\theta}}{\partial \theta} - \frac{\tau_{\theta\theta}}{r} = 0 \tag{6.18}$$

$$-\frac{1}{r}\frac{\partial p}{\partial \theta} + \frac{1}{r^2}\frac{\partial}{\partial r}(r^2\tau_{r\theta}) + \frac{1}{r}\frac{\partial \tau_{\theta\theta}}{\partial \theta} = 0 \tag{6.19}$$

in which the inertia effect is assumed negligible, which is justifiable for the very viscous molten polymer under consideration.

Integration of Eq. (6.18) with respect to r gives

$$p(r,\theta) - p(r_0,\theta) = \tau_{rr}(r,\theta) - \tau_{rr}(r_0,\theta) + \int_{r_0}^{r} \frac{1}{r}\left(\tau_{rr} - \tau_{\theta\theta} + \frac{\partial \tau_{r\theta}}{\partial \theta}\right) dr \tag{6.20}$$

in which r_0 is an arbitrary reference position at the upstream end of the die. Note that Eq. (6.20) also holds at the bubble–liquid interface. The force balance at the bubble–liquid interface due to interfacial tension σ gives

$$S_{n,g}(R) = S_{n,l}(R) - \sigma/R \tag{6.21}$$

where R is the radius of the threadlike thin cylinder of bubble (see Fig. 6.29), and $S_{n,g}$ and $S_{n,l}$ are components, perpendicular to the bubble surface, of the total stress at the gas phase and liquid phase, respectively.

For the liquid phase, $S_{n,l}$ may be represented by:

$$S_{n,l} = (-p + \tau_{rr})\sin^2\theta + (-p + \tau_{\theta\theta})\cos^2\theta + \tau_{r\theta}\sin 2\theta \tag{6.22}$$

Note that for a threadlike thin cylinder of gas bubble moving along the *centerline* of the converging channel (see Fig. 6.29), the region of our interest around the bubble is essentially the same as the centerline of the flow channel (i.e., $\theta \cong 0$) and therefore the physical variables of our concern (e.g., stresses, velocity) may be evaluated at the centerline, i.e., at $\theta = 0$. Thus, at $\theta = 0$, Eq. (6.22) may be rewritten

$$S_{n,l} = -p + \tau_{\theta\theta} \tag{6.23}$$

in which p is the pressure of the liquid phase.

For the gas phase, the normal stress $S_{n,g}$ may be represented by

$$S_{n,g} = -p_g \tag{6.24}$$

in which p_g is the pressure inside the gas bubble. Thus, use of Eqs. (6.23) and (6.24) in Eq. (6.21) gives

$$p = p_g + \tau_{\theta\theta} - \sigma/R \tag{6.25}$$

Substitution of Eq. (6.25) into Eq. (6.20) gives at $\theta = 0$

$$p_g(r) - p_g(r_0) = [\tau_{rr}(r,0) - \tau_{rr}(r_0,0)] - [\tau_{\theta\theta}(r,0) - \tau_{\theta\theta}(r_0,0)] + \int_{r_0}^{r} \frac{1}{r}\left(\tau_{rr} - \tau_{\theta\theta} + \frac{\partial \tau_{r\theta}}{\partial \theta}\right)_{\theta=0} dr + \sigma\left(\frac{1}{R} - \frac{1}{R_0}\right) \tag{6.26}$$

in which R_0 is the value of R evaluated at r_0.

Using the Coleman–Noll second-order fluid model [60] [see Eq. (2.53) in Chapter 2] in Eq. (6.26), Yoo and Han [59] obtained†

$$p_g(r) - p_g(r_0) = -\frac{\eta_0}{2}[4f(0) + f''(0)]\left[\frac{1}{r^2} - \frac{1}{r_0^2}\right] + \nu[f(0)f''(0) + 6[f(0)]^2]\left[\frac{1}{r^4} - \frac{1}{r_0^4}\right] + \sigma\left[\frac{1}{R} - \frac{1}{R_0}\right] \tag{6.27}$$

in which $f(0)$ is related to the volumetric flow rate per unit length of channel width $\bar{Q}$ by

$$\bar{Q} = -2\int_0^{\alpha} v_r(r,\theta) r\, d\theta = f(0)\left(\frac{2\alpha\cos 2\alpha - \sin 2\alpha}{1 - \cos 2\alpha}\right) \tag{6.28}$$

and $f''(0)$ is given by [61]:

$$f''(0) = -4f(0)/(1 - \cos 2\alpha) \tag{6.29}$$

in which α is the half-angle of the converging channel, η_0 and ν are the material constants in the second-order fluid model, and $p_g(r_0)$ is the pressure

† The practical limitations of the second-order model notwithstanding, this model yields explicit mathematical expressions that enable us to offer a theoretical interpretation of the experimental results presented above. Rheological models that portray the deformation history of the fluid (i.e., integral-type models) are expected to yield a more precise theoretical prediction.

inside the gas bubble at $r = r_0$, an arbitrary reference position in the upstream end of the converging channel.

From Eq. (6.27), we can now predict the pressure inside a threadlike thin gas bubble traveling along the centerline of the flow channel in terms of the material constants and volumetric flow rate of the suspending medium, provided that the profile of the bubble radius $R(r)$ is known. It can be shown that the contribution of the interfacial tension [the last term on the right-hand side of Eq. (6.27)] to the pressure p_g becomes negligibly small in the flow of very viscous molten polymers (i.e., those with large values of η_0 and v) or at high volumetric flow rates [i.e., for large values of $f(0)$ and $f''(0)$].

Figure 6.30 gives a theoretical prediction of the effect of volumetric flow rate on the profile of $p_g(r)$, using Eq. (6.27), in a gas bubble as it travels along the centerline of channel. The numerical values of the material constants used in obtaining Fig. 6.30 are: $\eta_0 = 3.97 \times 10^3 \text{N sec/m}^2$; $-v = 2.48 \times 10^3 \text{ N sec}^2/\text{m}^2$. It is seen in Fig. 6.30 that the pressure inside the bubble first decreases (hence the volume of the bubble will increase) and then increases (hence the volume of the bubble will decrease), going through a minimum as the bubble approaches the die exit. In other words, the predicted

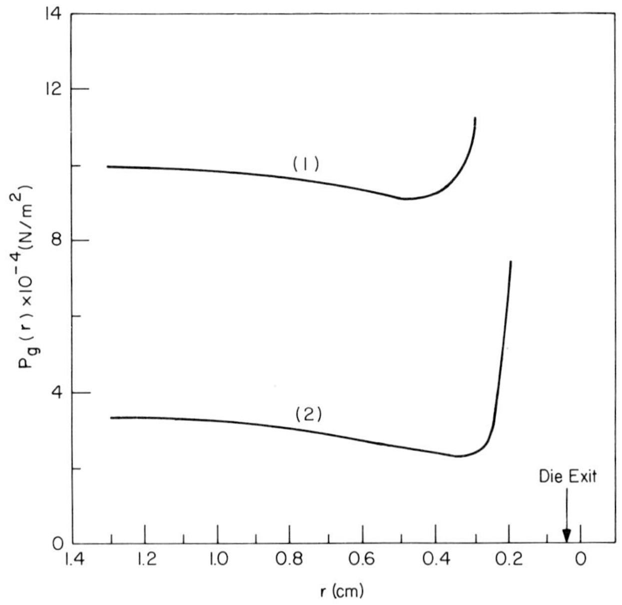

Fig. 6.30 Theoretically predicted pressure in a gas bubble traveling along the centerline of a converging channel at various volumetric flow rates of polystyrene melt (cm^3/min/channel width) [59]: (1) $\bar{Q} = 2.3$ and p_g ($r_0 = 1.3$ cm) $= 10^5$ N/m^2; (2) $\bar{Q} = 0.77$ and p_g ($r_0 = 1.3$ cm) $=$ 0.33×10^5 N/m^2. The die converging angle is 30°.

pressure profile p_g corroborates the *essential feature* of the experimental results given in Figs. 6.26–6.28. It is of interest to note that for a Newtonian suspending medium [i.e., for $\nu = 0$ in Eq. (6.27)], the pressure p_g will not go through a minimum, which then indicates that the gas bubble will continuously expand as it approaches the die exit. Therefore it can be concluded that the existence of a minimum in p_g is due to the *elastic* property of the suspending medium.

To summarize, it is shown above that a viscoelastic suspending medium can give rise to bubble dynamics quite different from a Newtonian medium. With the viscoelastic medium, it is possible to suppress the bubble growth inside a converging die. In foam extrusion of practical interest, one would like to have the foaming occur outside the extrusion die in order to achieve uniform cell structure in the products. Therefore the control of bubble formation and growth inside an extrusion die is of great importance.

6.3 BUBBLE DYNAMICS IN STRUCTURAL FOAM INJECTION MOLDING

6.3.1 Experimental Observation of Bubble Dynamics in Foam Injection Molding

All structural foam injection molding processes are either short-shot/low-pressure or full-shot/relatively high-pressure processes. High and low pressure refers to the clamping force needed to keep the mold closed during mold filling. The major difference between the two processes lies in the mold filling operation. In low pressure molding, a short shot is employed, and the mold filling is completed by the expansion of gases. In high-pressure molding, a gas-charged molten polymer is injected to fill the mold cavity completely, and after sufficient skin thickness is produced by cooling, the mold cavity is enlarged, allowing the still molten material in the center to foam and expand to fill the newly created space. Low-pressure molding has the advantage that conventional molding equipment can be used, possibly with minor modifications. On the other hand, with high-pressure molding, one can produce parts with smooth and discrete skins of controlled thickness and with uniform cell structure, which cannot be obtained in low pressure molding because of uncontrolled foaming. In either case, the success of structural foam molding depends largely on the uniformity of the final cell size and its distribution. Therefore, it is important to control the bubble growth in order to produce products of uniform quality.

In order to investigate bubble dynamics in foam injection molding, Han and co-workers [39, 62] constructed mold cavities with glass windows,

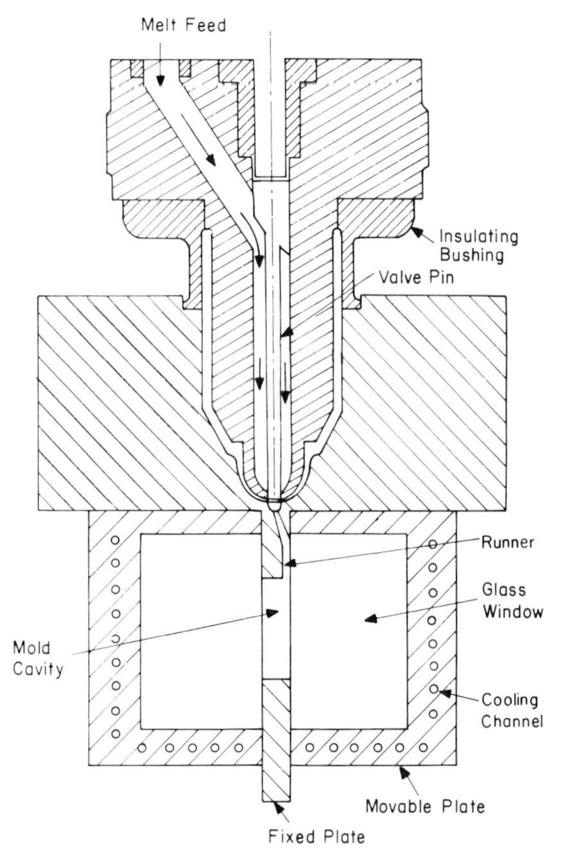

Fig. 6.31 Schematic of a rectangular mold cavity with glass windows, allowing one to observe bubble dynamics during the mold filling of gas-charged molten polymer [39].

permitting them to make visual observations of the dynamic behavior of gas bubbles during the mold filling and subsequent processes. Figure 6.31 gives a schematic of the plan view of the mold, including the injection nozzle, used by Villamizar and Han [39].

Photographs displaying the formation and growth of gas bubbles in the gas-charged molten polymer during the isothermal mold filling operation are given in Fig. 6.32 for polystyrene (PS) containing 0.1 wt. % of $NaHCO_3$, and in Fig. 6.33 for high-density polyethylene (HDPE) containing 0.2 wt. % of Celogen CB. (In isothermal operations, the mold temperature is the same as the temperature of the melt being injected.) Note that according to Villamizar and Han [39], the particular molding operation was a *short* shot under *isothermal* conditions, that is, injection of material was discontinued

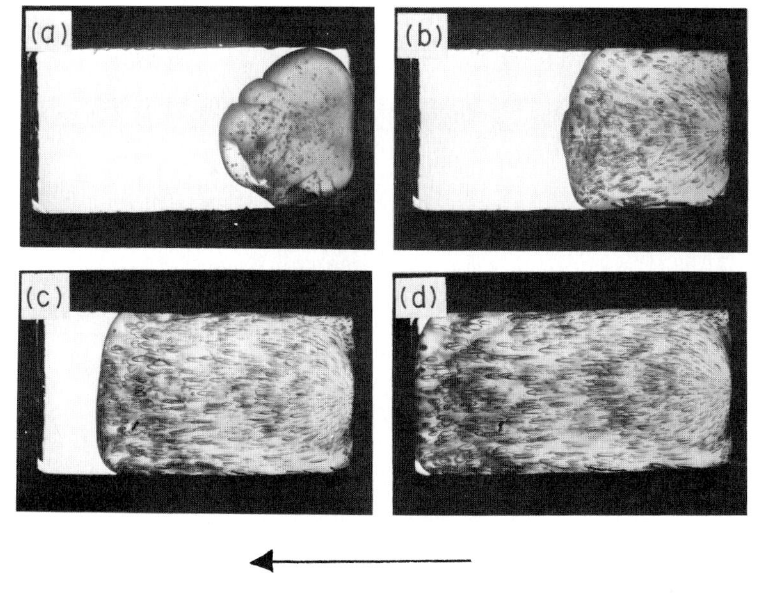

The Flow Direction

Fig. 6.32 Photographs showing the growth of gas bubbles (CO_2) during the mold filling of polystyrene containing 0.1 wt.% of $NaHCO_3$, under the isothermal condition ($T = 200°C$) [39]. Time after the injection started: (a) 20 sec; (b) 30 sec; (c) 45 sec; (d) 1 min 35 sec. The injection pressure is 1.378×10^6 N/m^2 (200 psi).

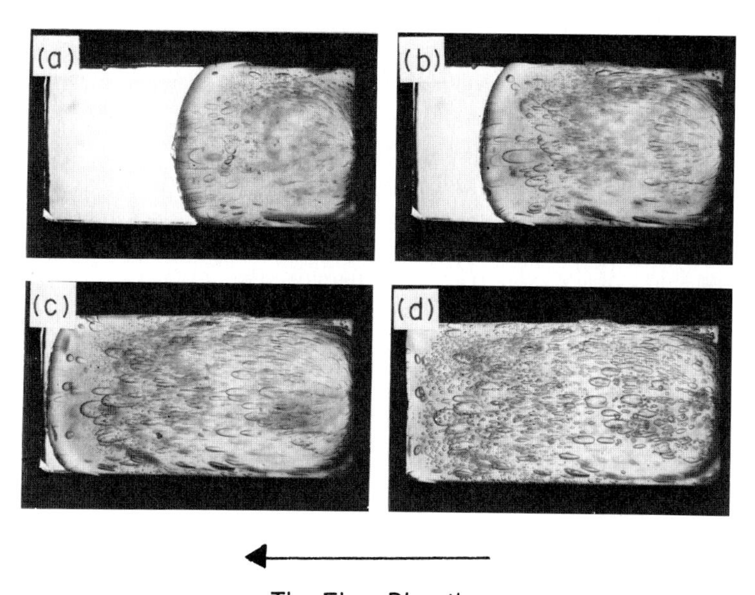

The Flow Direction

Fig. 6.33 Photographs showing the growth of gas bubbles (N_2) during the mold filling of high-density polyethylene containing 0.2 wt.% of Celogen CB, under the isothermal condition ($T = 180°C$) [39]: Time after the injection started: (a) 10 sec; (b) 23 sec; (c) 45 sec; (d) 1 min 20 sec.

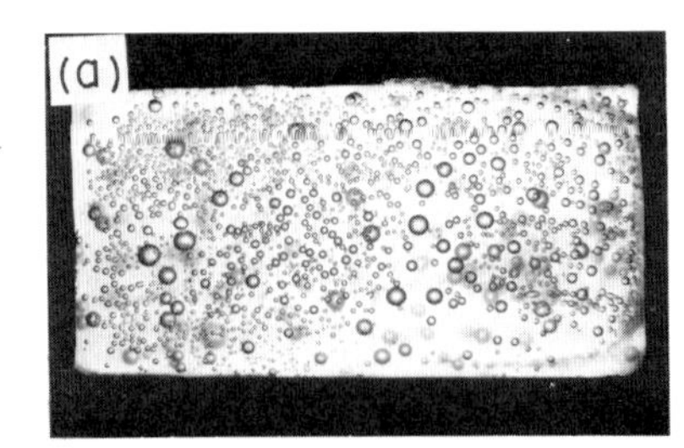

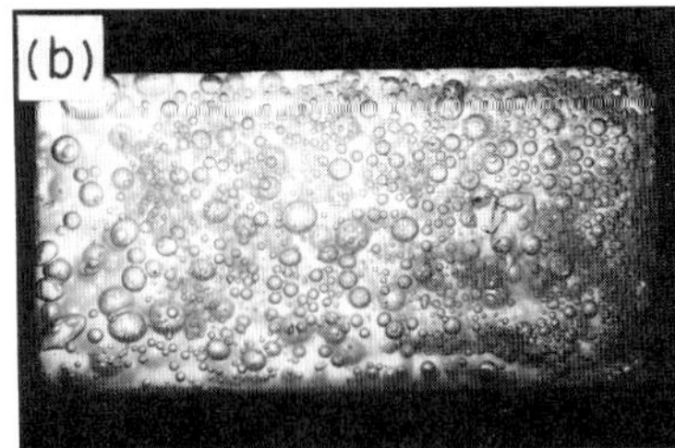

Fig. 6.34 Photographs showing the equilibrium size distribution (after the mold filling is completed) of gas bubbles (N_2) in HDPE containing 0.2 wt.% of Celogen CB, under the isothermal condition [39]: (a) $T = 220°C$; (b) $T = 260°C$.

when four-fifths of the mold cavity was filled, and therefore the bubbles grew as the gas-charged molten polymer expanded to fill the rest of the mold cavity.

It is of interest to note, in Figs. 6.32 and 6.33, that the bubbles are of different sizes, which may be attributable to the following reasons: (1) the bubbles, which start to grow at the gate of the mold cavity, continue to grow as the advancing front moves forward, and therefore the bubbles furthest from the gate should be larger than those near the gate; (2) the gas-forming component is not uniformly distributed in the molten polymer, so that gas bubbles nucleate first in the portion of the molten polymer where a greater concentration of blowing agent exists.

Moreover, the shape of the bubbles in Figs. 6.32 and 6.33 is nonspherical as the advancing melt front moves forward and fills the mold cavity. However, when a sufficiently long time (about 3–4 min) elapses after the mold cavity is filled, the bubbles become spherical under *isothermal* conditions, as shown in Fig. 6.34. Of course, the size and the number of the bubbles present in the mold cavity at any time depend on, among other things, the melt temperature, the injection pressure, the mold temperature, and the concentration of blowing agent.

Figure 6.35 gives a schematic, describing the bubble growth during the mold filling step under isothermal conditions. It is seen in Fig. 6.35 that there are two different mechanisms of bubble growth. One mechanism is due to the sudden decrease in external (melt) pressure during the mold filling, and the other mechanism is due to diffusion of gas from the polymer melt. The initial growth of gas bubbles is represented by the steeply sloping portion of the curves and the diffusion-controlled growth is represented by the much less steep portion.

Figure 6.36 shows that when the gas-charged molten polymer is injected into a *cold* mold cavity, bubble formation occurs in the core area of the mold cavity, and there is little bubble formation near the walls. This observation

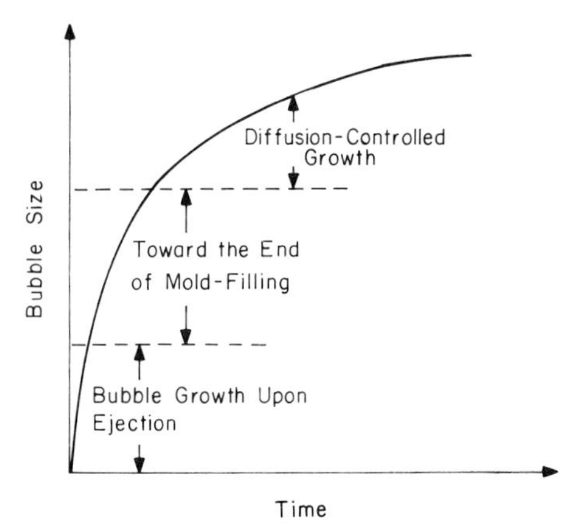

Fig. 6.35 Schematic describing various steps of bubble growth during the foam injection molding of gas-charged molten polymer [39].

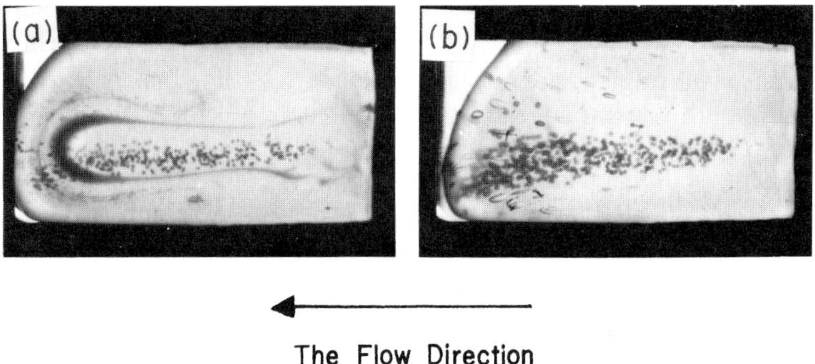

Fig. 6.36 Photographs showing the effect of mold temperature ($T = 70°C$) on bubble growth during the nonisothermal mold filling of polystyrene melt containing 0.1 wt.% of $NaHCO_3$ [39]: (a) injection melt temperature of 200°C; (b) injection melt temperature of 240°C.

may be explained by the fact that as soon as the hot melt contacts the cold wall, its viscosity increases very rapidly, and therefore the formation of bubbles near the wall is greatly suppressed, whereas in the central area of the mold cavity, the formation and growth of gas bubbles proceeds slowly as the molten polymer gradually solidifies from the wall to the core region. It can be concluded therefore that the *mold temperature* significantly influences the cell size and distribution in molded foam specimens.

Hence, as a practical solution for obtaining uniform cell size and distribution in a molded foam product, a high mold temperature is recommended. It should be remembered, however, that the use of a high mold temperature requires a longer molding cycle time.

The sudden release of pressure in a gas-charged molten polymer at the gate of the mold cavity gives rise to bubble formation and growth. However, as the injection pressure (i.e., the pressure within the runner) is increased, there will be less chance for bubble formation within the runner. Therefore the injection pressure must play an important role in determining the formation and growth of bubbles within the mold cavity.

The higher the injection pressure, the fewer, but more uniform, will be the bubbles. This may be attributable to the following facts. First, at high injection pressures, most of the bubbles start to form as the stress (built up in the melt while flowing through the runner) relaxes within the mold cavity as the melt front moves forward, whereas at low injection pressures the formation of gas bubbles might already have started inside the runner before the melt was actually ejected from the gate. Once bubbles are formed within the runner, they will keep growing as the melt front moves forward to fill the cavity. Therefore, a low injection pressure will give rise to large bubble sizes and a less uniform distribution of bubbles compared to a high injection pressure. Second, a high injection pressure requires a shorter mold fill time than a low injection pressure. The shorter mold fill time means that there is less time for the bubbles to grow while the cavity is being filled. Hence a high injection pressure will give rise to smaller bubbles and a more uniform bubble distribution than a low injection pressure. Third, a high injection pressure would require a longer time for the melt to relax its stress within the mold cavity than a low injection pressure does. This observation is based on the fact that the higher the injection pressure, the more energy will be stored in the viscoelastic polymer melt. Hence the injection pressure influences the growth rate of gas bubbles in a viscoelastic polymer melt.

The injection melt temperature plays an important role in determining the bubble dynamics during mold filling, as shown in Fig. 6.37. It is seen that at 180°C there is a relatively small number of bubbles during mold filling. This is attributable to the increase in melt viscosity as the temperature is decreased. It is of interest to note in Fig. 6.37 that jetting occurs initially, which is attributable to the low melt temperature. What is of particular interest in Fig. 6.37 is that the jetting that occurred initially changed into a radial flow pattern before the jet hit the opposite wall of the gate. This change in flow pattern during mold filling may be attributable to the initially very rapid expansion of gas and the consequent expansion of material from the gas-charged molten jet stream.

Recall that bubble growth during mold filling is controlled by two mechanisms (see Fig. 6.35), namely (a) a sudden decrease in external pressure (i.e., a hydrodynamic effect), and (b) the diffusion of gas from the polymer melt (i.e., a diffusion-controlled growth). The injection melt temperature affects both mechanisms. First, a low melt temperature gives rise to a slower

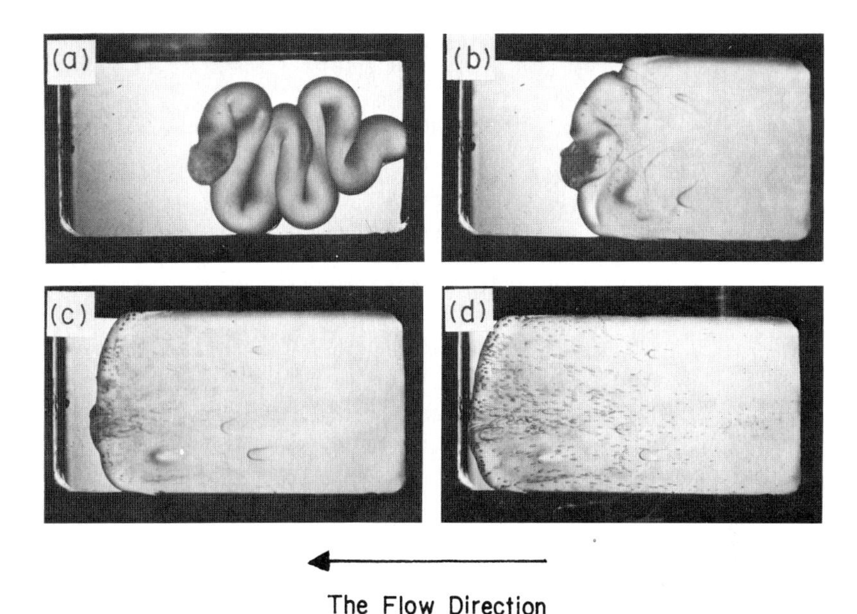

Fig. 6.37 Photographs showing the growth of gas bubbles (CO_2) during the isothermal ($T = 180°C$) mold filling of polystyrene containing 0.1 wt.% of $NaHCO_3$ [39]. Time after the injection started: (a) 30 sec; (b) 40 sec; (c) 55 sec; (d) 1 min 25 sec.

relaxation of the stress of the melt and hence a slower evolution of gas from the gas-charged melt stream. Second, the lower the melt temperature, the greater will be the viscosity of the melt and the lower will be the diffusivity of gases in the molten polymer [see Eq. (6.10)].

The concentration of blowing agent can significantly influence the number of bubbles formed, and the bubble size and its distribution during mold filling. Figure 6.38 shows photographs describing the bubble dynamics during mold filling of PS containing 0.5 wt% of $NaHCO_3$. A comparison of Fig. 6.38 with Fig. 6.32 shows clearly that the use of 0.5 wt.% of $NaHCO_3$ appears to have generated an excessive amount of gas, more than is needed from the solubility point of view. An excessive amount of blowing agent may bring about a coalescence of small bubbles, giving rise to a small number of large bubbles. This appears to have been the case in Fig. 6.38. Therefore one should carefully control the amount of blowing agent in order to obtain the mechanical properties desired of the foamed products.

It should be noted that, as will be discussed below, a molten polymer containing a reasonable amount of blowing agent has a lower viscosity than a polymer without a blowing agent. Remember that a low viscosity melt, upon ejection from the gate of a mold cavity, gives rise to faster bubble

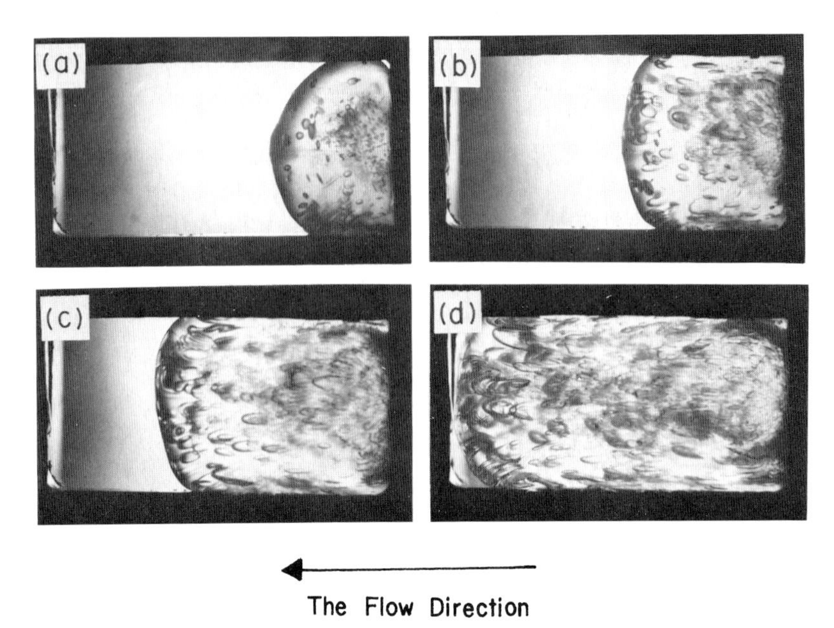

Fig. 6.38 Photographs showing the growth of gas bubbles (CO_2) during the isothermal ($T = 200°C$) mold filling of polystyrene containing 0.5 wt.% of $NaHCO_3$ [39]. Time after the injection started: (a) 10 sec; (b) 20 sec; (c) 30 sec; (d) 1 min 10 sec.

growth than a high viscosity melt, as discussed above in connection with the injection melt temperature. Therefore, both the greater number of bubbles and the larger size of bubbles in Fig. 6.38 may also be attributable to the reduced melt viscosity in the presence of an excessive amount of blowing agent in the molten polymer.

Figure 6.39 gives a plot of bubble radius versus time for growth in a typical *isothermal* mold filling operation. It is seen that while the mold is being filled, the growth rate (the slope of the curve) keeps changing and that after the mold is filled, the bubble growth rate tends to be constant. Figure 6.40 gives a similar plot in a typical *nonisothermal* mold filling operation. It is seen that the bubbles stop growing after about 50 sec. In this case, according to Villamizar and Han [39], the particular bubble chosen was located at the center of the mold cavity. It should be remembered, in reference to Fig. 6.36, that the rate at which a bubble grows depends very much on where it is located with respect to the *cold* mold wall, and therefore Figs. 6.39 and 6.40 are given only to demonstrate the difference in bubble growth rate between isothermal and nonisothermal mold filling operations.

In an actual foam molding operation, there are many bubbles, some of which grow continuously, whereas others coalesce to form larger bubbles.

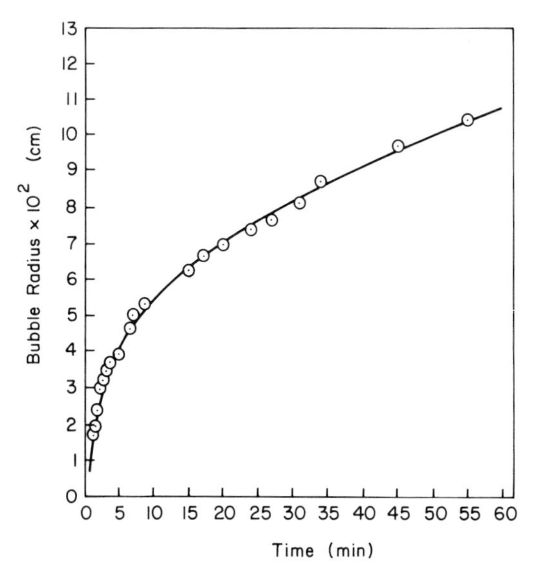

Fig. 6.39 Bubble radius versus time in the isothermal ($T = 200°C$) mold filling of polystyrene containing at 0.1 wt. % of $NaHCO_3$, at injection pressure of 4.13×10^6 N/m^2 (600 psi) [39].

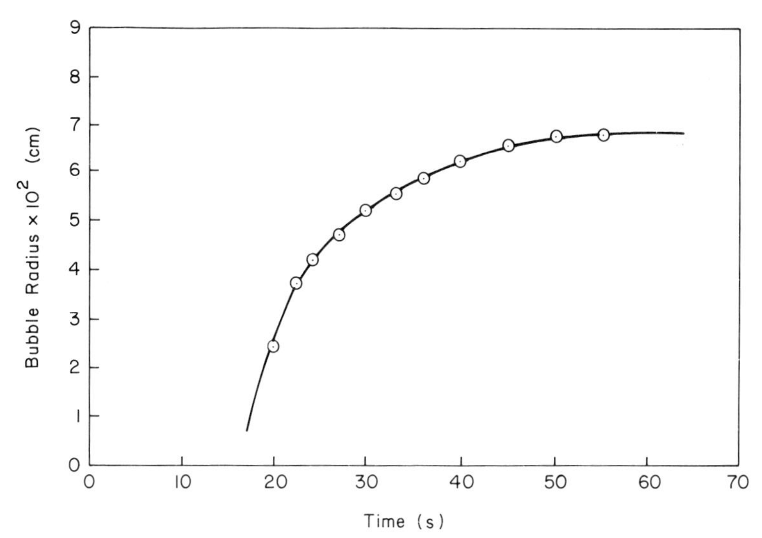

Fig. 6.40 Bubble radius versus time in the nonisothermal mold filling of polystyrene containing 0.1 wt. % of $NaHCO_3$, at injection pressure of 4.13×10^6 N/m^2 (600 psi) [39]. The injection melt temperature is 200°C and the mold temperature is 30°C.

Moreover, there is a spectrum of bubble sizes, as is clearly shown in some of the results discussed above. However, on the basis of the experimental observations presented above of bubble growth under different processing conditions, a general trend of bubble growth rate as affected by processing variables, is summarized schematically in Fig. 6.41. This figure is intended only to indicate, *qualitatively*, in what direction some of the important processing variables influence bubble growth during the foam injection molding operation.

At this juncture, it is worth pointing out that the rate of air escaping from the mold cavity through the vent lines is an important factor in controlling

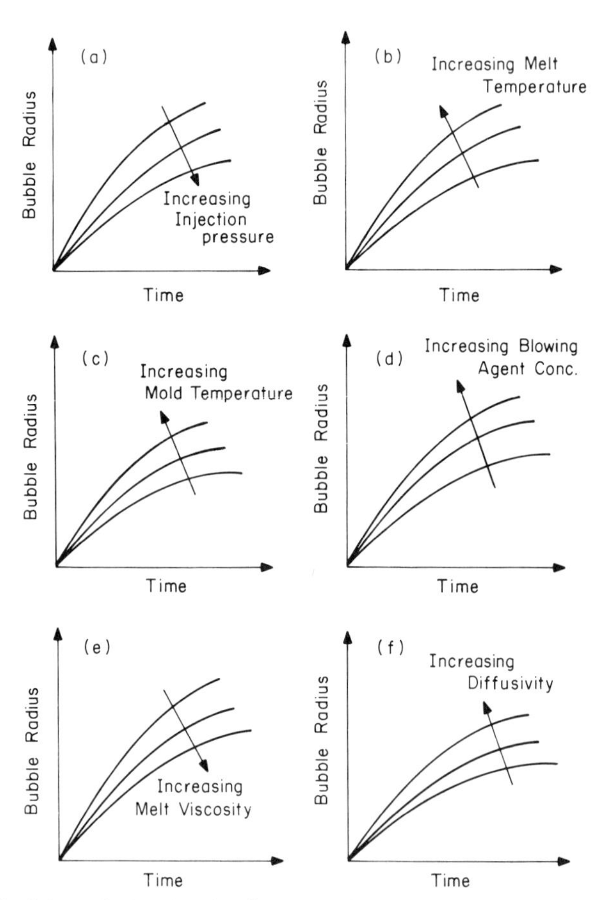

Fig. 6.41 Schematic showing the effects of various physical and processing variables on the growth of gas bubbles during mold filling [39]: (a) the effect of injection pressure; (b) the effect of injection melt temperature; (c) the effect of mold temperature; (d) the effect of blowing agent concentration; (e) the effect of melt viscosity; (f) the effect of diffusion coefficient.

the bubble formation and growth during mold filling. At low injection rates, the air in the mold cavity may have sufficient time to escape. However, at high injection rates, the air trapped in the cavity may not escape fast enough, thus exerting a pressure on the melt front and delaying the formation and growth of gas bubbles until the injection nozzle is closed. From the economic point of view, of course, fast injection is far more desirable than slow injection. Today, the structural foam molding industry tends to employ a mold filling time in the range of 1–5 sec. It should be pointed out, however, that a concern has been expressed that too fast an injection rate may cause a rough surface [17].

6.3.2 Theoretical Consideration of Bubble Growth in Foam Injection Molding

Consider a single gas bubble placed in a rectangular mold cavity, as schematically shown in Fig. 6.42. On the basis of the experimental observations described above and also, to a certain extent, for mathematical convenience, we shall consider bubble growth under *isothermal* conditions with the following assumptions: (1) the size of the bubble is sufficiently small during the period of its growth, so that the assumption of sphericity is warranted; (2) the bubble is located at the centerline of the rectangular cavity and bubble growth is not affected by the wall of the mold cavity (i.e., during the entire period of its growth, the radius of the bubble is very small compared to the width of the mold cavity); (3) the gas-charged molten polymer is incompressible; (4) the viscosity of the melt is so large that buoyancy is negligible during the entire period of bubble growth; (5) the bubble starts to grow after the injection nozzle is closed and thus the bubble is stationary. Using spherical coordinates (r, θ, φ) with the origin at the center of the bubble, the velocity field of the liquid medium is given by

$$v_r = v_r(r, t), \qquad v_\theta = v_\varphi = 0 \tag{6.30}$$

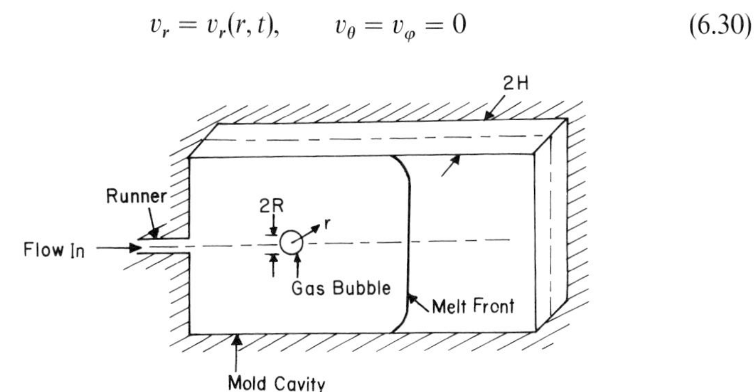

Fig. 6.42 Schematic of a rectangular mold cavity in which a single gas bubble grows during mold filling [62].

The equations of continuity and motion, respectively, may then be written

$$\partial(r^2 v_r)/\partial r - 0 \tag{6.31}$$

$$\rho\left(\frac{\partial v_r}{\partial t} + v_r \frac{\partial v_r}{\partial r}\right) = -\frac{\partial p}{\partial r} + \frac{1}{r^2}\frac{\partial}{\partial r}(r^2 \tau_{rr}) - \frac{(\tau_{\theta\theta} + \tau_{\varphi\varphi})}{r} \tag{6.32}$$

where ρ is the density of the liquid phase.

Noting that $v_r(r = R) = dR/dt = \dot{R}$, where R is the bubble radius, Eq. (6.31) may be rewritten

$$v_r = R^2 \dot{R}/r^2 \tag{6.33}$$

Substitution of Eq. (6.33) into Eq. (6.32), and use of spherical symmetry ($\tau_{\theta\theta} = \tau_{\varphi\varphi}$), give:

$$\rho\left(\frac{2R\dot{R}^2 + R^2\ddot{R}}{r^2} - \frac{2R^4\dot{R}^2}{r^5}\right) = -\frac{\partial p}{\partial r} + \frac{\partial \tau_{rr}}{\partial r} + \frac{2(\tau_{rr} - \tau_{\theta\theta})}{r} \tag{6.34}$$

where $\ddot{R} = d^2R/dt^2$. Integrating Eq. (6.34) from the bubble wall ($r = R$) to the mold wall ($r = H$) (see Fig. 6.42), we have

$$\rho(\tfrac{3}{2}\dot{R}^2 + R\ddot{R}) = p(R) - p(H) + \tau_{rr}(H) - \tau_{rr}(R) + 2\int_R^H \frac{(\tau_{rr} - \tau_{\theta\theta})}{r} dr \tag{6.35}$$

The force balance at the bubble–liquid interface ($r = R$) requires that

$$-p(R) + \tau_{rr}(R) = -p_g + 2\sigma/R \tag{6.36}$$

where p_g is the pressure inside the bubble and σ is the interfacial tension. Note in Eq. (6.36) that the normal stress τ_{rr} of the gas phase at the bubble wall is assumed to be negligibly small compared to the magnitude of the other terms, and that the surface viscosity and surface elasticity at the bubble–liquid interface are negligibly small.

Using the Zaremba–DeWitt model [see Eq. (2.42) in Chapter 2] in Eq. (6.35), with the aid of Eq. (6.36), Han and Yoo [62] obtained[†]:

$$\begin{aligned}\rho(\tfrac{3}{2}\dot{R}^2 + R\ddot{R}) = p_g - \frac{2\sigma}{R} - \frac{12\eta_0}{\lambda_1}\int_0^t e^{(s-t)/\lambda_1}\left[\frac{R^2(s)\dot{R}(s)}{R^3(t) - R^3(s)}\right]\ln\frac{R(t)}{R(s)}ds \\ - 3\tau_{rr,0}[e^{-t/\lambda_1}]\left[\ln\frac{H}{R}\right] - [p(H) - \tau_{rr}(H)]\end{aligned} \tag{6.37}$$

[†] The practical limitations of the Zaremba–DeWitt model notwithstanding, this model yields explicit mathematical expressions that enable us to offer a theoretical interpretation of the experimental results presented above. Rheological models that portray the deformation history of the fluid (i.e., integral-type models) are expected to yield a more precise theoretical prediction.

in which η_0 and λ_1 are the material constants in the Zaremba–DeWitt model, and $\tau_{rr,0} = \tau_{rr}\,(t = 0)$.

Note that p_g in Eq. (6.37) varies with time t, and therefore we must find an expression that relates p_g to bubble radius R. Such a relationship can be derived from a mass balance performed on the gas dissolved in the melt, i.e., by Fick's law of diffusion:

$$\frac{\partial C_A}{\partial t} + v_r \frac{\partial C_A}{\partial r} = D\left[\frac{1}{r^2}\frac{\partial}{\partial r}\left(r^2 \frac{\partial C_A}{\partial r}\right)\right] \tag{6.38}$$

in which D is the diffusivity of the gas in the polymer melt, and C_A is the concentration (weight fraction) of the gas dissolved in the melt.

As the gas bubble grows, the pressure inside the bubble p_g will decrease, giving rise to a concentration gradient at the bubble wall which, in turn, affects the mass transfer process. Hence, Eqs. (6.37) and (6.38) are intimately coupled through the relation between the solute concentration just outside the bubble wall, $C_w(t)$, and the concentration within the bubble, i.e., $p_g(t)$. Let us assume that this relation may be described by Henry's law,

$$C_A(R, t) = C_w(t) = K_p p_g(t) \tag{6.39}$$

where K_p is the Henry's law constant. We may assume further that the gas follows the ideal gas law,

$$p_g(t) = (R_g T/M)\rho_g(t) = A\rho_g(t) \tag{6.40}$$

where R_g is the universal gas constant, M is the molecular weight of the gas, ρ_g is the density of the gas in the bubble, T is the temperature, and A is defined as $R_g T/M$. Combining Eqs. (6.39) and (6.40), one obtains

$$\rho_g(t) = C_w(t)/K_p A \tag{6.41}$$

Note that at the bubble wall ($r = R$), the mass flux is represented by

$$\frac{d}{dt}(\rho_g R^3) = 3\rho D R^2 \left(\frac{\partial C_A}{\partial r}\right)_{r=R} \tag{6.42}$$

in which ρ is the density of the liquid phase. Therefore, in order to obtain information on bubble growth rate, we must solve Eqs. (6.37), (6.38), and (6.42) simultaneously.

Let us now briefly explain the physical significance of each term on the right-hand side of the hydrodynamic equation, Eq. (6.37). The first term represents the pressure inside the bubble, which acts to provide a driving force for bubble growth. Note that through this term, the hydrodynamic equation is coupled with the diffusion equation [i.e., Eq. (6.38)]. The second term describes the surface tension force acting at the bubble–liquid interface, providing a resistance to bubble growth. The third term represents the stress developed in the liquid phase due to the growing motion of the bubble wall. It should be noted that the magnitude of this term strongly depends on the viscosity (η_0) and elastic parameter (λ_1) of the liquid phase. This term also provides a resistance to bubble growth. The fourth term arises from the stress imposed on the liquid phase initially (i.e., at $t = 0$). Note that the polymer melt is subjected to a high stress field during injection, and a certain amount of stress may still remain unrelaxed in the liquid phase when the bubble starts to grow during mold filling. Also, note that the magnitude of the initial stress $\tau_{rr,0}$ depends strongly on the injection pressure. Finally, the last term represents the pressure (in more rigorous terms, the total normal stress) acting on the mold wall. Note that in the short-shot molding process, the pressure at the mold wall is approximately the same as the mold pressure (i.e., the pressure in the unfilled space of the mold cavity during injection). The mold pressure is expected to increase as the injection speed (or injection pressure) is increased. For example, at a high injection pressure, the air originally present in the mold cavity does not have enough time to escape through the mold vent, and thus the pressure inside the cavity increases. Hence, the mold pressure is a function of both the size of the mold vent and the injection pressure.

It is of interest to note that in a Newtonian medium, Eq. (6.37) reduces to

$$\rho(\tfrac{3}{2}\dot{R}^2 + R\ddot{R}) = p_g - 2\sigma/R - 4\eta_0\dot{R}/R - [p(H) - \tau_{rr}(H)] \qquad (6.43)$$

Han and Yoo [62] carried out the numerical computation of the system equations, Eqs. (6.37), (6.38), and (6.42). In their computation, the value of 1 μm was used for initial radius R_0 and, as a reference system, the carbon dioxide/polystyrene system was used to determine the physical/rheological parameters needed for computation. These values were then perturbed in order to investigate their influence on bubble growth, strictly from the point of view of theoretical interest.

Basically, there are two sets of variables that control the rate of bubble growth during mold filling under isothermal conditions. They are the physical/transport properties of the gas/liquid system (i.e., the solubility and diffusivity of the gas in the polymer melt, the viscosity and elasticity of the polymer melt, the interfacial tension between the gas and the polymer melt) and the processing variables (i.e., injection pressure, melt temperature, blowing agent concentration). Note that processing conditions may affect some of the physical/transport properties and therefore the two sets of variables are intimately related to each other, insofar as their influence on the bubble growth is concerned.

(a) Effect of Injection Pressure on Bubble Growth

Figure 6.43 shows the effect of injection pressure on bubble growth during mold filling. Note in Fig. 6.43 that, at a low injection pressure (curve 1) the initial stress $\tau_{rr,0}$ is assumed to be zero and the mold pressure is assumed to be at the ambient (1 atm), whereas at a high injection pressure (curve 3) the initial stress $\tau_{rr,0}$ and the mold pressure are chosen such that the gas bubble is in equilibrium with the opposing force in the liquid phase at $t = 0$.

It is seen in Fig. 6.43 that at the low injection pressure (solid curve 1) the plot of $\log R$ versus $\log t$ shows a constant slope of about 0.5, suggesting that the growth mechanism is diffusion controlled. However, as the injection pressure is increased (solid curves 2 and 3), initially the bubble growth is retarded and then the bubble size approaches that obtained at the low injection pressure (solid curve 1). The retardation of bubble growth at the

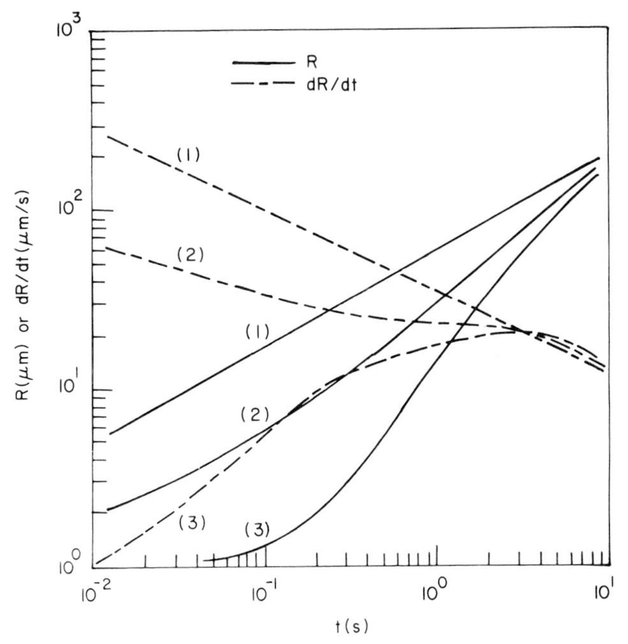

Fig. 6.43 Effect of injection pressure on bubble growth during mold filling [62]: (1) At low injection pressure for which $\tau_{rr,0} = 0$ and the mold pressure is at 1 atm (1.01×10^5 N/m^2); (2) At medium injection pressure for which $\tau_{rr,0} = 3.2 \times 10^3$ N/m^2 and the initial mold pressure is at 1.77 atm (1.79×10^5 N/m^2); (3) At high injection pressure for which $\tau_{rr,0} = 6.4 \times 10^3$ N/m^2 and the initial mold pressure is at 2.54 atm (2.58×10^5 N/m^2). Other system parameters are: $\rho = 0.88$ g/cm^3; $K_p = 4.26 \times 10^{-9}$ m^2/N; $D = 5.5 \times 10^{-6}$ cm^2/sec; $\sigma = 2.8 \times 10^{-2}$ N/m (28 dyn/cm); $C_0 = 2 \times 10^{-3}$ g/g; $\eta_0 = 4 \times 10^3$ N sec/m^2; $\lambda_1 = 0.90$ sec.

beginning is believed to be due to the presence of stress built up in the melt while being injected at high injection pressure (i.e., at fast injection speed). Figure 6.43 shows also that initially the rate of bubble growth dR/dt at a high injection pressure (dotted curve 3) is quite small compared to that at a low injection pressure (dotted curve 1). Note that dR/dt at a high injection pressure first increases and then decreases, going through a maximum, whereas dR/dt at a low injection pressure decreases continuously. It can then be concluded that at high injection pressures, bubble growth is controlled by the hydrodynamic effect in its early stage and then by the diffusion process in its later stage. On the other hand, at low injection pressures, bubble growth is controlled by the diffusion process over the entire period of mold filling.

The hydrodynamic-controlled mechanism in the early stage of bubble growth at high injection pressure can be seen more clearly when the bubble pressure p_g is plotted against time t, as given by Fig. 6.44. It is seen that at the low injection pressure (curve 1) p_g reaches the ambient (1 atm) immediately after the bubble starts to grow, whereas at the high injection pressure (curve 3) p_g decreases rather slowly with time, approaching the ambient asymptotically. The slow decay of p_g with time at high injection pressure is a clear indication of the presence of a force in the medium that opposes the bubble growth.

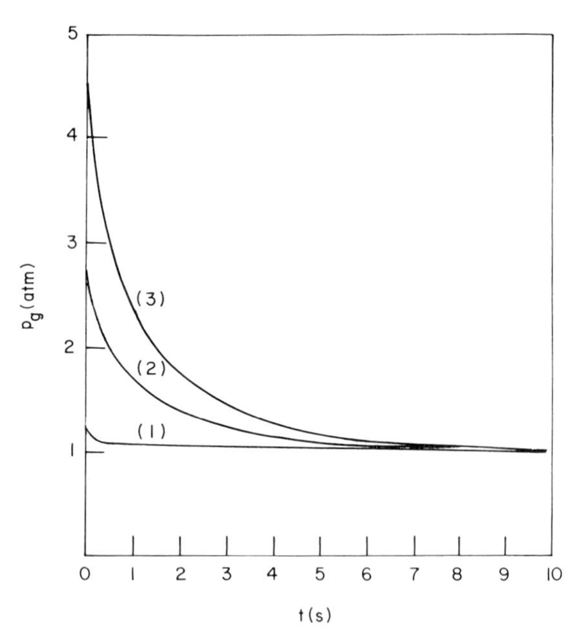

Fig. 6.44 Effect of injection pressure on bubble pressure during mold filling [62]: (1) At low injection pressure; (2) At medium injection pressure; (3) At high injection pressure. The numerical values of all system parameters are the same as in Fig. 6.43.

(b) Effect of Mold Vent on Bubble Growth

Since, from the strictly theoretical point of view, it is difficult, if not impossible, to take into account the effect of mold vent on bubble growth, Han and Yoo [62] have assumed that the mold pressure, the last term on the right-hand side of Eq. (6.37), decays exponentially with time t, as represented by $P(t) = P_0 \exp(-t/\lambda_m) + P_1$ in which P_1 is the asymptotic value of mold pressure (1 atm) at large t, P_0 is a constant value that depends on injection pressure, and λ_m is a time constant describing the decay rate of the mold pressure. Therefore λ_m can be considered as indirectly representing the size of the mold vent, i.e., the larger the value of λ_m, the slower the decay of the mold pressure, and hence the smaller the mold vent.

Figure 6.45 gives the effect of the size of mold vent on bubble growth. It is seen that the bubble radius R is considerably smaller with a large value of λ_m (small mold vent), as for solid curve 3, than with a small value of λ_m (large mold vent), as for solid curve 1. Furthermore, the rate of bubble growth dR/dt is lower with a small vent (dotted curve 3) than with a large vent

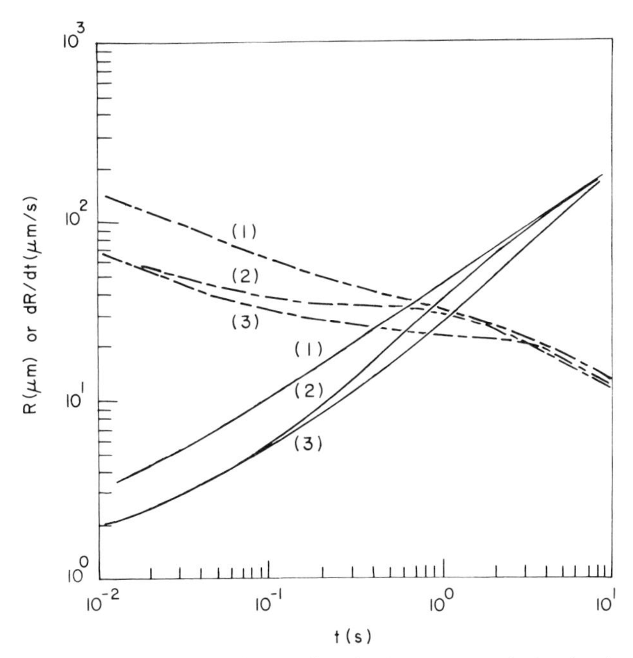

Fig. 6.45 Effect of the size of mold vent (i.e., the time constant λ_m for the air escape from mold cavity) on bubble growth during mold filling [62]: (1) $\lambda_m = 0$; (2) $\lambda_m = 0.5$ sec; (3) $\lambda_m = 2.0$ sec. It is assumed that $\tau_{rr,0} = 3.2 \times 10^3$ N/m^2 and the initial mold pressure is at 1.77 atm (1.79×10^5 N/m^2). Other system parameters are the same as in Fig. 6.43.

(dotted curve 1) in the early stage of bubble growth. This observation indicates that the mold vent can have a profound influence on the rate of bubble growth during mold filling, and consequently on the control of cell size in injection molded foam products.

(c) Effect of Melt Elasticity on Bubble Growth

Figure 6.46 shows the effect of melt elasticity (represented by the relaxation time constant λ_1) on bubble growth when the initial stress of the melt is zero (i.e., $\tau_{rr,0} = 0$) and the mold pressure is at the ambient pressure (i.e., at 1 atm). Such a situation will arise when the injection pressure (or injection speed) is low and the size of mold vent is sufficiently large. For comparison purposes, bubble growth in a Newtonian medium ($\lambda_1 = 0$) is also included in Fig. 6.46. It is seen that an increase in λ_1 gives rise to a large bubble size over the entire period of mold filling, and that the rate of bubble growth dR/dt in its early stage ($t \leq 0.2$ sec) is greater as the value of λ_1 is increased. In other words, in the early stage of bubble growth, the melt elasticity appears to enhance bubble growth when the injection pressure is sufficiently low,

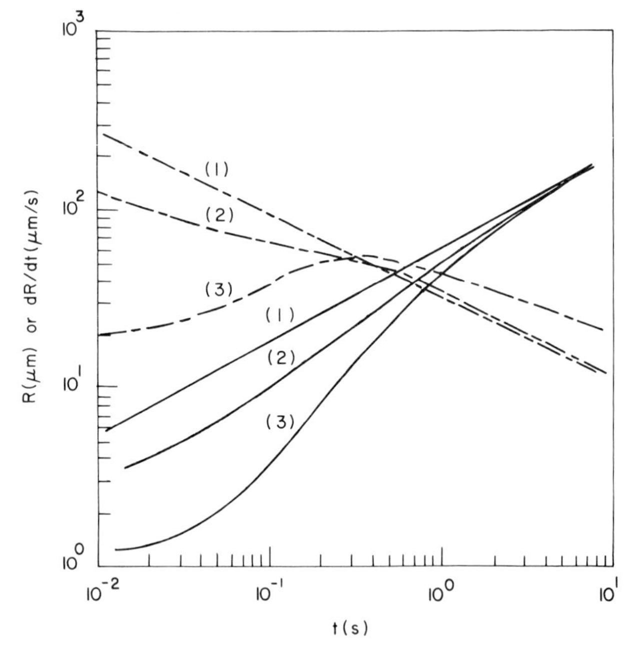

Fig. 6.46 Effect of medium elasticity on bubble growth during mold filling [62]: (1) $\lambda_1 = 5.0$ sec; (2) $\lambda_1 = 0.09$ sec; (3) $\lambda_1 = 0$ (Newtonian fluid). It is assumed that $\tau_{rr,0} = 0$ and the mold pressure is at 1 atm. Other system parameters are the same as in Fig. 6.43.

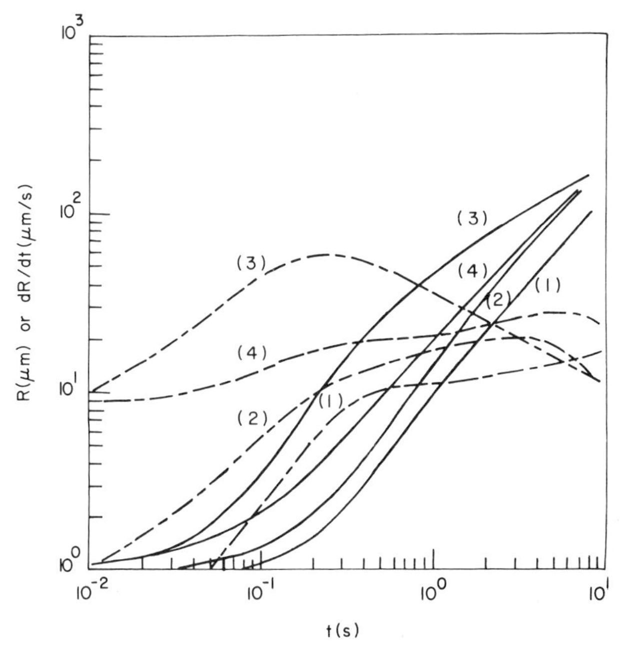

Fig. 6.47 Effect of the medium elasticity on bubble growth during mold filling [62]: (1) $\lambda_1 = 5.0$ sec; (2) $\lambda_1 = 0.9$ sec; (3) $\lambda_1 = 0.09$ sec; (4) $\lambda_1 = 0$ (Newtonian fluid). It is assumed that $\tau_{rr,0} = 6.4 \times 10^3$ N/m^2, the initial mold pressure is at 2.54 atm (2.56×10^5 N/m^2), and the time constant λ_m for the air escape from the mold cavity is 2.0 sec. Other system parameters are the same as in Fig. 6.43.

yielding a negligible amount of residual stress (recoverable shear strain) available at the mold gate.

However, the situation would be quite different when a significant amount of stress is present in the melt as it leaves the mold gate. Such a situation is illustrated in Fig. 6.47. It is seen that at the early stage, the rate of bubble growth in the viscoelastic medium can be greater, or less, than that in the Newtonian medium, depending on the extent of medium elasticity (i.e., depending on the value of λ_1).

Let us consider the reason why a viscoelastic medium can give rise to a greater rate of bubble growth in its early stage than the Newtonian medium, when the medium has zero initial stress (i.e., $\tau_{rr,0} = 0$). Figure 6.48 gives the stress in a medium generated by the motion of a growing gas bubble, in which S_{VE} is the quantity for a viscoelastic medium ($\lambda_1 = 5$ sec) computed from the third term on the right-hand side of Eq. (6.37), and S_{NE} is the quantity for the Newtonian medium ($\lambda_1 = 0$) computed from the third term on the right-hand side of Eq. (6.43). It is seen in Fig. 6.48 that S_{VE} is smaller by a few orders of

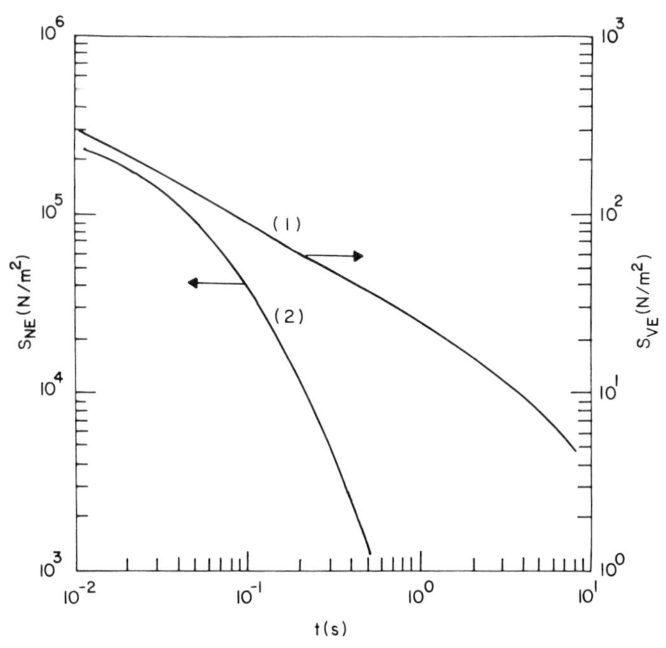

Fig. 6.48 Stress generated by the motion of a growing gas bubble in the liquid medium [62]: (1) In a viscoelastic medium; (2) In a Newtonian medium. It is assumed that $\tau_{rr,0} = 0$ and the mold pressure is at 1 atm. Other system parameters are the same as in Fig. 6.43, except that $\lambda_1 = 0$ for the Newtonian medium.

magnitude than S_{NE}. Therefore it can be concluded that the enhanced rate of bubble growth in a viscoelastic medium, compared to that in the Newtonian medium, is attributable to the small value of S_{VE} compared to S_{NE}, both representing the forces opposing bubble growth.

However, as the injection pressure is increased, the viscoelastic medium carries the residual stress (i.e., as yet unrelaxed stress) as it leaves the gate of the mold. This residual stress is represented by $\tau_{rr,0}$ in the fourth term on the right-hand side of Eq. (6.37). Note that the higher the injection pressure, the greater the value of $\tau_{rr,0}$ will be. This stress will then relax as the melt fills the mold cavity. The rate of stress decay depends on the relaxation time constant λ_1 [see Eq. (6.37)]. Note, however, that to all intents and purposes, $\tau_{rr,0}$ can be assumed to be zero for the Newtonian medium, regardless of the level of injection pressure, because a Newtonian fluid does not have the capacity to store energy. It is then clear that, for viscoelastic fluids, as the injection pressure is increased, the magnitude of the initial stress [represented by the fourth term in Eq. (6.37)] can exceed the magnitude of the stress buildup in the medium due to bubble growth [represented by the third term in Eq. (6.37)]. Under such a circumstance, the rate of bubble growth can be slower in a viscoelastic medium than in a Newtonian medium (see Fig. 6.47). Therefore it can be concluded that the role that the melt elasticity plays in controlling the rate of bubble growth during mold filling depends strongly on the level of injection pressure (or injection speed).

(d) Effect of Melt Viscosity on Bubble Growth

Figure 6.49 shows the effect of melt viscosity (zero shear viscosity) of polymer melt on bubble growth. It is seen that the rate of bubble growth is decreased as the melt viscosity is increased, especially in the early stage of mold filling. This is understandable in view of the fact that the viscosity η_0 appears only in the hydrodynamic equation [Eq. (6.37)] and that the bubble growth is controlled primarily by the hydrodynamic effect in the early stage of mold filling, especially when the injection pressure is sufficiently high. Note that the medium viscosity is expected to have little influence on bubble growth when it is controlled primarily by the diffusion process.

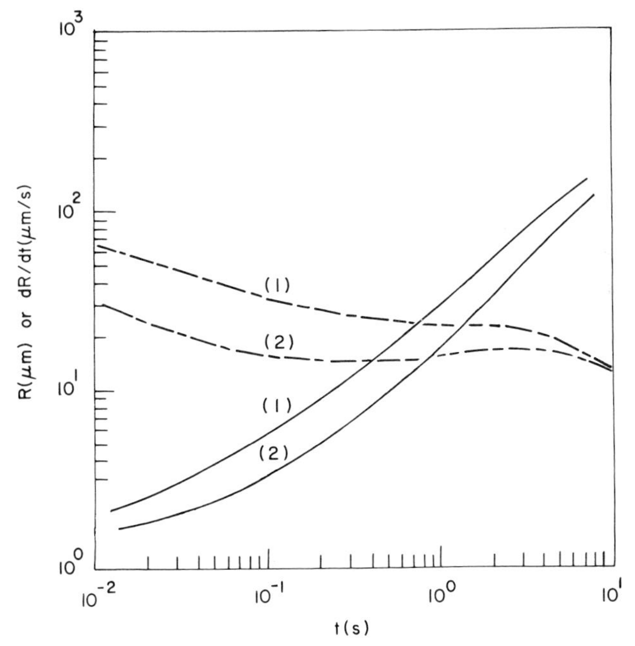

Fig. 6.49 Effect of melt viscosity on bubble growth during mold filling [62]: (1) $\eta_0 = 4 \times 10^3$ N sec/m^2; (2) $\eta_0 = 4 \times 10^4$ N sec/m^2. It is assumed that $\tau_{rr,0} = 3.2 \times 10^3$ N/m^2 and the initial pressure is at 1.77 atm (1.79×10^5 N/m^2). Other system parameters are the same as in Fig. 6.43.

(e) Effect of Blowing Agent Concentration on Bubble Growth

Figure 6.50 shows the effect of blowing agent concentration on bubble growth. It is seen that at the asymptotic stage ($t \geq 5$ sec) where bubble growth is controlled by the diffusion process, the rate of bubble growth dR/dt is proportional to the blowing agent concentration (i.e., doubling the blowing agent concentration also doubles the growth rate). However, when the rate of bubble growth is controlled by the hydrodynamic effect, dR/dt at the early stage is increased much more rapidly with an increase in blowing agent concentration. For example, at $t = 0.01$ sec, doubling the blowing agent concentration increases dR/dt by a factor of 3.9. This again is an indication that the liquid medium's hydrodynamic resistance to bubble growth would be much greater at low blowing agent concentrations than at high ones.

It is worth mentioning that, as will be discussed below, an increase in blowing agent concentration decreases the viscosity of a gas-charged molten polymer. As shown in Fig. 6.49, a decrease in melt viscosity in turn increases the rate of bubble growth. However, we have not included in Fig. 6.50 the

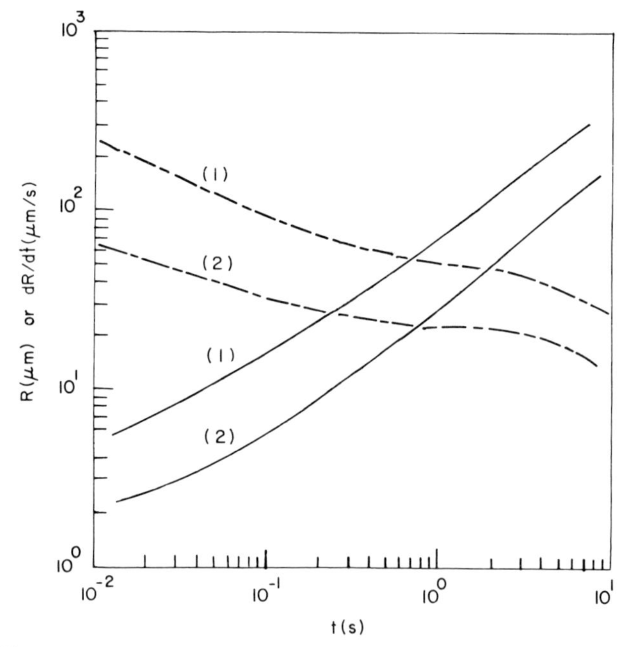

Fig. 6.50 Effect of the concentration of blowing agent on bubble growth during mold filling [62]: (1) $C_0 = 4 \times 10^{-3}$ g/g; (2) $C_0 = 2 \times 10^{-3}$ g/g. Other parameters are the same as in Fig. 6.49, except that $\eta_0 = 4 \times 10^3$ N sec/m^2.

effect of a decrease in melt viscosity, brought about by an increase in blowing agent concentration.

(f) Effect of Diffusion Coefficient on Bubble Growth

Figure 6.51 shows that an increase in diffusion coefficient enhances the rate of bubble growth. It is seen that at the asymptotic stage (say, $t \geq 5$ sec) of bubble growth, the rate of bubble growth dR/dt is approximately proportional to the square root of the diffusion coefficient. This is theoretically expected when bubble growth is controlled by the diffusion process [24]. However, as may be seen in Fig. 6.51, at the early stage of bubble growth (say, $t = 0.01$ sec), such a relationship is not obeyed. For instance, whereas the square root of the ratio of the two diffusion coefficients used in Fig. 6.51 is about 2.2, the ratio of the two growth rates is about 3.1. This is also an indication that in the early stage of mold filling, bubble growth is strongly influenced by the hydrodynamic effect.

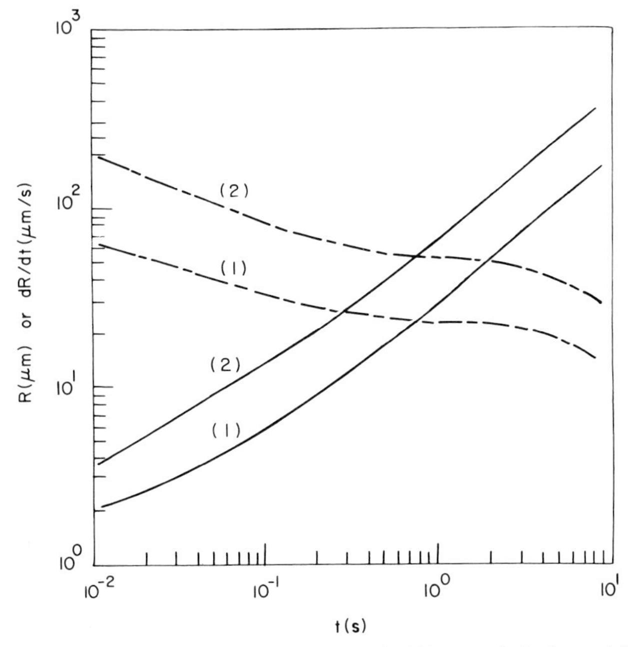

Fig. 6.51 Effect of diffusion coefficient of gas on bubble growth during mold filling [62]: (1) $D = 5.5 \times 10^{-6}$ cm^2/sec; (2) $D = 2.75 \times 10^{-5}$ cm^2/sec. Other parameters are the same as in Fig. 6.49, except that $\eta_0 = 4 \times 10^3$ N sec/m^2.

6.4 THE RHEOLOGICAL BEHAVIOR OF GAS-CHARGED MOLTEN POLYMERS

Measurement of the rheological properties of a polymer melt containing a gaseous component is of fundamental importance to a better control of foam processes and to the design of processing equipment (e.g., screw and die designs). As one may surmise, the rheological properties of a gas-charged molten polymer would vary with the type of foaming agent, the amount of foaming agent, and also with the type and amount of other additives (e.g., plasticizers, modifiers, nucleators, etc.).

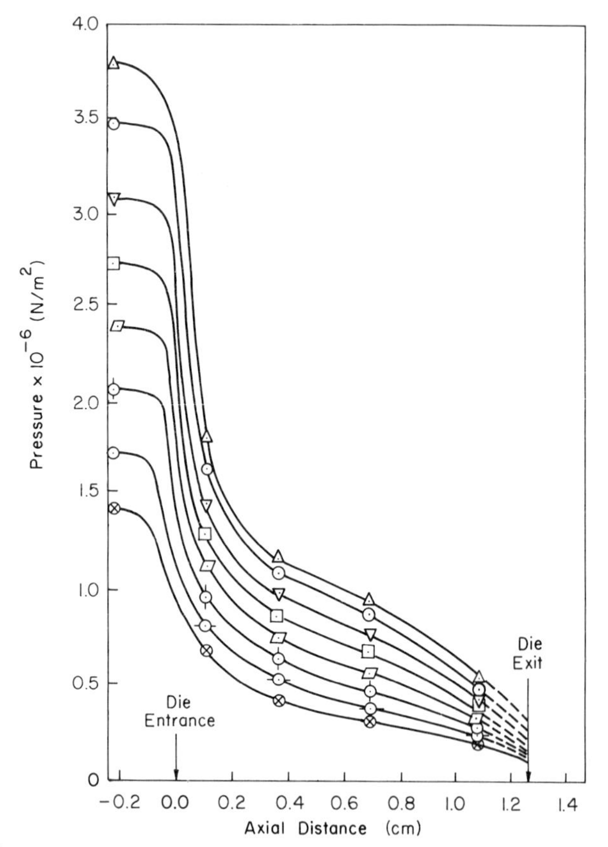

Fig. 6.52 Axial pressure profiles of high-density polyethylene ($T = 200°C$) containing a blowing agent, azodicarbonamide (Stepan Chemical Co., Kempore 125) generating nitrogen, in extrusion through a cylindrical die having an L/D ratio of 4, at various apparent shear rates (s^{-1}) [55]: (⊗) 53.3; (-○-) 75.8; (⏀) 103.7; (▱) 141.9; (□) 189.1; (▽) 249.1; (○) 337.8; (△) 412.2. From C. D. Han *et al.*, *J. Appl. Polym. Sci.* **20**, 1583. Copyright © 1976. Reprinted by permission of John Wiley & Sons, Inc.

On the other hand, only a relatively small amount of information about the rheological properties of gas-charged molten polymers is available in the literature [56, 63–66]. This may be attributable in part to the fact that rotational-type rheometers (see Chapter 2), widely used in many research laboratories, cannot be used with gas-charged polymers. This is because, with a rotational-type rheometer being an *open* system, it is *not* possible to suppress the growth of gas bubbles in a molten polymer. It is then clear that only a *closed* system (e.g., capillary- or slit-die rheometer) is suitable.

Figure 6.52 gives axial profiles of wall normal stress of high-density polyethylene containing 0.3 wt.% of a blowing agent (Celogen CB generating nitrogen) in a cylindrical die having an L/D ratio of 4. It is seen that the pressure profile starts to deviate from a straight line somewhere in the capillary as the melt approaches the die exit. Figure 6.53 shows axial profiles of wall normal stress of high-density polyethylene *with* and *without* a blowing agent in a cylindrical tube. Two things are worth noting in Fig. 6.53. One is that within the entire die length, the polymer *with* a blowing agent gives

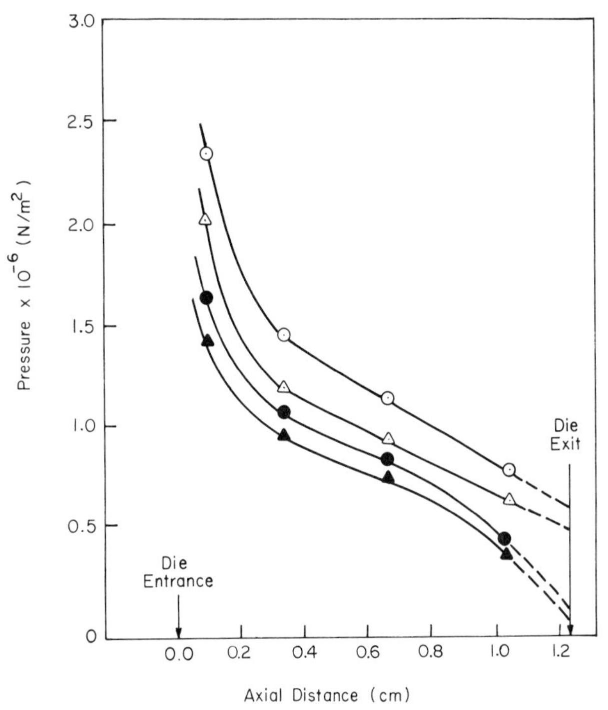

Fig. 6.53 Axial pressure profiles of high-density polyethylene ($T = 200°C$) in extrusion through a cylindrical die having an L/D ratio of 4, without (*open* symbols) and with a blowing agent (*closed* symbols), at various apparent shear rates (s^{-1}) [55]: (△, ▲) 249.2; (⊙, ●) 337.8. From C. D. Han *et al., J. Appl. Polym. Sci.* **20**, 1583. Copyright © 1976. Reprinted by permission of John Wiley & Sons, Inc.

rise to lower wall normal stresses than the polymer *without* a agent. The other is that after the melt passes the entrance region, the polymer *with* a blowing agent shows a curvature in its axial profile, whereas the polymer *without* a blowing agent shows a linear pressure profile. (See also Figs. 6.14 and 6.15.)

The fact that the pressure gradient may *not* be constant in the capillary for the melt containing a blowing agent should caution those who wish to determine the rheological properties, using a conventional (i.e., plunger-type) viscometer. This is because, when foaming occurs in the capillary, the following expression for the wall shear stress τ_w:

$$\tau_w = (-\partial p/\partial z) D/4 \tag{6.44}$$

in which $(-\partial p/\partial z)$ is the pressure gradient and D is the capillary diameter, and the Rabinowich–Mooney correction for the shear rate, commonly used for homopolymers, is *not* valid for determining the *bulk* viscosity of gas–polymer melt mixtures.

Another interesting observation that one can make in Fig. 6.52 is the entrance pressure drop for the polymer melt *with* a blowing agent. Figure 6.54 shows plots of entrance pressure drop versus apparent shear rate for high-density polyethylene melts *with* and *without* a blowing agent. It is seen that the melt *with* a blowing agent gives rise to much greater entrance pressure drops than the melt *without* a blowing agent. The difference in the entrance pressure drop between the two curves in Fig. 6.54 may be attributable to the

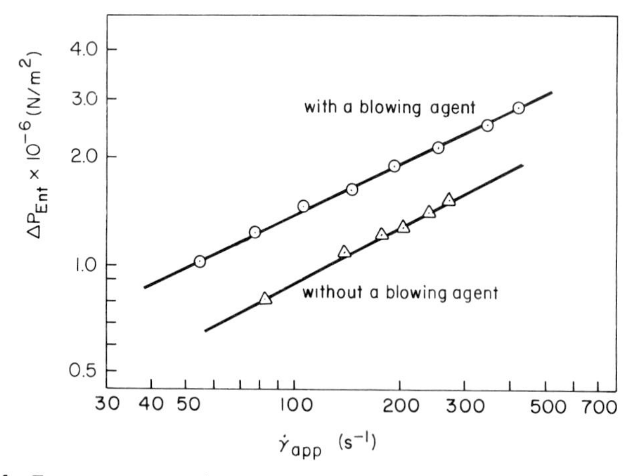

Fig. 6.54 Entrance pressure drop versus apparent shear rate for high-density polyethylene ($T = 200°$C) with and without a blowing agent [55]. From C. D. Han *et al.*, *J. Appl. Polym. Sci.* **20**, 1583, Copyright © 1976. Reprinted by permission of John Wiley & Sons, Inc.

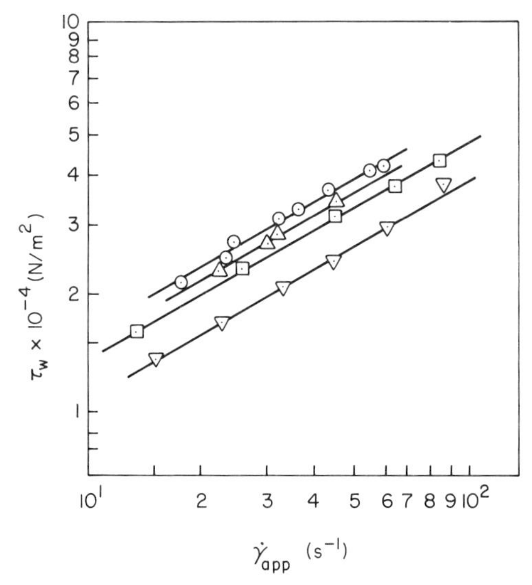

Fig. 6.55 Shear stress versus apparent shear rate for polystyrene (PS) ($T = 200°C$) with and without a blowing agent [56]: (⊙) pure PS; (△) PS containing 0.2 wt.% of $NaHCO_3$; (⊡) PS containing 0.4 wt.% of $NaHCO_3$; (▽) PS containing 0.8 wt.% of $NaHCO_3$.

possible evolution of the dissolved gas from the molten polymer as the melt passes through the die entrance.

One can, however, use a long capillary or slit die (i.e., one with a sufficiently large L/D ratio) and sufficiently high volumetric flow rates, so that constant pressure profiles (i.e., constant values of pressure gradient, $-\partial p/\partial z$) can be obtained over a large portion of the die, excluding both the entrance and exit regions. Under such conditions, Eq. (6.44) may be used to determine the wall shear stress τ_w and, thus, the *apparent* viscosity η (see Chapter 2). Han and Villamizar [56] used such an approach and reported viscosities of gas-charged molten polymers.

Figure 6.55 gives plots of wall shear stress τ_w versus *apparent* shear rate $\dot{\gamma}_{app}$ for polystyrene *with* and *without* $NaHCO_3$, and Fig. 6.56 gives plots of τ_w versus $\dot{\gamma}_{app}$ for polystyrene *with* and *without* Celogen CB. According to Han and Villamizar [56], the apparent shear rate was calculated, using the assumption that the blowing agent is completely dissolved in the melt over the range of pressure for which the pressure gradient $-\partial p/\partial z$ is reasonably constant, and that the density of *pure* polymer melt is not affected by dissolution of blowing agent. It is of interest to note in Figs. 6.55 and 6.56 that the wall shear stress τ_w is decreased and hence, the *apparent* shear viscosity η_{app} is decreased, as the concentration of blowing agent is increased.

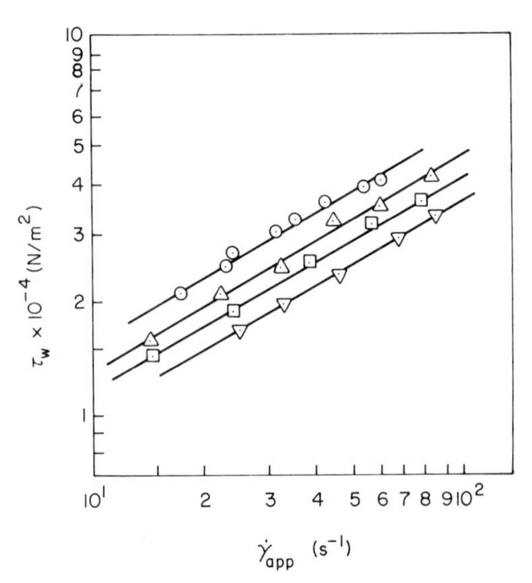

Fig. 6.56 Shear stress versus apparent shear rate for polystyrene (PS) ($T = 200°C$) with and without a blowing agent [56]: (⊙) pure PS; (△) PS containing 0.2 wt.% of Celogen CB; (⊡) PS containing 0.4 wt.% of Celogen CB; (▽) PS containing 0.8 wt.% of Celogen CB.

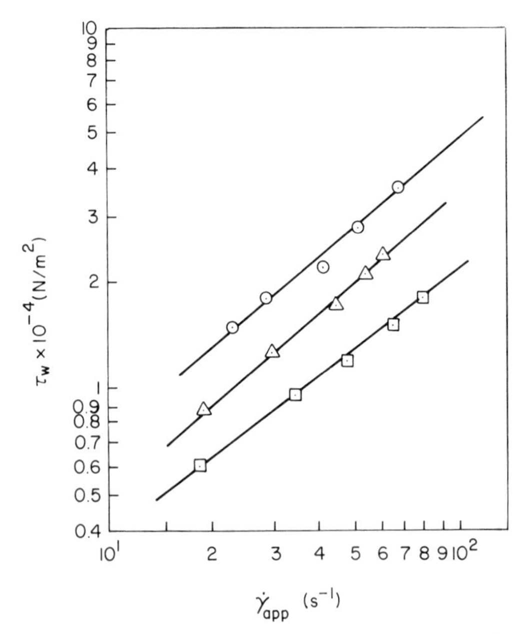

Fig. 6.57 Shear stress versus apparent shear rate for high-density polyethylene containing 0.8 wt.% of Celogen CB, at various melt temperatures (°C) [56]: (⊙) 180; (△) 220; (⊡) 260.

Figure 6.57 gives plots of τ_w versus $\dot{\gamma}_{app}$ for high-density polyethylene containing 0.8 wt. % of Celogen at different melt extrusion temperatures. It is seen that the wall shear stress (hence the apparent shear viscosity) is decreased as the extrusion temperature is increased.

Figure 6.58 gives the effect of the concentration of the physical blowing agent Freon 114 on the apparent viscosity of a low-density polyethylene, and shows that an increase in blowing agent concentration decreases the viscosity of the gas-charged polymer. Figures 6.59 and 6.60 give the effects of blowing agent concentration and temperature, respectively, on the apparent viscosity of polystyrene melts containing the physical blowing agent Freon 12. It is seen that the viscosity of the gas-charged polystyrene melt is decreased by an increase in melt temperature, and by an increase in blowing agent concentration.

With regard to the use of plunger-type instrument in determining the viscosity of gas-charged molten polymers, note that such an instrument provides information only of the pressure in the reservoir section of the capillary. In view of the fact that gas bubbles may nucleate in the entrance

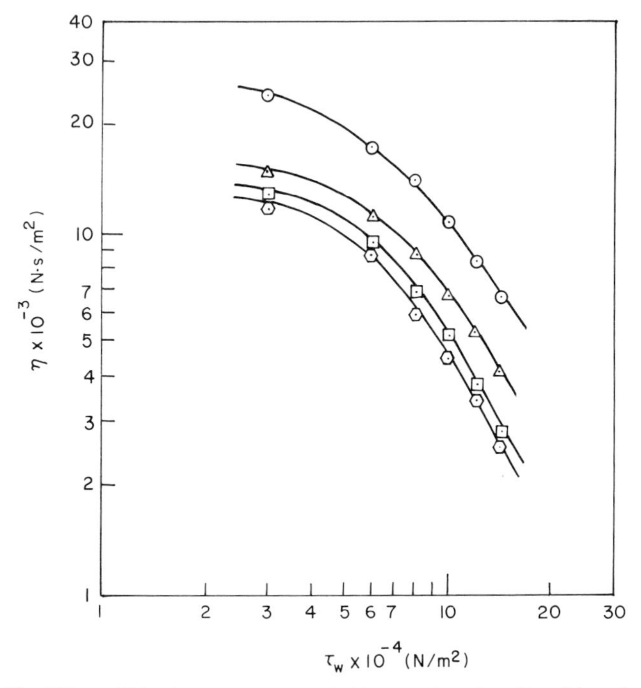

Fig. 6.58 Effect of blowing agent concentration on the viscosity of low-density polyethylene (LDPE) ($T = 120°C$) [63]; (⊙) no blowing agent; (△) 6 wt. % of Freon 114; (⊡) 15 wt. % of Freon 114; (hexagon with dot) 20 wt. % of Freon 114.

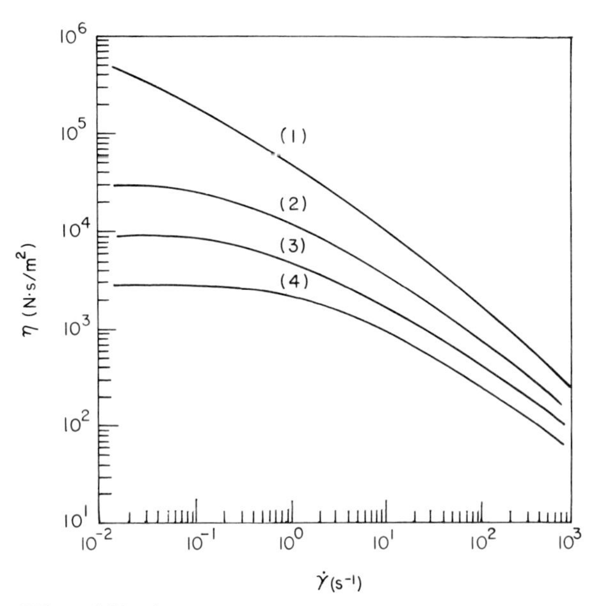

Fig. 6.59 Effect of blowing agent concentration on the viscosity of polystyrene (PS) ($T = 163°$C) [63]: (1) no blowing agent; (2) 4 wt. % of Freon 12; (3) 8 wt. % of Freon 12; (4) 12 wt. % of Freon 12.

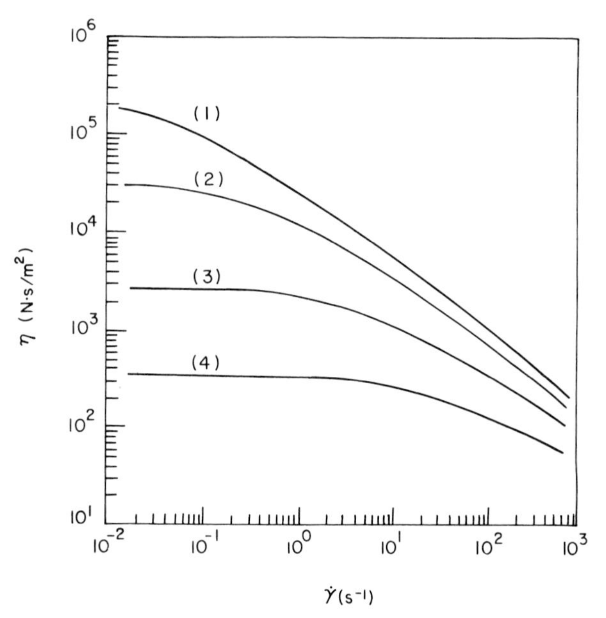

Fig. 6.60 Effect of temperature on the viscosity of polystyrene containing 4 wt.% of Freon 12 [63]: (1) $T = 135°$C; (2) $T = 163°$C; (3) $T = 204°$C; (4) $T = 246°$C.

region, where a large pressure drop usually occurs (see Fig. 6.52), and that microbubbles may start to grow before the polymer melt actually reaches the die exit, the use of the pressure measured in the reservoir section of the capillary would require an elaborate theoretical analysis. It should be pointed out that when gas bubbles start to come off from a polymer melt during its flow, the problem can no longer be treated in the same way as single-phase flow.

Let us now give consideration to the theoretical prediction of the *bulk* rheological properties of gas–liquid dispersed systems. Rigorous analysis of the problem will be extremely difficult because many gas bubbles dispersed in a liquid will grow as they flow through a conduit (i.e., subjected to a pressure-driven flow). The growth of gas bubbles is inevitable in such a flow field (e.g., in a capillary) as shown in Figs. 6.16 and 6.23. The problem becomes more complicated when a soluble gas, however small the solubility might be, is dispersed in the liquid. Note that the solubility of gas in a liquid is a function of the applied pressure and temperature. (See Figs. 6.10–6.12.)

Therefore, in order to facilitate our discussion here, let us consider the situation where the following assumptions may be valid: (1) the gas is *not* soluble in the liquid phase; (2) the gas is dispersed into the liquid, forming spherical bubbles of very small size; (3) the volume fraction of the gas phase is sufficiently low, allowing no coalescence of gas bubbles suspended in the liquid phase; (4) the pressure in the flow system is sufficiently high, so that *no* significant bubble growth occurs while the gas–liquid mixture flows through a conduit (e.g., in a cylindrical or slit die); (5) the shape of the gas bubbles remains spherical throughout the flow. We can then apply some of the phenomenological theories developed for predicting *bulk* rheological properties of emulsions, as discussed in Chapter 4.

Since the viscosity of the gas phase is negligibly small compared to that of the liquid phase (i.e., the viscosity ratio k approaches zero), the Taylor theory [67] for emulsions [see Eq. (4.18)] yields

$$\eta = \eta_0(1 + \phi) \tag{6.45}$$

where η_0 is the viscosity of the liquid phase and ϕ is the volume fraction of the gas phase. Such a prediction was first noted by Taylor himself [68]. Also, as k approaches zero, the Frankel–Acrivos theory [69] for emulsion [see Eqs. (4.31) and (4.32)] yields

$$\eta = \eta_0(1 + \phi) \tag{6.46}$$

$$\tau_{11} - \tau_{22} = 6.4(\eta_0^2\dot{\gamma}^2 a/\sigma)\phi \tag{6.47}$$

where $\dot{\gamma}$ is shear rate, a is the radius of a spherical gas bubble, and σ is the interfacial tension.

It is seen that the above phenomenological theories predict an increase of the bulk viscosity of the gas–liquid mixture over the viscosity of the suspending medium itself. Such theoretical predictions are *not* in agreement with the experimental observations discussed above in reference to gas-charged molten polymers (see Figs. 6.55–6.60). However, it should be remembered that the gases used as blowing agents are soluble in the molten polymer, and therefore the experimentally observed decrease in *bulk* viscosity of gas-charged molten polymers may be attributable to the plasticizing effect of the gas which, when dissolved, might have acted as an internal plasticizer.

It is of interest to note that the Frankel–Acrivos theory predicts the *bulk* elasticity, $\tau_{11} - \tau_{22}$, of a gas–liquid mixture. The *bulk* elasticity comes from the presence of interfacial tension σ between gas phase and the Newtonian liquid phase.

The bulk elasticity of a gas–liquid mixture can be determined experimentally from measurements of the wall normal stress in a slit or capillary die at sufficiently high shear rate (or shear stress), yielding a *constant* pressure gradient up to the die exit. In other words, the exit pressure method may be used to determine the normal stress difference only when the exit pressure is *greater* than the critical pressure for bubble inflation of the gas–liquid mixture under test. The principle of the exit pressure method is described briefly in Chapter 2 and elsewhere [70, 71].

6.5 PROCESSING–STRUCTURE–PROPERTY RELATIONSHIPS IN POLYMERIC FOAMS

In the past, a number of investigators [72–91] have discussed the mechanical properties (e.g., tensile strength, tear strength, tensile modulus) of polymeric foams. In a manner similar to filled polymer and polymer blends, the physical/mechanical properties of a polymeric foam depend, among many other factors, on foam morphology and the volume fraction of void. As one may surmise, the dispersion of a gaseous component has a profound influence on the mechanical properties of polymeric foam, indicating that processing conditions (i.e., the deformation and thermal histories of the gas-charged molten polymers) play an important role in determining structure–property relationships in polymeric foam. Unfortunately, most of the published literature discusses the structure–property relationships of foam specimen without referring to the history of specimen preparation (i.e., the processing conditions employed to obtain the specimens). Within the scope of this monograph, we shall discuss processing–structure–property relationships in polymeric foams, using the limited amount of experimental data available in the literature.

Han *et al.* [55] investigated the effect of processing variables and die geometry on the foam quality and the tensile properties of foam sheets,

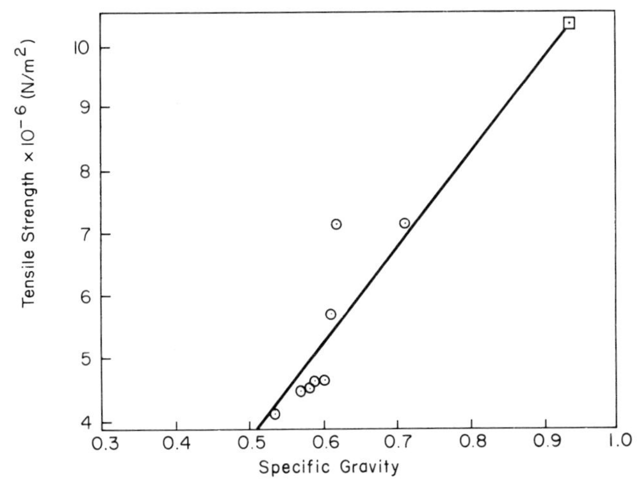

Fig. 6.61 Tensile strength versus specific gravity for low-density polyethylene (LDPE) foams [55]: (⊡) pure LDPE; (⊙) LDPE with 0.5 wt. % of azodicarbonamide (generating N_2). From C. D. Han *et al., J. Appl. Polym. Sci.* **20**, 1583. Copyright © 1976. Reprinted by permission of John Wiley & Sons, Inc.

produced with sheet-forming dies. They report that foam quality, as judged by appearance, was found to be better at a small die entrance angle, and that foam quality at a melt temperature of 160°C was better than that at the higher melt temperature of 200°C, when a low-density polyethylene was extruded, using azodicarbonamide as chemical blowing agent.

Figure 6.61 shows a linear relationship between the ultimate tensile strength and the specific gravity of the foam. Figure 6.62 shows that the quality of foam, as judged by appearance, can be quantitatively determined by measuring percent elongation. Interpreting the above, one can say that the tensile strength increases with specific gravity, and that the percent elongation is strongly dependent on the foam quality. Varenelli [78] has also

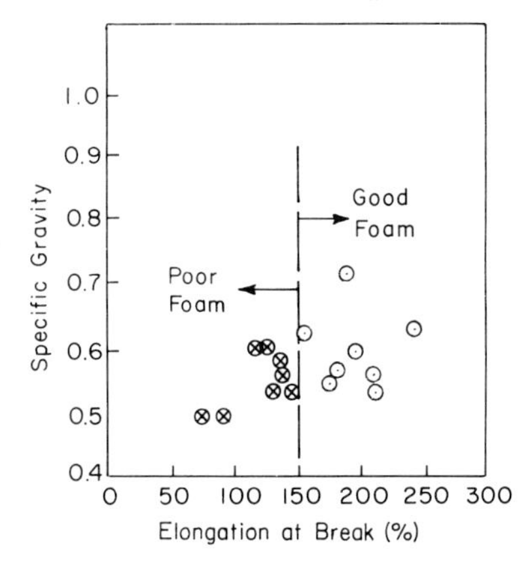

Fig. 6.62 Specific gravity versus elongation at break for low-density polyethylene foams [55]. The blowing agent used is 0.5 wt.% of azodicarbonamide. From C. D. Han *et al., J. Appl. Polym. Sci.* **20**, 1583. Copyright © 1976. Reprinted by permission of John Wiley & Sons, Inc.

shown that the modulus and tensile strength of flexible PVC foams are both related to the density in a linear fashion, as given in Fig. 6.63.

Some serious attempts were made to predict theoretically the mechanical properties of polymeric foam [73–75, 83]. Gent and Thomas [73–75] proposed a model to relate the Young's modulus of a latex rubber foam to the volume fraction of voids. Lederman [83] also derived a relationship between Young's modulus and foam density and tested the theory using experimental data for natural rubber latex foam. Their models are, however, applicable only to a completely open-cell structure (hence low-density polymeric foams) and thus fail to predict correctly the modulus of high-density polymeric foams.

In predicting Young's modulus of high-density polymeric foams, the theories that were originally developed by Kerner [92] and MacKenzie [93] for the shear modulus of filled composites, were modified. By modifying Kerner's theory, the following empirical expression for Young's modulus of a foam E_f is obtained [91]:

$$E_f/E_m = 12\phi/(23 - 11\phi) \tag{6.48}$$

and by modifying MacKenzie's theory, the following expression for Young's modulus is obtained [91]:

$$E_f/E_m = 4\phi(1 + 2\phi)/(13 + \phi - 2\phi^2) \tag{6.49}$$

in which E_m is the modulus of the matrix polymer and ϕ is volume fraction of void (i.e., the gas phase). Note that in the theories of Kerner and MacKenzie,

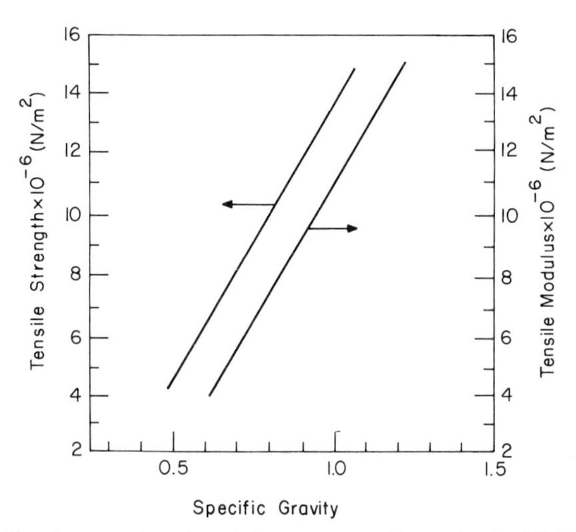

Fig. 6.63 Tensile strength and modulus versus specific gravity for PVC foams [78].

the composite is assumed to be filled with spherical, nonreinforcing particles and therefore, Eqs. (6.48) and (6.49) do not take into account the cell size and its distribution, and the nonsphericity of the cellular structure.

Figure 6.64 gives plots of reduced tensile modulus versus volume fraction of void for various polymeric foams, in which the theoretical predictions by Eqs. (6.48) and (6.49) are given. Note that the experimental data are fit rather well by a very simple empirical relationship, $E_f/E_m = \phi^2$.

Shetty and Han [94] performed an experimental study of sandwich foam coextrusion and discussed the effect of processing conditions on the foam morphology and mechanical properties of the foam sheets produced. In their study, a sheet-forming die with a feedblock was used for extruding two different polymer melts simultaneously, one with a blowing agent and the other without a blowing agent. A low-density polyethylene (LDPE) was used for the outer (skin) layer and an ethylene–vinyl acetate (EVA), with azodicarbonamide as blowing agent, was used for the inner (core) layer. The schematic of the coextruded foam resembles the one given in Fig. 6.4.

Figures 6.65–6.67 give photographs displaying the foam morphology of the coextruded foam sheets produced. It is seen that the bubble diameters vary over the range 0.25–3.0 mm.

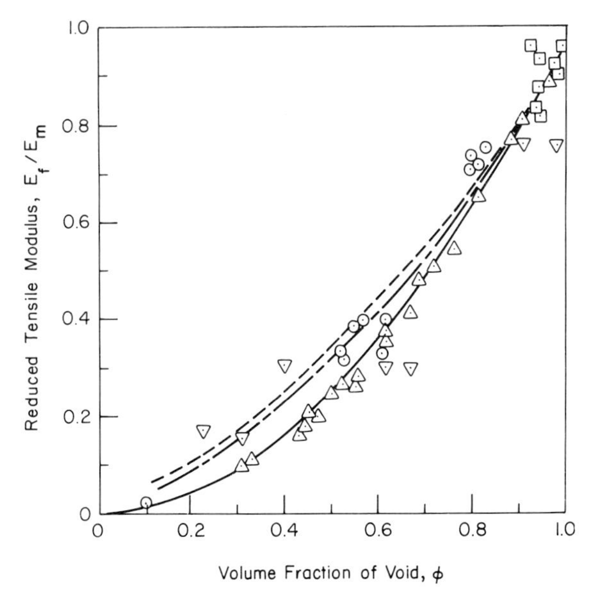

Fig. 6.64 Reduced tensile modulus versus volume fraction of void for various polymeric foams [89]: (⊙) Noryl and polycarbonate; (▽) general purpose polystyrene; (△) PP and PAN; (⊡) PP and PBT. The solid line (–) is predicted by $E_f/E_m = \phi^2$, the broken line (-··-) is predicted by Eq. (6.49) and the dashed line (---) is predicted by Eq. (6.48).

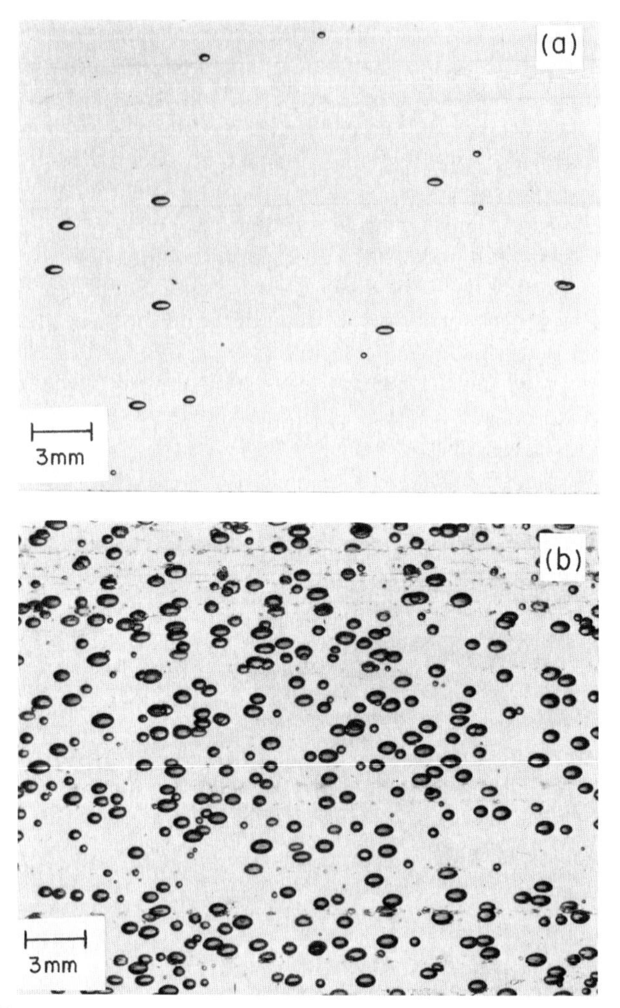

Fig. 6.65 Photographs displaying the morphology of three-layer sandwiched foams, with various layer thickness ratios of skin (LDPE) to foamed core (EVA plus blowing agent) components [94]: (a) 0.61; (b) 0.18. The extrusion temperature is 200°C, and the blowing agent employed is 0.3 wt.% of azodicarbonamide. From R. Shetty and C. D. Han, *J. Appl. Polym. Sci.* **22**, 2573. Copyright © 1978. Reprinted by permission of John Wiley & Sons, Inc.

Comparing Figs. 6.65a and 6.66a, and Figs. 6.65b and 6.66b, it is seen, as expected, that at the same processing temperature, the number of nucleation sites is increased with an increase in concentration of the blowing agent. Further, the cell size at a lower concentration of blowing agent is smaller than the cell size at a higher concentration of blowing agent. At a lower con-

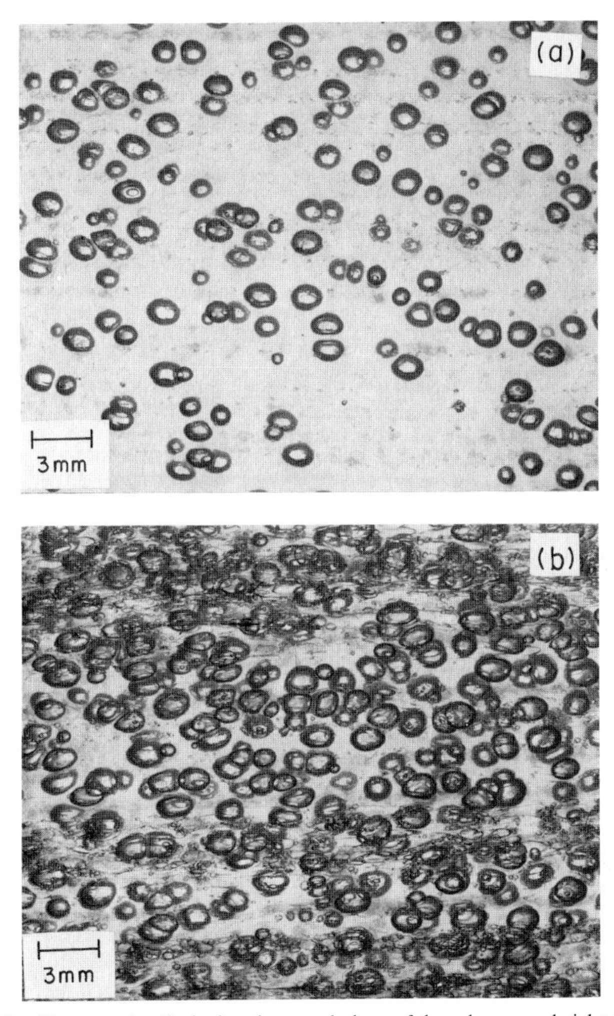

Fig. 6.66 Photographs displaying the morphology of three-layer sandwiched foams, with various layer thickness ratios of skin (LDPE) to foamed core (EVA plus blowing agent) components [94]: (a) 0.55; (b) 0.16. The extrusion temperature is 200°C, and the blowing agent employed is 0.6 wt.% of azodicarbonamide. From R. Shetty and C. D. Han, *J. Appl. Polym. Sci.* **22**, 2573. Copyright © 1978. Reprinted by permission of John Wiley & Sons, Inc.

centration of blowing agent, the number of nucleation sites is less, and hence the total void volume is less, resulting in an increase in melt viscosity. This effectively leads to a suppression of bubble growth and hence to a reduction in cell size. On the other hand, at higher concentrations of blowing agent, we have a decrease in melt viscosities, resulting in increased bubble growth and cell size.

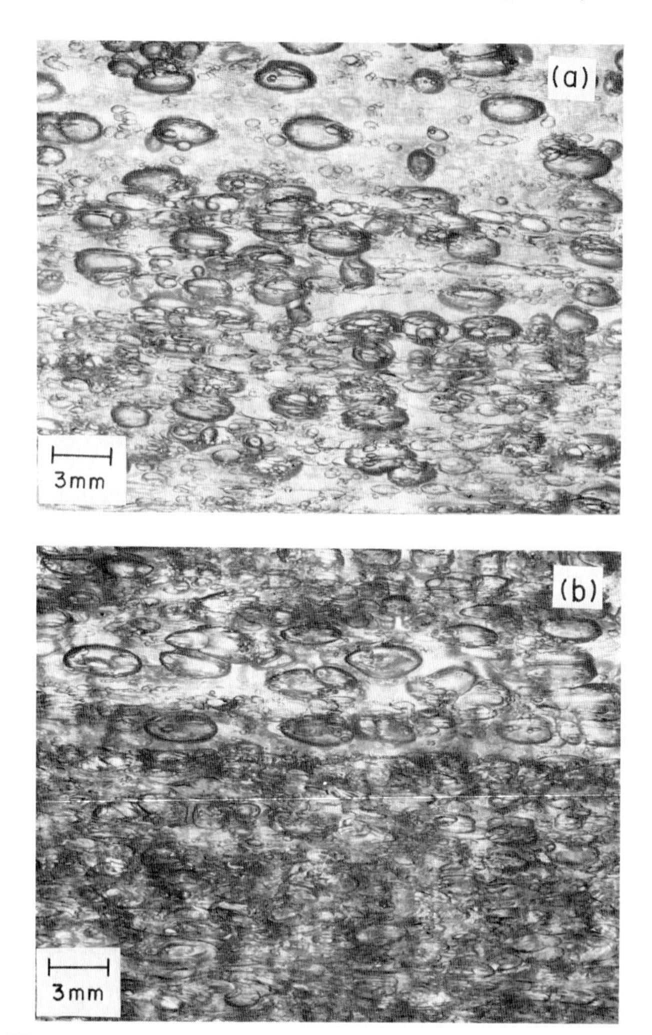

Fig. 6.67 Photographs displaying the morphology of three-layer sandwiched foams, with various layer thickness ratios of skin (LDPE) to foamed core (EVA plus blowing agent) components [94]: (a) 0.50; (b) 0.18. The extrusion temperature is 240°C, and the blowing agent employed is 0.6 wt. % of azodicarbonamide. From R. Shetty and C. D. Han, *J. Appl. Polym. Sci.* **22**, 2573. Copyright © 1978. Reprinted by permission of John Wiley & Sons, Inc.

Comparing Figs. 6.66a and 6.67a, and Figs. 6.66b and 6.67b, it is seen that the number of nucleation sites is approximately the same, but that there is an increase in cell size as the melt extrusion temperature is increased. At higher temperatures, the viscosities of the components used in processing are expected to decrease. This allows for an increase in bubble growth and hence an increase in cell size. At lower temperatures, the viscosities of the components increase, resulting in a decrease in cell size.

Comparing Figs. 6.65a and 6.65b, Figs. 6.66a and 6.66b, and Figs. 6.67a and 6.67b, it is seen that there is an increase in cell size as the skin-to-core ratio is decreased. The wall shear stress in the die at a smaller skin-to-core ratio is reduced because the skin component (LDPE) has a viscosity greater than the core component (EVA). This facilitates bubble growth and results in bigger cell sizes. On the other hand, at higher skin-to-core ratios, the wall shear stress is increased. This effectively suppresses bubble growth and results in a reduction of cell size.

From the above discussion, the following conclusions can be drawn: (1) An increase in the concentration of blowing agent results in an increase in the number of nucleation sites and cell size. (2) An increase in the processing temperature results in an increase in cell size. (3) An increase in the skin-to-core ratio results in a decrease in cell size in the foamed core.

Figure 6.68 shows the relationship between the tensile modulus and the skin-to-core ratio at various temperatures for the LDPE/EVA (foam)/LDPE system. It is seen that the tensile modulus of the sandwich foam sample

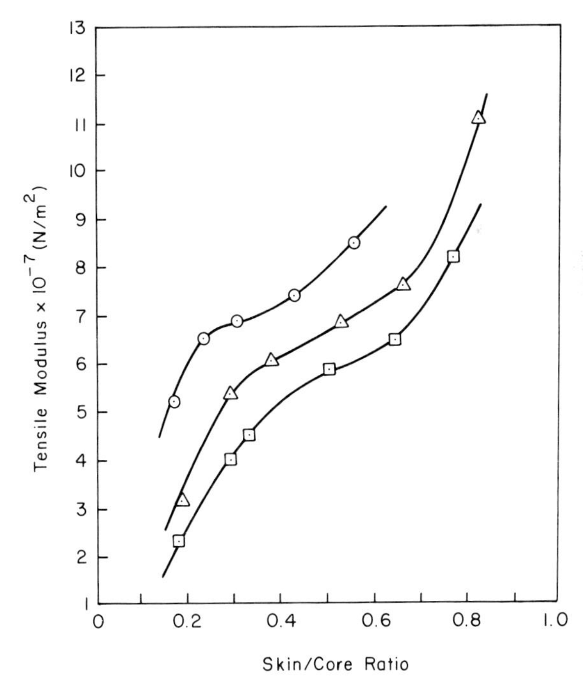

Fig. 6.68 Tensile modulus versus layer thickness ratio of the skin (LDPE) to core (EVA plus blowing agent) components for samples of three-layer sandwiched foams prepared at various extrusion temperatures (°C) [94]: (⊙) 200; (△) 220; (⊡) 240. The blowing agent used is 0.6 wt. % of azodicarbonamide. From R. Shetty and C. D. Han, *J. Appl. Polym. Sci.* **22**, 2573. Copyright © 1978. Reprinted by permission of John Wiley & Sons, Inc.

increases as the thickness ratio of the skin-to-core component increases. This can be explained by the fact that as the skin-to-core ratio is increased, the wall shear stress in the die is increased. This suppresses the growth of bubbles and results in smaller bubble sizes. On the other hand, as the skin-to-core ratio is decreased, the wall shear stress in the die is decreased. This allows for greater bubble growth, resulting in an increase in the size of bubbles. In extreme cases, where the skin is very thin, the bubbles may start to coalesce and may even rupture the skin (see, for instance, Fig. 6.67). Effectively, this increases the total volume of the voids in the foam sample, which causes a decrease in its tensile modulus.

It is also seen in Fig. 6.68 that the tensile modulus of the sandwich foam samples decreases as the processing temperature is increased. Note that at higher temperatures, there is a reduction in the melt viscosities of the components being processed. This allows for an increase in bubble growth and hence an increase in the size of the bubbles. The total volume of voids in the sandwiched foam is thus increased, resulting in a decrease in the tensile modulus. On the other hand, when the processing temperature is decreased, there is an increase in the melt viscosities of the components being processed. This results in a suppression of bubble growth and leads to a decrease in the size of the bubbles. The total void volume having thus been reduced, the tensile modulus of the foamed sample increases. From the above observation, we can conclude that (1) an increase in processing temperature effectively causes a decrease in the tensile modulus and (2) an increase in thickness ratio (skin/core) causes an increase in the tensile modulus of foamed sheets.

The cell size of polymeric foams can be varied by orders of magnitude through foaming techniques. The cell size is the structure parameter that can be modified by the proper choice of processing conditions (see Figs. 6.65–6.67), and it has a pronounced effect on the mechanical properties of foams. Figure 6.69 shows the effect of cell size on the mechanical properties of a rigid polystyrene foam. It is seen that both tensile and tear strengths increase with decreasing cell size. The reason for this phenomenon may be attributed to the ability of smaller cells to distribute tensile and compression forces much better than larger ones. Skochdopole and Rubens [77] reported experimental data, demonstrating the relationships of cell size, cell geometry, and polymer phase composition to the physical/mechanical properties of polyethylene foams.

It is worth mentioning that the geometry of cells in a foam also influences the mechanical properties. It has been found that the compressive stress in the direction of orientation is greater than that of foams of similar structure but having no orientation. The cells are elongated in the direction of orientation, and as a result, these cells buckle more than the unoriented ones. Since buckling requires much more stress than flexing, the stress-strain curve is higher if the load is applied in the oriented direction.

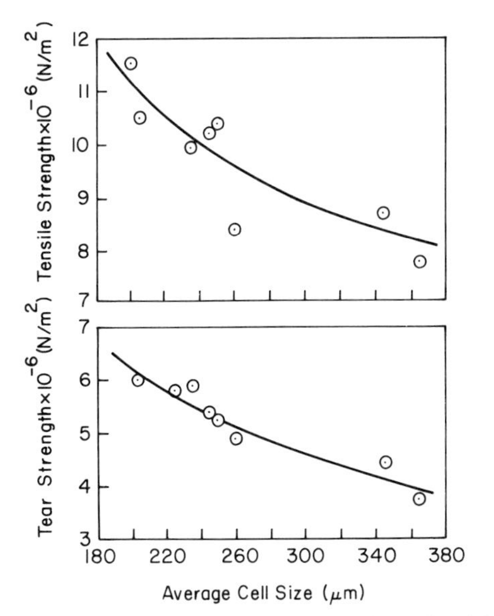

Fig. 6.69 Tensile and tear strengths versus average cell size for rigid polystyrene foams [79].

Problems

6.1 Consider an elongated, threadlike gas bubble placed on the longitudinal axis of a long cylindrical tube, as shown schematically in Fig. 6.70. Let us assume that (1) the radius of the gas bubble is very small compared to the tube radius; (2) bubble growth is far greater in the radial direction than in the longitudinal direction. The velocity field may be described in cylindrical coordinates (r, θ, z) by:

$$v_z = v_z(r, t), \qquad v_r = v_r(r, t), \qquad v_\theta = 0 \tag{6.50}$$

the equation of continuity by

$$(1/r)\,\partial(rv_r)/\partial r = 0 \tag{6.51}$$

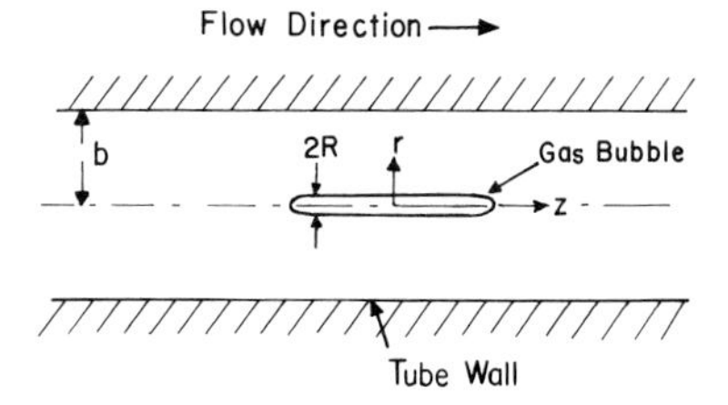

Fig. 6.70 Schematic showing a threadlike thin gas bubble traveling along the centerline of a cylindrical tube.

the equations of motion by

$$\rho\left(\frac{\partial v_r}{\partial t} + v_r \frac{\partial v_r}{\partial r}\right) = -\frac{\partial p}{\partial r} + \frac{1}{r}\frac{\partial}{\partial r}(r\tau_{rr}) - \frac{\tau_{\theta\theta}}{r} \tag{6.52}$$

$$\rho\left(\frac{\partial v_z}{\partial t} + v_r \frac{\partial v_z}{\partial r}\right) = -\frac{\partial p}{\partial z} + \frac{1}{r}\frac{\partial}{\partial r}(r\tau_{rz}) \tag{6.53}$$

and the mass balance equation for the concentration of the gas dissolved in the medium by

$$\frac{\partial C_A}{\partial t} + v_r \frac{\partial C_A}{\partial r} = D\left[\frac{1}{r}\frac{\partial}{\partial r}\left(r \frac{\partial C_A}{\partial r}\right)\right] \tag{6.54}$$

The force balance at the bubble wall requires that

$$-p_g + \sigma/R = -p(R) + \tau_{rr}(R) \tag{6.55}$$

in which σ is the interfacial tension, p_g is the pressure inside the gas bubble, R is the bubble radius, and p is the pressure in the liquid medium.

(a) Assuming that the stress and velocity fields are little disturbed by the presence of the bubble, derive the following expression describing the growth of the gas bubble,

$$\rho\left[(\dot{R}^2 + R\ddot{R})\ln\frac{b}{R} + \frac{(\dot{R}R)^2}{2}\left(\frac{1}{b^2} - \frac{1}{R^2}\right)\right]$$

$$= p_g - p(b) + \tau_{rr}(b) - \frac{\sigma}{R} + \int_R^b \frac{(\tau_{rr} - \tau_{\theta\theta})}{r}\,dr \tag{6.56}$$

in which ρ is the density of the medium, $\dot{R}$ and $\ddot{R}$ are the time derivatives of R with respect to time t, and b is the tube radius.

(b) The conservation of diffusing species at the bubble surface yields (with the end effect of the bubble neglected):

$$\frac{d}{dt}(\rho_g R^2) = 2RD\rho \left.\frac{dC_A}{dr}\right|_{r=R} \tag{6.57}$$

in which ρ_g is the density of the gas. Assuming that the gas phase follows the ideal gas behavior [see Eq. (6.40)], derive the following expression for the concentration of the diffusing species:

$$\frac{d}{dt}(\rho_g R^2) = \frac{8R^2 D\rho^2 (C_0 - C_w)^2}{3(R^2\rho_g - R_0^2\rho_{go})} \tag{6.58}$$

by applying the integral method suggested by Rosner and Epstein [36]. In Eq. (6.58) ρ_g represents the density of the gas evaluated at C_w (the bubble wall concentration) and ρ_{go} that evaluated at C_0 (the gas concentration in the medium far away from the bubble surface).

(c) Assuming that the rheological behavior of the liquid medium is represented by the Zaremba–DeWitt model [see Eq. (2.42) in Chapter 2], solve Eqs. (6.56) and (6.58) to predict the rate of bubble growth.

6.2 Assume that as soon as the mold cavity is filled, bubble growth may be described by the diffusion process, and that the rate of cooling of the mold cavity controls the rate of bubble growth. Consider then that ten small bubbles of uniform size are located between the two walls (5 × 20 cm) of a thin rectangular mold cavity, just as the mold filling was completed and the cooling began. The initial radius R_0 of the bubbles is very small compared to the thickness H of the rectangular mold cavity ($R_0 \ll H$), whose dimensions are: thickness (H) 0.5 cm, width (W) 5 cm, and length 20 cm, as schematically given in Fig. 6.42.

Using the following expressions describing the dependence of the diffusion coefficient D and Henry's law constant K_p on temperature T,

$$D(T) = D_0 \exp\left[\frac{-E_d}{R_g}\left(\frac{1}{T} - \frac{1}{T_0}\right)\right] \tag{6.59}$$

$$K_p(T) = K_{po} \exp\left[\frac{-E_s}{R_g}\left(\frac{1}{T} - \frac{1}{T_0}\right)\right] \tag{6.60}$$

in which R_g is the universal gas constant, and E_d and E_s are activation energies, derive a theoretical expression predicting the rate of bubble growth. State clearly all the assumptions that you may make in answering the question.

6.3 Figure 6.53 gives pressure profiles of a high-density polyethylene (HDPE), with and without a blowing agent (azodicarbonamide), being extruded through a cylindrical tube. It is seen that the pressure profile of the gas-charged HDPE (closed symbols) begins to show curvature (concave down) at about the middle of the tube, whereas the pressure profile of the pure HDPE (open symbols) shows a constant slope. (i) Explain what might have occurred in the tube when the gas-charged HDPE approached the die exit. (ii) Why, do you think, does the pressure gradient of the gas-charged HDPE show concave downward, which is opposite to that shown in Fig. 6.14?

6.4 Figure 6.52 gives pressure profiles of a gas-charged molten high-density polyethylene being extruded through a cylindrical tube. Derive theoretical expressions describing: (1) the variation in the density of the mixture inside the tube; (2) the volumetric flow rate of the mixture inside the tube; (3) the variation in the volume of free gas inside the tube. State clearly all the assumptions that you may make in answering the questions.

6.5 Show that Eq. (6.37) in this chapter will yield

$$\rho(\tfrac{3}{2}\dot{R}^2 + R\ddot{R}) = p_g - 2\sigma/R - 4\eta_0\dot{R}/R \quad 4\nu\ddot{R}/R$$
$$+ (4\beta + 2\nu)\dot{R}^2/R^2 - [p(H) - \tau_{rr}(H)] \qquad (6.61)$$

when the suspending medium is represented by the Coleman–Noll second-order fluid [see Eq. (2.53)]

Comparison of Eq. (6.61) with Eq. (6.37) reveals that the effect of initial stress in the melt on bubble growth cannot be described by the Coleman–Noll second-order fluid. This is a good example illustrating a serious drawback of the second-order fluid in describing the transient behavior of viscoelastic fluids.

REFERENCES

[1] E. A. Meinecke and R. C. Clark, "Mechanical Properties of Polymeric Foams." Technomic Publ., Westport, Connecticut, 1973.
[2] J. H. Saunders and R. H. Hansen, *in* "Plastic Foams" (K. C. Frisch and J. H. Saunders, eds.), Vol. 1, p. 23. Dekker, New York, 1972.
[3] C. J. Benning, "Plastic Foams." Wiley (Interscience), New York, 1969.
[4] K. C. Frisch and J. H. Saunders, eds., "Plastic Foams." Dekker, New York, 1972.
[5] B. A. Hunter and D. L. Schoene, *Ind. Eng. Chem.* **44**, 119 (1952).
[6] H. R. Lasman, "Modern Plastics Encyclopedia." McGraw-Hill, New York, 1965.
[7] R. G. LaCallade, *Plast. Eng.* **32**(6), 40 (1976).
[8] Ugine Kuhlmanm, Fr. Patent 1,498,620 (1967).
[9] Union Carbide Co., U.S. Patents 3,268,638 (1966); 3,436,440 (1969).
[10] USM Co., U.S. Patents 3,674,401 (1973); 3,697,204 (1973); 3,776,989 (1974).
[11] ICI Corp., U.S. Patents 3,559,280 (1972); 3,690,797 (1973); 3,773,156 (1974).
[12] D. F. Oxley and D. J. H. Sandiford, *Plast. Polym.* **39**, 288 (1971).
[13] R. G. Angell, *J. Cell. Plast.* **3**(11), 490 (1967).
[14] J. L. Throne, *J. Cell. Plast.* **12**, 161 (1976); 264 (1976).
[15] W. D. Harris, *Plast. Eng.* **32**(5), 26 (1976).
[16] A. Nicolay and G. Menges, 1975 ANTEC Preprints of SPE, p. 77 (1975).
[17] L. H. Gross and R. G. Angell, 1976 ANTEC Preprints of SPE, p. 162 (1976).
[18] R. H. Hansen and W. M. Martin, *Ind. Eng. Chem., Res. Dev.* **3**, 137 (1964).
[19] R. H. Hansen and W. M. Martin, *J. Polym. Sci., Part B* **3**, 325 (1965).
[20] C. W. Stewart, *J. Polym. Sci., Part A-2* **8**, 937 (1970).
[21] S. Wu, *J. Phys. Chem.* **74**, 632 (1970).
[22] A. N. Gent and D. A. Tompkins, *J. Appl. Phys.* **40**, 2520 (1969).
[23] Lord Rayleigh, *Philos. Mag.* **34**, 94 (1917).
[24] P. S. Epstein and M. S. Plesset, *J. Chem. Phys.* **18**, 1505 (1950).
[25] M. S. Plesset and S. A. Zwick, *J. Appl. Phys.* **23**, 95 (1952).
[26] L. E. Scriven, *Chem. Eng. Sci.* **10**, 1 (1959).
[27] E. J. Barlow and W. E. Langlois, *IBM J.* **6**, 329 (1962).
[28] L. A. Marique and G. Houghton, *Can. J. Chem. Eng.* **40**, 122 (1962).
[29] W. J. Yang and H. C. Yeh, *AIChE J.* **12**, 927 (1966).

[30] J. R. Street, *Trans. Soc. Rheol.* **12**, 103 (1968).
[31] J. L. Duda and J. S. Vrentas, *AIChE J.* **15**, 351 (1969).
[32] H. S. Fogler and J. D. Goddard, *Phys. Fluids* **13**, 1135 (1970).
[33] I. Tanasawa and W. J. Yang, *J. Appl. Phys.* **41**, 4526 (1970).
[34] E. Ruckenstein and E. J. Davis, *J. Colloid Interface Sci.* **31**, 142 (1970).
[35] J. R. Street, A. L. Fricke, and L. P. Reiss, *Ind. Eng. Chem., Fundam.* **10**, 54 (1971).
[36] D. E. Rosner and M. Epstein, *Chem. Eng. Sci.* **27**, 69 (1972).
[37] R. Y. Ting, *AIChE J.* **21**, 810 (1975).
[38] E. Zana and L. G. Leal, *Ind. Eng. Chem., Fundam.* **14**, 175 (1975).
[39] C. A. Villamizar and C. D. Han, *Polym. Eng. Sci.* **18**, 699 (1978).
[40] S. Y. Hobbs, *Polym. Eng. Sci.* **16**, 270 (1976).
[41] D. M. Newitt and K. E. Weale, *J. Chem. Soc.* p. 1541 (1948).
[42] J. L. Lundberg, M. B. Wilk, and M. J. Huyett, *J. Appl. Phys.* **31**, 1131 (1960).
[43] J. L. Lundberg, M. B. Wilk, and M. J. Huyett, *J. Polym. Sci.* **57**, 275 (1962).
[44] J. L. Lundberg, E. J. Mooney, and C. E. Rogers, *J. Polym. Sci., A-2* **7**, 947 (1969).
[45] J. L. Lundberg, M. B. Wilk, and M. J. Huyett, *Ind. Eng. Chem., Fundam.* **2**, 37 (1963).
[46] P. L. Durrill and R. G. Griskey, *AIChe J.* **12**, 1147 (1966).
[47] P. L. Durrill and R. G. Griskey, *AIChE J.* **15**, 106 (1969).
[48] D. D. Liu and J. M. Prausnitz, *Ind. Eng. Chem., Fundam.* **15**, 330 (1976).
[49] D. P. Maloney and J. M. Prausnitz, *AIChE J.* **22**, 74 (1976).
[50] L. I. Stiel and D. F. Harnish, *AIChE J.* **22**, 117 (1976).
[51] Y. L. Cheng and D. C. Bonner, *J. Polym. Sci., Polym. Phys. Ed.* **15**, 593 (1977).
[52] D. C. Bonner, *Polym. Eng. Sci.* **17**, 65 (1977).
[53] J. L. Duda and J. S. Vrentas, *J. Polym. Sci. A-2* **6**, 675 (1968).
[54] H. J. Karam, personal communication (1979).
[55] C. D. Han, Y. W. Kim, and K. D. Malhotra, *J. Appl. Polym. Sci.* **20**, 1583 (1976).
[56] C. D. Han and C. A. Villamizar, *Polym. Eng. Sci.* **18**, 687 (1978).
[57] A. Acrivos and T. S. Lo, *J. Fluid Mech.* **86**, 641 (1978).
[58] C. D. Han, "Rheology in Polymer Processing," Chap. 6. Academic Press, New York, 1976.
[59] H. J. Yoo and C. D. Han, *Polym. Eng. Sci.* **21**, 69 (1981).
[60] B. D. Coleman and W. Noll, *Rev. Mod. Phys.* **33**, 239 (1961).
[61] H. J. Yoo and C. D. Han, *J. Rheol.* **25**, 115 (1981).
[62] C. D. Han and H. J. Yoo, *Polym. Eng. Sci.* **21**, 518 (1981).
[63] H. J. Karam, personal communication (1979).
[64] L. L. Blyler and T. K. Kwei, *J. Polym. Sci., Part C* No. 35, 165 (1971).
[65] D. M. Bigg, J. R. Preston, and D. Brenner, *Polym. Eng. Sci.* **16**, 706 (1976).
[66] Y. Oyanagi and J. L. White *J. Appl. Polym. Sci.* **23**, 1013 (1979).
[67] G. I. Taylor, *Proc. R. Soc. London, Ser. A* **138**, 41 (1932).
[68] G. I. Taylor, *Proc. R. Soc. London, Ser. A* **226**, 34 (1954).
[69] N. A. Frankel and A. Acrivos, *J. Fluid Mech.* **44**, 65 (1970).
[70] C. D. Han, *Trans. Soc. Rheol.* **18**, 163 (1974).
[71] C. D. Han, "Rheology in Polymer Processing," Chap. 5. Academic Press, New York, 1976.
[72] J. A. Talalay, *Ind. Eng. Chem.* **46**, 1530 (1954).
[73] A. N. Gent and A. G. Thomas, *J. Appl. Polym. Sci.* **1**, 107 (1959).
[74] A. N. Gent and A. G. Thomas, *J. Appl. Polym. Sci.* **2**, 354 (1960).
[75] A. N. Gent and A. G. Thomas, *Rubber Chem. Technol.* **36**, 597 (1963).
[76] V. Matonis, *SPE J.* **20**, 1024 (1964).
[77] R. E. Skochdopole and L. C. Rubens, *J. Cell. Plast.* **1**, 91 (1965).

[78] A. D. Varenelli, *Wire Prod.* **36**(7), 861 (1961).
[79] P. W. Croft, *Plastics* **29**(10), 47 (1964).
[80] C. J. Benning, *J. Cell. Plast.* **3**, 125 (1967).
[81] M. N. Paul, *in* "Handbook of Foamed Plastics" (R. J. Bender, ed.), p. 306. Lake Publ. Corp., Libertyville, Illinois, 1965.
[82] V. Matonis, *SPE J.* **20**, 1024 (1964).
[83] J. M. Lederman, *J. Appl. Polym. Sci.* **15**, 693 (1971).
[84] R. E. Whittaker, *J. Appl. Polym. Sci.* **15**, 1205 (1971).
[85] B. S. Mehta and E. A. Colombo, *J. Cell. Plast.* **12**(1), 59 (1976).
[86] J. F. Zappala, *J. Cell. Plast.* **7**(11), 309 (1971).
[87] S. Y. Hobbs, *Polym. Eng. Sci.* **15**, 854 (1075).
[88] S. Y. Hobbs, *J. Cell. Plast.* **12**, 258 (1976).
[89] R. C. Progelhof and K. Eilers, SPE Divtec, Woburn, Massachusetts, September 1977.
[90] J. L. Throne, SPE PATEC, p. 185. Los Angeles, California, February 1979.
[91] R. C. Progelhof and J. L. Throne, *Polym. Eng. Sci.* **19**, 493 (1979).
[92] E. H. Kerner, *Proc. R. Soc. London, Ser. B* **69**, 808 (1956).
[93] J. K. MacKenzie, *Proc. Phys. Soc. London, Sect. B* **63**, 2 (1950).
[94] R. Shetty and C. D. Han, *J. Appl. Polym. Sci.* **22**, 2573 (1978).

Part II

Stratified Multiphase Flow in Polymer Processing

Stratified multiphase flow, commonly referred to in the polymer processing industry as coextrusion, has emerged as an attractive means of economically producing multilayered composite films, for instance, which otherwise would be very difficult to achieve by conventional lamination. The success of a coextrusion process depends very much on the die design, which should provide the means for combining different melt streams and of controlling the thickness of the layers of a composite film. Success also depends on the choice of the proper combination of rheological properties of the polymers involved.

It is also well documented in the literature that under certain conditions of stratified multiphase flow, products having irregular interfaces are obtained. This phenomenon is termed "interfacial instability." It is important to have a better understanding of the mechanism of the occurrence of interfacial instability in coextrusion, thus enabling one to avoid the occurrence of the undesirable flow instability by selecting polymer combinations on the basis of their rheological properties and by selecting optimal processing conditions.

In the next two chapters we shall discuss some fundamental aspects of coextrusion operations in cylindrical, rectangular, and annular dies, by presenting a unified approach to a better understanding of coextrusion processes from both the materials and the processing points of view. Emphasis will be placed on the development of a criterion (or criteria) for determining the processability of two or more thermoplastic polymeric materials in the coextrusion operation. Such a criterion will lead to a better understanding of the effect of the molecular characteristics of the polymers on the mechanical/physical properties of the coextruded composite produced, the effect of the rheological properties of individual polymers on coextrudability, and the effect of processing conditions on the quality of the coextruded composite produced.

7

Stratified Flow (Coextrusion) of Polymeric Systems

7.1 INTRODUCTION

There are two basic types of coextrusion process, which make use of dies having a *circular* cross section. As schematically shown in Fig. 7.1, one type coextrudes two polymers *side by side* through a cylindrical die, and the other coextrudes two (or more) polymers *concentrically* through a cylindrical die. Each of these two processes has specific objectives to be achieved in producing a final product of desired properties.

The fiber industry has long used side-by-side coextrusion with spinnerette holes having circular cross sections in producing bicomponent (conjugate) fibers [1–7]. As long as the fiber produced in this manner retains a flat interface between the two components, it provides unique properties resembling those of natural wools, the so-called crimped (or wool) fibers. The crimping characteristics result from the different thermal expansion coefficients of the individual components, leading to the buckling of the filament while it is either being cooled or coagulated along the length of the spinline after exiting from the spinnerette. Figure 7.2 gives a photograph of the longitudinal view of polypropylene/polyester bicomponent fibers at drawn state, showing their crimping characteristics. Such fibers result in a wool-like effect in knit or woven cloth.

The fiber industry has also introduced the concept of *concentric*, stratified two-phase flow in a circular die, for the production of fibers having antistatic or flame-retardant properties. In such a process, two melt streams, one *with* and one *without* an additive (e.g., antistatic or flame-retardant components),

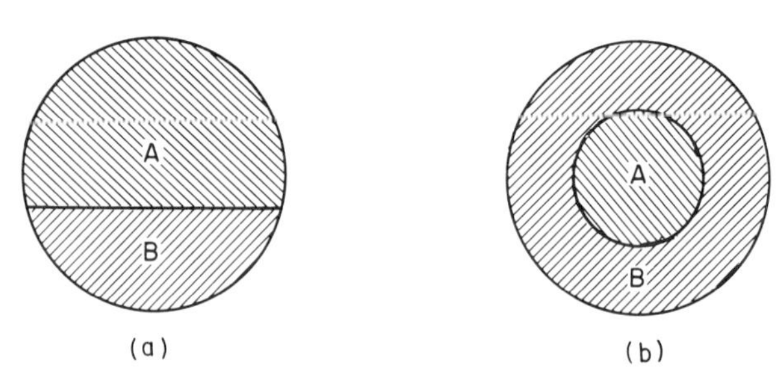

Fig. 7.1 Schematic illustrating two different types of coextrusion operations in a circular die: (a) side-by-side coextrusion; (b) sheath-core coextrusion.

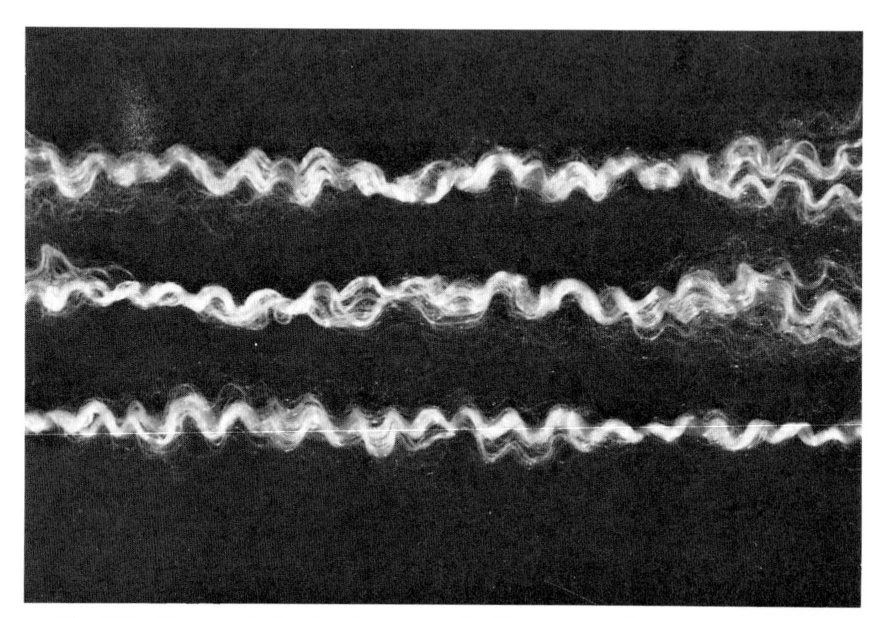

Fig. 7.2 Photograph showing the crimps of a bicomponent fiber of polypropylene and polyethylene terephthalate.

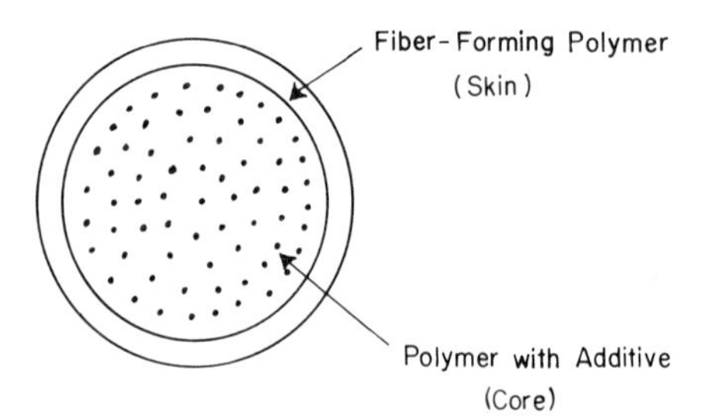

Fig. 7.3 Schematic of the sheath-core fiber cross section in which the core component contains an additive.

are coextruded concentrically through spinnerette holes having circular cross sections. Figure 7.3 gives a schematic of the fiber cross section, in which the core contains an additive, shielded completely by the homopolymer (the skin) *without* an additive.

There are other applications, which make use of stratified two-phase flow in a circular die. For instance, one can produce foamed rods shielded by a smooth skin component. Such a foamed product offers advantages in that it can be produced with a smaller amount of resin and, due to its flexible nature, can be used for certain purposes, otherwise difficult to achieve. From the processing point of view, the production of foamed rods, for instance, involves the combination of a foaming process (dispersed flow) and a coextrusion process (stratified flow). An understanding of the foaming operation, discussed in Chapter 6, will help one better control such a combined process.

As a matter of fact, stratified two-phase flow in a cylindrical tube has been a subject of considerable interest, in the late 1950's and early 1960's, in the transportation of heavy viscous crude oil through long pipelines [8–14]. The addition of a small amount of water to the crude oil flowing through the pipeline considerably reduced the pressure gradient, resulting in a reduction in pumping cost. This is because the water, which is far less viscous than the heavy crude oil, migrates to the inner wall of the pipeline and acts as a lubricant. This observation stimulated both theoretical and experimental studies dealing with the stratified laminar flow of oil–water mixtures in a cylindrical tube. It should be mentioned, however, that those studies dealt with two Newtonian liquids.

Coextrusion processes have gained an importance in manufacturing multilayer thin films and sheets that have the combined properties of the two or more polymers involved [15–22]. The coextrusion multilayer film process is more economical than the conventional laminating process, and yields much thinner layers than would be possible by lamination. In producing multilayer films, rectangular dies for flat films and annular dies for blown films are employed.

In the coextrusion of films, it is desirable that the strength of one material balances any weakness of the other in order to produce a structure having nearly uniform properties. To be successful, coextruded materials must show compatibility or affinity in order to have good adhesion, and industry has succeeded in finding materials possessing outstanding adhesion. For instance, polyethylene ionomer/nylon is one commercially available film-forming system. It is particularly well suited to the coextrusion process, since the two materials exhibit a chemical affinity for each other, and therefore have excellent adhesion. In the selection of materials, other characteristics are also important to consider. For instance, the nylon provides strength and an oxygen barrier, whereas the polyethylene provides a water barrier. The combined characteristics are exactly suited to films for packaging certain foodstuffs.

There are many other applications where the coextrusion process may be used effectively for producing a whole new range of products, which combine the best properties of each layer of the sheet. Figure 7.4 gives schematics describing some important applications of the sheet-forming coextrusion process. In reference to Fig. 7.4, one can achieve surface protection by coextruding a protective layer onto a normally degradable polymer. Typically, environment-resistant polymers (e.g., pigmented PVC) can be extruded over lower cost base materials (e.g., unpigmented PVC). Such sheets are used in exterior siding materials for houses, which require good weather resistance. Hard, scratch-resistant layers on softer materials is another application. Multicolor products can be made from sheets having different colors on the two sides by coextruding two differently pigmented resins through a sheet-forming die. Other applications are the covering of scrap material with uniformly colored material on one or both sides (i.e., two-layer or three-layer sheet coextrusion), thus enabling one to reclaim discolored or contaminated material. Another interesting application is the coextrusion of base resins on a foam core to obtain so-called sandwiched foam products, for greater rigidity, low cost, and good insulating properties. Other applications are the combination of rigid, brittle outer layers with an impact-resistant central layer, and the combination of two materials having little adhesion, but with an interlayer of adhesive.

In wire coating coextrusion, two (or more) molten polymers, concentrically arranged, are forced to flow through an annulus where the inner

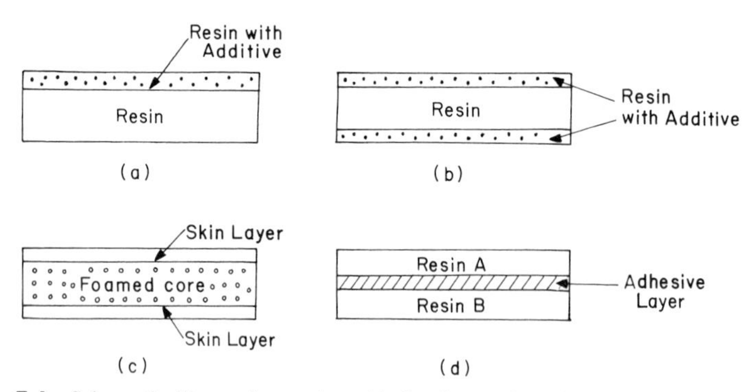

Fig. 7.4 Schematic illustrating various kinds of extrudate that may be obtained from coextrusion through a rectangular (or slit) die: (a) two-layer product in which one layer contains an additive (e.g., pigment); (b) three-layer product in which two outer layers contain an additive; (c) three-layer product representing the sandwiched foams in which the core component contains the porous structure; (d) three-layer product in which the middle layer is an adhesive.

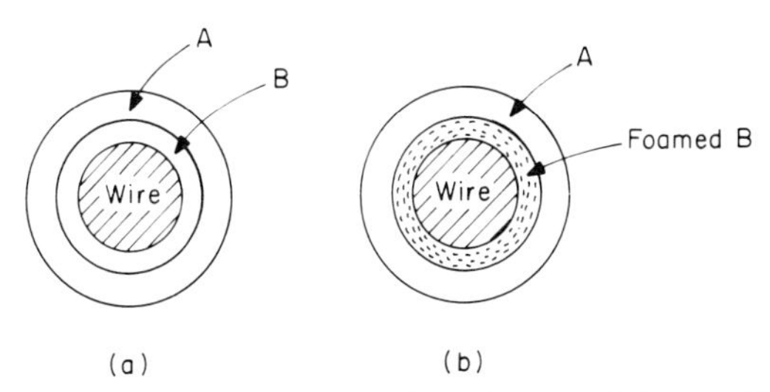

Fig. 7.5 Schematic showing the cross section of coextruded wire: (a) with two homopolymers; (b) with two polymers in which one contains the cellular structure (foamed core).

cylinder (i.e., the wire) moves at a constant speed [23]. In order to improve the insulating characteristics of the coated material, one sometimes uses a foamed inner layer surrounded by a base resin skin component, as schematically shown in Fig. 7.5.

In more recent years, stratified two-phase flow in a cylindrical tube (and in a rectangular duct) has been dealt with from the point of view of polymer processing operations. Some researchers [24–26] made theoretical studies of the mechanisms of interface deformation, and others [27–32] carried out experimental investigations. With a few exceptions [33, 34], most of the theoretical studies have dealt with the stratified two-phase flow of Newtonian liquids, with particular attention to the equilibrium shape of the interface when two such fluids were forced to flow side by side. It has been found, both theoretically and experimentally, that, in general, the more viscous component tends to be convex at the interface, and that the less viscous component tends to preferentially wet the wall of the die. Furthermore, the fluid viscosity appears to be more important than the fluid elasticity, at least in the initial deformation of the interface between the two components. However, under certain flow conditions, very irregular interfaces have been observed and such a phenomenon has been explained in terms of both the fluid viscosity and elasticity [30, 35].

As one may surmise, control of coextrusion operations in practice requires a better understanding of the flow problems associated with the processes. One can conceive of several interesting and fundamental flow problems whose solution could guide the industry concerned to design better dies, to choose optimum processing conditions, and to select proper combinations of polymer systems from the standpoint of their rheological properties.

We shall discuss in this chapter the flow problems involved with the stratified flow (coextrusion) of polymeric liquids through a cylindrical tube, or in a rectangular die, or in an annular die. Emphasis will be placed on the effects of the rheological properties of the individual components on the interface from its initial shape, and the processing characteristics (e.g., velocity and stress distributions) as affected by rheological properties and processing conditions.

7.2 ANALYSIS OF SHEATH-CORE COEXTRUSION IN A CYLINDRICAL TUBE

Consider two-phase concentric flow in a cylindrical tube, as schematically shown in Fig. 7.6, with the following assumptions: (1) no external forces exerted on the flow; (2) the two fluids have the same density (i.e., negligible buoyancy effect); (3) the two fluids are incompressible; (4) the two fluids follow a power-law model.

Using cylindrical coordinates, the z component of the equations of motion for tubular flow may be written

$$-\frac{\partial p}{\partial z} + \frac{1}{r}\frac{\partial}{\partial r}(r\tau_{rz}) = 0 \tag{7.1}$$

For the power-law fluid defined as

$$\tau_{rz} = K(\partial v_z/\partial r)^n \tag{7.2}$$

in which K and n are power-law constants, integrating Eq. (7.1) gives for phase A

$$v_{z,A} = -(\zeta/2K_A)^{1/n_A}\{n_A/(n_A+1)\}r^{(n_A+1)/n_A} + c_3 \tag{7.3}$$

and, for phase B

$$v_{z,B} = -(\zeta/2K_B)^{1/n_B}\{n_B/(n_B+1)\}r^{(n_B+1)/n_B} + c_4 \tag{7.4}$$

in which C_3 and C_4 are integration constants and ζ is the pressure gradient defined as

$$\zeta = -\partial p/\partial z \tag{7.5}$$

Note that n_A and K_A are the power-law constants for phase A, and n_B and K_B are those for phase B.

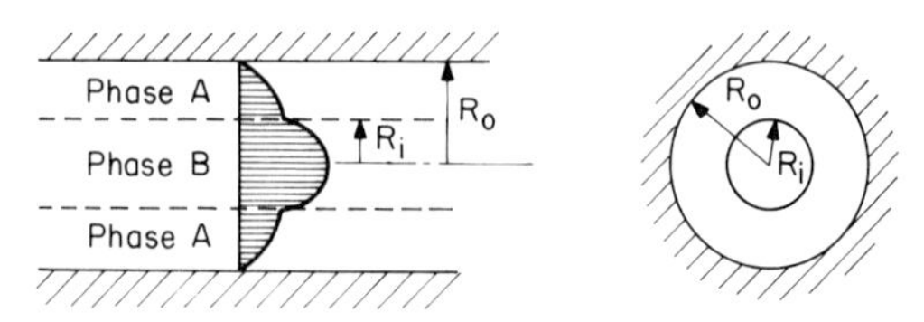

Fig. 7.6 Schematic showing the velocity profile in the sheath-core coextrusion through a cylindrical tube.

Using the boundary conditions

$$\text{at } r = R_0, \quad v_{z,A} = 0 \tag{7.6}$$

$$\text{at } r = R_i, \quad v_{z,A} = v_{z,B} \tag{7.7}$$

we have, from Eqs. (7.3) and (7.4), the velocity profiles

$$v_{z,A} = \left(\frac{\zeta}{2K_A}\right)^{\alpha_A} \frac{1}{(\alpha_A + 1)} [R_0^{\alpha_A+1} - r^{\alpha_A+1}] \tag{7.8}$$

$$v_{z,B} = \left(\frac{\zeta}{2K_B}\right)^{\alpha_B} \left(\frac{1}{\alpha_B + 1}\right) [R_i^{\alpha_B+1} - r^{\alpha_B+1}] + \left(\frac{\zeta}{2K_A}\right)^{\alpha_A} \frac{1}{(\alpha_A + 1)} [R_0^{\alpha_A+1} - R_i^{\alpha_A+1}] \tag{7.9}$$

where

$$\alpha_A = 1/n_A; \qquad \alpha_B = 1/n_B \tag{7.10}$$

The volumetric flow rates Q_A and Q_B may be obtained by

$$Q_A = 2\pi \int_{R_i}^{R_0} v_{z,A} r\, dr = A\zeta^{\alpha_A}\left[1 - \left(\frac{\alpha_A + 3}{\alpha_A + 1}\right)\lambda^2 + \left(\frac{2}{\alpha_A + 1}\right)\lambda^{\alpha_A+3}\right] \tag{7.11}$$

$$Q_B = 2\pi \int_0^{R_i} v_{z,B} r\, dr = A\zeta^{\alpha_A}\left(\frac{\alpha_A + 3}{\alpha_A + 1}\right)(\lambda^2 - \lambda^{\alpha_A+3}) + B\zeta^{\alpha_B}\lambda^{\alpha_B+1} \tag{7.12}$$

respectively, where $\lambda = R_i/R_0$ and

$$A = \left(\frac{R_0}{2K_A}\right)^{\alpha_A}\left(\frac{\pi R_0^3}{\alpha_A + 3}\right); \qquad B = \left(\frac{R_0}{2K_B}\right)^{\alpha_B}\left(\frac{\pi R_0^3}{\alpha_B + 3}\right) \tag{7.13}$$

The total volumetric flow rate $Q = Q_A + Q_B$ may be given as

$$Q = A\zeta^{\alpha_A}[1 - \lambda^{\alpha_A+3}] + B\zeta^{\alpha_B}\lambda^{\alpha_B+3} \tag{7.14}$$

Note that Eqs. (7.11) and (7.12) may be used to predict the volumetric flow rates (Q_A and Q_B) of the individual components when the rheological parameters (K_A, n_A, K_B, n_B), the tube radius (R_0), the interface position (λ), and the pressure gradient (ζ) are specified.

Figure 7.7 gives some representative velocity profiles computed by the use of Eqs. (7.8) and (7.9), simulating the sheath-core coextrusion of low-density polyethylene (LDPE) and polystyrene (PS). It should be pointed out

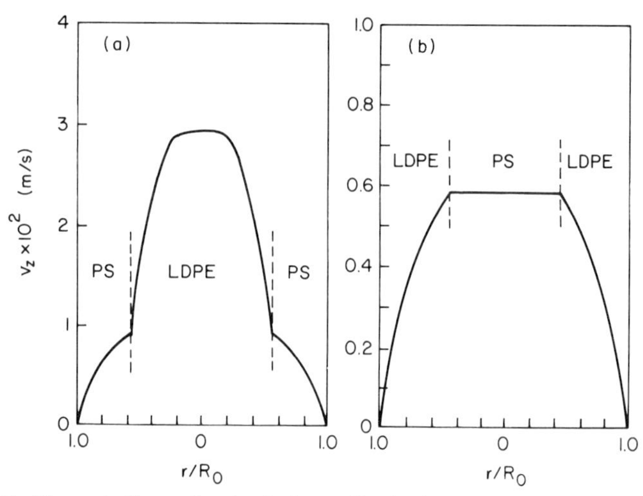

Fig. 7.7 Theoretically predicted velocity profiles in the sheath-core coextrusion through a cylindrical tube: (a) PS (sheath)/LDPE (core) system, $-\partial p/\partial z = 2.75 \times 10^7$ N/m^3, $Q_{PS} = 68.2$ cm^3/min, $Q_{LDPE} = 38.4$ cm^3/min; (b) LDPE (sheath)/PS (core) system, $-\partial p/\partial z = 6.56 \times 10^6$ N/m^2, $Q_{LDPE} = 29.4$ cm^3/min, $Q_{PS} = 68.2$ cm^3/min. The power-law constants for the polymers used are (a) For PS: $K = 2.105 \times 10^4$ N sec^n/m^2, $n = 0.28$; (b) For LDPE: $K = 0.437 \times 10^4$ N sec^n/m^2, $n = 0.48$.

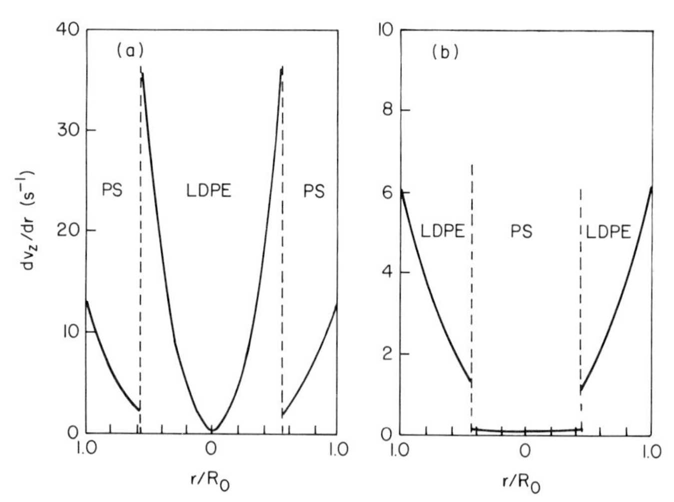

Fig. 7.8 Theoretically predicted profiles of velocity gradient in the sheath-core coextrusion through a cylindrical tube: (a) PS (sheath)/LDPE (core) system; (b) LDPE (sheath)/PS (core) system.

that in computing the velocity profiles, the interface position in two layers is not known *a priori*. Therefore, the dimensionless interface position λ must be determined by trial and error until the computed value of the volumetric flow ratio, Q_A/Q_B, agrees with the experimentally observed one, when the pressure gradient ζ and the other parameters are specified.

Figure 7.8 gives the distribution of velocity gradient, and Fig. 7.9 the distribution of melt viscosity for the LDPE/PS system. The viscosity varies across the tube radius because the polymer melts exhibit shear-thinning behavior described by the power-law model. If two Newtonian fluids were coextruded, the viscosities of each phase would have *flat* profiles across the tube radius with a finite jump at the interface, irrespective of the shapes of the velocity and velocity gradient profiles.

At this juncture it is worth examining some of the assumptions made in the above analysis, namely: (1) isothermal flow; (2) no slippage at the interface between the layers [see Eq. (7.7)]; (3) no slippage at the tube wall [see Eq. (7.6)]. Violation of any of these assumptions should be judged on the basis of unequivocal experimental evidence. We admit that experimental verification of the violation of any of these assumptions is a very difficult task, indeed. Even when one finds, experimentally, that some of these assumptions are violated with a *certain* class of fluids, one must *not* conclude that all fluids violate the assumptions. This is because in polymer processing operations we deal with very diversified fluids, and one class of fluids can behave very differently from another under identical flow conditions. Keeping

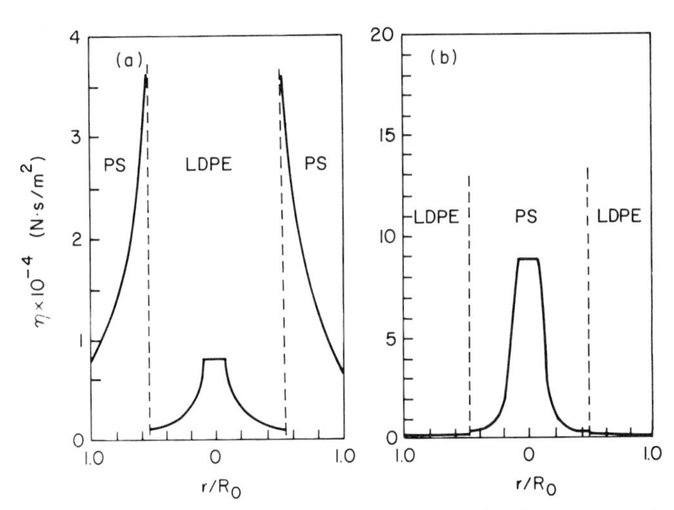

Fig. 7.9 Theoretically predicted profiles of viscosity in the sheath-core coextrusion through a cylindrical tube: (a) PS (sheath)/LDPE (core) system; (b) LDPE (sheath)/PS (core) system.

this in mind, let us now examine the situations where violations of each of the three assumptions delineated above may occur. We do this solely to demonstrate that different solutions can be obtained by modifying the original assumptions.

In the past, there has been some concern expressed about viscous heating when very viscous molten polymers flow through a conduit. When viscous heating becomes significant, the temperature rises in the melts and the velocity and stress distributions in the tube are affected. Uhland [36] has analyzed the concentric two-phase flow of power-law fluids in a cylindrical tube, taking into account viscous dissipation. In his analysis, the equations of motion and the equation of energy were coupled by the temperature and shear-rate-dependent viscosity, and numerical solutions were obtained using specific values of system parameters and thermal boundary conditions. Uhland's analysis shows that (1) the temperature of the fluid increases in the direction of flow due to viscous dissipation; (2) the temperature and velocity fields are largely determined by the flow properties of the two fluids and the thermal boundary conditions; (3) most energy is dissipated in the outer layer near the die wall; (4) the interface position changes along the die due to the variations in the temperature and velocity profiles; (5) the shear stress is considerably influenced by energy dissipation and depends on thermal boundary conditions. These findings are in agreement with our intuitive expectations. It should be pointed out, however, that the degree of significance of these findings may vary with the type of fluid system, flow conditions (e.g., extrusion rate), design of the die (e.g., the length of the die), the viscosity ratio of the two fluids, etc. Therefore a generalization of the findings on the basis of a specific fluid system cannot be made to other fluid systems.

The possibility of polymeric liquids slipping at the tube wall has been discussed in the literature [37, 38]. Again, the extent of wall slippage, if any, would depend on the type of fluid and also on the condition of the tube wall surface. For instance, the external lubricants, often used in the processing of polyvinylchloride (PVC), are believed to reduce the pressure drop considerably in extrusion. Such an experimental observation has been attributed to wall slippage of lubricants which tend to migrate to the tube wall.

7.3 ANALYSIS OF FLAT-FILM COEXTRUSION IN A RECTANGULAR DUCT

There are two basic methods for coextruding flat film and sheet, one using a multimanifold die, the other using a single manifold die with a feedblock system at the die inlet. In using a multimanifold die, the various polymer layers are each fed to separate manifolds and are brought out to the full width before being merged into one layer just before the die lip. It is therefore possible to use polymers with widely differing rheological properties, without having to worry about interfacial instability between the layers. However, this method requires special and expensive dies with elaborate flow control adjustments across the width of each layer. Furthermore, the process is not suited for making films of many layers (more than three layers, say) and for heat sensitive inner layers whose contact with the die surface may be undesirable.

On the other hand, the feedblock method uses a low-cost die, and can extrude products having many layers. In this method, a conventional flat-film or sheet-forming die is preceded by a device, called a feedblock, in which the different layers are assembled. Figure 7.10 gives a schematic of a feedblock. A major disadvantage of the feedblock method is that it is difficult to produce uniform layers from materials of widely differing rheological properties, because of interface migration (and possibly interfacial instability) as the materials flow through the die.

We can conceive of any number of layers in a flat film, and the possible combinations between different polymers in a given multilayer film will increase rapidly as the number of layers increases. In order to discuss the essential processing characteristics of multilayer flat-film coextrusion, we shall consider the following two cases [39–41]: (a) two-layer (type A/B)

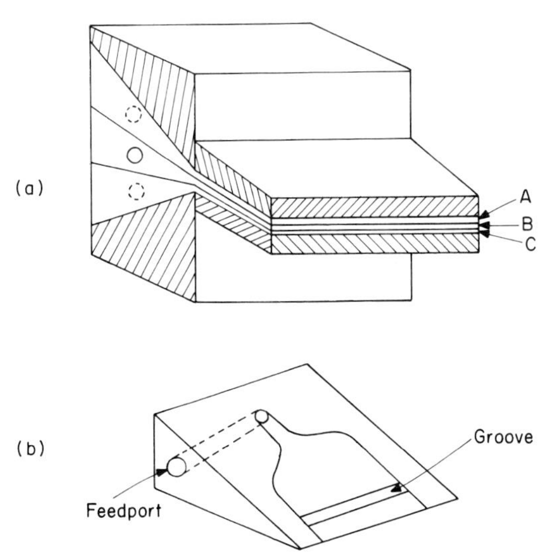

Fig. 7.10 Schematic of a flat-film coextrusion die assembly: (a) pictorial view of the die assembly including the feedblock; (b) a feedblock section.

Fig. 7.11 Schematic illustrating possible layer rearrangements in flat-film coextrusion: (a) two-layer film; (b) three-layer film.

coextrusion, and (b) three-layer (type $A/B/C$) coextrusion, as schematically shown in Fig. 7.11. Some of the reasons why we discuss these two cases will become clear as we proceed with our analysis and discuss the deformation of the interface in each case.

7.3.1 Two-Layer Flat-Film Coextrusion

Consider stratified two-phase flow, as depicted in Fig. 7.12, where phase A is assumed to be less viscous than phase B, with the maximum in velocity thus occurring in phase A. Assume further that (1) the flow is at steady state and is fully developed in the region of our interest, (2) the depth h of the flow channel is very small compared to the width w (i.e., $h \ll w$), (3) the effect of external forces (e.g., the gravitational effect) is negligible, (4) there is no slippage at the die wall and at the interface between the two liquid layers, and (5) the interface shape is flat and the interface position stays constant in the direction of flow.

Referring to Fig. 7.12, the z component of the equations of motion is given by

$$-\partial p/\partial z + \partial \tau_{yz}/\partial y = 0 \tag{7.15}$$

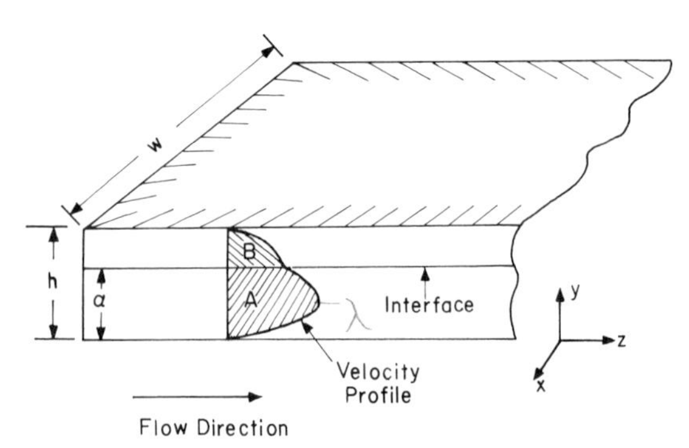

Fig. 7.12 Schematic showing the velocity profile in the two-layer coextrusion through a rectangular channel.

For the flow situation under consideration (see Fig. 7.12), the velocity field is given by

$$v_z = v_z(y); \qquad v_y = v_x = 0 \tag{7.16}$$

Integrating Eq. (7.15) gives for phase A

$$\tau_{yz,A} = -\zeta(y - \lambda), \qquad 0 \leq y \leq \alpha \tag{7.17}$$

and, for phase B

$$\tau_{yz,B} = -\zeta(y - \lambda), \qquad \alpha \leq y \leq h \tag{7.18}$$

in which ζ is the pressure gradient defined by

$$\zeta = -\partial p_A/\partial z = -\partial p_B/\partial z = \text{const} \tag{7.19}$$

Note that Eq. (7.19) implies that the pressure gradients in both phases are the same. This has been verified experimentally by Yu and Han [39]. Note that λ is an integration constant, which corresponds to the position at which the maximum in velocity (and hence the minimum in shear stress) occurs.

In order to obtain the velocity profile and then to calculate the volumetric flow rate, we choose a power-law model defined as

$$\tau_{yz,A} = K_A \dot{\gamma}^{n_A} \tag{7.20}$$

where K_A and n_A are the power-law constants for phase A, and $\dot{\gamma}$ is the velocity gradient defined as

$$\dot{\gamma} = |dv_{z,A}/dy|, \qquad 0 \leq y \leq \alpha \tag{7.21}$$

Similarly, for phase B we have

$$\tau_{yz,B} = K_B \dot{\gamma}^{n_B} \tag{7.22}$$

where K_B and n_B are the power-law constants for phase B, and $\dot{\gamma}$ is defined by

$$\dot{\gamma} = -dv_{z,B}/dy, \qquad \alpha \leq y \leq h \tag{7.23}$$

Now, combining Eqs. (7.17) and (7.20) and integrating the resulting expression, we obtain

$$v_{z,A} = \left(\frac{\zeta}{K_A}\right)^{\alpha_A} \left(\frac{1}{\alpha_A + 1}\right) \{\lambda^{\alpha_A + 1} - |\lambda - y|^{\alpha_A + 1}\}, \qquad 0 \leq y \leq \alpha \tag{7.24}$$

Similarly, combining Eqs. (7.18) and (7.22) and integrating the resulting expression, we obtain

$$v_{z,B} = \left(\frac{\zeta}{K_B}\right)^{\alpha_B} \left(\frac{1}{\alpha_B + 1}\right) \{(h - \lambda)^{\alpha_B + 1} - (y - \lambda)^{\alpha_B + 1}\}, \qquad \alpha \leq y \leq h \tag{7.25}$$

where

$$\alpha_A = 1/n_A, \qquad \alpha_B = 1/n_B \tag{7.26}$$

It should be noted that Eqs. (7.24) and (7.25) contain a constant λ, yet to be determined with the aid of boundary condition:

$$\text{at } y = \alpha, \qquad \tau_{yz,A} = \tau_{yz,B} \tag{7.27}$$

That is, the equation

$$\left(\frac{\zeta}{K_A}\right)^{\alpha_A}\left(\frac{1}{\alpha_A+1}\right)\{\lambda^{\alpha_A+1} - |\lambda-\alpha|^{\alpha_A+1}\}$$
$$= \left(\frac{\zeta}{K_B}\right)^{\alpha_B}\left(\frac{1}{\alpha_B+1}\right)\{(h-\lambda)^{\alpha_B+1} - (\alpha-\lambda)^{\alpha_B+1}\} \tag{7.28}$$

must be solved for λ. The solution of Eq. (7.28) requires a trial-and-error procedure, using some kind of successive iteration scheme. Note, however, that, in determining the parameter λ from Eq. (7.28), the interface position α (see Fig. 7.12) has to be specified.

The volumetric flow rates, Q_A and Q_B, may be obtained from:

$$Q_A = w\left(\frac{\zeta}{K_A}\right)^{\alpha_A}\left(\frac{1}{\alpha_A+1}\right)\left\{\lambda^{\alpha_A+1}\alpha - \frac{\lambda^{\alpha_A+2}}{\alpha_A+2} - \frac{(\alpha-\lambda)^{\alpha_A+2}}{\alpha_A+2}\right\} \tag{7.29}$$

$$Q_B = w\left(\frac{\zeta}{K_B}\right)^{\alpha_B}\left(\frac{1}{\alpha_B+1}\right)\left\{(h-\lambda)^{\alpha_B+1}(h-\alpha) - \frac{(h-\lambda)^{\alpha_B+2}}{\alpha_B+2} + \frac{(\alpha-\lambda)^{\alpha_B+2}}{\alpha_B+2}\right\} \tag{7.30}$$

Figure 7.13 gives velocity profiles for the polypropylene (PP)/polystyrene (PS) system, which were obtained with the aid of Eqs. (7.24) and (7.25), using volumetric flow rates, and pressure gradients determined experimentally in a rectangular channel. Figure 7.14 gives plots of viscosity versus shear stress for the PP and PS employed. It is seen that the polymer melts obey a power-law model.

It should be pointed out that in computing velocity profiles, such as those given in Fig. 7.13, one needs information about the position of interface α (i.e., the layer thicknesses) in the die. Yu and Han [39] used computed values of the volumetric flow ratio, Q_A/Q_B, as a guide for determining the values of α, by comparing them with the experimentally determined values of Q_A/Q_B. Note that Eqs. (7.29) and (7.30) yield

$$\frac{Q_A}{Q_B} = \frac{(\zeta/K_A)^{\alpha_A}}{(\zeta/K_B)^{\alpha_B}}\left(\frac{\alpha_B+1}{\alpha_A+1}\right) \tag{7.31}$$
$$\left\{\frac{\lambda^{\alpha_A+1}\alpha - [\lambda^{\alpha_A+2}/(\alpha_A+2)] - [(\alpha-\lambda)^{\alpha_A+2}/(\alpha_A+2)]}{(h-\lambda)^{\alpha_B+1}(h-\alpha) - [(h-\lambda)^{\alpha_B+2}/(\alpha_B+2)] + [(\alpha-\lambda)^{\alpha_B+2}/(\alpha_B+2)]}\right\}$$

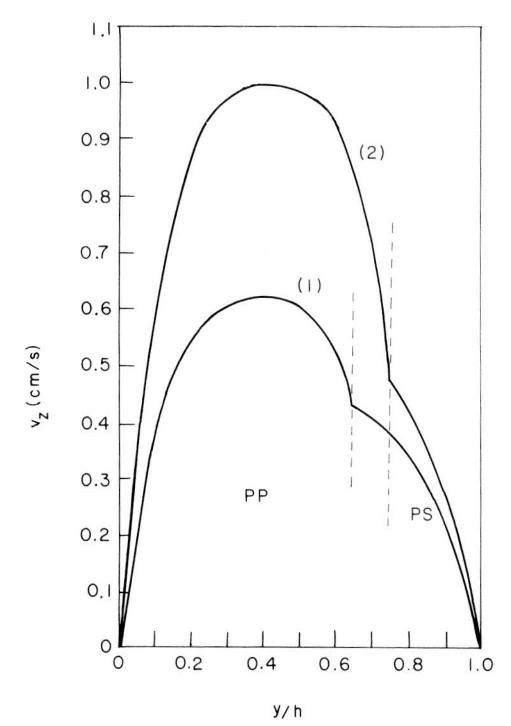

Fig. 7.13 Theoretically predicted velocity profiles in the two-layer (PP/PS) coextrusion through a rectangular channel: (1) $-\partial p/\partial z = 2.46 \times 10^7$ N/m^3; $Q = 19.8$ cm^3/min; (2) $-\partial p/\partial z = 2.74 \times 10^7$ N/m^3; $Q = 26.9$ cm^3/min; The power-law constants used are (a) For PP: $K = 0.724 \times 10^4$ N sec^n/m^2, $n = 0.451$; (b) For PS: $K = 2.127 \times 10^4$ N sec^n/m^2, $n = 0.301$.

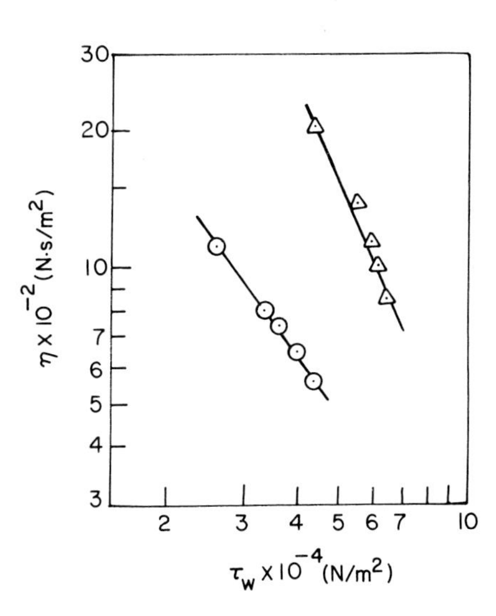

Fig. 7.14 Viscosity versus shear stress for the polymers ($T = 200°$C) used in the computations that yielded the velocity profiles given in Fig. 7.13: (△) PS; (⊙) PP.

It is seen that the volumetric flow ratio, Q_A/Q_B, is not equal to the layer thickness ratio, $\alpha/(h-\alpha)$, as one might expect. Q_A/Q_B is a complicated function which depends on the pressure gradient ζ, the power-law constants of each phase (n_A, K_A, n_B, and K_B), and the parameters α and λ. Therefore, for a given fluid system and flow condition, where the pressure gradient ζ and flow rates (Q_A and Q_B) are specified, the interface position α may be determined from Eq. (7.31), when the predicted value of Q_A/Q_B agrees with the experimentally determined one, provided Eq. (7.28) is satisfied. In other words, Eqs. (7.28) and (7.31) may be used for determining values of α and λ.

7.3.2 Three-Layer Flat-Film Coextrusion

Consider stratified, three-phase flow in a rectangular channel as depicted in Fig. 7.15. Using the same assumptions made in the analysis of two-phase flow presented above, we have:

$$\tau_{yz,A} = -\zeta(y-\lambda), \qquad 0 \le y \le \alpha \tag{7.32}$$

$$\tau_{yz,B} = -\zeta(y-\lambda), \qquad \alpha \le y \le \beta \tag{7.33}$$

$$\tau_{yz,C} = -\zeta(y-\lambda), \qquad \beta \le y \le h \tag{7.34}$$

where $\lambda = c_1 = c_2 = c_3$ (integration constants).

Following the same procedure described above in the analysis of two-phase flow, we have

$$v_{z,A} = \left(\frac{\zeta}{K_A}\right)^{\alpha_A}\left(\frac{1}{\alpha_A+1}\right)\left\{\lambda^{\alpha_A+1} - (\lambda-y)^{\alpha_A+1}\right\} \tag{7.35}$$

$$v_{z,B} = \left(\frac{\zeta}{K_B}\right)^{\alpha_B}\left(\frac{1}{\alpha_B+1}\right)\left\{(\lambda-\alpha)^{\alpha_B+1} - (\lambda-y)^{\alpha_B+1}\right\} + \left(\frac{\zeta}{K_A}\right)^{\alpha_A}\left(\frac{1}{\alpha_A+1}\right)\left\{\lambda^{\alpha_A+1} - (\lambda-\alpha)^{\alpha_A+1}\right\} \tag{7.36}$$

$$v_{z,C} = \left(\frac{\zeta}{K_C}\right)^{\alpha_C}\left(\frac{1}{\alpha_C+1}\right)\left\{(h-\lambda)^{\alpha_C+1} - (\lambda-y)^{\alpha_C+1}\right\} \tag{7.37}$$

in which

$$\alpha_A = 1/n_A, \qquad \alpha_B = 1/n_B, \qquad \alpha_C = 1/n_C \tag{7.38}$$

and n_A, K_A, n_B, K_B, n_C, K_C are power-law constants for the respective phases.

With the aid of boundary condition

$$\text{at } y = \beta, \qquad v_{z,B} = v_{z,C} \tag{7.39}$$

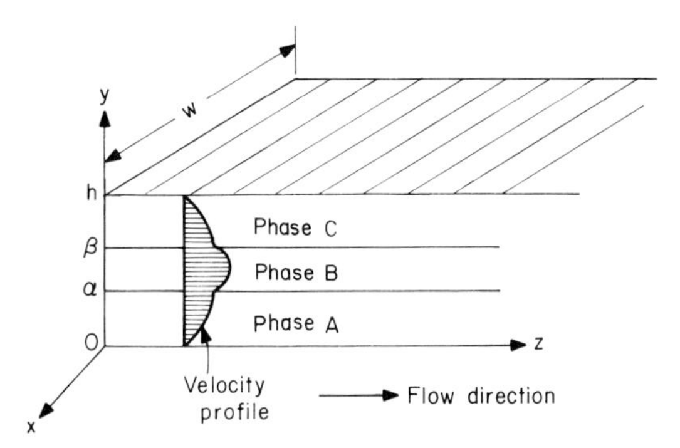

Fig. 7.15 Schematic showing the velocity profile in the three-layer coextrusion through a rectangular channel.

Eqs. (7.36) and (7.37) yield

$$\left(\frac{\zeta}{K_B}\right)^{\alpha_B}\left(\frac{1}{\alpha_B+1}\right)\{(\lambda-\alpha)^{\alpha_B+1}-(\lambda-\beta)^{\alpha_B+1}\}$$
$$+\left(\frac{\zeta}{K_A}\right)^{\alpha_A}\left(\frac{1}{\alpha_A+1}\right)\{\lambda^{\alpha_A+1}-(\lambda-\alpha)^{\alpha_A+1}\}$$
$$=\left(\frac{\zeta}{K_C}\right)^{\alpha_C}\left(\frac{1}{\alpha_C+1}\right)\{(h-\lambda)^{\alpha_C+1}-(\lambda-\beta)^{\alpha_C+1}\} \tag{7.40}$$

This equation contains three parameters: λ, α, and β. Since we do not know the interface positions, α and β, *a priori*, two additional relationships are needed, so that α and β, together with λ, can be determined. As discussed above, such relationships may be obtained by using volumetric flow ratios, say, Q_A/Q_B and Q_B/Q_C.

The volumetric flow rates Q_A, Q_B, and Q_C, may be obtained from:

$$Q_A = w\left(\frac{\zeta}{K_A}\right)^{\alpha_A}\left(\frac{1}{\alpha_A+1}\right)\left\{\lambda^{\alpha_A+1}\alpha+\frac{(\lambda-\alpha)^{\alpha_A+2}}{\alpha_A+2}-\frac{\lambda^{\alpha_A+2}}{\alpha_A+2}\right\} \tag{7.41}$$

$$Q_B = w\left(\frac{\zeta}{K_B}\right)^{\alpha_B}\left(\frac{1}{\alpha_B+1}\right)\left\{(\lambda-\alpha)^{\alpha_B+1}(\beta-\alpha)+\frac{(\lambda-\beta)^{\alpha_B+2}}{\alpha_B+2}-\frac{(\lambda-\alpha)^{\alpha_B+2}}{\alpha_B+2}\right\}$$
$$+w\left(\frac{\zeta}{K_A}\right)^{\alpha_A}\left(\frac{1}{\alpha_A+1}\right)\{\lambda^{\alpha_A+1}-(\lambda-\alpha)^{\alpha_A+1}\}(\beta-\alpha) \tag{7.42}$$

$$Q_C = w\left(\frac{\zeta}{K_C}\right)^{\alpha_C}\left(\frac{1}{\alpha_C+1}\right)\left\{(h-\lambda)^{\alpha_C+1}(h-\beta)+\frac{(\lambda-h)^{\alpha_C+2}}{\alpha_C+2}-\frac{(\lambda-\beta)^{\alpha_C+2}}{\alpha_C+2}\right\} \tag{7.43}$$

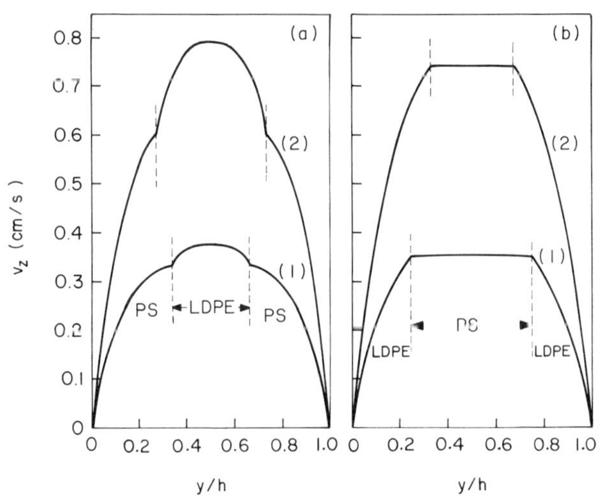

Fig. 7.16 Theoretically predicted velocity profiles in the three-layer coextrusion through a rectangular channel [41]: (a) PS/LDPE/PS system: (1) $-\partial p/\partial z = 4.37 \times 10^7$ N/m^3, $Q = 19.4$ cm^3/min; (2) $-\partial p/\partial z = 5.35 \times 10^7$ N/m^3, $Q = 35.2$ cm^3/min; (b) LDPE/PS/LDPE system: (1) $-\partial p/\partial z = 2.46 \times 10^7$ N/m^3, $Q = 29.4$ cm^3/min; (2) $-\partial p/\partial z = 3.31 \times 10^7$ N/m^3, $Q = 72.4$ cm^3/min. The power-law constants used are (a) For PS: $K = 0.775 \times 10^4$ N sec^n/m^2, $n = 0.32$; (b) For LDPE: $K = 0.294 \times 10^4$ N sec^n/m^2, $n = 0.45$.

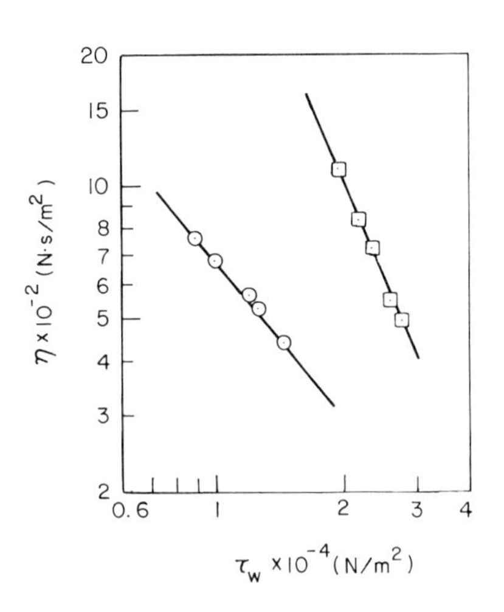

Fig. 7.17 Viscosity versus shear stress for the polymers ($T = 200$°C) used in the computations that yielded the velocity profiles given in Fig. 7.16: (⊡) PS; (⊙) LDPE.

Figure 7.16 gives velocity profiles for the symmetric three-layer systems, PS/LDPE/PS and LDPE/PS/LDPE, which were obtained with the aid of Eqs. (7.35)–(7.37), and volumetric flow rates and pressure gradients experimentally determined in a rectangular channel. Figure 7.17 gives plots of viscosity versus shear stress for the PS and LDPE employed. It is seen that the polymer melts obey a power-law model. Note that in this particular situation, phase A and phase C are the same material and have the same layer thickness. Therefore, the maximum in velocity occurs at the center of phase B, i.e., $\lambda = h/2$. Since, in this case, the parameter λ is known ($\lambda = h/2$) and $\alpha + \beta = h$, Eq. (7.40) was used to find the interface positions, α and β.

7.4 ANALYSIS OF WIRE COATING COEXTRUSION

In the wire coating coextrusion technique, two different polymers concentrically coat a wire in a single step. Tough, abrasion-resistant nylon, for example, can be coated over a much less expensive polyethylene core, or one can have a thin coat of color compound over unpigmented insulator, thus taking advantage of the properties of two components for reducing cost. A considerable saving in the cost of processing is achieved by applying the two coats in a single step. In this process, two different polymer melts from separate extruders are brought into a single cross-head and die, where they are made to flow in a concentric annular manner over an axially moving wire.

Compared to single-layer wire coating [42], coextrusion requires the investigation of three additional variables. They are: (1) the *relative* rheological properties (i.e., the viscosity and elasticity ratios) of the two polymers being coextruded; (2) the position of the layer interface with respect to the wire or the die wall; (3) the ratio of the coating thicknesses. Questions then arise as to how these variables affect the velocity and stress distributions, and the pressure losses, inside the die.

Let us consider stratified two-phase concentric flow through an annulus, where the inner surface of the two coaxial cylinders moves at a constant speed. This flow problem simulates the combined drag and pressure-driven flow of two polymers, forming two concentric layers, in a pressure-type wire coating die [23]. For the theoretical development, we make the following assumptions: (1) the flow is at steady state; (2) the flow is laminar and isothermal; (3) both the gravitational and inertia effects are negligible.

Referring to Fig. 7.18, where the velocity maximum is assumed to occur in phase A, the velocity field may be written as

$$v_z = v_z(r), \qquad v_r = v_\theta = 0 \tag{7.44}$$

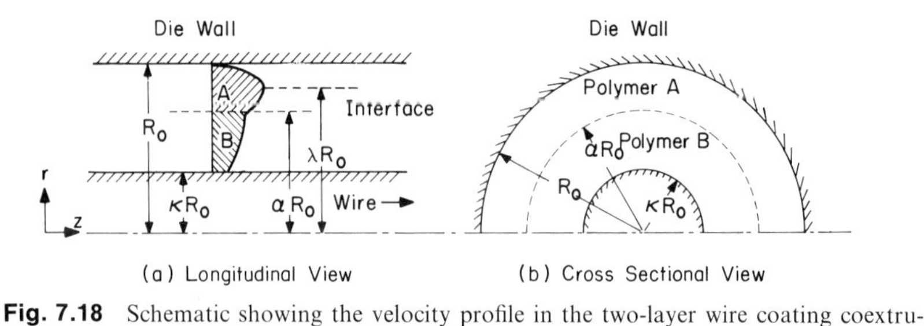

Fig. 7.18 Schematic showing the velocity profile in the two-layer wire coating coextrusion [23].

The z component of the equations of motion can be given for phase A as

$$-\frac{\partial p_A}{\partial z} + \frac{1}{r}\frac{\partial}{\partial r}(r\tau_{rz,A}) = 0, \qquad \text{for} \quad R_\mathrm{i} \leq r \leq R_0 \tag{7.45}$$

and for phase B as

$$-\frac{\partial p_B}{\partial z} + \frac{1}{r}\frac{\partial}{\partial r}(r\tau_{rz,B}) = 0, \qquad \text{for} \quad R_\mathrm{w} \leq r \leq R_\mathrm{i} \tag{7.46}$$

where R_0 is the radius of the die opening, R_w is the radius of the wire, and R_i is the radial position at which the interface occurs.

For fully developed flow, it can be shown that

$$-\partial p_A/\partial z = -\partial p_B/\partial z = \zeta \tag{7.47}$$

Integrating Eqs. (7.45) and (7.46) with respect to r, one obtains

$$\tau_{rz,A} = -\tfrac{1}{2}\zeta r + c/r, \qquad \text{for} \quad R_\mathrm{i} \leq r \leq R_0 \tag{7.48}$$

$$\tau_{rz,B} = -\tfrac{1}{2}\zeta r + c/r, \qquad \text{for} \quad R_\mathrm{w} \leq r \leq R_\mathrm{i} \tag{7.49}$$

in which c is a constant of integration. Let λ be the dimensionless radial coordinate at which $\tau_{rz(A)} = 0$ or the fluid velocity is at a maximum. Then, using the dimensionless radial coordinate $\xi = r/R_0$, Eqs. (7.48) and (7.49) can be rewritten

$$\tau_{rz,A} = \tfrac{1}{2}\zeta R_0\{(\lambda^2 - \xi^2)/\xi\}, \qquad \text{for} \quad \alpha \leq \xi \leq 1 \tag{7.50}$$

$$\tau_{rz,B} = \tfrac{1}{2}\zeta R_0\{(\lambda^2 - \xi^2)/\xi\}, \qquad \text{for} \quad \kappa \leq \xi \leq \alpha \tag{7.51}$$

Boundary conditions may be given as

$$\left.\begin{array}{lll} \text{at } r = R_\mathrm{w} \text{ (or } \xi = \kappa), & v_{z,B} = V_\mathrm{w} & \text{(7.52a)} \\ \text{at } r = R_\mathrm{i} \text{ (or } \xi = \alpha), & v_{z,A} = v_{z,B} & \text{(7.52b)} \\ \text{at } r = R_\mathrm{i} \text{ (or } \xi = \alpha), & \tau_{rz,A} = \tau_{rz,B} & \text{(7.52c)} \\ \text{at } r = R_0 \text{ (or } \xi = 1), & v_{z,A} = 0 & \text{(7.52d)} \end{array}\right\}$$

in which $\kappa = R_\mathrm{w}/R_0$, $\alpha = R_\mathrm{i}/R_0$, and V_w is the wire velocity.

In order to solve Eqs. (7.50) and (7.51) with the boundary conditions, Eq. (7.52), we choose a power-law model for both phases, i.e., for phase A

$$\tau_{rz,A} = -K_A \left|\frac{dv_z}{dr}\right|^{n_A - 1} \frac{dv_z}{dr} = -\frac{K_A}{(R_0)^{n_A}} \left|\frac{dv_z}{d\xi}\right|^{n_A - 1} \frac{dv_z}{d\xi} \tag{7.53}$$

and for phase B

$$\tau_{rz,B} = -K_B \left|\frac{dv_z}{dr}\right|^{n_B - 1} \frac{dv_z}{dr} = -\frac{K_B}{(R_0)^{n_B}} \left|\frac{dv_z}{d\xi}\right|^{n_B - 1} \frac{dv_z}{d\xi} \tag{7.54}$$

where K_A and n_A are the power-law constants for fluid A, and K_B and n_B are the power-law constants for fluid B.

Combining Eqs. (7.50) and (7.53), and integrating the resulting equation with the aid of boundary conditions, we get for fluid A,

$$v_{z,A}(\xi) = R_0 \left(\frac{\zeta R_0}{2K_A}\right)^{\alpha_A} \int_\xi^1 \left(\frac{s^2 - \lambda^2}{s}\right)^{\alpha_A} ds, \qquad \text{for} \quad \lambda \le \xi \le 1 \tag{7.55}$$

$$v_{z,A}(\xi) = R_0 \left(\frac{\zeta R_0}{2K_A}\right)^{\alpha_A} \int_\alpha^\xi \left(\frac{\lambda^2 - s^2}{s}\right)^{\alpha_A} ds + v_{z,A}(\alpha), \qquad \text{for} \quad \alpha \le \xi \le \lambda \tag{7.56}$$

where $\alpha_A = 1/n_A$, and combining Eqs. (7.51) and (7.54) and integrating the resulting equation with the aid of boundary conditions, we obtain for fluid B,

$$v_{z,B}(\xi) = R_0 \left(\frac{\zeta R_0}{2K_B}\right)^{\alpha_B} \int_\kappa^\xi \left(\frac{\lambda^2 - s^2}{s}\right)^{\alpha_B} ds + V_w, \qquad \text{for} \quad \kappa \le \xi \le \alpha \tag{7.57}$$

where $\alpha_B = 1/n_B$.

Using Eqs. (7.55) and (7.56), we obtain at $\xi = \lambda$

$$v_{z,A}(\alpha) = v_\alpha = R_0 \left(\frac{\zeta R_0}{2K_A}\right)^{\alpha_A} \left[\int_\lambda^1 \left(\frac{s^2 - \lambda^2}{s}\right)^{\alpha_A} ds - \int_\alpha^\lambda \left(\frac{\lambda^2 - s^2}{s}\right)^{\alpha_A} ds\right] \tag{7.58}$$

and at $\xi = \alpha$ (at the phase interface), with the aid of the boundary condition Eq. (7.52b), we obtain from Eqs. (7.57) and (7.58)

$$V_w = R_0 \left(\frac{\zeta R_0}{2K_A}\right)^{\alpha_A} \left[\int_\lambda^1 \left(\frac{s^2 - \lambda^2}{s}\right)^{\alpha_A} ds - \int_\alpha^\lambda \left(\frac{\lambda^2 - s^2}{s}\right)^{\alpha_A} ds\right] - R_0 \left(\frac{\zeta R_0}{2K_B}\right)^{\alpha_B} \int_\kappa^\alpha \left(\frac{\lambda^2 - s^2}{s}\right)^{\alpha_B} ds \tag{7.59}$$

This is the determining equation for the parameter λ, which depends on the wire velocity V_w, the pressure gradient ζ, the dimensionless wire radius κ, the radius of the die opening R_0, the rheological constants of the fluids A and B, and the interface position α.

The volumetric flow rates Q_A and Q_B for the phases A and B, respectively, can be obtained using the following expressions:

$$Q_A = 2\pi R_0^2 \int_\alpha^1 v_{z,A}(\xi)\xi\, d\xi$$

$$= \pi R_0^2 v_\alpha(\lambda^2 - \alpha^2) + \pi R_0^3 \left(\frac{\zeta R_0}{2K_A}\right)^{\alpha_A} \int_\alpha^1 \frac{|\lambda^2 - \xi^2|^{\alpha_A + 1}}{\xi^{\alpha_A}}\, d\xi \tag{7.60}$$

$$Q_B = 2\pi R_0^2 \int_\kappa^\alpha v_{z,B}(\xi)\xi\, d\xi$$

$$= \pi R_0^2 V_w(\alpha^2 - \kappa^2) + \pi R_0^3 \left(\frac{\zeta R_0}{2K_B}\right)^{\alpha_B} \int_\kappa^\alpha \frac{(\alpha^2 - \xi^2)(\lambda^2 - \xi^2)^{\alpha_B}}{\xi^{\alpha_B}}\, d\xi \tag{7.61}$$

in which v_α and V_w are defined by Eqs. (7.58) and (7.59), respectively. The total volumetric flow rate can be obtained by adding Eqs. (7.60) and (7.61), and making use of Eqs. (7.58) and (7.59) in the resulting equation, thus:

$$Q_A + Q_B = \pi R_0^3 \left(\frac{\zeta R_0}{2K_A}\right)^{\alpha_A} \int_\alpha^1 \frac{|\lambda^2 - \xi^2|^{\alpha_A + 1}}{\xi^{\alpha_A}}\, d\xi$$

$$+ \pi R_0^2 V_w(\lambda^2 - \kappa^2) + \pi R_0^3 \left(\frac{\zeta R_0}{2K_B}\right)^{\alpha_B} \int_\kappa^\alpha \frac{(\lambda^2 - \xi^2)^{\alpha_B + 1}}{\xi^{\alpha_B}}\, d\xi \tag{7.62}$$

The ratio of volumetric flow rates of phase A and phase B can be given as in Eq. (7.63) (p. 365). Equation (7.63) shows that for $n_A \neq n_B$, the ratio of the volumetric flow rates of the two phases is also a function of the pressure gradient ζ.

From the foregoing analysis, it is evident that in order to predict the shear stress distribution, the velocity profile, and the volumetric flow rates of each phase, one needs to determine the parameters α and λ. The correct values of α and λ are such that they should simultaneously satisfy Eqs. (7.59) and (7.63). Note that the wire speed V_w, the pressure gradient ζ, the dimensionless wire radius κ, the radius of the die opening R_0, the rheological constants of the two phases, and the ratio of the flow rates of the two phases Q_A/Q_B must be specified. In determining the correct values of α and λ, we may use the following guidance: (1) The position of maximum velocity λ has a tendency to lie in the less viscous phase; (2) The interfacial position α is strongly dependent on the ratio of the volumetric flow rates of the two phases. As Q_A/Q_B increases, α decreases and vice versa; (3) The position of the maximum fluid velocity λ is strongly dependent on the wire velocity V_w and the pressure gradient ζ.

A trial-and-error procedure may be adopted in which values of α and λ are assumed that will satisfy Eqs. (7.59) and (7.63) simultaneously. Once the

$$\frac{Q_A}{Q_B} = \frac{\displaystyle\int_{\alpha}^{1} \frac{|\lambda^2 - \xi^2|^{\alpha_A + 1}}{\xi^{\alpha_A}}\, d\xi + (\lambda^2 - \alpha^2)\left[\int_{\lambda}^{1} \left(\frac{\xi^2 - \lambda^2}{\xi}\right)^{\alpha_A} d\xi - \int_{\alpha}^{\lambda} \left(\frac{\lambda^2 - \xi^2}{\xi}\right)^{\alpha_A} d\xi\right]}{\displaystyle(\alpha^2 - \kappa^2)\left[\int_{\lambda}^{1} \left(\frac{\xi^2 - \lambda^2}{\xi}\right)^{\alpha_A} d\xi - \int_{\alpha}^{\lambda} \left(\frac{\lambda^2 - \xi^2}{\xi}\right)^{\alpha_A} d\xi\right] + \frac{\left(\frac{\zeta R_0}{2K_B}\right)^{\alpha_B} \int_{\kappa}^{\alpha} \frac{(\kappa^2 - \xi^2)(\lambda^2 - \xi^2)^{\alpha_B}}{\xi^{\alpha_B}}\, d\xi}{\left(\frac{\zeta R_0}{2K_A}\right)^{\alpha_A}}} \tag{7.63}$$

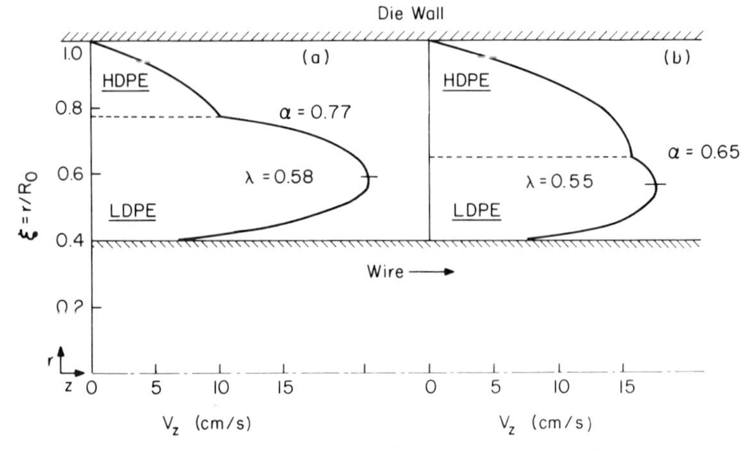

Fig. 7.19 The effect of layer thickness ratio on the velocity profile in the wire coating coextrusion of HDPE/LDPE system [23]. Melt flow rate, $Q = 48.2\ \text{cm}^3/\text{min}$; Wire speed, $V_w = 6.50$ cm/sec. Curve (a) for $Q_A/Q_B = 0.32$; Curve (b) for $Q_A/Q_B = 1.48$. The power-law constants used are (a) for HDPE: $K = 1.01 \times 10^4\ \text{N sec}^n/\text{m}^2$, $n = 0.45$; (b) for LDPE: $K = 0.30 \times 10^4\ \text{N sec}^n/\text{m}^2$, $n = 0.45$.

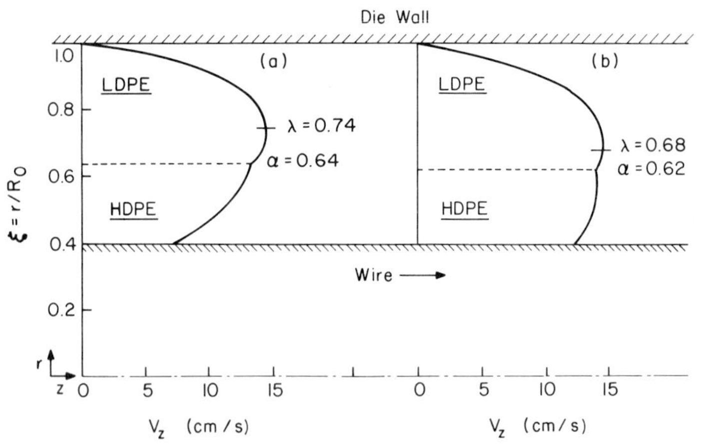

Fig. 7.20 The effect of wire velocity on the velocity profile in the wire coating coextrusion of LDPE/HDPE system [23]. Melt flow rate, $Q = 45.5\ \text{cm}^3/\text{min}$; $Q_A/Q_B = 2.23$. Curve (a) for $V_w = 6.50$ cm/sec; Curve (b) for $V_w = 12.4$ cm/sec.

values of α and λ are determined, one can calculate the shear stress distribution from Eqs. (7.50) and (7.51), the velocity distribution from Eqs. (7.55)–(7.57), and the volumetric flow rates Q_A and Q_B from Eqs. (7.60) and (7.61).

The equations relevant to the situation where the maximum in the velocity profile occurs in phase *B*, instead of in phase *A*, can be derived in a similar manner.

Figure 7.19 illustrates the effect on the velocity profile of the HDPE/LDPE system of changing the ratio of the thicknesses of the individual components. It is seen that at a constant wire speed and volumetric flow rate, the position of the maximum fluid velocity is not very sensitive to the ratio of the flow rates of the individual components. However, the interfacial position α depends strongly on the ratio of the flow rates of the individual components.

Figure 7.20 illustrates the effect of wire speed on the velocity profile of the LDPE/HDPE system at the same total volumetric flow rate and at the same ratio of the flow rates of the individual components. It is seen that both α and λ are lowered by an increase in the wire velocity. However, the decrease in λ is much more pronounced than the decrease in α.

The following conclusions can be drawn regarding the nature of the velocity profiles shown above: (1) the position of the maximum fluid velocity λ has a tendency to lie in the less viscous component and this tendency is more pronounced if the two components differ widely in their viscosities; (2) the position of the maximum fluid velocity λ is a strong function of the wire velocity but is not very sensitive to the ratio of the flow rates of each component; (3) the interfacial position α is a strong function of the ratio of the flow rates of each component but is not very sensitive to the wire velocity.

Figure 7.21 shows the effect of the ratio of the thicknesses of each component on the shear stress profile for the HDPE/LDPE system. It is seen that at a constant total flow rate and wire speed, the shear stress at the die wall and the shear stress at the wire surface are little affected by a change in the thickness ratio. However, the shear stress at the interface is dependent strongly upon the interfacial position α and hence the thickness ratio. As the interfacial position α approaches the position of the maximum fluid velocity λ, the shear stress at the interface decreases. Thus, by controlling the ratio of the coating thicknesses of the two components, the interfacial shear stress can be controlled.

Figure 7.22 illustrates the effect of the wire velocity on the shear stress profile of the LDPE/HDPE system. It is seen that an increase in the wire velocity at a constant total volumetric flow rate considerably decreases the shear stress at the wire. However, the shear stress at the die wall is little affected.

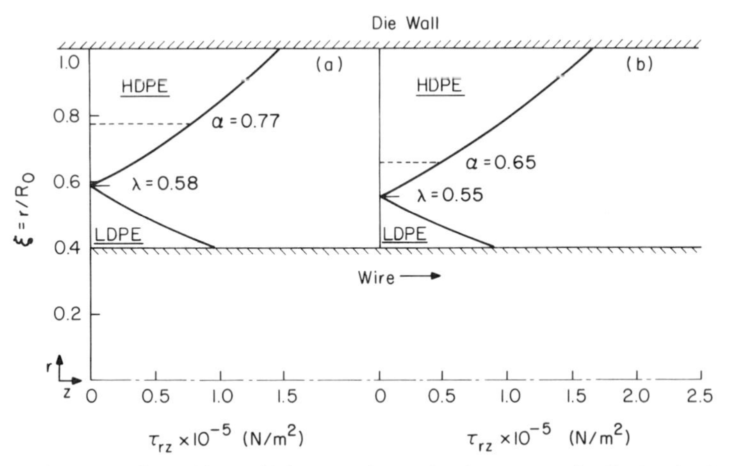

Fig. 7.21 The effect of layer thickness ratio on the shear stress distribution in the wire coating coextrusion of HDPE/LDPE system [23]. Extrusion conditions are the same as in Fig. 7.19. For curve (a), $(\tau_w)_{die} = 1.47 \times 10^5$ N/m^2 and $(\tau_w)_{wire} = 0.95 \times 10^5$ N/m^2; For curve (b), $(\tau_w)_{die} = 1.68 \times 10^5$ N/m^2 and $(\tau_w)_{wire} = 0.87 \times 10^5$ N/m^2.

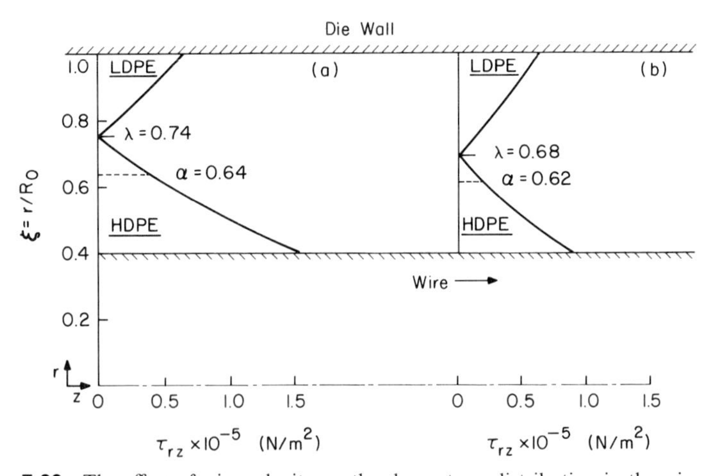

Fig. 7.22 The effect of wire velocity on the shear stress distribution in the wire coating coextrusion of LDPE/HDPE system [23]. Extrusion conditions are the same as in Fig. 7.20. For curve (a), $(\tau_w)_{die} = 0.71 \times 10^5$ N/m^2 and $(\tau_w)_{wire} = 1.53 \times 10^5$ N/m^2; For curve (b), $(\tau_w)_{die} = 0.64 \times 10^5$ N/m^2 and $(\tau_w)_{wire} = 0.88 \times 10^5$ N/m^2.

7.5 ANALYSIS OF BLOWN-FILM COEXTRUSION

There are basically two methods for producing multilayer blown films; namely: (1) two or more melt streams are fed separately into the feedblock and, at the die inlet, form concentric layers, which are then forced to flow through the annular space; and (2) two or more melt streams are fed into the feedport ring and the layers are generated by the rotation of the inner and outer mandrels of the tubing die separately or together [43–45]. The former method requires as many extruders as the number of film layers needed and therefore is not practical for more than two or three layers. However, the latter method can provide up to a few hundred layers by designing the feedrings properly and by controlling the rotational speeds of the inner and outer mandrels. We shall now discuss each of these coextrusion processes in greater detail.

7.5.1 Feedblock Method

In this method, two (or more) melt streams are fed into the feedblock and are combined within the body of the die, and the resultant film is made up of two (or more) plies adhering strongly to each other. Figure 7.23 gives a schematic of a three-layer blown-film die.

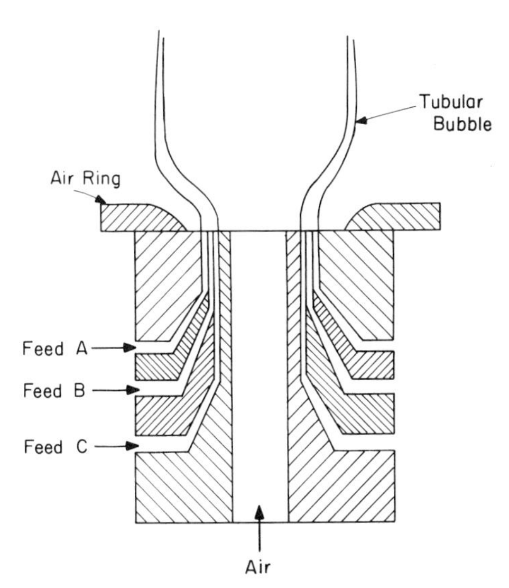

Fig. 7.23 Schematic of a blown-film die assembly.

Analysis of a two-layer blown-film coextrusion process may be carried out in the same manner as that of a wire coating coextrusion process having the wire stationary. Setting the wire velocity equal to zero (i.e., $V_w = 0$) in Eqs. (7.57), (7.59), (7.61), and (7.62), we reduce Eq. (7.63) to

$$\frac{Q_A}{Q_B} = \left(\frac{\zeta R_0}{2K_A}\right)^{\alpha_A} \left\{ \frac{\int_\alpha^1 \frac{|\lambda^2 - \xi^2|^{\alpha_A + 1}}{\xi^{\alpha_A}} d\xi + (\lambda^2 - \alpha^2)\left[\int_\lambda^1 \left(\frac{\xi^2 - \lambda^2}{\xi}\right)^{\alpha_A} d\xi - \int_\alpha^\lambda \left(\frac{\lambda^2 - \xi^2}{\xi}\right)^{\alpha_A} d\xi\right.}{\left[\left(\frac{\zeta R_0}{2K_B}\right)^{\alpha_B} \int_\kappa^\alpha \frac{(\kappa^2 - \xi^2)(\lambda^2 - \xi^2)^{\alpha_B}}{\xi^{\alpha_B}} d\xi\right]} \right. \tag{7.64}$$

Let us now consider a three-layer blown-film coextrusion process. In order to facilitate our analysis we assume that the maximum in the velocity profile occurs in the middle layer, phase B, as schematically shown in Fig. 7.24. This particular layer configuration is very similar to that of the three-layer *flat-film* coextrusion considered above, except that in the former the materials flow through an annulus whereas in the latter the materials flow through a rectangular channel.

Using the same assumptions made in the analysis of wire coating coextrusion presented above, we have

$$\tau_{rz,A} = -\tfrac{1}{2}\zeta R_0\{(\xi^2 - \lambda^2)/\xi\}, \quad \text{for} \quad \alpha \leq \xi \leq 1 \tag{7.65}$$

$$\tau_{rz,B} = -\tfrac{1}{2}\zeta R_0\{(\xi^2 - \lambda^2)/\xi\}, \quad \text{for} \quad \beta \leq \xi \leq \alpha \tag{7.66}$$

$$\tau_{rz,C} = -\tfrac{1}{2}\zeta R_0\{(\xi^2 - \lambda^2)/\xi\}, \quad \text{for} \quad \kappa \leq \xi \leq \beta \tag{7.67}$$

in which λ is an integration constant and ζ is the pressure gradient defined by Eq. (7.47), and $\xi = r/R_0$ is the dimensionless radial position. We assume further that the velocity field is given by Eq. (7.44).

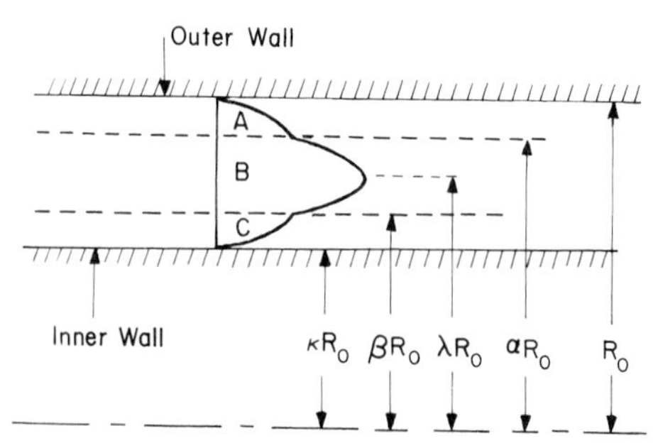

Fig. 7.24 Schematic showing the velocity profile in three-layer coextrusion through an annular die.

Now, boundary conditions may be given as follows:

$$\text{at } \xi = \kappa, \qquad v_{z,C} = 0, \tag{7.68a}$$

$$\text{at } \xi = \beta, \qquad v_{z,B} = v_{z,C}, \tag{7.68b}$$

$$\text{at } \xi = \beta, \qquad \tau_{rz,B} = \tau_{rz,C}, \tag{7.68c}$$

$$\text{at } \xi = \alpha, \qquad v_{z,A} = v_{z,B}, \tag{7.68d}$$

$$\text{at } \xi = \alpha, \qquad \tau_{rz,A} = \tau_{rz,B}, \tag{7.68e}$$

$$\text{at } \xi = 1, \qquad v_{z,A} = 0. \tag{7.68f}$$

Following essentially the same procedure described above in the analysis of two-layer wire coating coextrusion, we have

$$v_{z,A} = R_0\left(\frac{\zeta R_0}{2K_A}\right)^{\alpha_A} \int_\xi^1 \left(\frac{s^2 - \lambda^2}{s}\right)^{\alpha_A} ds, \qquad \text{for} \quad \alpha \le s \le 1, \tag{7.69}$$

$$v_{z,B} = v_\alpha + R_0\left(\frac{\zeta R_0}{2K_B}\right)^{\alpha_B} \int_\xi^\alpha \left(\frac{s^2 - \lambda^2}{s}\right)^{\alpha_B} ds, \qquad \text{for} \quad \lambda \le s \le \alpha \tag{7.70}$$

$$v_{z,B} = v_\beta + R_0\left(\frac{\zeta R_0}{2K_B}\right)^{\alpha_B} \int_\beta^\xi \left(\frac{\lambda^2 - s^2}{s}\right)^{\alpha_B} ds, \qquad \text{for} \quad \beta \le s \le \lambda \tag{7.71}$$

$$v_{z,C} = R_0\left(\frac{\zeta R_0}{2K_C}\right)^{\alpha_C} \int_\kappa^\xi \left(\frac{\lambda^2 - s^2}{s}\right)^{\alpha_C} ds, \qquad \text{for} \quad \kappa \le s \le \beta \tag{7.72}$$

in which v_α in Eq. (7.70) may be determined with the aid of boundary condition, Eqs. (7.68d), as

$$v_\alpha = R_0\left(\frac{\zeta R_0}{2K_A}\right)^{\alpha_A} \int_\alpha^1 \left(\frac{s^2 - \lambda^2}{s}\right)^{\alpha_A} ds \tag{7.73}$$

and v_β in Eq. (7.71) may be determined by equating the right-hand sides of Eqs. (7.70) and (7.71) at $\xi = \lambda$ (at which the maximum in velocity occurs), yielding

$$v_\beta = R_0\left(\frac{\zeta R_0}{2K_A}\right)^{\alpha_A} \int_\alpha^1 \left(\frac{s^2 - \lambda^2}{s}\right)^{\alpha_A} ds + R_0\left(\frac{\zeta R_0}{2K_B}\right)^{\alpha_B} \left[\int_\lambda^\alpha \left(\frac{s^2 - \lambda^2}{s}\right)^{\alpha_B} ds - \int_\beta^\lambda \left(\frac{\lambda^2 - s^2}{s}\right)^{\alpha_B} ds\right] \tag{7.74}$$

It should be noted that Eqs. (7.69)–(7.74) contain the parameter λ, yet to be determined. Assuming that the interface positions, α and β, are specified; the value of λ may be determined by use of boundary condition, Eq. (7.68b), yielding:

$$R_0\left(\frac{\zeta R_0}{2K_A}\right)^{\alpha_A}\int_\alpha^1\left(\frac{s^2-\lambda^2}{s}\right)^{\alpha_A}ds$$
$$+R_0\left(\frac{\zeta R_0}{2K_B}\right)^{\alpha_B}\left[\int_\lambda^\alpha\left(\frac{s^2-\lambda^2}{s}\right)^{\alpha_B}ds-\int_\beta^\lambda\left(\frac{\lambda^2-s^2}{s}\right)^{\alpha_B}ds\right]$$
$$=R_0\left(\frac{\zeta R_0}{2K_C}\right)^{\alpha_C}\int_\kappa^\beta\left(\frac{\lambda^2-s^2}{s}\right)^{\alpha_C}ds \tag{7.75}$$

Since the interface positions, α and β, are not known, one needs two additional relationships which will permit one to determine α and β. As discussed above, such relationships may be obtained by using volumetric flow ratios, say, Q_A/Q_B and Q_B/Q_C.

The volumetric flow rates, Q_A, Q_B, and Q_C, may be obtained from:

$$Q_A=\pi R_0^3\left(\frac{\zeta R_0}{2K_A}\right)^{\alpha_A}\int_\alpha^1\frac{(s^2-\alpha^2)(s^2-\lambda^2)^{\alpha_A}}{s^{\alpha_A}}ds \tag{7.76}$$

$$Q_B=\pi R_0^3\left(\frac{\zeta R_0}{2K_B}\right)^{\alpha_B}\int_\beta^\alpha\frac{|\lambda^2-s^2|^{\alpha_B+1}}{s^{\alpha_B}}ds+\pi R_0^2 v_\alpha(\alpha^2-\lambda^2)$$
$$+\pi R_0^2 v_\beta(\lambda^2-\beta^2) \tag{7.77}$$

$$Q_C=\pi R_0^3\left(\frac{\zeta R_0}{2K_C}\right)^{\alpha_C}\int_\kappa^\beta\frac{(\beta^2-s^2)(\lambda^2-s^2)^{\alpha_C}}{s^{\alpha_C}}ds \tag{7.78}$$

in which v_α and v_β are defined by Eqs. (7.73) and (7.74), respectively. It can be shown that the total volumetric flow rate, $Q=Q_A+Q_B+Q_C$, may be given as:

$$Q=\pi R_0^3\left(\frac{\zeta R_0}{2K_A}\right)^{\alpha_A}\int_\alpha^1\frac{(\lambda^2-s^2)^{\alpha_A+1}}{s^{\alpha_A}}ds+\pi R_0^3\left(\frac{\zeta R_0}{2K_B}\right)^{\alpha_B}\int_\beta^\alpha\frac{|\lambda^2-s^2|^{\alpha_B+1}}{s^{\alpha_B}}ds$$
$$+\pi R_0^3\left(\frac{\zeta R_0}{2K_C}\right)^{\alpha_C}\int_\kappa^\beta\frac{(\lambda^2-s^2)^{\alpha_C+1}}{s^{\alpha_C}}ds \tag{7.79}$$

Note that, for $\alpha_A=\alpha_B=\alpha_C$ (i.e., $n_A=n_B=n_C$) and $K_A=K_B=K_C$, Eq. (7.79) reduces to the expression for the volumetric flow rate in single-phase annular flow.

Figures 7.25 and 7.26 give the theoretically predicted velocity and shear stress distributions, respectively, in a two-layer blown-film die, where high-

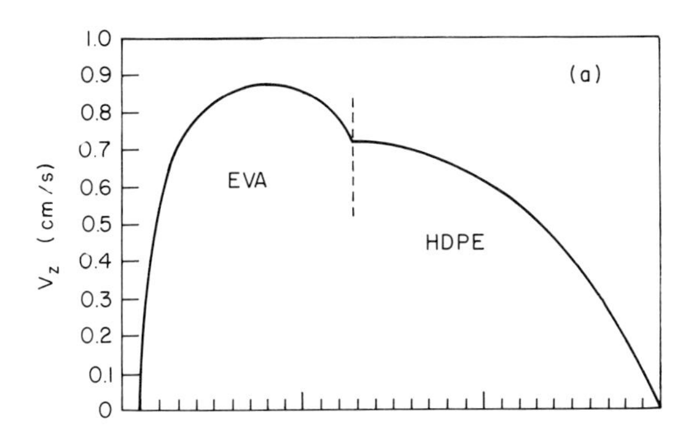

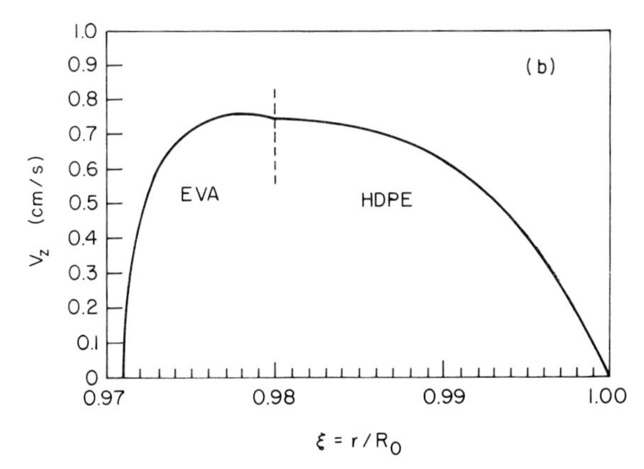

Fig. 7.25 Theoretically predicted velocity profiles in the two-layer coextrusion of ethylene-vinyl acetate (EVA)/high-density polyethylene (HDPE) system through an annular die: (a) $-\partial p/\partial z = 1.08 \times 10^8$ N/m^3, $Q_{\mathrm{EVA}} = 21.8$ cm^3/min, $Q_{\mathrm{HDPE}} = 22.1$ cm^3/min, (b) $-\partial p/\partial z = 1.08 \times 10^8$ N/m^3, $Q_{\mathrm{EVA}} = 22.1$ cm^3/min, $Q_{\mathrm{HDPE}} = 28.1$ cm^3/min. The power-law constants used are: (a) For EVA: $K = 2.9 \times 10^3$ N sec^n/m^2, $n = 0.39$; (b) For HDPE: $K = 1.04 \times 10^4$ N sec^n/m^2, $n = 0.49$.

density polyethylene (HDPE) forms the outer layer and ethylene–vinyl acetate (EVA) forms the inner layer.

Figures 7.27 and 7.28 give the theoretically predicted velocity and shear stress distributions, respectively, in a three-layer blown-film die. Note that for the same pressure drop to occur, the volumetric flow rate for the EVA/HDPE/EVA system, where the HDPE is sandwiched by the EVA, is

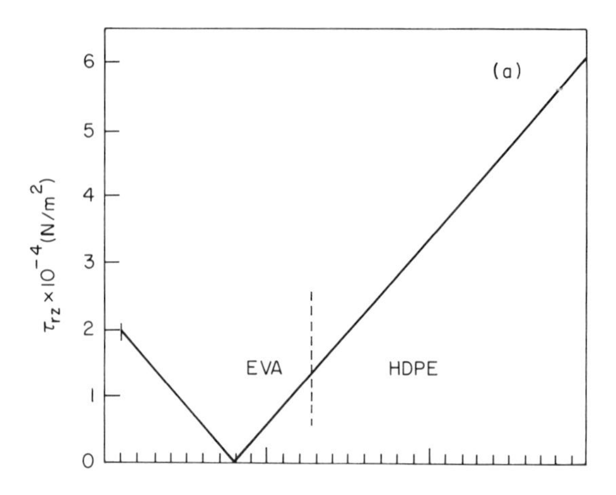

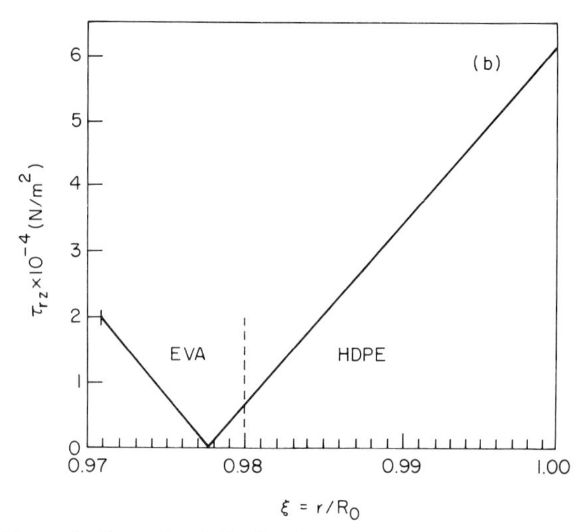

Fig. 7.26 Theoretically predicted distributions of shear stress in the two-layer coextrusion of EVA/HDPE system through an annular die. Processing conditions are the same as in Fig. 7.25.

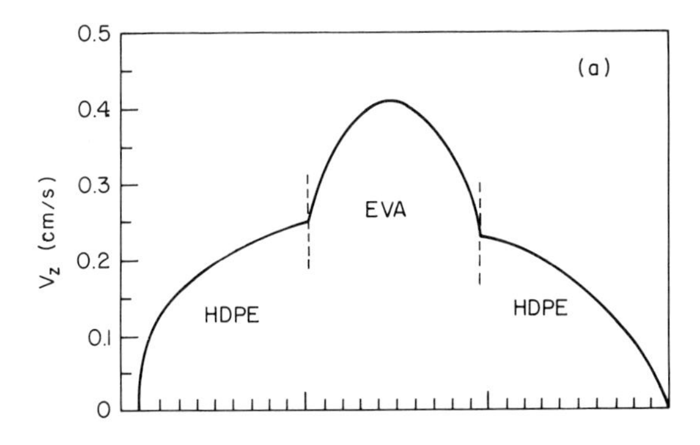

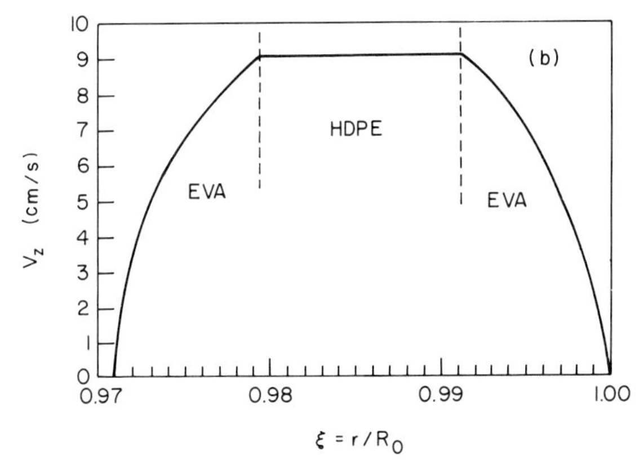

Fig. 7.27 Theoretically predicted velocity profile in the three-layer coextrusion through an annular die: (a) HDPE/EVA/HDPE system: $-\partial p/\partial z = 1.08 \times 10^8$ N/m^3, $Q_{\text{HDPE}} = 8.46$ cm^3/min; $Q_{\text{EVA}} = 8.43$ cm^3/min; (b) EVA/HDPE/EVA system: $-\partial p/\partial z = 1.08 \times 10^8$ N/m^3, $Q_{\text{EVA}} = 268$ cm^3/min, $Q_{\text{HDPE}} = 269$ cm^3/min. The power-law constants used are the same as in Fig. 7.25.

about 32 times that for the HDPE/EVA/HDPE system where the EVA is sandwiched by the HDPE. The difference in the volumetric flow rates between the two system is due to the difference in viscosities between the EVA and the HDPE. This suggests then that the use of the lower viscosity resin as the skin component will allow one to increase the production rate in the manufacture of sandwiched three-layer blown films.

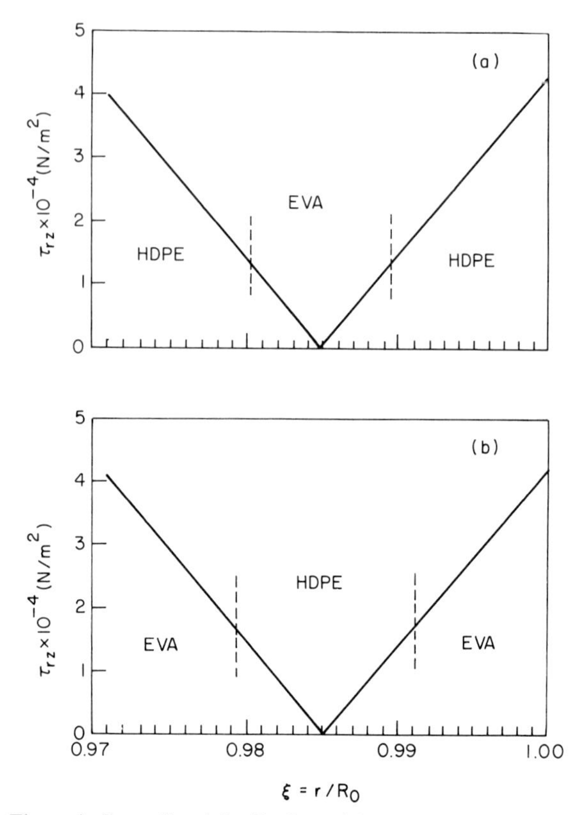

Fig. 7.28 Theoretically predicted distributions of shear stress in the three-layer coextrusion through an annular die: (a) HDPE/EVA/HDPE system; (b) EVA/HDPE/EVA system. Processing conditions are the same as in Fig. 7.27.

7.5.2 Rotational-Mandrel Method

Figure 7.29 gives a schematic of the die that may be used for making multilayer blown films [45]. The principles of operation are as follows. Molten polymers from two feed supplies, *A* and *B*, are fed separately into the die distribution manifolds, while the inner mandrel rotates at a constant speed. One polymer is introduced into the torroidal manifold, while the other polymer is introduced into the annular manifold. The polymers are arranged into alternate layers by connecting feedslots in the feedport ring in such a manner that the individual polymers are introduced at a number of feedslots uniformly spaced around the ring as schematically shown in Fig. 7.30. The number of layers to be generated depends, for the given number of feedslots, on the rotational speed of the inner mandrel and the

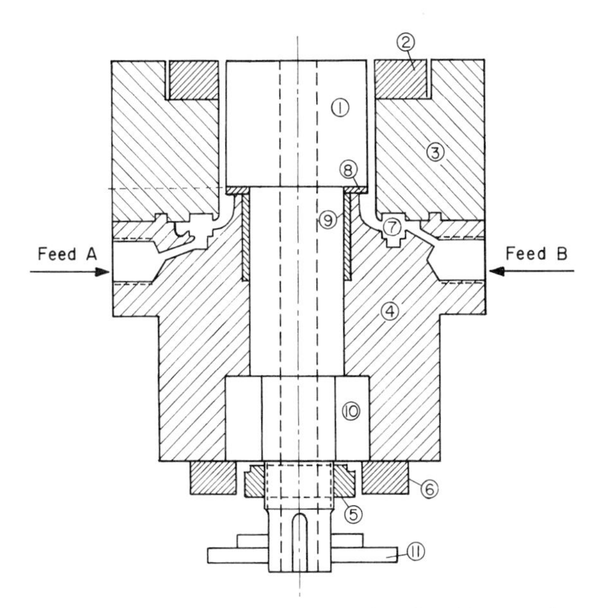

Fig. 7.29 Schematic of the blown-film coextrusion die with a rotating mandrel [45]: (1) rotating shaft, (2) centering ring, (3) upper die; (4) lower die; (5) threaded ring; (6) bearing retainer; (7) feed ring; (8) thrust bearing; (9) bearing; (10) tinker bearing; (11) sprocket.

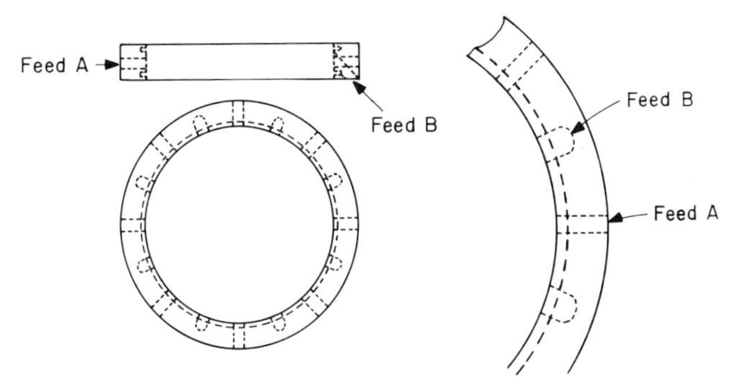

Fig. 7.30 Schematic describing the details of the feedport for the die described in Fig. 7.29 [45].

feed ratio of the two polymers. Polymer melts, flowing through the annular space, form alternating layers of A and B as shown schematically in Fig. 7.31. Idealized layer patterns generated in this process are shown schematically in Fig. 7.32.

Some of the problems of practical interest in dealing with this coextrusion process are: (1) the effect of the rotational speed of the inner (or outer)

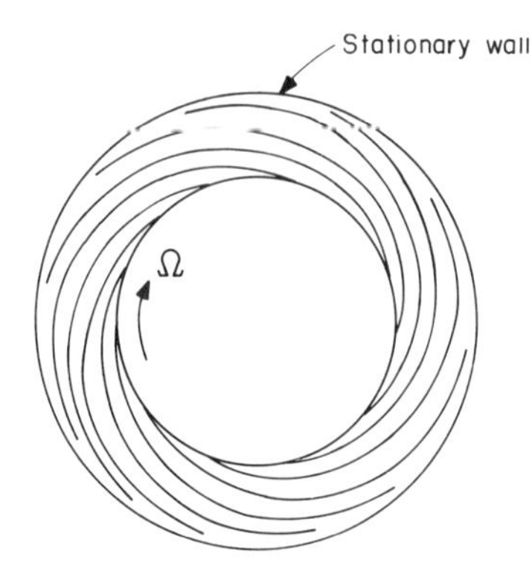

Fig. 7.31 Schematic of the layers that may be generated in the blown-film coextrusion die with a rotating inner mandrel [45].

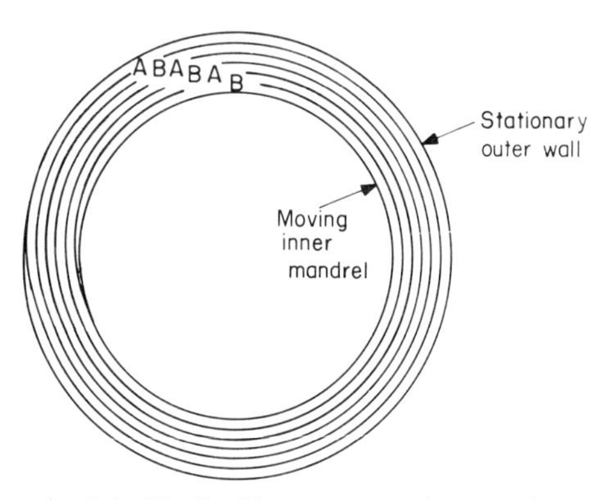

Fig. 7.32 Schematic of the idealized layer patterns that may be generated in the blown-film coextrusion die with a rotating inner mandrel.

mandrel on the number of layers generated and the striation thickness of composite films, and (2) the effect of the viscosity ratio of two (or more) molten polymers on the striation thickness of composite films. In the following we shall present a theoretical analysis simulating the coextrusion process.

Consider the steady, stratified multilayer helical flow of power-law *non-Newtonian* fluids through an annular space consisting of two concentric cylinders, with the inner cylinder rotating at a predetermined angular speed. Referring to Fig. 7.33, and using cylindrical coordinates, let the inner cylinder rotate with an angular velocity Ω_1 and the outer cylinder be stationary. The polymer melt flowing upward has a pressure gradient along the z axis.

The flow field may be described by helical motion, which has the combined features of forward motion due to pressure drop and angular motion

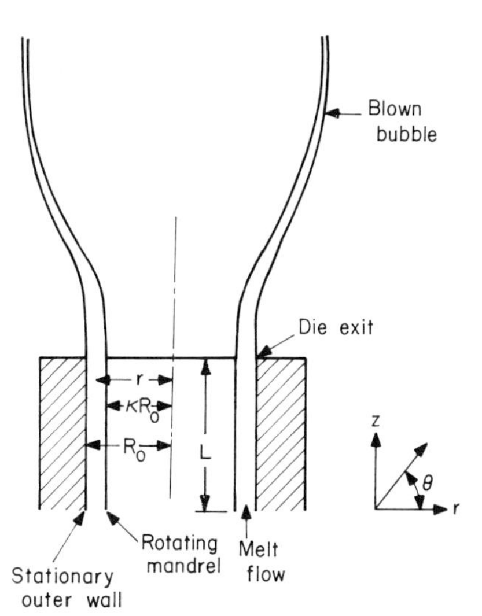

Fig. 7.33 Schematic of a tubular blown-film die [45].

due to rotation. For steady helical flow, the velocity field may be written as:

$$v_r = 0, \qquad v_\theta = r\omega(r), \qquad v_z = u(r) \tag{7.80}$$

in which $\omega(r)$ is the velocity component of angular movement, $u(r)$ is the velocity component of forward movement, and r is an arbitrary radial distance from the center axis.

Referring to Fig. 7.33, the $r\theta$ and rz component of the equations of motion for steady helical flow are given by:

$$\partial\tau_{r\theta}/\partial r + 2\tau_{r\theta}/r = 0 \tag{7.81}$$

$$\partial\tau_{rz}/\partial r + \tau_{rz}/r - \partial p/\partial z - \rho g_z = 0 \tag{7.82}$$

in which ρ is the density of the fluid and g_z is the z component of gravitational acceleration.

Integrating Eqs. (7.81) and (7.82), respectively, with respect to r, one obtains

$$\tau_{r\theta} = c_1/r^2 \tag{7.83}$$

$$\tau_{rz} = \tfrac{1}{2}r\zeta + c_2/r \tag{7.84}$$

in which ζ is the axial pressure gradient defined as

$$\zeta = \partial p/\partial z + \rho g_z \tag{7.85}$$

For the analysis to follow, the power-law fluid model will be considered:

$$\boldsymbol{\tau}^i = \eta_i(\mathrm{II}_i)\mathbf{d}^i \qquad \{i = 1 \text{ to } (N+1)\} \tag{7.86}$$

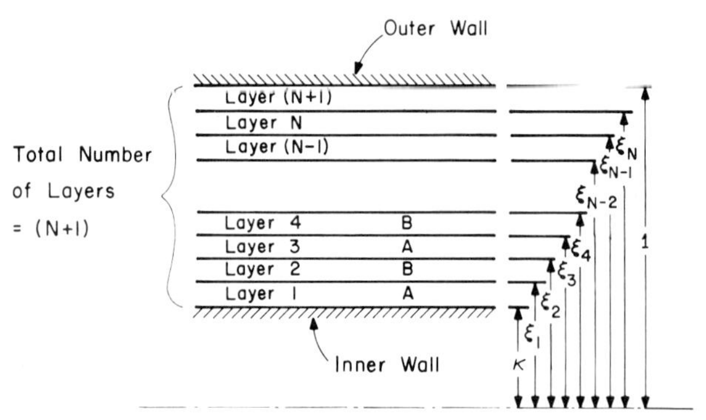

Fig. 7.34 Schematic of the dimensionless radial positions of the interfaces in an annulus [45].

where

$$\eta_i(\mathrm{II}_i) = K_i(4\mathrm{II}_i)^{(n_i - 1)/2} \tag{7.87}$$

The subscript i refers to the layer being considered, the total number of layers being $(N + 1)$ (see Fig. 7.34), with *odd* values of i denoting polymer A and *even* values of i denoting polymer B.

For the velocity field given by Eq. (7.80), one has

$$\|\mathbf{d}^i\| = \tfrac{1}{2}\begin{Vmatrix} 0 & r\omega_i' & u_i' \\ r\omega_i' & 0 & 0 \\ u_i' & 0 & 0 \end{Vmatrix} \tag{7.88}$$

in which $\omega_i' = d\omega_i/dr$ and $u_i' = du_i/dr$. Note that the second invariant II_i of $\mathbf{d}^i$ is given by

$$\mathrm{II}_i = (r^2\omega_i'^2 + u_i'^2)/4 \tag{7.89}$$

and is a function only of r.

Now the use of Eq. (7.86) in Eqs. (7.83) and (7.84) gives

$$\tau_{r\theta}^i = \alpha/\xi^2 = \xi(d\omega_i/d\xi)\eta_i(\mathrm{II}_i) \tag{7.90}$$

$$\tau_{rz}^i = \frac{\zeta R_0}{2}\left(\frac{\xi^2 - \lambda^2}{\xi}\right) = \frac{1}{R_0}\frac{du_i}{d\xi}\eta_i(\mathrm{II}_i) \tag{7.91}$$

in which $\xi = r/R_0$, α and λ may be expressed in terms of the integration constants, c_1 and c_2, given in Eqs. (7.83) and (7.84), respectively, and R_0 is the radius of the outer cylinder (see Fig. 7.33).

Equations (7.90) and (7.91) hold true for all layers between the two coaxial cylinders. Integration of Eqs. (7.90) and (7.91) with appropriate boundary conditions will give expressions that describe the velocity components of forward movement, $u_i(\xi)$, and angular motion, $\omega_i(\xi)$.

The following boundary conditions are now imposed:

(i) at $r = \kappa R_0$(inner wall), where $0 < \kappa < 1$

$$\omega_i(\kappa R_0) = \Omega_I, \qquad u_i(\kappa R_0) = 0 \tag{7.92}$$

(ii) at $r = R_0$(outer wall),

$$\omega_i(R_0) = 0, \qquad u_i(R_0) = 0 \tag{7.93}$$

(iii) at the phase interface, ω and u are continuous, i.e.,

$$\omega_A(\xi_j) = \omega_B(\xi_j), \qquad u_A(\xi_j) = u_B(\xi_j) \qquad \text{for all} \quad j = 1 \text{ to } N \tag{7.94}$$

Note that, in Eq. (7.94), ξ_j is the location of the interfaces (see Fig. 7.34).

As may be surmised, when we have an *odd* number of layers generated we will have the same polymer forming layers on the inner and outer walls of the annulus. On the other hand, when we have an *even* number of layers generated, the polymers forming layers at the inner and outer walls are different.

We shall consider the case where an *odd* number of layers is generated. Integrating Eqs. (7.90) and (7.91) gives:

(i) From the inner wall to the first interface ($\xi = \xi_1$):

$$\omega_A(\xi_1) = \Omega_I + \alpha \int_\kappa^{\xi_1} \frac{ds}{s^3 \eta_A[\mathrm{II}_A(s)]} \tag{7.95}$$

$$u_A(\xi_1) = \frac{\zeta R_0^2}{2} \int_\kappa^{\xi_1} \frac{(s^2 - \lambda^2)\, ds}{s \eta_A[\mathrm{II}_A(s)]} \tag{7.96}$$

(ii) Between two successive interfaces (i.e., from ξ_j to ξ_{j+1}):

$$\omega_i(\xi_{j+1}) = \omega_i(\xi_j) + \alpha \int_{\xi_j}^{\xi_{j+1}} \frac{ds}{s^3 \eta_i[\mathrm{II}_i(s)]} \tag{7.97}$$

$$u_i(\xi_{j+1}) = u_i(\xi_j) + \frac{\zeta R_0^2}{2} \int_{\xi_j}^{\xi_{j+1}} \frac{(s^2 - \lambda^2)\, ds}{s \eta_i[\mathrm{II}_i(s)]} \tag{7.98}$$

Equations (7.97) and (7.98) hold true for all values of j from 1 to $N - 1$. Also, as mentioned earlier, i designates the layer being considered, with *odd* values of i denoting polymer A and *even* values of i denoting polymer B. It should be noted that the numerical value of i is greater than the numerical value of j by unity. These relationships hold true in all the following derivations.

(iii) From the last interface ($\xi = \xi_N$) to the outer wall ($\xi = 1$):

$$0 = \omega_A(\xi_N) + \alpha \int_{\xi_N}^{1} \frac{ds}{s^3 \eta_A[\mathrm{II}_A(s)]} \tag{7.99}$$

$$0 = u_A(\xi_N) + \frac{\zeta R_0^2}{2} \int_{\xi_N}^{1} \frac{(s^2 - \lambda^2)\, ds}{s \eta_A[\mathrm{II}_A(s)]} \tag{7.100}$$

Adding Eqs. (7.95), (7.97), and (7.99) gives

$$-\Omega_I = \alpha \int_{\kappa}^{\xi_1} \frac{ds}{s^3 \eta_A[\mathrm{II}_A(s)]} + \alpha \sum_{j=1}^{N-1} \left[\int_{\xi_j}^{\xi_{j+1}} \frac{ds}{s^3 \eta_i[\mathrm{II}_i(s)]} \right] + \alpha \int_{\xi_N}^{1} \frac{ds}{s^3 \eta_A[\mathrm{II}_A(s)]} \tag{7.101}$$

Adding Eqs. (7.96), (7.98), and (7.100) gives

$$0 = \int_{\kappa}^{\xi_1} \frac{(s^2 - \lambda^2)\, ds}{s \eta_A[\mathrm{II}_A(s)]} + \sum_{j=1}^{N-1} \left[\int_{\xi_j}^{\xi_{j+1}} \frac{(s^2 - \lambda^2)\, ds}{s \eta_i[\mathrm{II}_i(s)]} \right] + \int_{\xi_N}^{1} \frac{(s^2 - \lambda^2)\, ds}{s \eta_A[\mathrm{II}_A(s)]} \tag{7.102}$$

The total volumetric flow rate Q may be determined from

$$Q = \sum_{j=0}^{j=N} \left[\int_{r_j}^{r_{j+1}} 2\pi u_i(r) r\, dr \right] = 2\pi R_0^2 \left[\sum_{j=0}^{j=N} \left\{ \int_{\xi_j}^{\xi_{j+1}} u_i(s) s\, ds \right\} \right] \tag{7.103}$$

where, $\xi_0 = \kappa$ and $\xi_{N+1} = 1$.

Using Eqs. (7.96), (7.98), and (7.100) in Eq. (7.103) gives

$$\begin{aligned} Q = {} & \frac{\zeta \pi R_0^4}{2} \left[\int_{\kappa}^{\xi_1} \frac{(s^2 - \kappa^2)}{\eta_A[\mathrm{II}_A(s)]} \frac{(s^2 - \lambda^2)}{s} ds \right] \\ & + \sum_{j=1}^{N-1} \left[\frac{\zeta \pi R_0^4}{2} \int_{\xi_j}^{\xi_{j+1}} \frac{(s^2 - \xi_j^2)}{\eta_i[\mathrm{II}_i(s)]} \frac{(s^2 - \lambda^2)}{s} ds + \pi R_0^2 u_i(\xi_j)(\xi_{j+1}^2 - \xi_j^2) \right] \\ & + \frac{\zeta \pi R_0^4}{2} \left[\int_{\xi_N}^{1} \frac{(s^2 - \xi_N^2)}{\eta_A[\mathrm{II}_A(s)]} \frac{(s^2 - \lambda^2)}{s} ds \right] \end{aligned} \tag{7.104}$$

The analysis is similar when the total number of layers generated is *even*, except that the polymers forming layers at the inner and outer walls are different.

Let Q_A and Q_B be the volumetric flow rates of the individual polymers, and let h be the annular gap. Let h_A and h_B be the sum of the thicknesses of layers of polymers A and B, respectively. Assuming the layer thicknesses to be in the same ratio as the volumetric flow rates, we have

$$h_A = \{Q_A/(Q_A + Q_B)\}h; \qquad h_B = \{Q_B/(Q_A + Q_B)\}h \tag{7.105}$$

If the number of layers $(N + 1)$ assumed is *even*, then the individual layer thicknesses l_A and l_B are given by

$$l_A = 2h_A/(N + 1); \qquad l_B = 2h_B/(N + 1) \tag{7.106}$$

If the number of layers $(N + 1)$ assumed is *odd*, then the individual layer thicknesses are given by

$$l_A = 2h_A/(N + 2); \qquad l_B = 2h_B/N \tag{7.107}$$

In assuming the number of layers generated, we are guided by the residence time of the polymers in the annulus and the number of overlappings due to the rotation of the inner mandrel, for any given processing condition.

The residence time for the polymers in the annulus is given by

$$t_R = \pi(R_0^2 - R_w^2)L/Q \tag{7.108}$$

where t_R is the residence time, R_0 is the radius of outer wall, R_w is the radius of inner wall, L is the length of the annulus, and Q is the total volumetric flow rate. If the angular speed of rotation of the inner wall is Ω_I, the number of rotations N_R completed by the polymer phases at the inner wall before exiting from the annulus is given by

$$N_R = \Omega_I t_R \tag{7.109}$$

The analysis presented above is very versatile in that it may be easily extended to accommodate any number of feedslots for each of the polymer melts. Han and Shetty [45] presented the computational procedure to obtain the volumetric flow rate in terms of system parameters. Information on the velocity and stress profiles, number of layers, and the layer thicknesses in the annular space (as functions of the rheological parameters of the fluid and processing conditions) is of extreme importance for controlling the blown film coextrusion process.

Figure 7.35 gives a plot of the theoretically predicted number of layers generated versus the angular speed of rotation Ω_I for the LDPE/EVA and the LDPE/PP systems. These plots show an increase in the number of layers

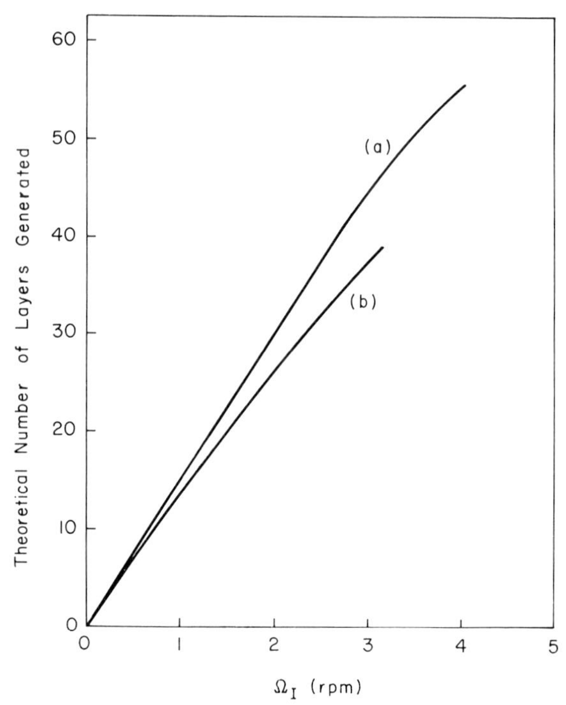

Fig. 7.35 Theoretically predicted number of layers versus angular speed of the inner mandrel in the blown-film coextrusion die [45]: (a) LDPE/EVA system; (b) LDPE/PP system. The power-law constants used are: (a) For LDPE: $K = 0.437 \times 10^4$ N sec^n/m^2, $n = 0.40$; (b) For EVA: $K = 0.26 \times 10^4$ N sec^n/m^2, $n = 0.61$; (c) For PP: $K = 0.641 \times 10^4$ N sec^n/m^2, $n = 0.46$.

generated with an increase in the angular speed of rotation of the inner mandrel. The theoretical prediction assumed eight feedslots for each polymer in the feedport. (See Fig. 7.30.)

Representative linear and angular velocity profiles are given in Figs. 7.36 and 7.37, respectively, for the LDPE/PP system, where we see that the slope of the velocity profiles changes at the interfaces, thereby indicating a discontinuity in the velocity gradient at each interface.

Figure 7.38 gives distributions of the axial velocity gradient for the LDPE/PP system. It is seen that the velocity gradient (or shear rate) is discontinuous at the interfaces. This is again due to the fact that there is an abrupt change in the melt viscosities of the two polymers across the interface. Figure 7.39 gives representative shear stress distributions for the LDPE/PP system. It is seen that the shear stress is continuous across the interfaces.

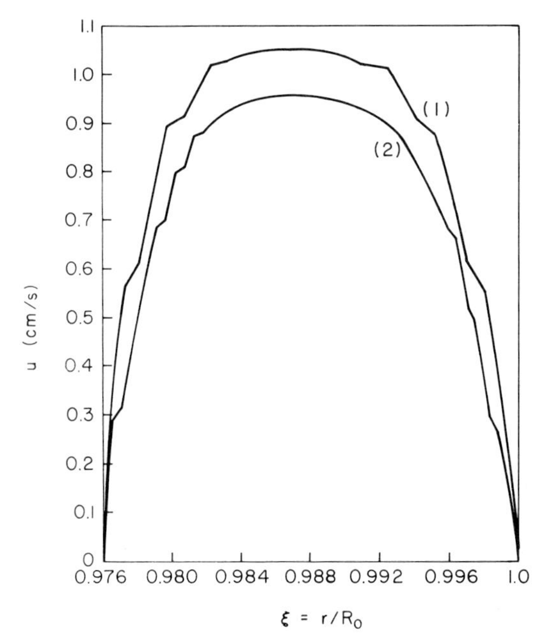

Fig. 7.36 Theoretically predicted distributions of axial velocity of LDPE/PP system in a blown-film coextrusion die with the rotating inner mandrel [45]: (1) $-\partial p/\partial z = 7.64 \times 10^7$ N/m^3, $\Omega_I = 1.27$ rpm; (2) $-\partial p/\partial z = 7.18 \times 10^7$ N/m^3, $\Omega_I = 2.78$ rpm. The power-law constants used are the same as in Fig. 7.35.

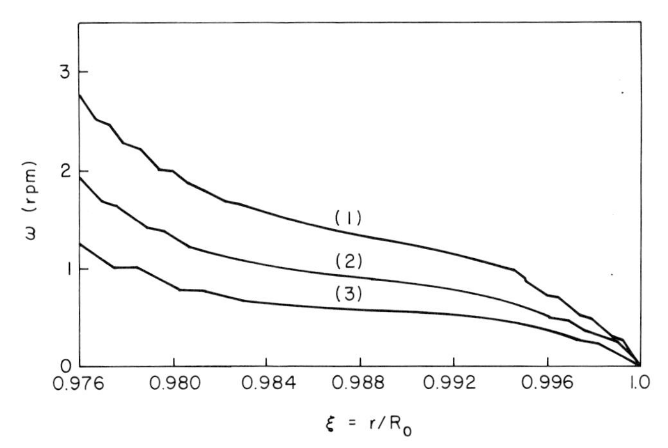

Fig. 7.37 Theoretically predicted distributions of angular velocity of LDPE/PP system in a blown-film coextrusion die with the rotating inner mandrel [45]: (1) $\Omega_I = 1.27$ rpm; (2) $\Omega_I = 1.95$ rpm; (3) $\Omega_I = 2.78$ rpm. Other processing conditions are the same as in Fig. 7.36.

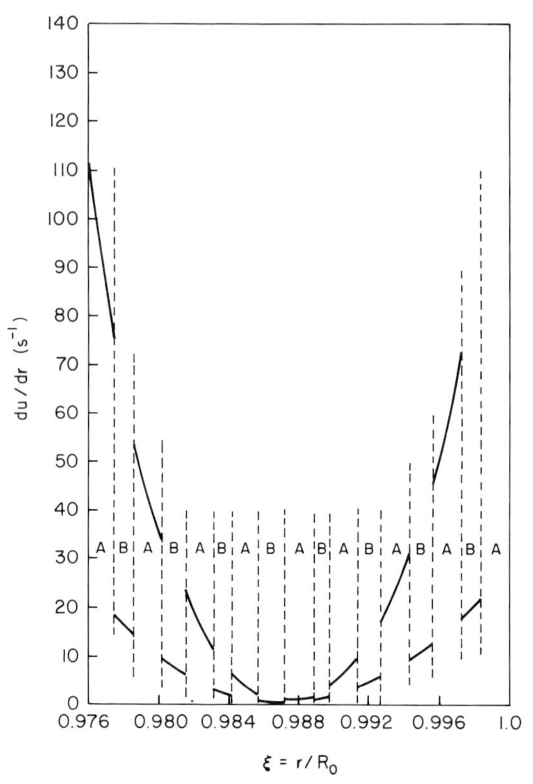

Fig. 7.38 Theoretically predicted distributions of axial velocity gradient of LDPE/PP system in a blown-film coextrusion die with the rotating inner mandrel [45]. The angular speed Ω_I is 1.27 rpm, and other processing conditions are the same as in Fig. 7.36.

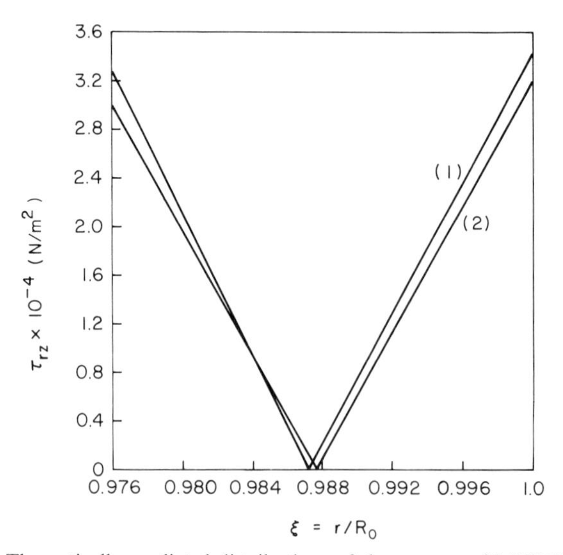

Fig. 7.39 Theoretically predicted distributions of shear stress of LDPE/PP system in a blown-film coextrusion die with the rotating inner mandrel [45]. Processing conditions are the same as in Fig. 7.36.

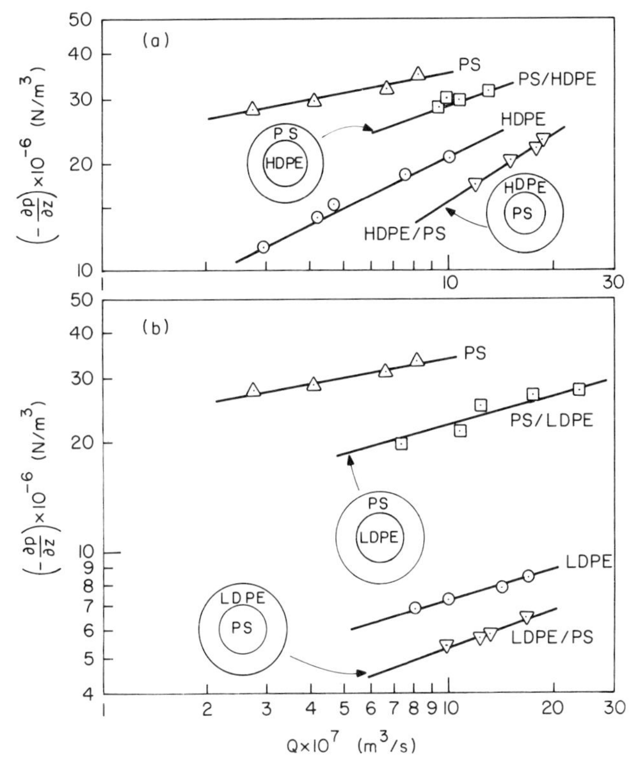

Fig. 7.40 Pressure gradient versus volumetric flow rate in the sheath-core coextrusion through a cylindrical tube [46]: (a) PS/HDPE and HDPE/PS systems; (b) PS/LDPE and LDPE/PS systems.

7.6 PRESSURE GRADIENT REDUCTION IN COEXTRUSION

From the processing point of view, information on pressure gradients is of practical importance because they are intimately related to the energy requirement and production rate.

Figure 7.40 gives plots of pressure gradient $(-\partial p/\partial z)$ versus the combined volumetric flow rate Q for the sheath-core coextrusion of polystyrene (PS)/low-density polyethylene (LDPE), and of polystyrene (PS)/high-density polyethylene (HDPE), in a cylindrical tube. Figure 7.41 gives the melt viscosity behavior of the individual components, PS, LDPE, and HDPE.

It is seen in Fig. 7.40 that the value of $(-\partial p/\partial z)$ of a two-phase system lies below that of the individual components when the less viscous component forms the outer layer, but between those of the individual components when

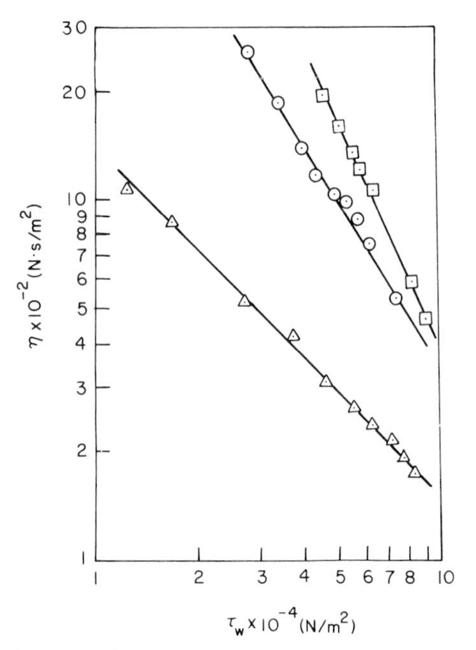

Fig. 7.41 Viscosity versus shear stress for the polymers ($T = 200°$C) used in the coextrusion experiments that yielded the pressure gradients given in Fig. 7.40: (⊡) PS; (⊙) HDPE; (△) LDPE.

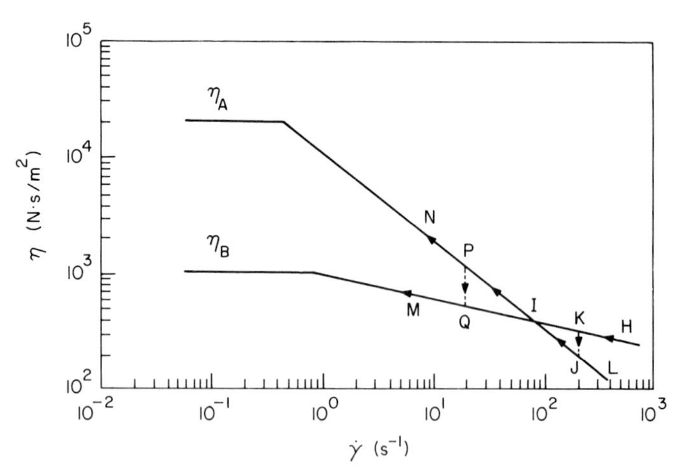

Fig. 7.42 Viscosity versus shear rate for two fluids whose viscosities cross over at a particular shear rate [46].

the less viscous component forms the inner layer. What is of great interest is that the pressure gradient $(-\partial p/\partial z)$ of a two-phase system can become lower than that of the less viscous of the two individual components.

We show below that for a certain combination of rheological properties of the two components being coextruded, a seemingly anomalous pressure gradient reduction can indeed be predicted [46]. For given values of the volumetric flow rates, Q_A and Q_B, we can compute the pressure gradient, $\zeta = (-\partial p/\partial z)$, using Eqs. (7.11) and (7.12). Let us now consider two examples.

Example 1 Let $R_0 = 0.3$ cm, $Q = 3$ cm^3/sec, $K_A = 1.1 \times 10^4$ N sec$^{0.25}$/m^2, $n_A = 0.25$, $K_B = 9.4 \times 10^2$ N sec$^{0.8}$/m^2, and $n_B = 0.80$. Let us assume further that the zero-shear viscosities (η_0) of the individual components are given by: $(\eta_0)_A = 2 \times 10^4$ N sec/m^2 and $(\eta_0)_B = 10^3$ N sec/m^2. Figure 7.42 gives plots of viscosity versus shear rate for the two components being coextruded, and it is seen that viscosities cross over at a certain shear rate. The computed pressure gradients are plotted in Fig. 7.43, where we see that when component A flows outside component B, the pressure gradient goes through a minimum, whereas when component B flows outside component A, the pressure gradient goes through a maximum.

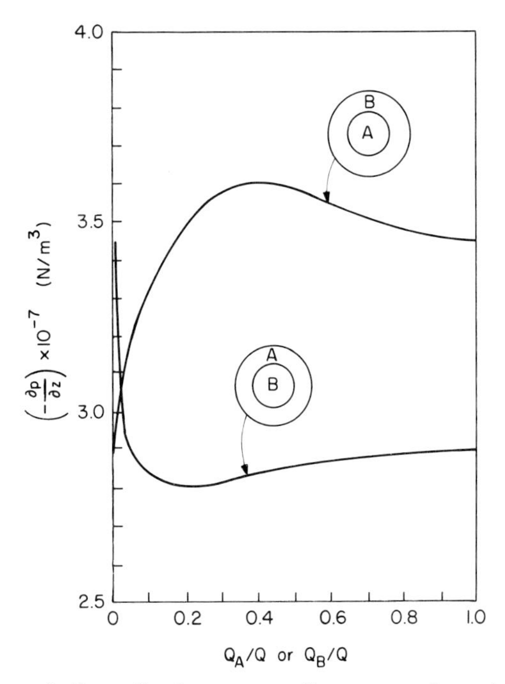

Fig. 7.43 Theoretically predicted pressure gradient versus volumetric flow ratio for the two fluids whose viscosity curves are given in Fig. 7.42 [46].

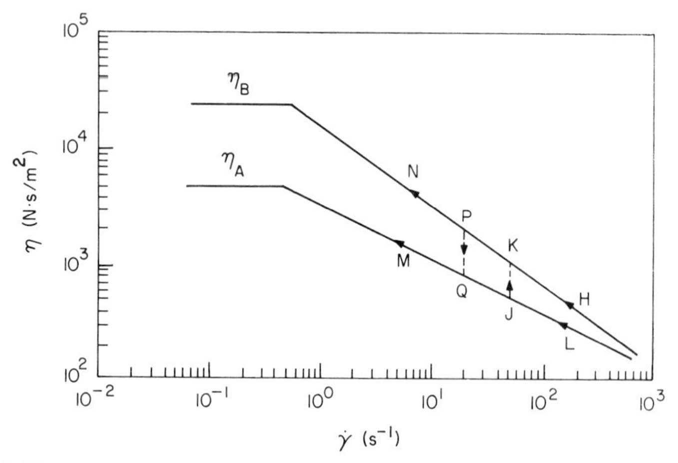

Fig. 7.44 Viscosity versus shear rate for two fluids whose viscosities do not cross over in the entire range of shear rates one is interested in [46].

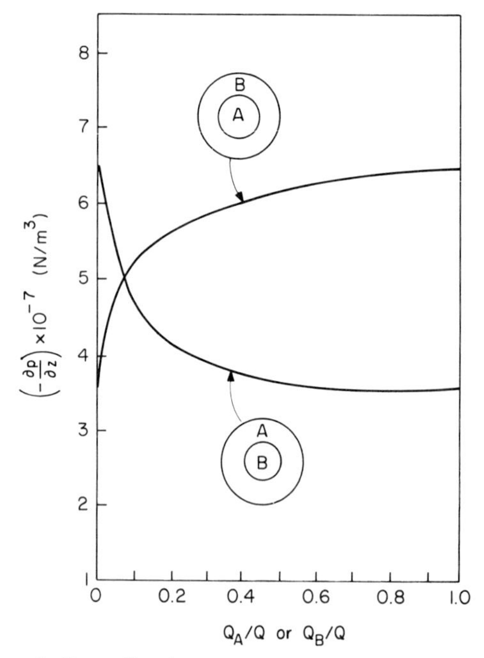

Fig. 7.45 Theoretically predicted pressure gradient versus volumetric flow ratio for the two fluids whose viscosities are given in Fig. 7.44 [46].

Example 2 Let $R_0 = 0.3$ cm, $Q = 3\ \mathrm{cm}^3/\mathrm{sec}$, $K_A = 4.4 \times 10^3\ \mathrm{N\,sec}^{0.48}/\mathrm{m}^2$, $n_A = 0.48$, $K_B = 2.1 \times 10^4\ \mathrm{N\,sec}^{0.28}/\mathrm{m}^2$, $n_B = 0.28$, $(\eta_0)_A = 6 \times 10^3\ \mathrm{N\,sec/m}^2$, and $(\eta_0)_B = 3 \times 10^4\ \mathrm{N\,sec/m}^2$. Figure 7.44 gives plots of viscosity versus shear rate for the two components being coextruded and it is seen that viscosities do *not* cross over in the entire range of shear rates we are interested in. The computed pressure gradients are plotted in Fig. 7.45, where we see that the pressure gradient exhibits neither a maximum nor a minimum.

With reference to Fig. 7.43, the situation where a minimum pressure gradient occurs has the viscosity distribution as shown in Fig. 7.46a, and the situation where a maximum pressure gradient occurs has the viscosity distribution as shown in Fig. 7.46b. In order to facilitate our discussion, let us examine the shear-dependent viscosities, given in Fig. 7.42, of the two fluids being coextruded. The pressure gradient goes through a minimum when the viscosity crossover occurs at the interface of the two fluids, following the path LIM in Fig. 7.42, whereas it goes through a maximum when the viscosity crossover occurs at the interface, following the path HIN in Fig. 7.42. The

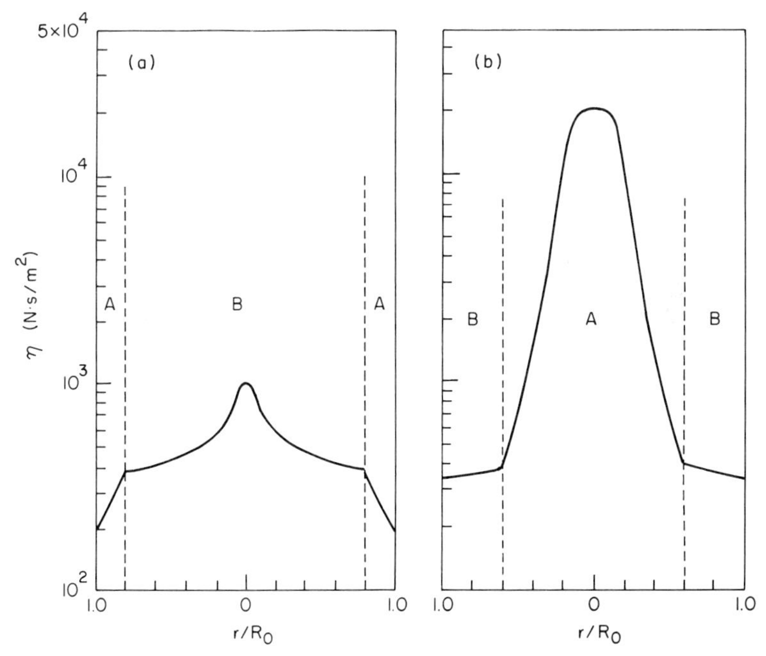

Fig. 7.46 Theoretically predicted viscosity profiles in sheath-core tubular coextrusion considered in Example 1, with reference to Fig. 7.43 [46]:

(a) $(-\partial p/\partial z)_{A/B}$ is at minimum, $\lambda = 0.8$, $(Q_A/Q = 0.166)$;
(b) $(-\partial p/\partial z)_{B/A}$ is at maximum, $\lambda = 0.6$, $(Q_B/Q = 0.44)$.

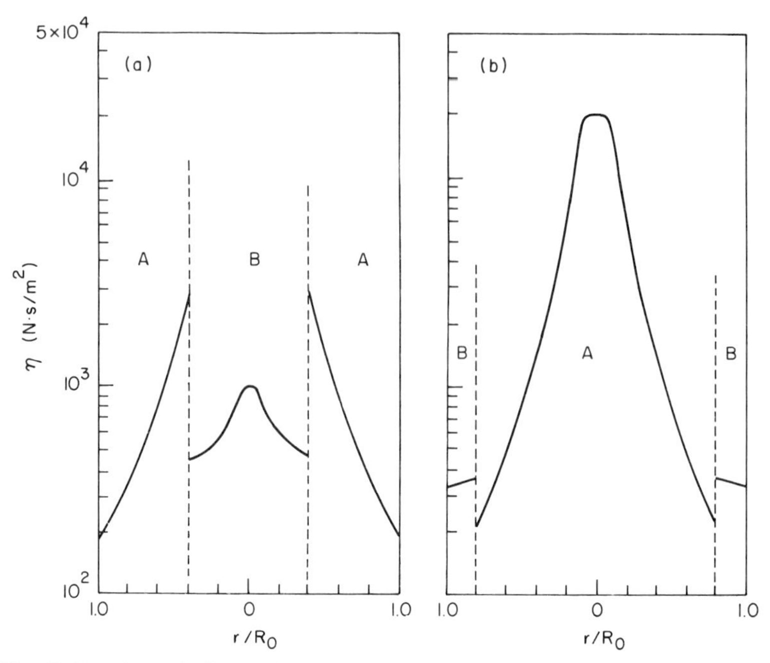

Fig. 7.47 Theoretically predicted viscosity profiles in sheath-core tubular coextrusion considered in Example 1, with reference to Fig. 7.43 [46]:

(a) $(-\partial p/\partial z)_{A/B} < (-\partial p/\partial z)_A < (-\partial p/\partial z)_B$, $\lambda = 0.4$ $(Q_A/Q = 0.76)$;
(b) $(-\partial p/\partial z)_{B/A} > (-\partial p/\partial z)_B > (-\partial p/\partial z)_A$, $\lambda = 0.8$ $(Q_B/Q = 0.13)$.

actual viscosity profiles inside a tube were computed, using the expression:

$$\eta = K(\partial v_z/\partial r)^{n-1} \tag{7.110}$$

with the aid of Fig. 7.42. Note that the velocity gradient $\partial v_z/\partial r$ was obtained from Eqs. (7.8) and (7.9) for the respective phases.

Figure 7.47 gives the viscosity distributions for the situations where the viscosity crossover does *not* occur at the interface of the two fluids inside a tube. Figure 7.47a may represent the situation where the path LIPQM in Fig. 7.42 is followed, and Fig. 7.47b may represent the situation where the path HKJIN in Fig. 7.42 is followed.

With reference to Fig. 7.45, we have the viscosity distributions, as shown in Fig. 7.48, for particular values of interface position. Note that there is *no* viscosity crossover in the two fluids being coextruded, as shown in Fig. 7.44. Figure 7.48a may represent the situation where the path LJKN in Fig. 7.44 is followed, and Fig. 7.48b may represent the situation where the path HPQM in Fig. 7.44 is followed.

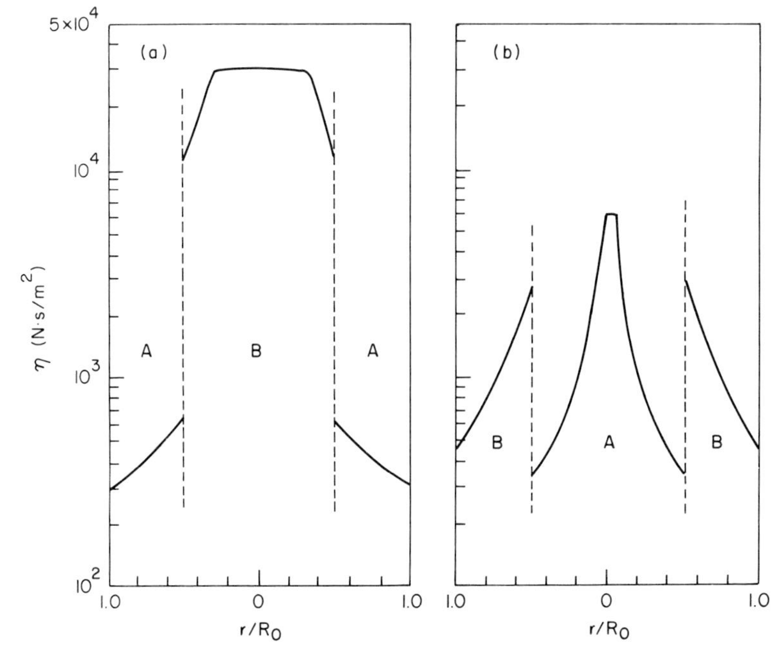

Fig. 7.48 Theoretically predicted viscosity profiles in sheath-core tubular coextrusion considered in Example 2, with reference to Fig. 7.45 [46]:

(a) $(-\partial p/\partial z)_{A/B}$ decreases monotonically, $\lambda = 0.5$ $(Q_A/Q = 0.62)$;
(b) $(-\partial p/\partial z)_{B/A}$ increases monotonically, $\lambda = 0.5$ $(Q_B/Q = 0.59)$.

It is demonstrated above that depending on the choice of the power-law constants, K and n, for the individual components, at otherwise identical conditions, it is theoretically possible that the pressure gradients in two-phase concentric flow through a cylindrical tube can become smaller than the pressure gradient of the less viscous component of the two.

More experiments are needed to verify the theoretical predictions. Experiments are important because once the theory is verified, one can apply it to coextrusion operations in order to minimize pressure drops, and hence to increase production rate by manipulating the rheological properties of the two polymers to be coextruded. From the practical point of view, the manipulation of rheological properties can be realized by changing the polymer's molecular weight, and/or its distribution, by using a small amount of "unharmful" additive, and by varying the melt extrusion temperature. It is worth pointing out that the viscosity ratio of two polymers can be reversed by increasing (or decreasing) the melt extrusion temperature when the viscosity of one polymer is far more sensitive to temperature than the other.

7.7 INTERFACE DEFORMATION IN STRATIFIED TWO-PHASE FLOW

7.7.1 Side-by-Side Coextrusion in a Cylindrical Tube

In this type of coextrusion, two polymers are fed separately into the feedblock and they meet at the die inlet. The heart of the processing equipment is the proper design of the feedblock, which controls the two melt streams to ensure that the interface is flat on arrival at the die inlet. Figure 7.49 gives a schematic of the design of the feedblock, which has a *knife-edged* flow divider. The two melt streams, fed separately from their respective extruders into the feedblock, meet each other at the die inlet, and the combined stream flows through the capillary and leaves the die exit.

Figure 7.50 shows the extrudate cross section of a polystyrene (PS)/low-density polyethylene (LDPE) system, coextruded through three circular dies with different L/D ratios, at approximately the same flow rate. As can be

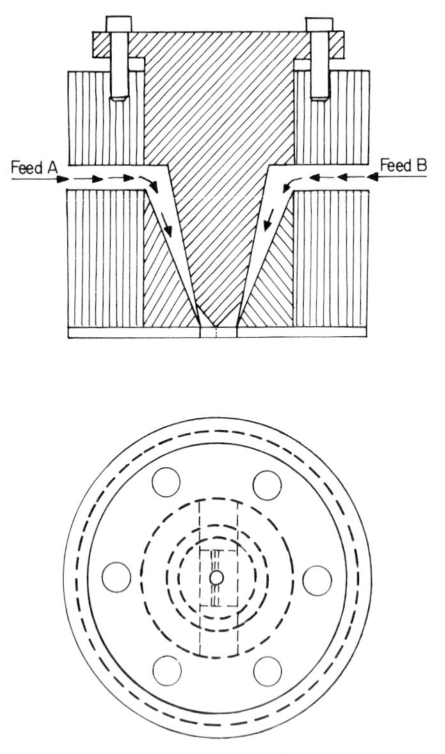

Fig. 7.49 Schematic of the coextrusion die assembly which has a knife-edged flow divider.

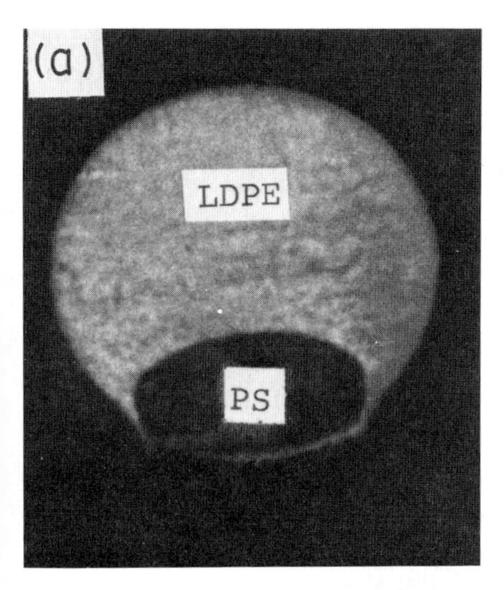

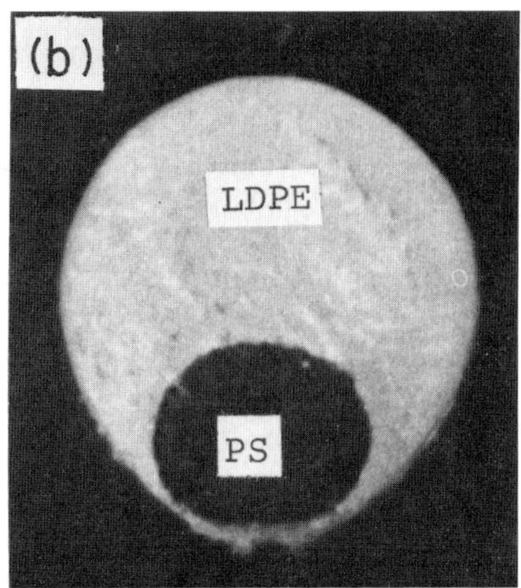

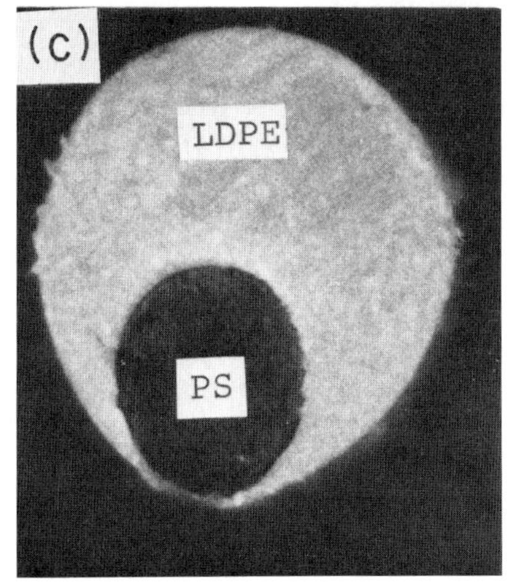

Fig. 7.50 Effect of the capillary length-to-diameter (L/D) ratio on the interface shape of two polymers (LDPE and PS) at 200°C, coextruded side by side through a circular die [28]: (a) L/D = 4, $Q_{PS} = 8.3$ cm^3/min, $Q_{LDPE} = 33.2$ cm^3/min; (b) L/D = 11, $Q_{PS} = 8.3$ cm^3/min, $Q_{LDPE} = 30.8$ cm^3/min; (c) L/D = 18, $Q_{PS} = 8.3$ cm^3/min, $Q_{LDPE} = 30.8$ cm^3/min. From C. D. Han, *J. Appl. Polym. Sci.* **17**, 1289, Copyright © 1973. Reprinted by permission of John Wiley & Sons, Inc.

seen, the shape of the interface becomes almost completely circular at L/D ratios of 11 and 18, giving rise to a sheath-core configuration. This is in spite of the fact that, when the polymers met at the die entrance, the interface was flat.

Let us now try to offer a quantitative interpretation of the experimentally observed interface shapes given in Fig. 7.50 in terms of the rheological properties of the fluids concerned. Figure 7.51 gives plots of melt viscosity versus shear stress and Fig. 7.52 of first normal stress difference versus shear stress for PS and LDPE melts. Over the range of extrusion conditions tested (i.e., apparent shear rate of 90–600 $\sec^{-1}$, and wall shear stress of 0.3×10^5–0.9×10^5 N/m^2), we have the following information:

$$\eta_{\mathrm{LDPE}}/\eta_{\mathrm{PS}} = 0.14\text{–}0.34$$

$$(\tau_{11} - \tau_{22})_{\mathrm{LDPE}}/(\tau_{11} - \tau_{22})_{\mathrm{PS}} = 2.53\text{–}2.37$$

It is seen that the LDPE is less viscous, but more elastic, than the PS. It is therefore difficult to conclude which of the two rheological properties, viscosity or elasticity, is primarily responsible for the nature of the interface shape observed. What is required are melt systems in which the two fluids have different viscosities and the same elasticity or different elasticities and the same viscosity.

Figure 7.53 shows the extrudate cross section of a polystyrene (PS)/high-density polyethylene (HDPE) system, extruded through a capillary having an L/D ratio of 11. Note that the two polymers were fed side by side (i.e., with a flat interface) at the die inlet, using a knife-edged flow divider (see

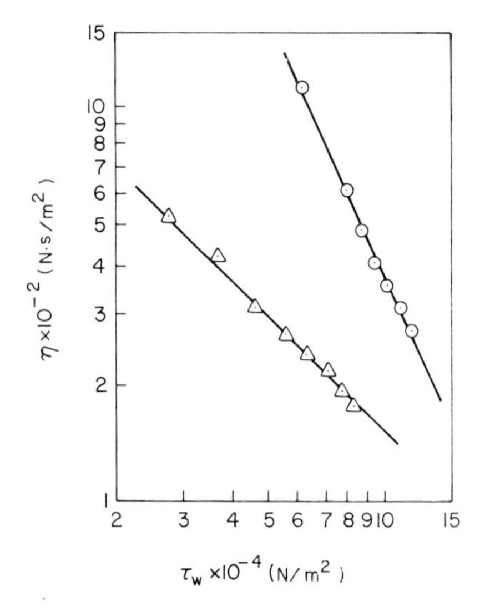

Fig. 7.51 Viscosity versus shear stress for the polymers ($T = 200°$C) used in the coextrusion experiments that yielded the interface shapes given in Fig. 7.50: (⊙) PS; (△) LDPE.

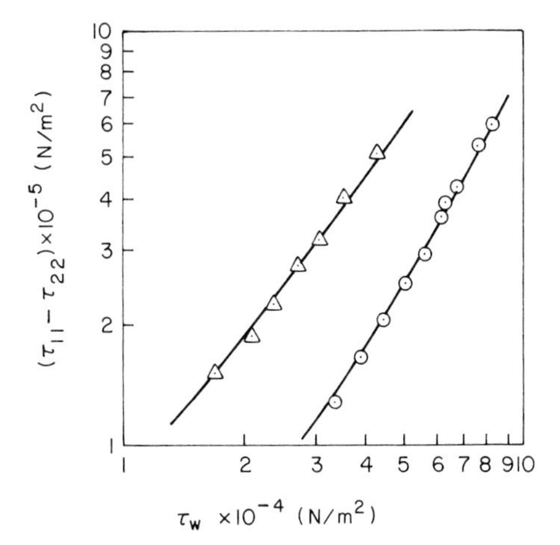

Fig. 7.52 First normal stress difference versus shear stress for the polymers ($T = 200°C$) used in the coextrusion experiments that yielded the interface shapes given in Fig. 7.50: (⊙) PS; (△) LDPE.

Fig. 7.49). It is seen in Fig. 7.53 that at 200°C the PS pushes into the HDPE whereas at 240°C the HDPE pushes into the PS, giving rise to an interface of curtate cycloid shape. In other words, the PS/HDPE system exhibits an interface curvature reversal as the melt temperature is increased from 200 to 240°C. The interface shapes given in Fig. 7.53 may be explained with the aid of the rheological properties of the melts coextruded.

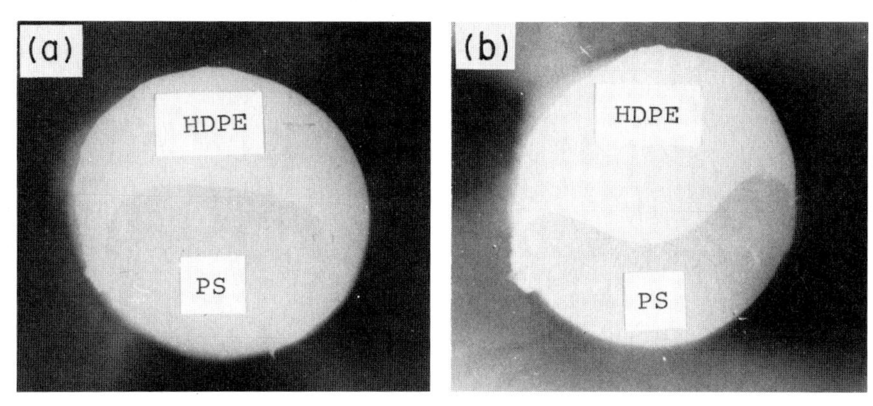

Fig. 7.53 Interface shape of two polymers (HDPE and PS) at 200°C, coextruded side by side through a circular die [31]; (a) at 200°C; (b) at 240°C. The die used has an L/D ratio of 11. From C. D. Han, *J. Appl. Polym. Sci.* **20**, 2609. Copyright © 1976. Reprinted by permission of John Wiley & Sons, Inc.

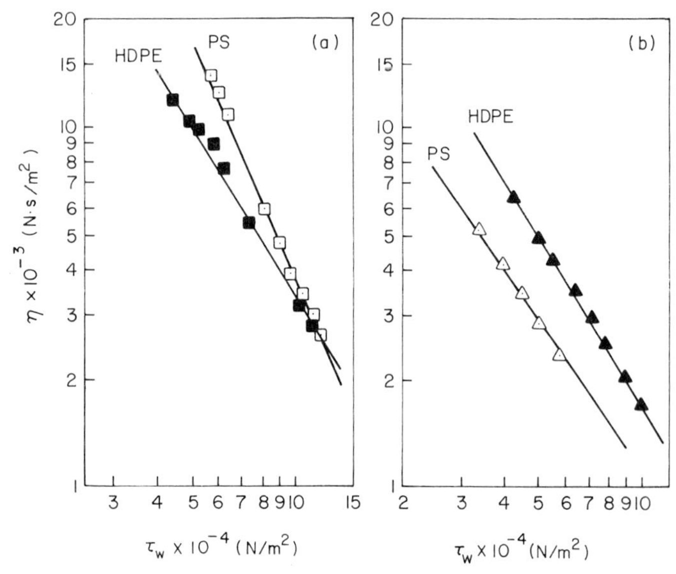

Fig. 7.54 Viscosity versus shear stress for the polymers used in the coextrusion experiments that yielded the interface shapes given in Fig. 7.53 [31]: (a) at 200°C; (b) at 240°C. From C. D. Han, *J. Appl. Polym. Sci.* **20**, 2609. Copyright © 1976. Reprinted by permission of John Wiley & Sons, Inc.

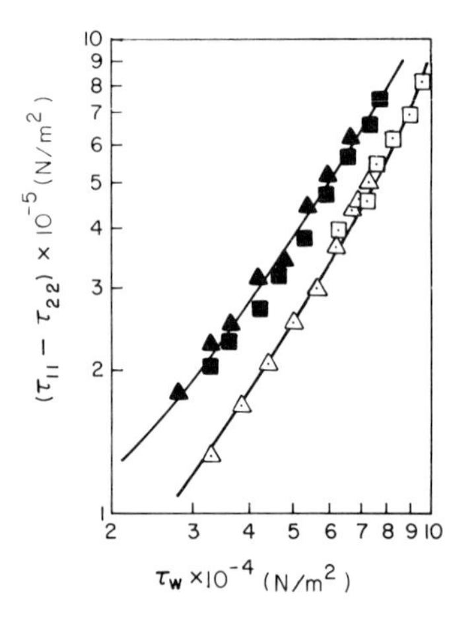

Fig. 7.55 First normal stress difference versus shear stress for the polymers used in the coextrusion experiments that yielded the interface shapes given in Fig. 7.53 [31]: Open symbols for PS: (□) 200°C; (△) 240°C. Closed symbols for HDPE: (■) 200°C; (▲) 240°C. From C. D. Han, *J. Appl. Polym. Sci.* **20**, 2609. Copyright © 1976. Reprinted by permission of John Wiley & Sons, Inc.

Figure 7.54 gives plots of melt viscosity versus shear stress and Fig. 7.55 of first normal stress difference versus shear stress for PS and HDPE melts at 200°C and 240°C. It is seen that over the range of extrusion conditions tested (i.e., wall shear stress below 0.9×10^5 N/m²) the PS is more viscous than the HDPE at 200°C, whereas the PS is less viscous than the HDPE at 240°C. On the other hand, the ratio of melt elasticities of the two components remains essentially constant, as may be seen in Fig. 7.55. It can be concluded, therefore, that the interface curvature reversal given in Fig. 7.53 may be attributable to the viscosity ratio reversal of the two components as the melt temperature is increased from 200°C to 240°C.

It is of interest to note that at a given melt temperature, viscosity ratio reversal can be obtained when the two components have viscosities such that crossovers occur in the plots of melt viscosity versus shear stress. As a matter of fact, some researchers [29, 30] reported that interface curvature reversal occurred at the shear stress (or shear rate) where the viscosity crossover of the two melts also occurred. Therefore, on the basis of experimental evidence, it can be concluded that the viscosity ratio of two components predominates over the elasticity ratio in determining the interface shape.

Some theoretical studies have also been reported on the stratified flow of two immiscible *Newtonian* fluids in a cylindrical tube. In the flow of viscoelastic polymeric melts, a theoretical analysis of the two-phase flow in a cylindrical tube accompanying a change in interface shape can become extremely complicated, and it is no wonder that little on this subject has been reported in the literature.

MacLean [24] and Everage [25] used the variational method, together with the principle of minimum viscous dissipation, in determining the energetically preferred interface configuration when two Newtonian (or power-law) fluids were forced to flow side by side through a cylindrical tube. Such an approach requires the specification of an interface configuration and velocity field in order to evaluate the integral which represents the rate of viscous dissipation per unit length of tube. In the situation where a curved interface keeps changing as the flow continues through a cylindrical tube, the determination of the velocity field and the interface configuration is a nontrivial matter. Assuming that interface shapes may be described by arcs of a circle, which is reasonable in view of the experimental observations given in Figs. 7.50 and 7.53, Everage [25] evaluated the integral, which represents the rate of viscous dissipation per unit length of tube, and compared the result with the one corresponding to the sheath-core configuration (i.e., concentric interface shape). He reached the conclusion that for a given pressure drop, the sheath-core configuration is the energetically preferred flow configuration.

As may be seen in Figs. 7.50 and 7.53, the extent of encapsulation (or the degree of interface deformation) from the initial flat interface depends on the viscosity ratio and the residence time (duration of flow) in a tube. Rigorous analysis of this problem in a cylindrical tube becomes quite complicated due to the necessity of using the cylindrical coordinate system.

Note that the shear stress is greatest at the tube wall. Therefore the rate of viscous dissipation per unit length of tube will be smaller when the less

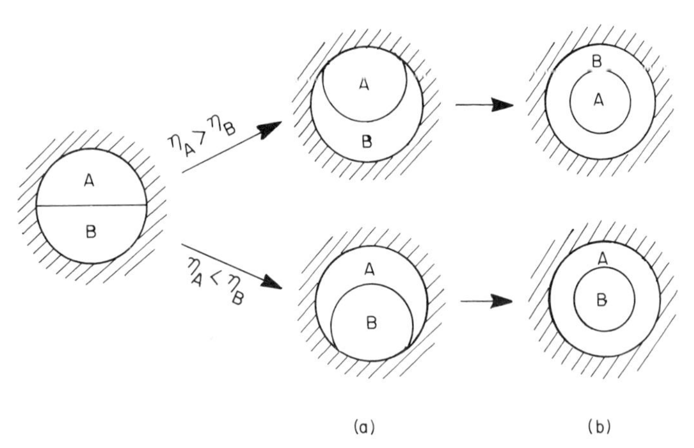

Fig. 7.56 Schematic showing the evolution of the change of interface shape in side-by-side coextrusion through circular dies: (a) a small L/D ratio; (b) a sufficiently large L/D ratio.

viscous component of the two wets the tube wall, surrounding the more viscous component staying in the core. This indeed has been the case as evidenced by the experimental results shown in Figs. 7.50 and 7.53. It can be concluded therefore that when two liquids are forced to flow side by side through a cylindrical tube, the more viscous component tends to convex at the interface and the less viscous component tends to preferentially wet the wall of the tube, eventually leading to the sheath-core configuration (i.e., complete encapsulation by the less viscous component) if sufficient time for flow is allowed (i.e., with a sufficiently long tube). Figure 7.56 gives a schematic describing the equilibrium shape of an interface in a sufficiently long cylindrical tube.

What role do the elastic properties of the two components involved play in determining the interface shape? There is some experimental evidence that when the viscosity ratio of the two components is the same, the more elastic component tends to wrap around the less elastic component [30], and that melt elasticity differences tend to cause rugged interface shapes [30, 35]. This problem will be dealt with theoretically in greater detail in Chapter 8.

7.7.2 Sheath-Core Coextrusion in a Cylindrical Tube

In this type of coextrusion, two polymers are fed separately into the feedblock and they meet at the die inlet, forming concentrical layers. Figure 7.57 gives a schematic describing the way two melt streams may be fed from separate extruders. In sheath-core coextrusion, the core component may

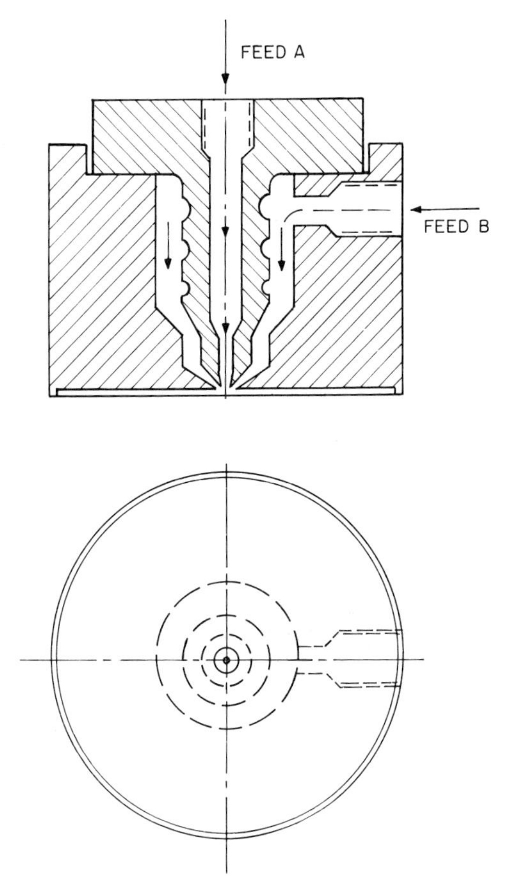

Fig. 7.57 Schematic of the coextrusion die assembly that may be used for concentric and/or eccentric flows through a circular die [31]. From C. D. Han, *J. Appl. Polym. Sci.* **19**, 1875. Copyright © 1975. Reprinted by permission of John Wiley & Sons, Inc.

contain an additive (e.g., a flame-retardant component, or a blowing agent which gives off gas bubbles) or the skin component may contain a reinforcing agent or a pigment. One can think of numerous applications of this type of coextrusion.

Pictures of extrudate cross section showing the shape of the interface are given in Fig. 7.58 for a polystyrene/high-density polyethylene (PS/HDPE) system. Note in these pictures that the two melts were fed *concentrically* to the die inlet. It is seen that when the PS was fed *inside* the other component, the interfacial shape was little affected whether the die was short (L/D = 4) or long (L/D = 18). However, when the PS was fed *outside* the other component, the long die (L/D = 18) tended to yield a slightly eccentric interface.

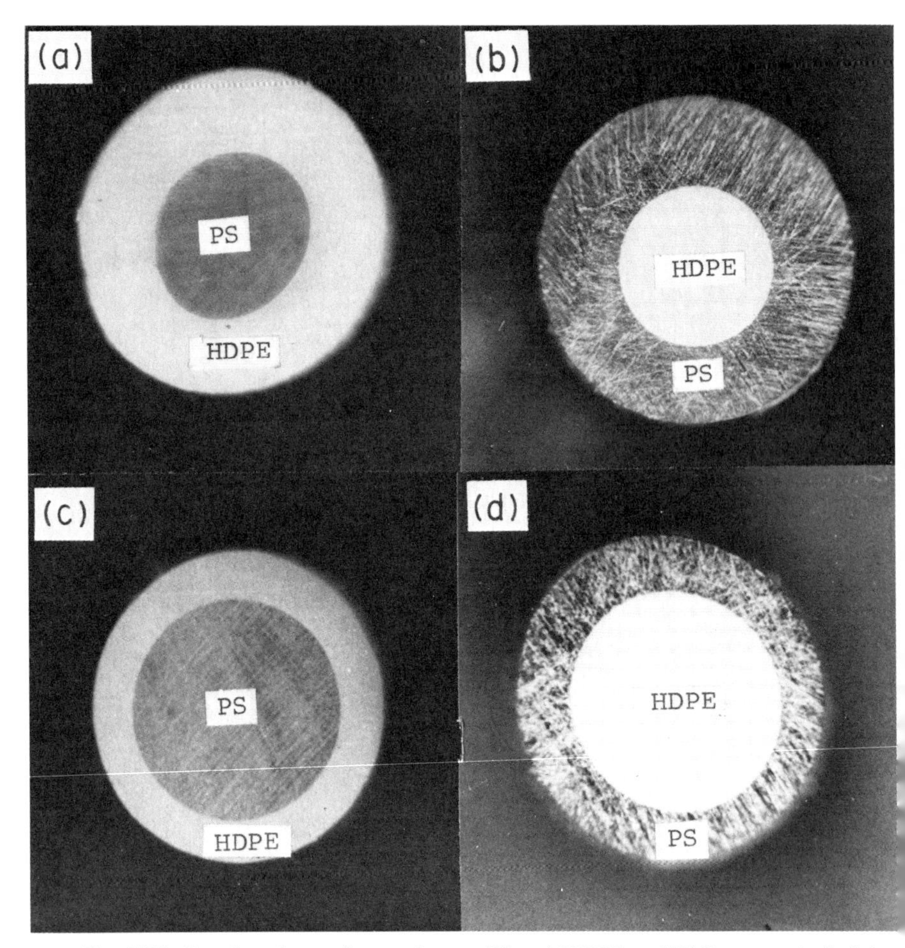

Fig. 7.58 Interface shape of two polymers (PS and HDPE) at 200°C, *concentrically* fed into a circular die [31]: (a) HDPE (sheath)/PS (core) with an L/D ratio of 4; (b) PS (sheath)/HDPE (core) with an L/D ratio of 4; (c) HDPE (sheath)/PS (core) with an L/D ratio of 18; (d) PS (sheath)/HDPE (core) with an L/D ratio of 18. From C. D. Han, *J. Appl. Polym. Sci.* **19**, 1875. Copyright © 1975. Reprinted by permission of John Wiley & Sons, Inc.

The difference between these instances may be attributable to the fact that the less viscous component tends to wrap around the more viscous component.

Perhaps a more dramatic observation, which may support the contention set forth above, can be seen in Fig. 7.59 for the HDPE/PS system. Note in these pictures that the two melts were fed *eccentrically* to the die inlet. It is seen that, when the PS (more viscous) was fed outside the other component

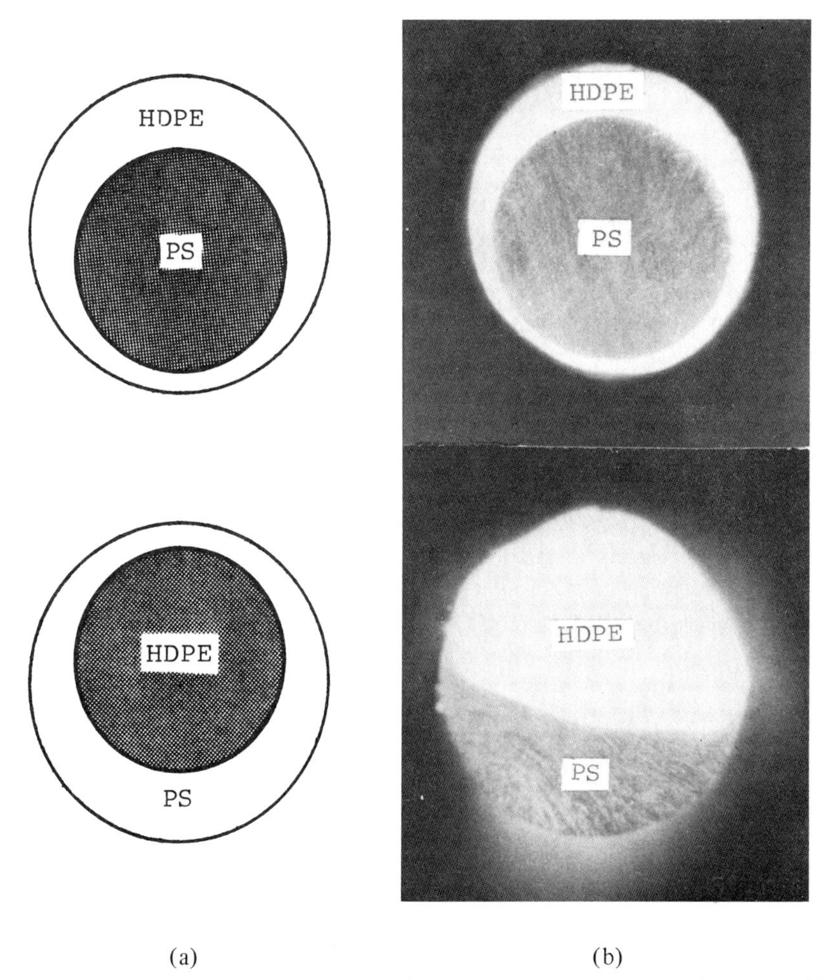

(a) (b)

Fig. 7.59 Interface shape of the two polymers (PS and HDPE) at 200°C, *eccentrically* fed into a circular die (L/D = 18) [31]: (a) interface at the die inlet; (b) interface in the extrudate. From C. D. Han, *J. Appl. Polym. Sci.* **19**, 1875, Copyright © 1975. Reprinted by permission of John Wiley & Sons, Inc.

(less viscous), the shape of the interface in the extrudate was completely different from that at the die inlet.

On the basis of the rheological properties of the two polymers, we have the following information:

$$\eta_{HDPE}/\eta_{PS} = 0.52\text{–}0.89$$
$$(\tau_{11} - \tau_{22})_{HDPE}/(\tau_{11} - \tau_{22})_{PS} = 1.59\text{–}1.39$$

Note that the component which is less viscous is more elastic than the other component.

It appears then that the interface shape is more strongly influenced by a difference between the viscosities of two components than by a difference between their elasticities. For instance, it is seen in Fig. 7.59 that, when the HDPE (less viscous) was fed *outside* the PS (more viscous), hardly any change is seen in the interface shape. This may be attributable to the fact that in the fully developed flow region, it is the viscosity, and not the elasticity, which governs the pressure drop across the capillary. On the other hand, when the HDPE was fed *inside* the PS, the less viscous component (i.e., HDPE) tended to wrap around the more viscous component, regardless of the substantial difference between the elasticities of the two components.

7.7.3 Coextrusion of Two Concentric Layers in an Annular Die

Coextrusion of two concentric layers in an annular die gives rise to layer rearrangement that is far more complex than seems possible at first thought. Consider the situation where two polymers are fed concentrically to the die inlet. Depending on the viscosity ratio of the two polymers, the layer configuration would look like the ones schematically shown in Fig. 7.60, when the residence time of the flow is sufficiently large. This speculation (or expectation) is based on the principle of minimum viscous dissipation, requiring that the less viscous fluid be in contact with both walls of the die, with the more viscous fluid pushed to the center of the annular space. In other words, in both two-layer blown-film coextrusion and two-layer wire coating coextrusion, interface migration will occur so as to minimize the energy of viscous dissipation.

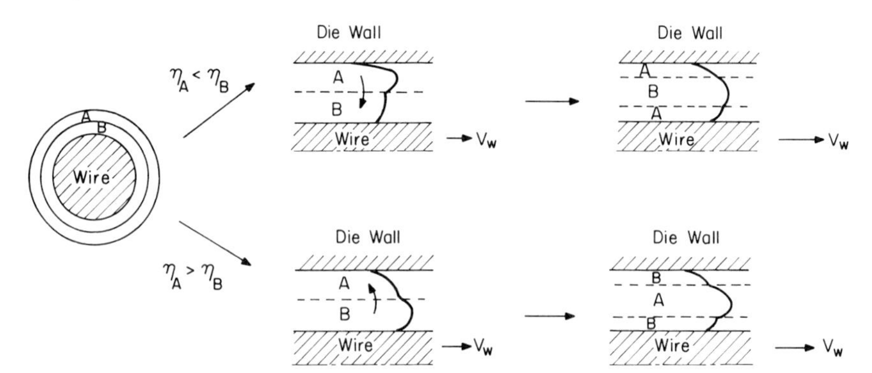

Fig. 7.60 Schematic showing the ultimate layer arrangement in long-duration flow when two polymers having different viscosities are coextruded through a wire coating coextrusion die [23].

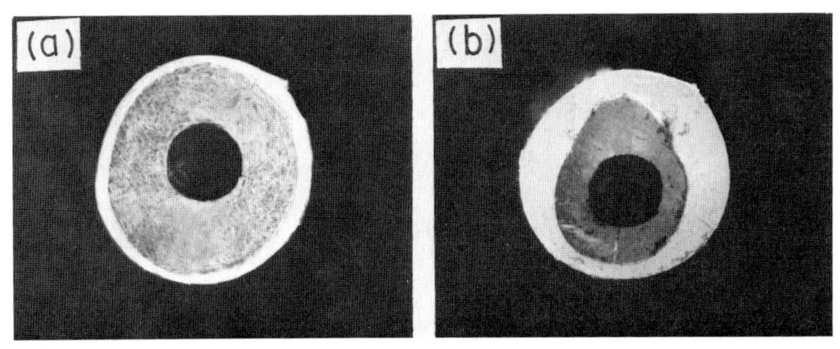

Fig. 7.61 Photographs of the cross sections of the coextruded wires with LDPE (outer layer)/STYRON 678 (inner layer) [23]: (a) $V_w = 18.58$ cm/sec, $Q_A/Q_B = 0.19$, $Q = 52.1$ cm^3/min. (b) $V_w = 8.33$ cm/sec, $Q_A/Q_B = 1.49$, $Q = 55.6$ cm^3/min.

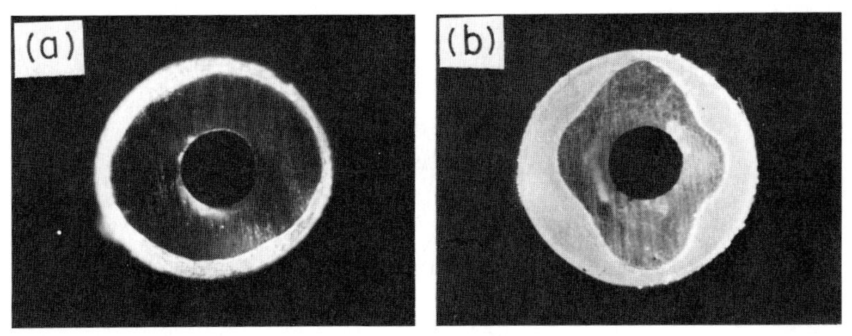

Fig. 7.62 Photographs of the cross sections of coextruded wires with KRATON (outer layer)/STYRON 686 (inner layer) [23]: (a) $V_w = 6.50$ cm/sec, $Q_A/Q_B = 0.29$, $Q = 44.4$ cm^3/min; (b) $V_w = 14.69$ cm/sec, $Q_A/Q_B = 1.12$, $Q = 120.5$ cm^3/min.

Figures 7.61 and 7.62 give photographs showing the interface shape of two layers in the cross section of coextruded two-layer wires. It should be noted that an increase in wire speed decreases the shear stress of the polymer at the wire surface, thus influencing the rate of interface movement. However, the basic feature of the layer rearrangement, schematically shown in Fig. 7.60, will remain the same, irrespective of the wire speed.

7.7.4 Side-by-Side Coextrusion in a Rectangular Duct

Flow visualization experiments of stratified two-phase flow in a rectangular duct would reveal the process by which one phase pushes into the other, that is, the evolution of the interface shape towards the steady-state configuration. Indeed, such experiments were carried out by Minagawa and White [32] and Khan and Han [34]. Figure 7.63 gives a schematic displaying

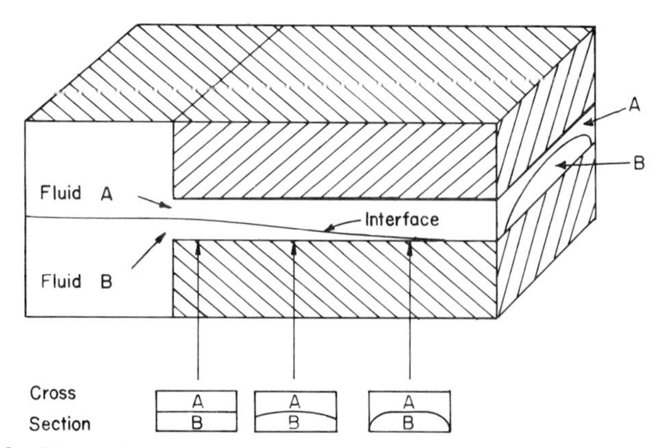

Fig. 7.63 Schematic showing the phase interface, which changes progressively as two polymer melts flow, side by side, through a rectangular channel [34]. From A. A. Khan and C. D. Han, *Trans. Soc. Rheol.* **20**, 595. Copyright © 1976. Reprinted by permission of John Wiley & Sons, Inc.

the evolution of the interface shape observed as the melts flow from the die extrance to the die exit. According to Han and co-workers [34, 39], when LDPE and PS were coextruded, the interface moved from its initial position on the LDPE side to its final position on the PS side, as viewed from the direction perpendicular to flow.

Figure 7.64 is a photograph of the cross section of a frozen extrudate sample of LDPE and PS. It is seen that the LDPE completely covers the short side of the rectangle, tending to wrap around the PS. Khan and Han [34] report that on the basis of the rheological properties of the polymers

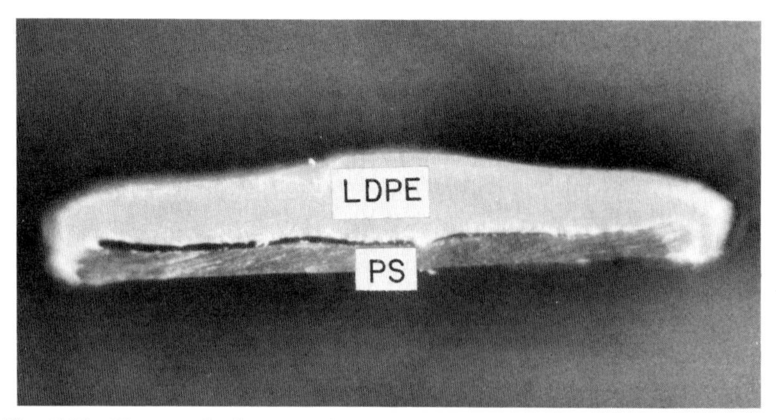

Fig. 7.64 Photograph of an extrudate cross section showing the phase interface of low-density polyethylene (LDPE) and polystyrene (PS), coextruded side by side in a rectangular die [34]. From A. A. Khan and C. D. Han, *Trans. Soc. Rheol.* **20**, 595. Copyright © 1976. Reprinted by permission of John Wiley & Sons, Inc.

they employed, the following relationships hold:

$$\eta_{\text{LDPE}}/\eta_{\text{PS}} = 0.08\text{–}0.21$$
$$(\tau_{11} - \tau_{22})_{\text{LDPE}}/(\tau_{11} - \tau_{22})_{\text{PS}} = 2.61\text{–}2.42$$

Note that for the LDPE/PS system, the component which is less viscous (LDPE) is more elastic than the other component (PS). It appears then that the less viscous component (LDPE in LDPE/PS system) tends to wrap around the more viscous component.

Assuming that the viscosity difference between the components being coextruded predominantly governs the equilibrium shape of the phase interface, we may speculate on the equilibrium interface shapes as schematically shown in Fig. 7.65, when a sufficiently large residence time for flow is provided, i.e., when a very long rectangular die is used.

Referring to Fig. 7.65, a finite die length is required to attain equilibrium flow (i.e., complete encapsulation) when two liquids having dissimilar rheological properties are forced to flow through a rectangular duct. Analysis of this problem (i.e., the evolution of the interface shape towards the steady-state configuration) requires consideration of the entrance region flow, which involves nonstationary boundaries due to interface motion. Rigorous analysis of the problem with regard to non-Newtonian viscoelastic fluids is

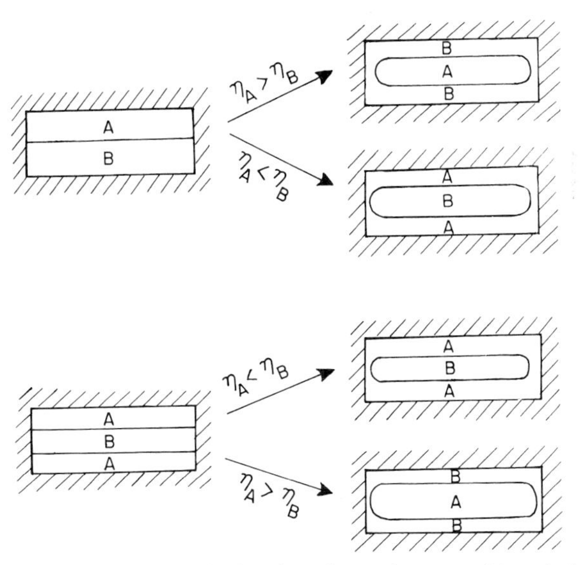

Fig. 7.65 Schematics showing the interface shapes that may ultimately be obtained in long-duration flow when two (or three) polymers having different viscosities are coextruded through a rectangular die.

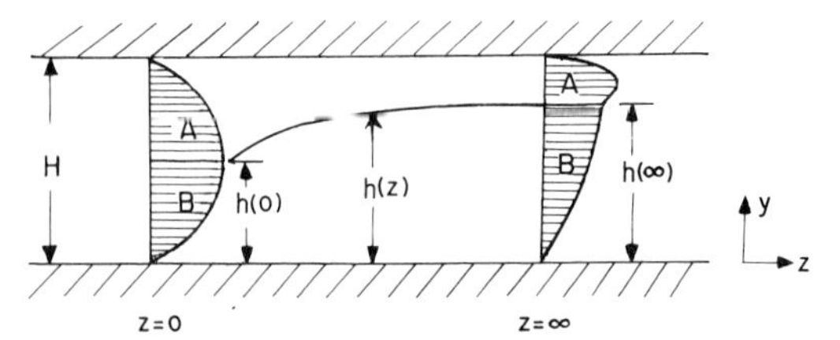

Fig. 7.66 Schematic showing the development of interface position along the axis when two polymers are coextruded through a rectangular channel [47]. From J. L. White and B. L. Lee, *Trans. Soc. Rheol.* **19**, 457. Copyright © 1975. Reprinted by permission of John Wiley & Sons, Inc.

extremely difficult, if not impossible, and thus, in the past, some approximate analyses have been reported with regard to Newtonian fluids flowing through parallel plates.

Postulating that when the pressure gradients and pressures in the two phases are unequal the lower viscosity phase encapsulates the higher viscosity phase, White and Lee [47] carried out an analysis of the entrance region flow of two Newtonian fluids flowing through long parallel plates, as schematically shown in Fig. 7.66. Using the simplifying assumptions that the interface velocity varies linearly with interface position, and that the pressure difference between the phases increases with displacement from equilibrium, they obtained the following expression for $h(z)$, the depth of

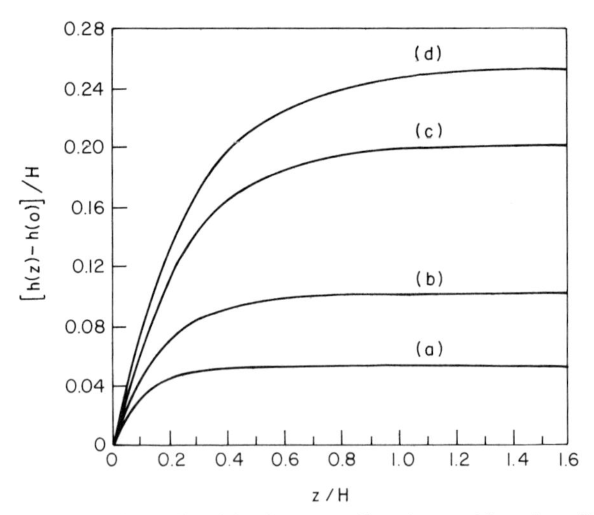

Fig. 7.67 Theoretically predicted development of interface position of stratified two-phase flow of Newtonian fluids through a rectangular channel, for various values of the viscosity ratio k [47]: (1) 3.13; (2) 8.12; (3) 42.18; (4) 96.83. From J. L. White and B. L. Lee, *Trans. Soc. Rheol.* **19**, 457, Copyright © 1975. Reprinted by permission of John Wiley & Sons, Inc.

fluid B at position z (see Fig. 7.66):

$$\frac{h(z)}{H} = \xi(\delta) = \frac{h(0)}{H} + \frac{h(\infty) - h(0)}{H}\left\{1 - \exp\left[-\left(\frac{8}{U_0 - U_\infty}\right)^{1/2}\delta\right]\right\} \quad (7.111)$$

where U_0 and U_∞ are the interface velocities at $z = 0$ and $z = \infty$, respectively, $\delta = z/H$, and $\xi(\infty)$ is to be obtained, for equal volumetric flow rates of the two phases (i.e., $Q_A = Q_B$), from the equation:

$$k^2[1 - \xi(\infty)]^4 + 4k\xi(\infty)[1 - \xi(\infty)][1 - 2\xi(\infty)] - [\xi(\infty)]^4 = 0 \quad (7.112)$$

in which k is the viscosity ratio of phase B to phase A, η_B/η_A.

Figure 7.67 gives the theoretically predicted interface positions of two Newtonian fluids flowing through parallel plates. It is seen that interface motion takes place rapidly, with the steady-state interface position attained within a plate length less than one plate separation (i.e., $h(z) \leq H$), and that the plate length required to attain equilibrium flow is predicted to increase with viscosity ratio.

The question still remains to be answered as to how important the fluid elasticities are, compared to the fluid viscosities, in deforming the interface between two viscoelastic fluids. Khan and Han [34] attempted to determine the initial direction of interface deformation in the stratified two-phase flow of viscoelastic fluids through a rectangular duct.

Consider the schematic of the flow geometry and the coordinate system given in Fig. 7.68. In order to determine the interface configuration perturbed from the initial flat position, a normal stress continuity condition will be used across the interface. This condition will permit us to determine the magnitude of the interface displacement due to the imbalance of normal stresses across it. In other words, if the normal stress across the interface

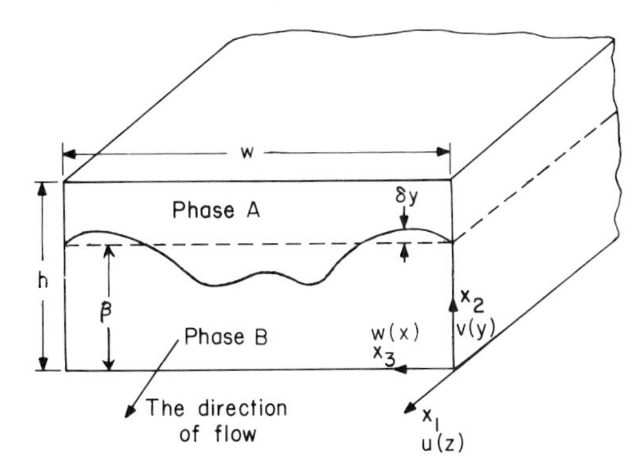

Fig. 7.68 Schematic of a rectangular channel in which deformation of interface occurs in stratified two-phase flow [34]. From A. A. Khan and C. D. Han, *Trans. Soc. Rheol.* **20**, 595. Copyright © 1976. Reprinted by permission of John Wiley & Sons, Inc.

is not balanced, the component having a greater normal stress will push into the other component.

If we assume that the surface tension forces are negligible compared to the viscous and elastic forces (this is justifiable for polymer melt flows), the difference in the total normal stress pointing in the y direction across the interface may be represented as

$$\Delta S_{yy}(\beta/h, x) = [(S_{yy})_A - (S_{yy})_B]_{y=\beta/h} \tag{7.113}$$

where $\Delta S_{yy}(\beta/h, x)$ is the difference in the total normal stress evaluated at the interface $y = \beta/h$, at the position x. If ΔS_{yy} is not zero, then the interface will move a distance δy in the y direction (see Fig. 7.68) such that the following equality is satisfied:

$$(\Delta S_{yy})_{y=\beta/h} + (\partial\, \Delta S_{yy}/\partial y)_{y=\beta/h}\, \delta y = 0 \tag{7.114}$$

or

$$\delta y = [-\Delta S_{yy}/(\partial\, \Delta S_{yy}/\partial y)]_{y=\beta/h} \tag{7.115}$$

By definition, the total normal stress S_{yy} can be represented by the sum of the hydrostatic pressure p and the deviatoric stress τ_{yy}:

$$S_{yy} = -p + \tau_{yy} \tag{7.116}$$

Hence, substituting Eq. (7.116) into Eq. (7.115) gives

$$\delta y = \left[\frac{-(p_A - p_B) + [(\tau_{yy})_A - (\tau_{yy})_B]}{(\partial p_A/\partial y) - (\partial p_B/\partial y) - \partial[(\tau_{yy})_A - (\tau_{yy})_B]/\partial y}\right]_{y=\beta/h} \tag{7.117}$$

Our task now is to evaluate the vertical displacement δy in terms of the rheological properties of the fluids under consideration.

For steady laminar flow through a rectangular channel, the equations of motion may be written

$$-\partial p/\partial z + \partial \tau_{zy}/\partial y + \partial \tau_{zx}/\partial x = 0 \tag{7.118}$$
$$-\partial p/\partial y + \partial \tau_{yy}/\partial y + \partial \tau_{yx}/\partial x = 0 \tag{7.119}$$
$$-\partial p/\partial x + \partial \tau_{xy}/\partial y + \partial \tau_{xx}/\partial x = 0 \tag{7.120}$$

where z, y, x are dimensionless coordinate variables defined by

$$z = x_1/h, \qquad y = x_2/h, \qquad x = x_3/h \tag{7.121}$$

in which h is the channel depth, and x_1, x_2, x_3 are rectangular Cartesian coordinates as shown in Fig. 7.68. In writing Eqs. (7.118)–(7.120), we have assumed that the inertia terms are negligible, which is justifiable for polymer melt flows, and that the gravitational forces are negligible, which is justifiable in commericial extrusion operation.

Now, at a fixed position in the z direction, Eq. (7.117) may be rewritten

$$\delta y = \left[\frac{-\int_0^x (\partial p_A/\partial x)\,dx + \int_0^x (\partial p_B/\partial x)\,dx + [(\tau_{yy})_A - (\tau_{yy})_B]}{(\partial \tau_{yx}/\partial x)_A - (\partial \tau_{yx}/\partial x)_B} \right]_{y=\beta/h} \tag{7.122}$$

in which Eq. (7.119) and the relationships

$$p_A = \int_0^x \left(\frac{\partial p_A}{\partial x} \right) dx; \qquad p_B = \int_0^x \left(\frac{\partial p_B}{\partial x} \right) \tag{7.123}$$

are used. Substitution of Eq. (7.120) into Eq. (7.122) to eliminate $\partial p/\partial x$ gives

$$\delta y = \left[\frac{-\int_0^x \{\partial[(\tau_{yx})_A - (\tau_{yx})_B]/\partial y\}\,dx + [(\tau_{yy} - \tau_{xx})_A - (\tau_{yy} - \tau_{xx})_B]}{\partial[(\tau_{yx})_A - (\tau_{yx})_B]/\partial x} \right]_{y=\beta/h} \tag{7.124}$$

It should be pointed out that the vertical displacement δy given by Eq. (7.124) is a first-order approximation [see Eq. (7.114)]. Therefore, any result based on such approximation is valid only when the displacement is very small compared to the width of the channel, i.e., $\delta y \ll h$ (see Fig. 7.68).

Using the Coleman–Noll second-order fluid model [see Eq. (2.53) in Chapter 2] and some simplifying assumptions made for the mathematical convenience, Khan and Han [34] carried out an analysis to obtain the displacement δy of the interface with the aid of Eq. (7.124).

Figure 7.69 shows the effect of fluid viscosity on interface deformation when elasticities of the two fluids are identical. It is seen that the higher viscosity component pushes into the other, that is, the less viscous component

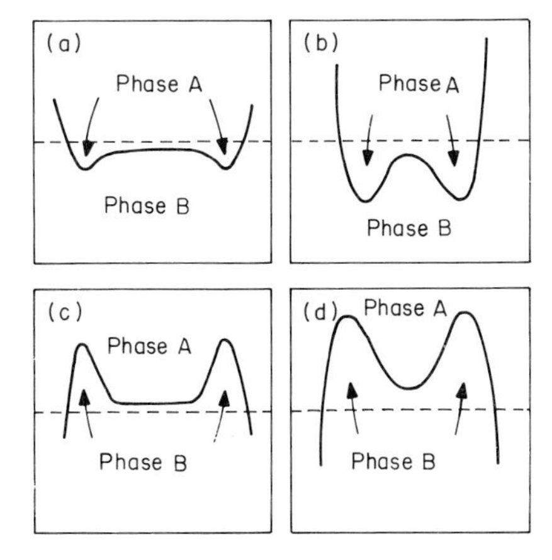

Fig. 7.69 Sketch of the interface shape in stratified flow, in a rectangular channel, of two second-order fluids which have the same elasticity but different viscosities, at various values of viscosity ratio, $k = (\eta_0)_B/(\eta_0)_A$ [34]: (a) 0.95; (b) 0.70; (c) 1.20; (d) 1.50. From A. A. Khan and C. D. Han, *Trans. Soc. Rheol.* **20**, 595. Copyright © 1976. Reprinted by permission of John Wiley & Sons, Inc.

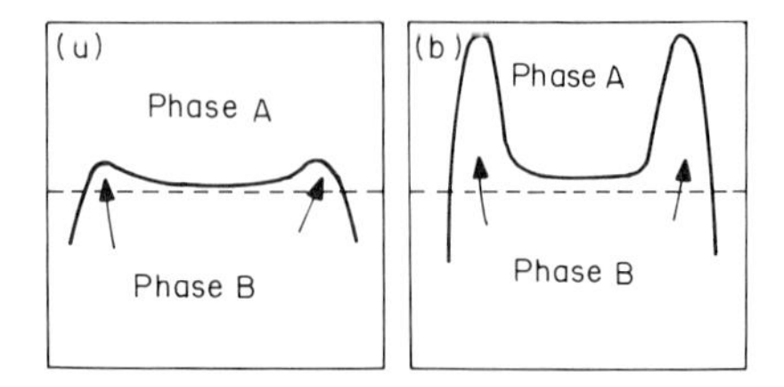

Fig. 7.70 Sketch of the interface shape in stratified flow, in a rectangular channel, of two second-order fluids which have the same viscosity but different elasticities, at various values of elasticity parameter ratio, ν_B/ν_A [34]: (a) 0.95; (b) 0.80. ν_k ($k = A, B$) is a material constant associated with the Coleman–Noll second-order fluid [see Eq. (2.53) in Chapter 2]. From A. A. Khan and C. D. Han, *Trans. Soc. Rheol.* **20**, 595. Copyright © 1976. Reprinted by permission of John Wiley & Sons, Inc.

tends to encapsulate the more viscous one. This is consistent with the experimental observation discussed above.

Figure 7.70 shows the effect of fluid elasticity on interface deformation when the viscosities of the two fluids are identical. Here, the less elastic component pushes into the more elastic one, i.e., the more elastic component tends to wrap around the less elastic one. This result seems to support the experimental observations of Southern and Ballman [30].

Figure 7.71 shows the interface deformation of two fluids, whose viscosities and elasticities differ from each other. In view of the fact that the interface shape is little affected by the differences in fluid elasticities in the two cases (a) and (b) in Fig. 7.71, one can conclude that viscosity plays a much stronger role than elasticity in determining the equilibrium shape of the interface between two fluids.

What role then do the elastic properties of the two components involved play in the stratified flow of two viscoelastic fluids? Some of the recent experimental studies [48, 49] report that one sometimes observes a very rugged (i.e., irregular) interface between the components under certain

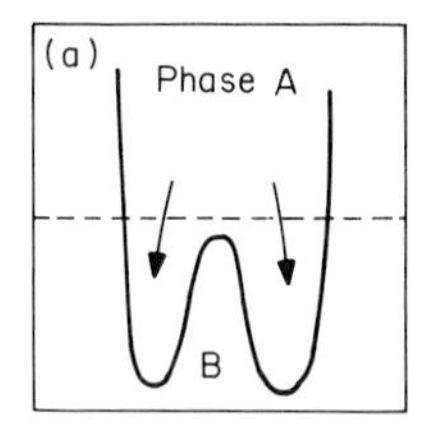

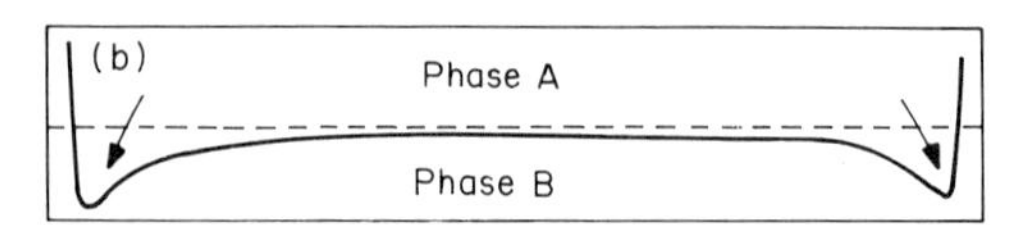

Fig. 7.71 Sketch of the interface shape in stratified flow, in a rectangular channel, of two second-order fluids which have different viscosities and elasticities [34]: (a) $k = 0.70$, $\nu_B/\nu_A = 0.70$; (b) $k = 0.80$, $\nu_B/\nu_A = 1.20$.

extrusion conditions and that the fluid elasticity plays an important role causing the rugged interface between the layers. In Chapter 8, we shall discuss interfacial instability that may occur in the stratified flow of viscoelastic fluids, as relevant to coextrusion processes.

It should be pointed out that the details of an analysis of rheologically complex flow problems, such as the one discussed above, often depends on the choice of a rheological model. Therefore the theoretical results presented above, which is based on the Coleman–Noll second-order fluid, must be interpreted with caution, because it is possible for one to reach different conclusions by using different rheological models.

Problems

7.1 For the three concentric layers of polymers A, B, and C being coextruded in a cylindrical tube, as schematically shown in Fig. 7.72, derive expressions for predicting the velocity distribution and volumetric flow rate for the following viscosity relationships:

(i) $\eta_A > \eta_B > \eta_C$
(ii) $\eta_A > \eta_C > \eta_B$
(iii) $\eta_B > \eta_A > \eta_C$

Assume that each polymer follows a power law in its shear stress-shear rate relationship.

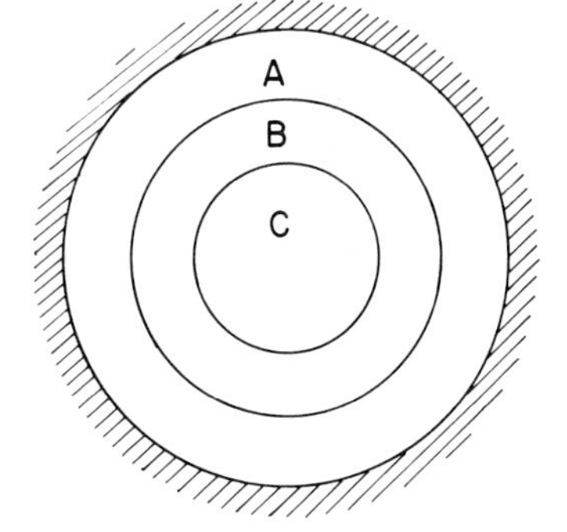

Fig. 7.72 Schematic illustrating three concentric layers of polymers being coextruded in a cylindrical tube.

7.2 For certain polymers, when the extrusion rate is high, viscous heating is appreciable. In such an instance, one must consider the temperature variation of the polymer melt during extrusion. Consider the three-layer coextrusion in a cylindrical tube, referred to in Problem 7.1 (see Fig. 7.72), and derive expressions for predicting the temperature and velocity distributions of the melts inside the tube for the following viscosity relationships:

(i) $\eta_A > \eta_B > \eta_C$; $E_A > E_B > E_C$
(ii) $\eta_A > \eta_C > \eta_B$; $E_A > E_C > E_B$
(iii) $\eta_B > \eta_A > \eta_C$; $E_B > E_A > E_C$

in which E_A, E_B, and E_C are the flow activation energies of polymers A, B, and C, respectively.

7.3 Consider three-layer isothermal coextrusion in a flat-film die, as shown schematically in Fig. 7.73. Assuming that each polymer follows a power law in its shear stress–shear rate relationship, derive expressions for predicting the velocity distribution and volumetric flow rate for the following viscosity relationships:

(i) $\eta_A > \eta_B > \eta_C$
(ii) $\eta_A > \eta_C > \eta_B$
(iii) $\eta_B > \eta_A > \eta_C$

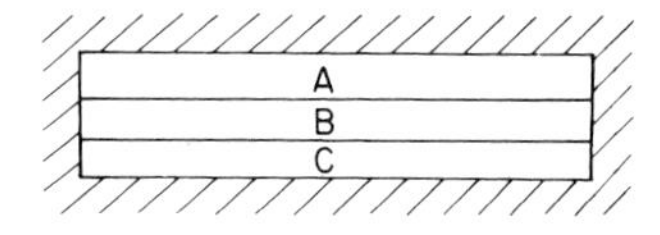

Fig. 7.73 Schematic illustrating three-layer flat film in a rectangular die.

7.4 Consider five-layer isothermal coextrusion in a flat-film die, schematically shown in Fig. 7.74. Assuming that each polymer follows a power law in its shear stress–shear rate relationship, derive expressions for predicting the velocity and shear stress distributions of the melts inside the die for the following viscosity relationships:

(i) $\eta_A > \eta_B > \eta_C$
(ii) $\eta_C > \eta_B > \eta_A$
(iii) $\eta_B > \eta_A > \eta_C$
(iv) $\eta_A > \eta_C > \eta_B$

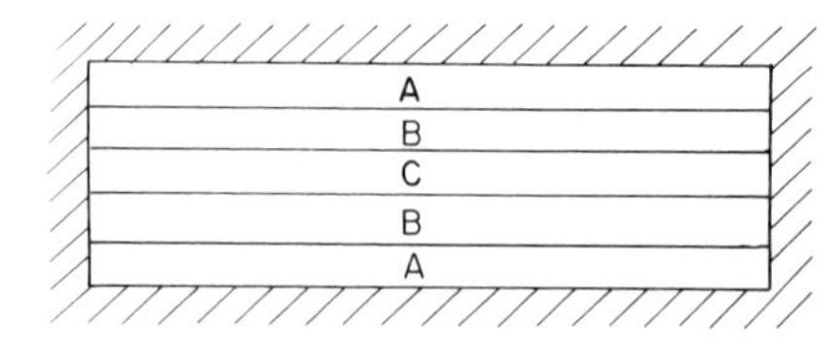

Fig. 7.74 Schematic illustrating five-layer flat film in a rectangular die.

7.5 In three-layer flat-film coextrusion with the layer arrangements as shown schematically in Fig. 7.75, the experimentally determined pressure gradients are shown schematically in Fig. 7.76. If we increase the number of layers to approximately 100, with the same layer thick-

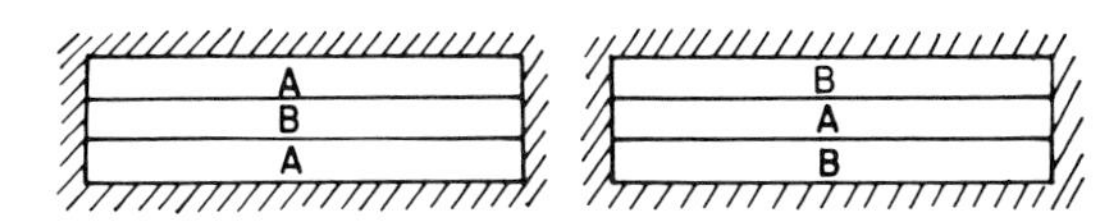

Fig. 7.75 Schematic illustrating three-layer flat films in a rectangular die.

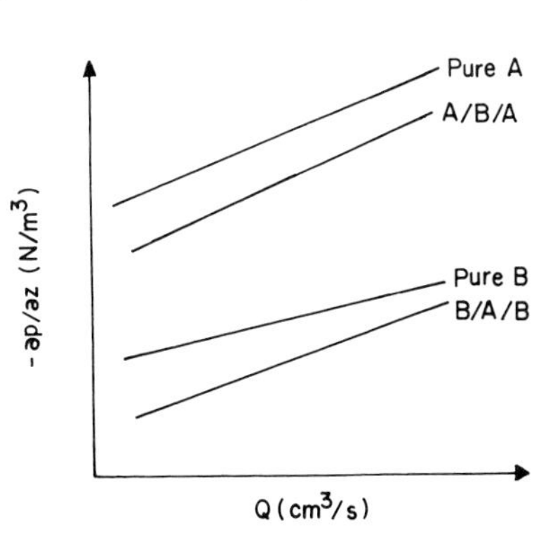

Fig. 7.76 Schematic illustrating pressure gradient versus volumetric flow rate in the three-layer coextrusion through a rectangular die.

ness in each layer, what do you think the pressure gradient will look like? Draw your estimated pressure gradient on the sketch given in Fig. 7.76, and give your reason for the answer. Assume that the two polymers (A and B) have the same volumetric flow rate and hence an equal number of layers.

7.6 White and Lee [47] considered the interface movement of two Newtonian fluids flowing through two parallel plates, as shown schematically in Fig. 7.66. You are now asked to consider the following rheological models:

(i) The power-law fluid,
(ii) The Zaremba–DeWitt model,
(iii) The Coleman–Noll second-order fluid.

Using the boundary conditions (in reference to Figure 7.66) given by:

(i) $v_{z,A}(H, z) = 0; \qquad v_{z,B}(0, z) = 0$
(ii) $v_{z,A}(y, z) = v_{z,B}(y, z) = U$

in which U is the velocity at the interface, derive the expressions for predicting:

(a) the velocity profiles $v_{z,A}(z)$ and $v_{z,B}(z)$;
(b) the volumetric flow rates, Q_A for phase A, and Q_B for phase B;
(c) the interfacial velocity U in terms of the flow rates Q_A and Q_B (you may assume that the shear stress is continuous at the interface);

(d) the expression which will allow one to determine the interface position $y(z)$, assuming that the interface moves when the pressure gradients and pressures in the two phases are unequal during flow.

7.7 Consider the two-layer wire coating coextrusion process in which the rheological properties of the two polymers, A and B, are such, that the maximum in velocity occurs in phase B, the phase in contact with the wire surface, as given schematically in Fig. 7.77. Assuming that a power-law model describes the rheological behavior of the individual polymers, A, and B, derive expressions for the velocity profiles and the volumetric flow rates of phases A and B in the coextrusion die. Assume that no interface rearrangement occurs in the die during coextrusion.

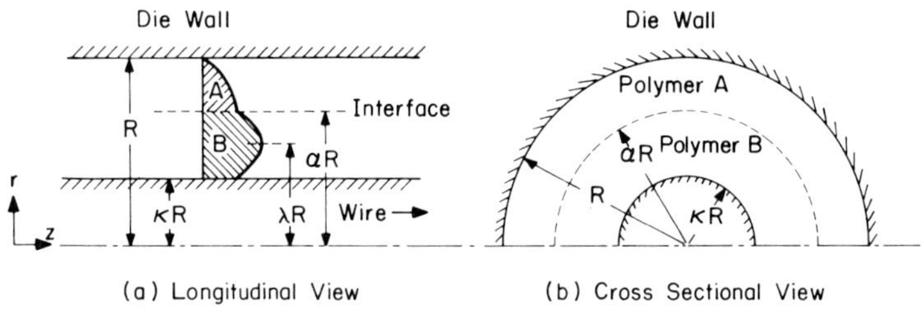

Fig. 7.77 Schematic showing the velocity profile in the two-layer wire coating coextrusion.

7.8 Consider the three-layer blown-film coextrusion process in which the three different polymers form the layer arrangement shown schematically in Fig. 7.78. Assuming that there is no interface rearrangement while the polymers are coextruded in the die, derive expressions for the velocity profiles and the volumetric flow rates of phases A, B, and

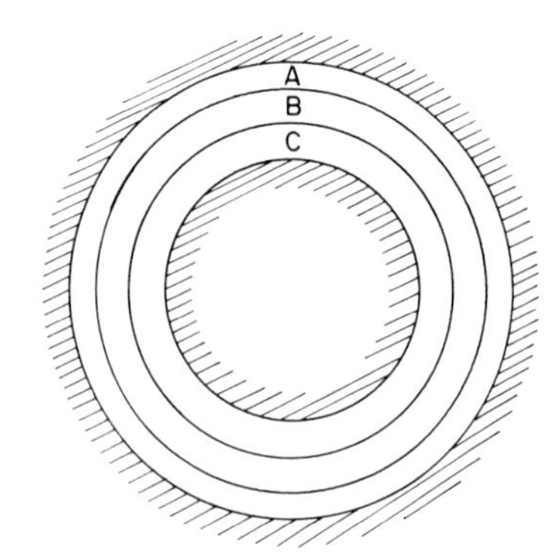

Fig. 7.78 Schematic showing three concentric layers of polymers being coextruded in an annular die.

C for the following cases:

(a) When the maximum in velocity occurs in phase A.
(b) When the maximum in velocity occurs in phase C.

You may assume that a power-law model describes the rheological behavior of the individual polymers.

7.9 In reference to Fig. 7.78 for Problem 7.8, consider now that the length of the die is sufficiently long to cause layer rearrangement during coextrusion. Predict the ultimate layer arrangement for the following viscosity relationships:

(a) $\eta_A = \eta_C; \quad \eta_A, \eta_C > \eta_B$
(b) $\eta_A > \eta_B > \eta_C$
(c) $\eta_A < \eta_B < \eta_C$

REFERENCES

[1] W. A. Sisson and F. F. Morehead, *Text. Res. J.* **23**, 152 (1953).
[2] E. M. Hicks, J. F. Ryan, R. B. Taylor, and R. L. Tichenor, *Text. Res. J.* **30**, 675 (1960).
[3] E. M. Hicks, E. A. Tippets, J. V. Hewett, and R. H. Brand, *in* "Man-Made Fibers" (H. Mark, S. M. Atlas, and E. Cernia, eds.), Vol. 1, p. 375. Wiley, New York, 1967.
[4] R. A. Buckley and R. J. Phillips, *Chem. Eng. Prog.* **66**(10), 41 (1969).
[5] W. A. Sisson, U.S. Patent 2,443,771 (1948).
[6] T. F. Ryan and R. L. Tichenor, U. S. Patent 2,988,420 (1961).
[7] T. Nagano, Jpn. Patent 48-17492 (1973).
[8] T. W. F. Russel, G. W. Hodgson, and G. W. Govier, *Can. J. Chem. Eng.* **37**, 9 (1959).
[9] T. W. F. Russel and M. E. Charles, *Can. J. Chem. Eng.* **37**, 18 (1959).
[10] M. E. Charles, G. W. Govier, and G. W. Hodgson, *Can. J. Chem. Eng.* **39**, 27 (1961).
[11] M. E. Charles and P. J. Redberger, *Can. J. Chem. Eng.* **40**, 70 (1962).
[12] M. E. Charles and L. U. Lilleleht, *J. Fluid Mech.* **22**, 217 (1965).
[13] M. E. Charles and L. U. Lilleleht, *Can. J. Chem. Eng.* **43**, 110 (1965).
[14] A. R. Gemmel and N. Epstein, *Can. J. Chem. Eng.* **40**, 215 (1962).
[15] W. J. Schrenk, K. J. Cleereman, and T. Alfrey, *SPE Trans.* **19**, 192 (1963).
[16] W. J. Schrenk and T. Alfrey, *SPE J.* **29**, 38 (1973).
[17] W. J. Schrenk, *Plast. Eng.* **1**, 65 (1974).
[18] H. O. Corbett, U.S. Patents 3,320,636 (1967); 3,398,431 (1968).
[19] P. H. Squires, U.S. Patent 3,476,627 (1969).
[20] D. Chisholm and W. J. Schrenk, U.S. Patent 3,557,265 (1971).
[21] D. F. Wiley, U.S. Patents 3,769,380 (1973); 3,882,219 (1975); 3,900,548 (1975).
[22] F. R. Nissel, U.S. Patents 3,919,865 (1975); 3,940,221 (1976); 3,959,431 (1976).
[23] C. D. Han and D. A. Rao, *Polym. Eng. Sci.* **20**, 128 (1980).
[24] D. L. MacLean, *Trans. Soc. Rheol.* **17**, 385 (1973).
[25] A. E. Everage, *Trans. Soc. Rheol.* **17**, 629 (1973).
[26] M. C. Williams, *AIChE J.* **21**, 1204 (1975).
[27] J. H. Southern and R. L. Ballman, *Appl. Polym. Symp.* No 20, 1234 (1973).
[28] C. D. Han, *J. Appl. Polym. Sci.* **17**, 1289 (1973).

[29] B. L. Lee and J. L. White, *Trans. Soc. Rheol.* **18**, 467 (1974).
[30] J. H. Southern and R. L. Ballman, *J. Polym. Sci., Part A-2* **13**, 863 (1975).
[31] C. D. Han, *J. Appl. Polym. Sci.* **19**, 1875 (1975); **20**, 2609 (1976).
[32] N. Minagawa and J. L. White, *Polym. Eng. Sci.* **15**, 825 (1975).
[33] J. L. White, R. C. Ufford, K. R. Dharod, and R. L. Price, *J. Appl. Polym. Sci.* **16**, 1313 (1972).
[34] A. A. Khan and C. D. Han, *Trans Soc. Rheol.* **20**, 595 (1976).
[35] A. A. Khan and C. D. Han, *Trans Soc. Rheol.* **21**, 101 (1977).
[36] E. Uhland, *Polym. Eng. Sci.* **17**, 671 (1977).
[37] M. Mooney, *J. Rheol.* **2**, 210 (1931).
[38] M. Natov and E. Djagarowa, *Macromol. Chem.* **100**, 126 (1967).
[39] T. C. Yu and C. D. Han, *J. Appl. Polym. Sci.* **17**, 1203 (1973).
[40] A. Y. Malkin, M. L. Friedman, K. D. Vachaghin, and G. V. Vinogradov, *J. Appl. Polym. Sci.* **19**, 375 (1975).
[41] C. D. Han and R. Shetty, *Polym. Eng. Sci.* **16**, 697 (1976).
[42] C. D. Han and D. A. Rao, *Polym. Eng. Sci.* **18**, 1019 (1978).
[43] W. J. Schrenk, K. J. Cleereman, and T. Alfrey, Jr., *SPE Trans.* **3**, 192 (1963).
[44] W. J. Schrenk and T. Alfrey, Jr., *SPE J.* **29**, 38 (1973).
[45] C. D. Han and R. Shetty, *Polym. Eng. Sci.* **18**, 187 (1978).
[46] C. D. Han and H. B. Chin, *Polym. Eng. Sci.* **19**, 1156 (1979).
[47] J. L. White and B. L. Lee, *Trans. Soc. Rheol.* **19**, 457 (1975).
[48] C. D. Han and R. Shetty, *Polym. Eng. Sci.* **18**, 180 (1978).
[49] W. J. Schrenk, N. L. Bradley, T. Alfrey, and H. Maack, *Polym. Eng. Sci.* **18**, 620 (1978).

8

Interfacial Instability in Stratified Multiphase Flow

8.1 INTRODUCTION

The manufacture of multiphase (or multicomponent) plastic constructions (e.g., multilayer films, multilayer coatings of wires and cables) is not without problems. Dissimilar materials seem inevitably to possess different rheological properties and often "in the wrong direction." The coextrusion processes for packaging film, blown containers, and insulated wire and cable are currently hampered by the occurrence of irregular interfaces. This leads to high scrap rates and ratios of layer thicknesses that are not economical, and/or preclude the use of materials, which have highly desirable properties, but are not readily processable. Figure 8.1 gives a schematic displaying the irregular interfaces between the layers that may occur in flat-film coextrusion, blown-film coextrusion, and wire coating coextrusion.

When two or more polymers are coextruded, it is important to produce smooth interfaces between the layers, because an irregular interface is detrimental to the quality of the product, as determined by its mechanical and/or optical properties [1,2]. As an example, Fig. 8.2 displays a photograph showing irregularities in the interface of a three-layer sheet coextruded in a commercial production unit, using two polymers of the same molecular structure, but of different molecular weights; the low molecular weight polymer layer being sandwiched between layers of the high molecular weight one. The photograph was taken looking through the coextruded sheet.

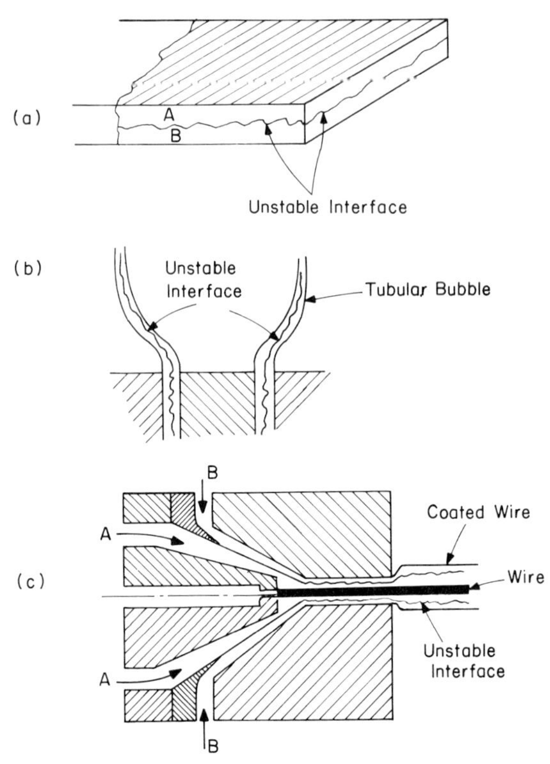

Fig. 8.1 Schematic describing the irregular interfaces that may occur in coextrusion through: (a) a rectangular die; (b) tubular film die; (c) pressure-type wire coating die.

Because of the transparency of the particular polymers used, we were able to photograph the irregular interfaces between the layers.

In order to help one understand the photograph, a schematic of the flow geometry used for the coextrusion is given in Fig. 8.3. In reference to Fig. 8.3, the photograph in Fig. 8.2 represents the z-x plane. Note that the irregular patterns in Fig. 8.2 were present at the interface between the layers and that the two outer surfaces were smooth.

In the past, the occurrence of irregular interfaces was reported by Southern and Ballman [3] and Lee and White [4], who carried out co-extrusion experiments using a *circular* die, by Schrenk *et al.* [2] and Han and Shetty [1], who carried out flat-film coextrusion experiments, and by Han and Rao [5] who conducted experiments of wire coating coextrusion. It is worth pointing out that *irregular* interfaces, perhaps better represented by the schematic given in Fig. 8.3, should *not* be confused with the movement of *regular* interfaces, discussed in Chapter 7, which can result from purely

Fig. 8.2 Photograph showing irregular interfaces of the three-layer coextruded sheet of polyvinyl chloride [1].

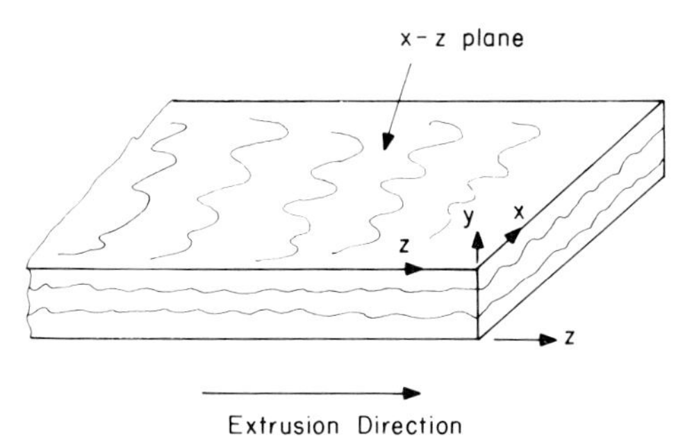

Fig. 8.3 Schematic showing the die geometry used in producing the coextruded sheet given in Fig. 8.2 [1].

viscous effects, interpreted in terms of the concept of minimum energy dissipation [6, 7]. It may be conjectured, therefore, that *irregular* interface shapes, such as those shown in Fig. 8.2, must have an origin quite different from that giving rise to *regular* interface shapes. We shall refer to the rugged or irregular interface as *interfacial instability*.

Interfacial instability in the stratified flow of two Newtonian fluids has been reported by earlier investigators [8, 9] who used an oil-and-water system in a rectangular duct. They concluded that viscosity stratification was a major cause of the interfacial instability observed. Using hydrodynamic stability analysis, Yih [10] investigated the effects of variations in the viscosity and density ratios of two Newtonian fluids on interfacial instability in plane Poiseuille and plane Couette flows. Hickox [11] also carried out a hydrodynamic analysis of interfacial instability in the axisymmetric, laminar flow of two Newtonian fluids, flowing concentrically in a straight cylindrical tube. Both Yih and Hickox concluded that even for low Reynolds numbers, interfacial instability can occur by viscosity stratification. Using a modified Oldroyd model, Li [12, 13] extended the Yih analysis to study interfacial instability in plane Poiseuille flow when viewed from the direction perpendicular to the flow. He showed that the fluid interface can be either regular or irregular, depending on the values of the viscosity ratio, elasticity ratio, density ratio, and depth ratio of the two fluids.

Using the Coleman–Noll second-order fluid as a rheological model, Khan and Han [14] have carried out a hydrodynamic stability analysis of interfacial instability in two-layer flat-film coextrusion, and have reported that interfacial instability is influenced by both the viscosity and elasticity ratios of the two fluids being coextruded, and the thickness ratio of the two layers.

In this chapter we shall first point out the importance of the rheological/physical variables that govern the occurrence of interfacial instability in coextrusion processes. Then we shall provide an operating guide that will help avoid the occurrence of interfacial instability, when proper precautions are taken in the selection of polymeric systems, on the basis of their rheological properties, and in the selection of optimal processing conditions.

8.2 EXPERIMENTAL OBSERVATIONS OF INTERFACIAL INSTABILITY IN COEXTRUSION

Our objective here is to present *experimental* evidence to correlate the processing conditions, at which interfacial instability begins, to the rheological properties of the individual polymers being coextruded. When the problem to be analyzed is very complex, such as the one at hand, it is revealing to first present experimental observations, albeit obtained over a limited range of processing conditions, and then to discuss theoretical analyses that explain the experimental results in more general terms.

In presenting experimental results, we shall consider three coextrusion processes separately, namely, flat-film coextrusion, wire coating coextrusion, and blown-film coextrusion. It will be shown that, irrespective of the type of coextrusion process under consideration, common rheological factors control the occurrence of interfacial instability.

8.2.1 Interfacial Instability in Flat-Film Coextrusion

Figure 8.4 displays photographs showing the irregular nature of the interface in three-layer flat-film samples collected at different extrusion conditions. The photographs show about 10 cm of the flat side of the film, in which the low-density polyethylene (LDPE), forming the core component, is sandwiched between the polystyrene (PS), forming the skin component, i.e., the system is PS/LDPE/PS. Table 8.1 gives the processing conditions used for generating the coextruded films shown in Fig. 8.4.

It is seen in Fig. 8.4 that as the skin-to-core thickness ratio is decreased, the severity of interfacial instability is increased. On the other hand, Table 8.1 shows that the shear stress at the die wall for Fig. 8.4c is lower than that for Figs. 8.4a and 8.4b, whereas the shear stress at the interface (between the PS and LDPE layers) is progressively increased from Fig. 8.4a to Fig. 8.4c. This observation appears to indicate that the severity of interfacial instability in coextrusion is directly correlatable to the shear stress at the interface.

It should be pointed out that the outer surface of the film shown in Fig. 8.4 is smooth, and therefore the experimentally observed irregular interfaces should *not* be confused with the rather well-known phenomenon of flow instability, or "melt fracture." Melt fracture refers to the flow phenomenon which yields extrudate with rough or twisted surface, and sometimes, surfaces which are very severely distorted. The phenomenon of melt fracture and its rheological interpretations is extensively discussed in a review article [15] and a monograph [16].

Han and Shetty [1] report that when the PS was sandwiched between layers of LDPE (i.e., the system is LDPE/PS/LDPE), an irregular interface was *not* observed, even at a flow rate greater than that at which the PS/LDPE/PS system gave rise to irregular interfaces. This observation indicates clearly that the occurrence of interfacial instability is not a function of the total volumetric flow rate of the combined stream alone, but also depends on which of the two polymers is the skin component, and which is the core component.

Plots of wall shear stress τ_w versus Q, the total volumetric flow rate of the combined streams, are given in Fig. 8.5 for the three-layer PS/LDPE/PS system. Note that in their study, Han and Shetty [1] determined τ_w from

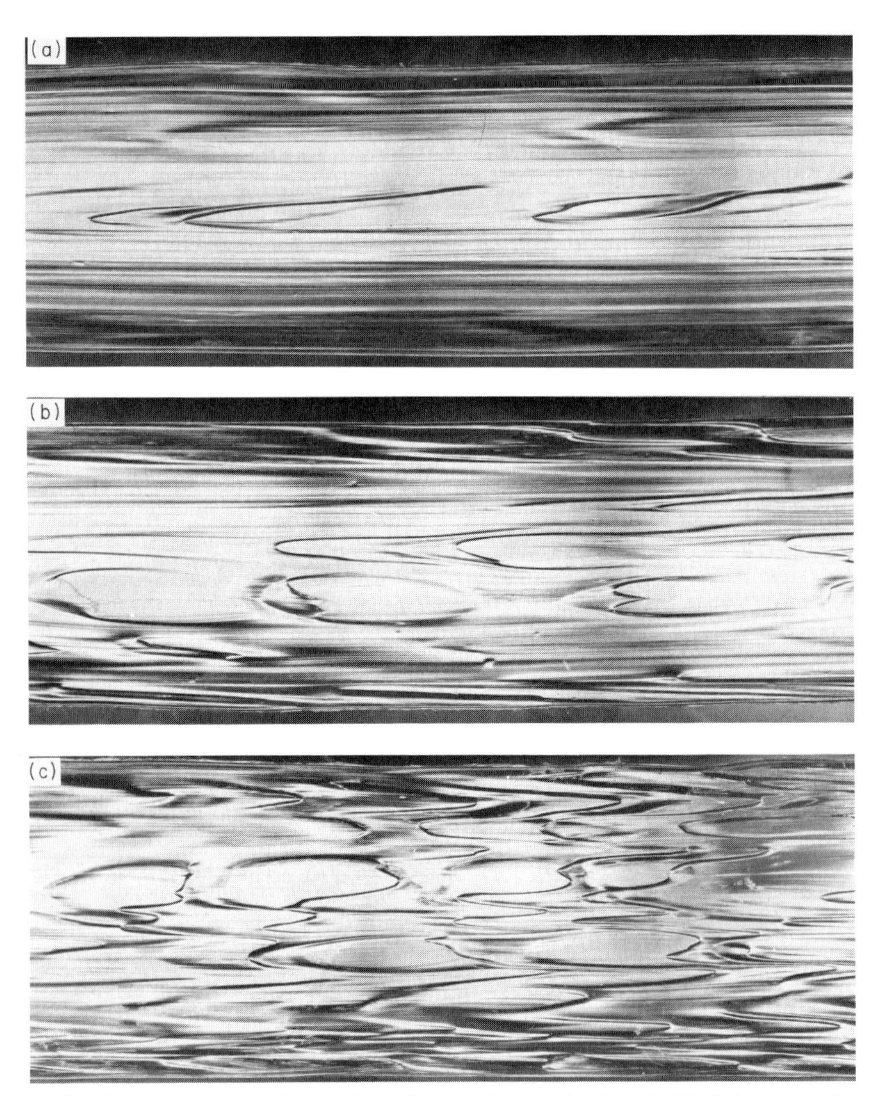

Fig. 8.4 Photographs showing irregular interfaces of the PS/LDPE/PS three-layer flat films coextruded at 200°C, at various values of the skin-to-core thickness ratio, h_{PS}/h_{LDPE} [1]: (a) 0.93; (b) 0.54; (c) 0.38.

TABLE 8.1

Processing Conditions That Produced the Coextruded Films (PS/LDPE/PS) Shown in Fig. 8.4

Photograph	Q_{PS} (cm^3/min)	Q_{LDPE} (cm^3/min)	$\tau_w \times 10^{-4}$ (N/m^2)	$(\tau_w)_{Int} \times 10^{-4}$ (N/m^2)	$\frac{h_{PS}}{h_{LDPE}}$
a	18.30	9.79	2.514	0.876	0.93
b	18.30	16.99	2.730	1.314	0.54
c	12.67	16.99	2.356	1.333	0.38

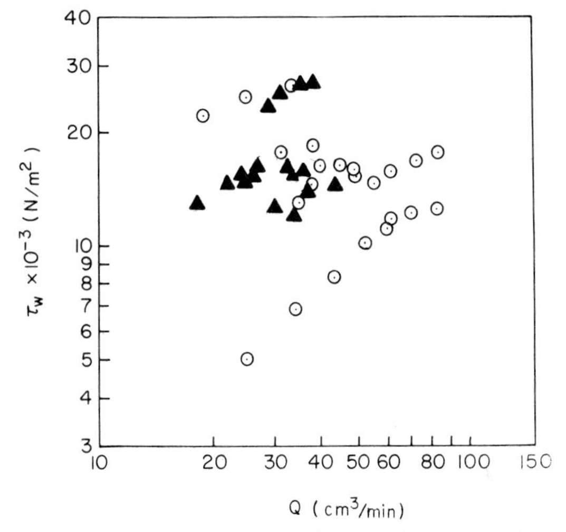

Fig. 8.5 Wall shear stress versus volumetric flow rate in the three-layer flat-film coextrusion of PS/LDPE/PS system [1]: (⊙) stable interface; (▲) unstable interface.

wall normal stress measurements. It is seen in Fig. 8.5 that such plots do *not* shed any light on determining the critical condition for the onset of interfacial instability.

Figure 8.6 gives plots of wall shear stress τ_w versus the ratio of layer thickness h_{PS}/h_{LDPE} for the polymer systems investigated. It is seen that a clear trend exists which separates the stable region from the unstable region, and that critical values of wall shear stress for an onset of interfacial instability increase with h_{PS}/h_{LDPE}. In other words, the ratio of layer thickness, h_{PS}/h_{LDPE}, plays an important role in the occurrence of interfacial instability.

On the basis of Fig. 8.6, a few practical means of avoiding irregular interfaces may be suggested. One is to increase the melt extrusion temperature, which in turn will reduce the viscosities of the melt streams, and hence the

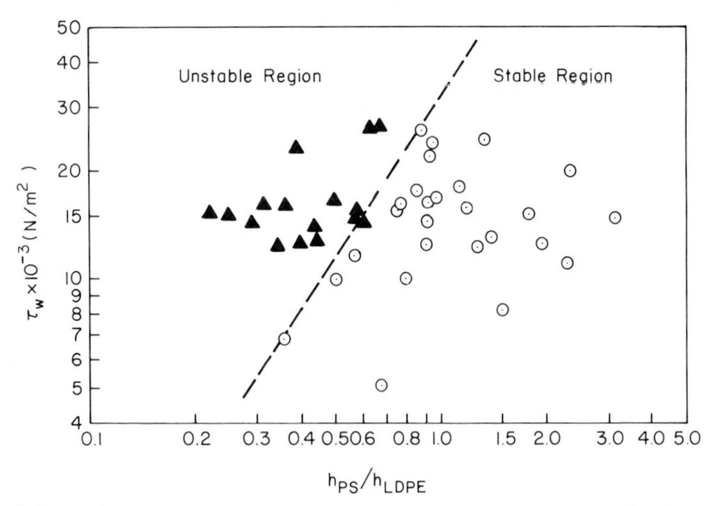

Fig. 8.6 Wall shear stress versus layer thickness ratio in the three-layer flat-film coextrusion of PS/LDPE/PS system [1]: (⊙) stable interface; (▲) unstable interface.

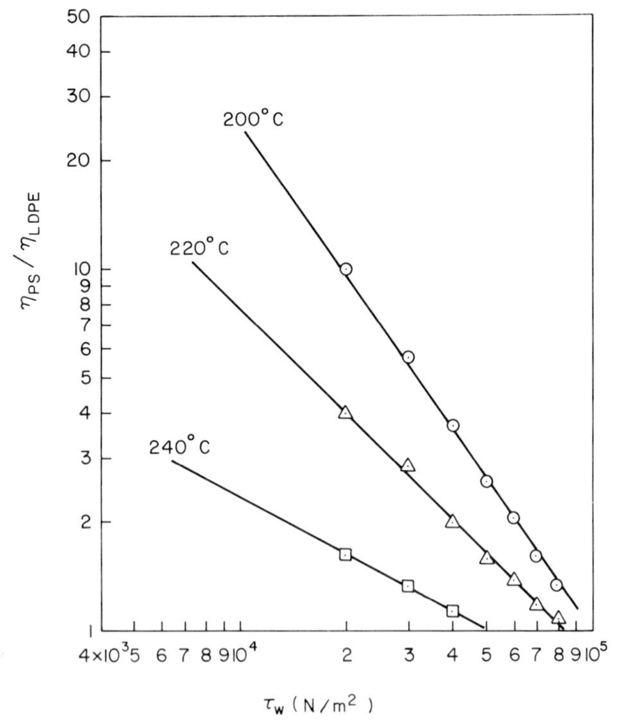

Fig. 8.7 Viscosity ratio of polystyrene (PS) to low-density polyethylene (LDPE) versus wall shear stress at different melt temperatures [1].

wall shear stress. Another means of avoiding irregular interface is to change the ratio of layer thickness by regulating the melt flow rate of the respective streams. In practice, however, the layer thickness may be mandated by a specific product specification and therefore one may not be completely free to change it.

The sensitivity of melt viscosity to temperature often varies from material to material. In the coextrusion operation, therefore, the viscosity ratio of the two polymers concerned is of fundamental and practical interest. Figure 8.7 gives plots of the viscosity ratio of PS to LDPE versus wall shear stress τ_w at 200, 220, and 240°C.

Figure 8.8 gives plots of viscosity ratio η_{PS}/η_{LDPE} versus layer thickness ratio h_{PS}/h_{LDPE} for the PS/LDPE/PS system. The plots are constructed using Figs. 8.6 and 8.7. It is seen that interfacial instability will occur only over certain ranges of η_{PS}/η_{LDPE} and h_{PS}/h_{LDPE}. For instance, when η_{PS}/η_{LDPE} is less than 4, the interface will be stable, *almost* independent of the values of h_{PS}/h_{LDPE}. Note in Fig. 8.7 that when the LDPE forms the skin component, the viscosity ratios of η_{LDPE}/η_{PS} are less than 4.0 over the range of τ_w investigated, $10^4 < \tau_w < 10^5$ N/m^2. Han and Shetty [1] report that the LDPE/PS/

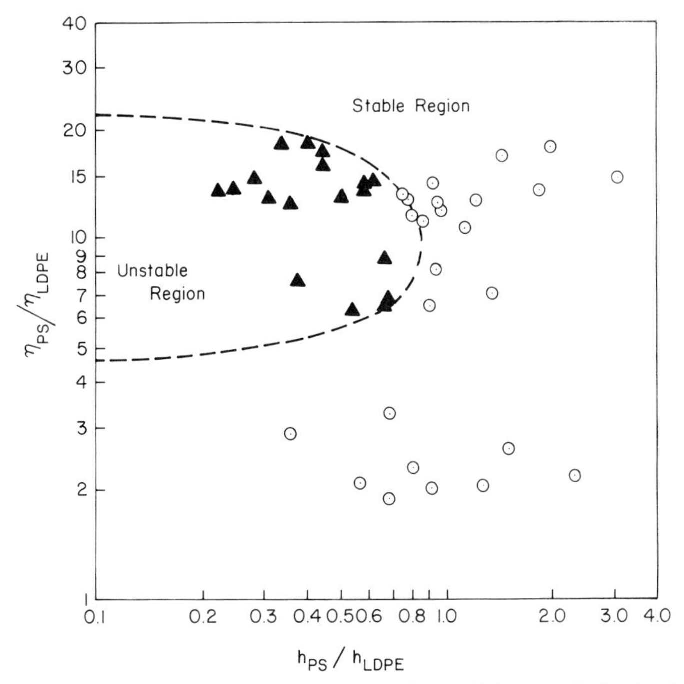

Fig. 8.8 Viscosity ratio of PS to LDPE versus layer thickness ratio in the three-layer flat-film coextrusion of PS/LDPE/PS system [1]: (⊙) stable interface; (▲) unstable interface.

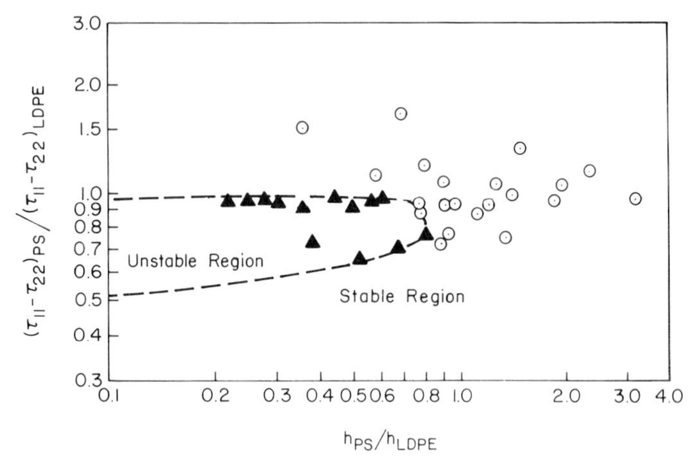

Fig. 8.9 Normal stress difference ratio of PS/LDPE versus layer thickness ratio in the three-layer flat-film coextrusion of PS/LDPE/PS system [1]: (⊙) stable interface; (▲) unstable interface.

LDPE system gives rise to regular interface, independent of the values of h_{LDPE}/h_{PS}.

Figure 8.9 gives plots of the ratio of the first normal stress difference of PS to LDPE, $(\tau_{11} - \tau_{22})_{PS}/(\tau_{11} - \tau_{22})_{LDPE}$, versus layer thickness ratio, h_{PS}/h_{LDPE} for the PS/LDPE/PS system. It is seen that interfacial instability occurs only over certain ranges of $(\tau_{11} - \tau_{22})_{PS}/(\tau_{11} - \tau_{22})_{LDPE}$ and h_{PS}/h_{LDPS}. However, when the layer arrangement was reversed, i.e., for the LDPE/PS/LDPE system, the ratios of first normal stress difference of LDPE to PS, $(\tau_{11} - \tau_{22})_{LDPE}/(\tau_{11} - \tau_{22})_{PS}$, were greater than 1.0 over the range of τ_w investigated, $10^4 < \tau_w < 10^5$ N/m^2, and only regular interfaces were observed [1].

In view of the fact that the polymer melts investigated possess both viscosity and elasticity, which are two independent rheological properties, the violation of the constraint with respect to either the viscosity ratio or the elasticity ratio or both would give rise to interfacial instability in coextrusion. Therefore, using Figs. 8.8 and 8.9, plots of η_{PS}/η_{LDPE} versus $(\tau_{11} - \tau_{22})_{PS}/(\tau_{11} - \tau_{22})_{LDPE}$ are prepared and given in Fig. 8.10, with h_{PS}/h_{LDPE} as parameter. It is of interest to note in Fig. 8.10 that the permissible ranges of viscosity and elasticity ratios which will lead to regular interfaces will become narrower as the skin-to-core layer thickness ratio becomes smaller. It should be pointed out, however, that the determination of the region of interfacial stability, given in Fig. 8.10, depends on the accuracy of the experimentally determined U-shaped curves given in Figs. 8.8 and 8.9.

What is shown above is an experimental correlation between the rheological properties of the polymers being coextruded and the processing

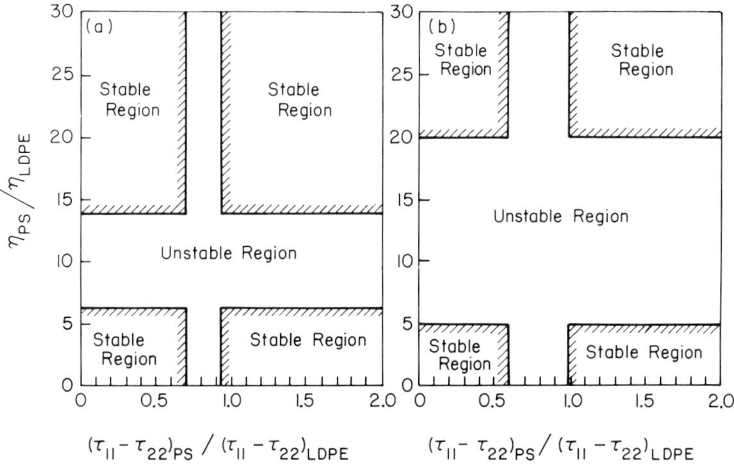

Fig. 8.10 Stability region in the three-layer flat-film coextrusion of the PS/LDPE/PS system, represented in terms of the viscosity and normal stress difference ratios, at various values of the skin-to-core thickness ratio [1]: (a) 0.7; (b) 0.3.

variables (e.g., flow rate) or product specification (e.g., layer thickness ratio). Note that the processing variables, such as melt extrusion temperature and extrusion rate, affect the rheological properties, which in turn affect the pressure drop in the die. In the absence of theoretical analysis, such experimental correlations may prove useful in avoiding the onset of interfacial instability. However, what is needed is an operating guide which is based on a sound, rigorous theoretical analysis.

8.2.2 Interfacial Instability in Wire Coating Coextrusion

The success of a wire coating coextrusion operation lies in obtaining a distortion-free coating with a smooth surface. An irregular interface might impair the electrical and mechanical properties of the insulation. Hence, it is important to study the effects of processing conditions (and the rheological and physical properties of different polymer combinations) on the stability of the interface during the coextrusion operation.

Cross sections of coextruded wires, displaying irregular interfaces, were obtained with the PS (STYRON 678)/LDPE and the PS (STYRON 686)/KRATON systems, and are given in Figs. 8.11 and 8.12. In these systems, the high viscosity component forms the outer layer. After careful examination of their experimental data, together with the visual observation of extrudate cross sections, Han and Rao [5] were able to establish some empirical correlations between the shear stress at the interface and the layer thickness

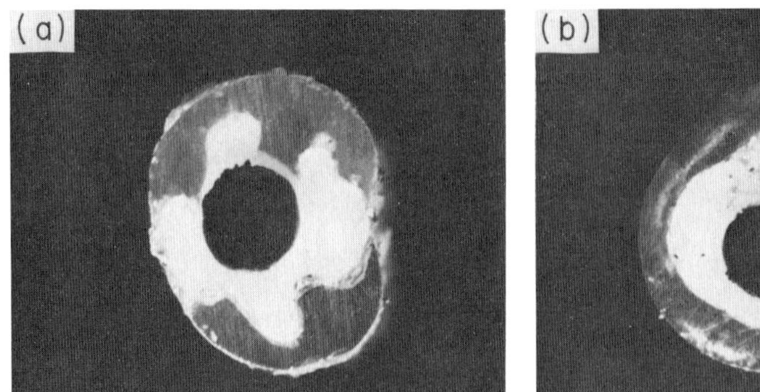

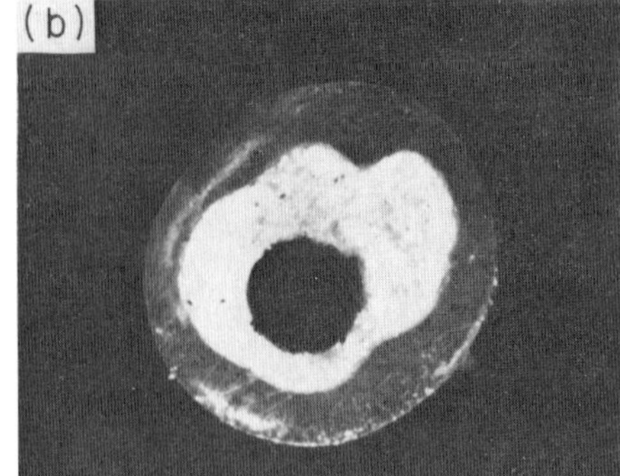

Fig. 8.11 Photographs of the cross sections of coextruded wires of PS (outer layer)/LDPE (inner layer) system: (a) wire speed = 8.1 cm/sec, $Q = 43.0\ \text{cm}^3/\text{min}$; (b) wire speed = 12.8 cm/sec, $Q = 57.3\ \text{cm}^3/\text{min}$.

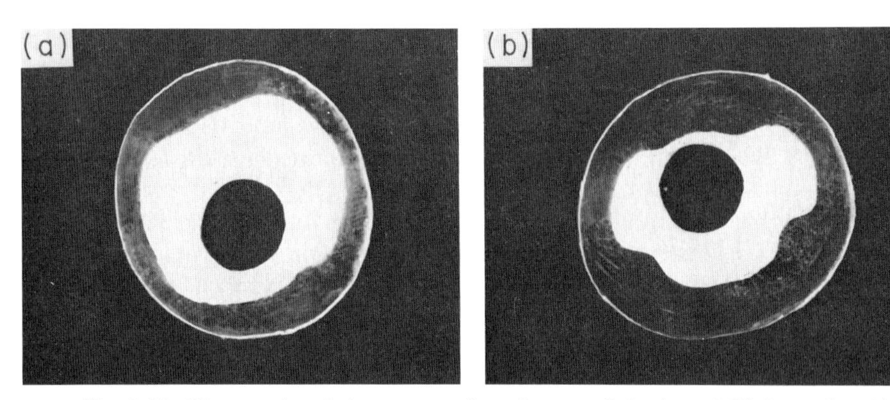

Fig. 8.12 Photographs of the cross section of coextruded wires of PS (outer layer)/ KRATON (inner layer) system [5]: (a) wire speed = 4.1 cm/sec, $Q = 48.0\ \text{cm}^3/\text{min}$; (b) wire speed = 6.5 cm/sec, $Q = 38.8\ \text{cm}^3/\text{min}$.

ratio, and between the viscosity ratio at the interface and the layer thickness ratio.

Figure 8.13 gives plots of axial pressure gradient versus the ratio of the volumetric flow rate of the outer component to that of the inner component, Q_A/Q_B, for the KRATON/PS (STYRON 686) and PS (STYRON 686)/ KRATON systems. It is seen that distinct regions of stability and instability exist, and that the axial pressure gradient and the ratio of the flow rates of the two phases play important roles in determining the occurrence of interfacial instability. For a given Q_A/Q_B ratio, there exists a *critical* pressure

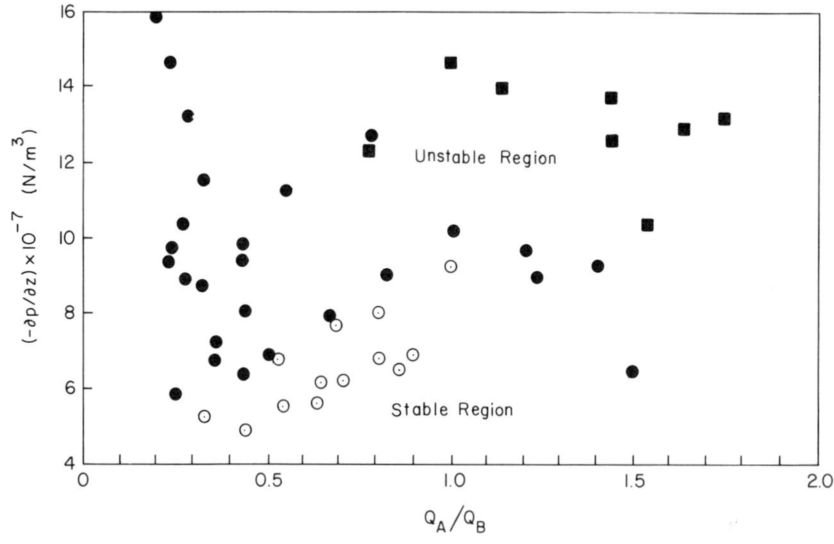

Fig. 8.13 Pressure gradient versus volumetric flow ratio in wire coating coextrusion [5]: (⊙) stable interface for the KRATON/PS system; (●) unstable interface for the KRATON/PS system; (■) unstable interface for the PS/KRATON system.

gradient above which interfacial instability occurs. The value of this critical pressure gradient first increases and then decreases as the Q_A/Q_B ratio is increased. It is seen also that, for a given Q_A/Q_B ratio, the transition from unstable flow to stable flow is possible if the pressure gradient is sufficiently decreased. This can be accomplished in one of the following ways: (1) by reducing the total volumetric flow rate at a constant wire speed, (2) by increasing the wire speed at a fixed total volumetric flow rate, (3) by increasing the extrusion temperature so as to reduce the melt viscosities of the materials, (4) by increasing the diameter of the die opening.

Note that the values of the interface position α and the maximum fluid velocity position λ may be determined by using the theoretical expressions developed in Chapter 7 [see Eqs. (7.59)–(7.63)] and the interfacial shear stress $(\tau_{rz})_{\text{Int}}$ may be calculated, using the expressions:

$$(\tau_{rz})_{\text{Int}} = \tfrac{1}{2}\zeta R_0\{(\alpha^2 - \lambda^2)/\alpha\}, \qquad \text{for} \quad \alpha \geq \lambda \tag{8.1}$$

$$(\tau_{rz})_{\text{Int}} = \tfrac{1}{2}\zeta R_0\{(\lambda^2 - \alpha^2)/\alpha\}, \qquad \text{for} \quad \alpha \geq \lambda \tag{8.2}$$

The layer thickness ratio of the outer phase to the inner phase, h_A/h_B may be determined using the expression [5]:

$$h_A/h_B = (1 - \alpha)/(\alpha - \kappa) \tag{8.3}$$

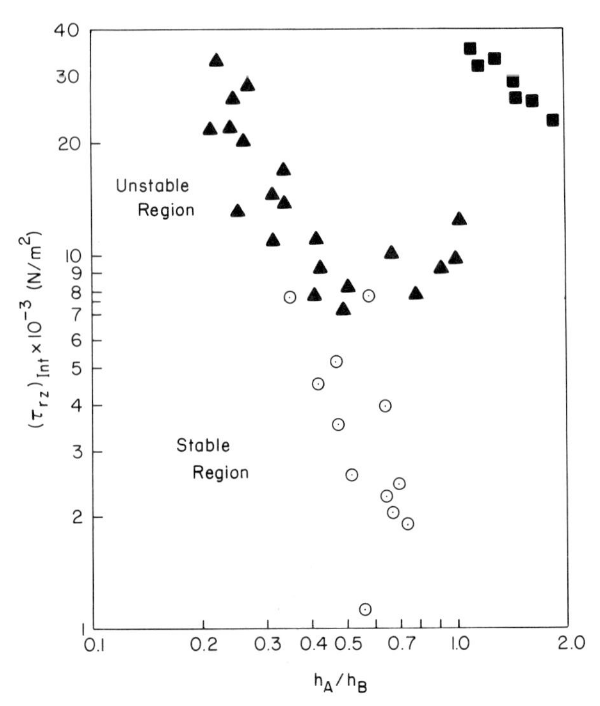

Fig. 8.14 Shear stress at the interface versus layer thickness ratio in wire coating co-extrusion [5]. Symbols are the same as in Fig. 8.13.

Plots of $(\tau_{rz})_{\text{Int}}$ versus h_A/h_B are given in Fig. 8.14 for the KRATON/PS (STYRON 686) and PS (STYRON 686) KRATON systems. A critical value of interfacial shear stress appears to exist.

Figure 8.15 gives plots of the viscosity ratio at the interface, $(\eta_A/\eta_B)_{\text{Int}}$, versus h_A/h_B ratio for the PS (STYRON 686)/KRATON, KRATON/PS (STYRON 686), HDPE/PS (STYRON 678), and LDPE/PS (STYRON 678) systems, and Fig. 8.16 gives plots of the first normal stress difference ratio at the interface, $[(\tau_{11} - \tau_{22})_A/(\tau_{11} - \tau_{22})_B]_{\text{Int}}$, versus h_A/h_B for the same systems.

It appears from Figs. 8.15 and 8.16 that interfacial instability occurs when the viscosity and normal stress difference ratios at the interface exceed critical values, regardless of the types of the two polymers being coextruded. The fluids investigated being viscoelastic fluids, the regions of stability (or instability) may be obtained by a composite plot as given in Fig. 8.17, prepared from Figs. 8.15 and 8.16. Note that the stability region given in Fig. 8.17 refers to the rheological properties evaluated at the phase interface, and therefore the thickness ratio h_A/h_B of the two layers does not come into the picture.

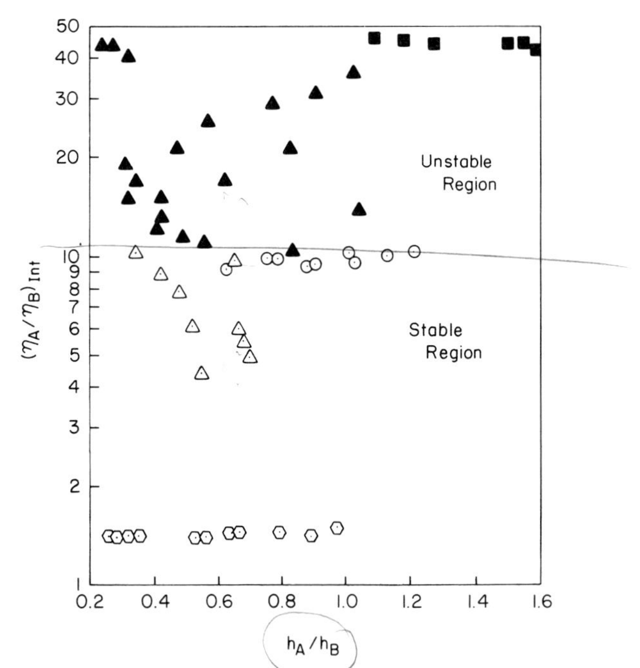

Fig. 8.15 Viscosity ratio at the interface versus layer thickness ratio in wire coating coextrusion [5]: (⊙) stable interface for the HDPE/PS system; (△) stable interface for the KRATON/PS system; (⬡) stable interface for the LDPE/PS system; (▲) unstable interface for the KRATON/PS system. (■) unstable interface for the PS/KRATON system.

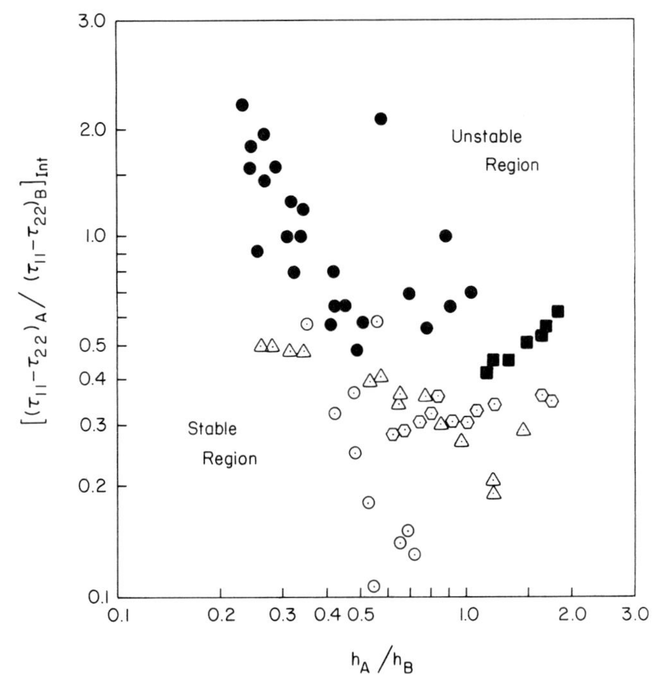

Fig. 8.16 Normal stress difference ratio at the interface versus layer thickness ratio in wire coating coextrusion [5]. Symbols are the same as in Fig. 8.15.

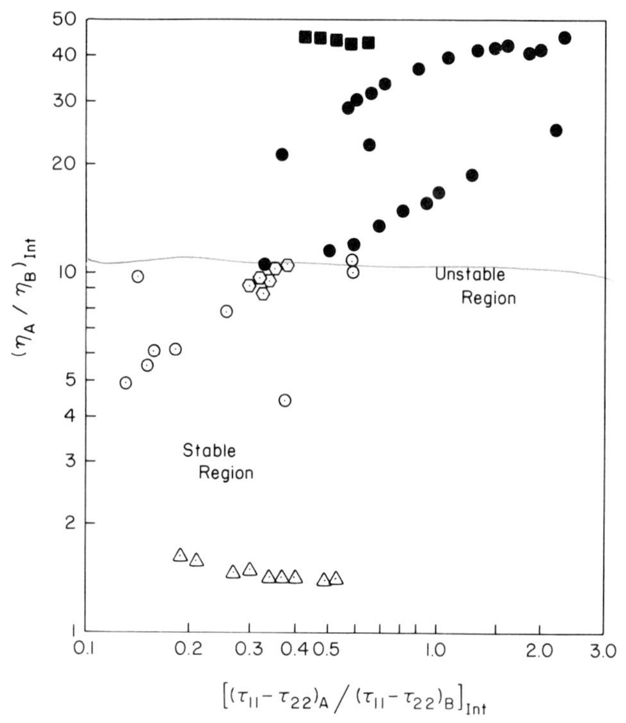

Fig. 8.17 Stability regions in two-layer wire coating coextrusion, expressed in terms of the viscosity and normal stress difference ratio at the interface [5]. Symbols are the same as in Fig. 8.15.

8.3 THEORETICAL CONSIDERATIONS OF INTERFACIAL INSTABILITY IN COEXTRUSION

The experimental observations presented above indicate clearly that many factors are involved in the occurrence of interfacial instability in coextrusion, including (1) the rheological properties of the individual polymers being coextruded, (2) the ratio of layer thicknesses, (3) the extrusion melt temperature, (4) the extrusion rate (or the shear stress at the die wall), and (5) the shear stress at the phase interface. Considering the large number of polymer combinations that may be possible for coextrusion, it can be easily surmised that no reasonable amount of experimentation can provide an operating guide that can guarantee regular interfaces in the coextruded product in every case. Therefore, a theoretical analysis (or analyses) will be of great value for achieving a better understanding of the phenomenon of interfacial instability.

Our primary objective here is to obtain stability regions in terms of the rheological parameters of the fluids involved (i.e., the ratios of viscosity and elasticity parameters) and the ratio of layer thicknesses. Note that, as discussed above in reference to the experimental observation, the effects of processing conditions (e.g., extrusion rate, melt temperature) on stability regions will reflect on the choice of both the viscosities and the elasticities of the fluids being coextruded.

Yih [10] has shown that for Newtonian flow, the problem of three-dimensional stability can be reduced to that of two-dimensional stability, i.e., Squire's theorem [17] can be invoked. Lockett [18] showed, however, that for viscoelastic fluids, Squire's theorem does not necessarily hold. We therefore have to consider three-dimensional flow in order to realistically represent the physical system. Although a full three-dimensional treatment of the disturbance would be desirable, one can easily imagine the complexity of the equations to be dealt with. Therefore, Khan and Han [14] considered, instead, two sets of two-dimensional disturbances: (1) disturbances spatially dependent only upon the x_1 and x_2 coordinates; (2) disturbances spatially dependent only upon the x_2 and x_3 coordinates.

Let us consider the situation where disturbances depend on the x_1 and x_2 coordinates. In other words, we consider plane Poiseuille flow when viewed from the direction perpendicular to flow, that is, into the $x_1 - x_2$ plane in Fig. 8.18. Let us further express perturbation variables u, v, w, p', and σ_{ij} as follows:

$$\begin{aligned} u_1 &= U(y) + u(z, y), \quad u_2 = v(z, y), \quad u_3 = w(z, y) \\ p &= \bar{p} + p'(z, y), \quad \tau_{ij} = \bar{\tau}_{ij} + \sigma_{ij}(z, y) \end{aligned} \tag{8.4}$$

The linearized equations of motion may be written in terms of the perturbation variables thus

$$\varepsilon_k \left[\frac{\partial u}{\partial \theta} + U_k \frac{\partial u}{\partial z} + \frac{\partial U_k}{\partial y} v \right] = -\frac{\partial p'}{\partial z} + \frac{\partial \sigma_{11}}{\partial z} + \frac{\partial \sigma_{12}}{\partial y} \tag{8.5}$$

$$\varepsilon_k \left[\frac{\partial v}{\partial \theta} + U_k \frac{\partial v}{\partial z} \right] = -\frac{\partial p'}{\partial y} + \frac{\partial \sigma_{12}}{\partial z} + \frac{\partial \sigma_{22}}{\partial y} \tag{8.6}$$

$$\varepsilon_k \left[\frac{\partial w}{\partial \theta} + U_k \frac{\partial w}{\partial z} \right] = \frac{\partial \sigma_{13}}{\partial z} + \frac{\partial \sigma_{23}}{\partial y} \tag{8.7}$$

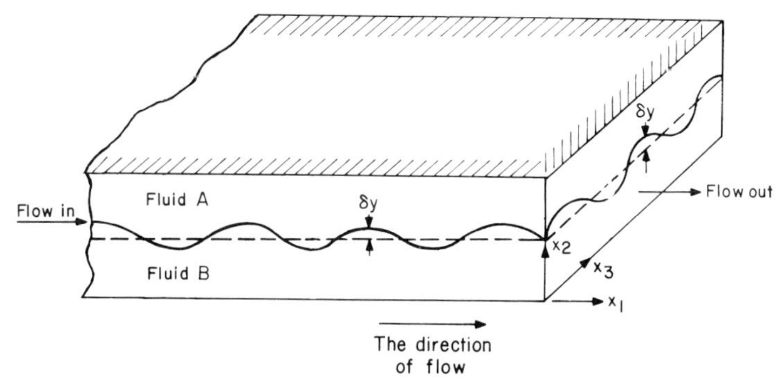

Fig. 8.18 Schematic of a rectangular channel in which unstable interface in stratified two-phase flow occurs in x_1-x_2 and x_2-x_3 planes. The dotted line represents a flat interface before a disturbance is introduced, and the solid curve represents a wavy interface after a disturbance is introduced [14]. From A. A. Khan and C. D. Han, *Trans. Soc. Rheol.* **21**, 101. Copyright © 1977. Reprinted by permission of John Wiley & Sons, Inc.

where ε_k is defined by

$$\varepsilon_k = \begin{cases} 1 & \text{for} \quad k = A \text{ (phase } A) \\ m_{\mathrm{d}} = \rho_B/\rho_A & \text{for} \quad k = B \text{ (phase } B) \end{cases} \tag{8.8}$$

Note in Eq. (8.4) that $\overline{\tau}_{ij}$ are the steady-state ijth components of the stress tensor τ, $\overline{p}$ is the steady-state hydrostatic pressure, and the following dimensionless variables are introduced:

$$\begin{aligned} z &= x_1/d_A; \qquad y = x_2/d_A; \qquad x = x_3/d_A \\ U &= \bar{U}/U_{\mathrm{m}}; \qquad p = \overline{p}/\rho_A U_{\mathrm{m}}^2; \qquad \tau_{ij} = \overline{\tau}_{ij}/\rho_A U_{\mathrm{m}}^2 \\ \theta &= tU_{\mathrm{m}}/d_A; \qquad \mathrm{Re} = \rho_A U_{\mathrm{m}} d_A/(\eta_0)_A; \qquad n = d_B/d_A \end{aligned} \tag{8.9}$$

in which U_{m} denotes the average velocity, $\bar{U}$ the steady-state velocity, ρ the fluid density, Re the Reynolds number, d_A and d_B the layer thicknesses of phases A and B respectively, n the layer thickness ratio, and $(\eta_0)_A$ the viscosity of phase A.

As is common in hydrodynamic linear stability analysis [10, 12], we assume that all perturbation quantities follow an exponential type of time dependence and a periodic type of spatial dependence, that is

$$q' = \hat{q}(y)\exp[i\alpha(z - c\theta)] \tag{8.10}$$

in which q' denotes an arbitrary perturbation variable, $\hat{q}(y)$ is the amplitude of a periodic function, α is a real positive number which is proportional to the reciprocal of wavelength, and $c = c_{\mathrm{r}} + ic_{\mathrm{i}}$ a complex number. It is apparent from Eq. (8.10) that stability is determined by the sign of the imaginary part of the complex number c. That is, positive values of c_i imply that the amplitude of the disturbance increases with time θ and thus the flow is unstable, and negative values of c_i indicate that the flow is stable.

We see also that the equation of continuity may be written in terms of the perturbation variables:

$$\partial u/\partial z + \partial v/\partial y = 0 \tag{8.11}$$

This equation permits us to use a stream function χ defined as

$$u = \partial\chi/\partial y, \qquad v = -\partial\chi/\partial z \tag{8.12}$$

If we now substitute the following expressions [see Eq. (8.10)]:

$$\begin{aligned} \chi &= \Phi(y)\exp[i\alpha(z - c\theta)] \\ p' &= f(y)\exp[i\alpha(z - c\theta)] \\ \sigma_{ij} &= F_{ij}(y)\exp[i\alpha(z - c\theta)] \end{aligned} \tag{8.13}$$

into Eq. (8.12) we obtain

$$\begin{aligned} u &= \Phi'(y)\exp[i\alpha(z - c\theta)] \\ v &= -i\alpha\Phi(y)\exp[i\alpha(z - c\theta)] \end{aligned} \tag{8.14}$$

in which Φ' denotes the derivative of Φ with respect to y. Introducing

$$w = \xi(y)\exp[i\alpha(z - c\theta)] \tag{8.15}$$

together with Eqs. (8.13) and (8.14) into Eqs. (8.5) to (8.7), and employing the Coleman–Noll second-order fluid model [see Eq. (2.53) in Chapter 2], Khan and Han [14] obtained the following system of differential equations that governs the stability of the interface on the x_1-x_2 plane:

$$\begin{aligned} &i\alpha\,\mathrm{Re}\,\varepsilon_k[-c\xi_k + U_k\xi_k] \\ &\quad = -\delta_k\varepsilon_k\alpha^2 + (M_1)_k\gamma_k\xi_k' i\alpha + (M_2)_k i\alpha^3 c\xi_k \\ &\qquad -(M_2)_k U_k i\alpha^3\xi_k + \delta_k\xi_k'' + (M_1 + M_2)_k i\alpha[\xi_k\,\partial\gamma_k/\partial y + \gamma_k\xi_k'] \\ &\qquad -(M_2)_k i\alpha c\xi_k'' + (M_2)_k i\alpha[\gamma_k\xi_k + U_k\xi_k''] \end{aligned} \tag{8.16}$$

and

$$\begin{aligned} &i\alpha\,\mathrm{Re}\,\varepsilon_k[\Phi_k\alpha^2(U_k - c) - \Phi_k''(U_k - c) + \Phi_k\,\partial\gamma_k/\partial y] \\ &\quad = \Phi_k^{iv}[-\delta_k + (M_2)_k i\alpha c - (M_2)_k U_k i\alpha] \\ &\qquad + \Phi_k''[2\delta_k - 2i\alpha^3 c(M_2)_k + 2(M_2)_k U_k i\alpha^3] \\ &\qquad + \Phi_k[-\delta_k\alpha^4 + (M_2)_k i\alpha^5 c - (M_2)_k U_k i\alpha^5] \end{aligned} \tag{8.17}$$

where

$$\gamma_k = \left(\frac{\partial U}{\partial y}\right)_k = \frac{\mathrm{Re}\,\zeta}{\delta_k} y + \frac{\mathrm{Re}\, b_1}{\delta_k}, \qquad k = A, B \tag{8.18}$$

$$(M_1)_k = \beta_k \,\mathrm{Re}/\rho_A d_A^2; \qquad (M_2)_k = \nu_k \,\mathrm{Re}/\rho_A d_A^2; \quad k = A, B \tag{8.19}$$

and

$$\delta_k = \begin{cases} 1 & \text{for phase } A \\ k = (\eta_0)_B/(\eta_0)_A & \text{for phase } B; \end{cases} \qquad b_1 = \frac{\zeta}{2}\left[\frac{n^2 - k}{n + k}\right], \tag{8.20}$$

in which ζ denotes the dimensionless pressure gradient $\partial p/\partial z$. β_k and ν_k in Eq. (8.19) are material constants associated with the Coleman–Noll second-order fluid [see Eq. (2.53) in Chapter 2].

By employing the perturbation technique, Khan and Han [14] obtained

$$(c_0)_\mathrm{i} = 0 \tag{8.21}$$

for the zeroth-order perturbation in α. Note that we are interested only in the imaginary component $(c_0)_\mathrm{i}$ of the complex number c_0, because we are interested in absolute stability which is determined by the sign of $(c_0)_\mathrm{i}$. The real component $(c_0)_\mathrm{r}$ of the complex number c_0 determines oscillatory instability, which is not of primary interest here. Equation (8.21) indicates that for infinitely large wavelengths, the stability is marginal and the system is neither stable nor unstable.

For the first-order perturbation in α, Khan and Han [14] report

$$(c_1)_\mathrm{i} = -bd/(a^2 + b^2) \tag{8.22}$$

where

$$\begin{aligned} a &= f_1(k, n, (M_2)_A, m_\mathrm{v}, \mathrm{Re}, m_\mathrm{d}) \\ b &= f_2(k, n, (M_2)_A, m_\mathrm{v}, \mathrm{Re}, m_\mathrm{d}) \\ d &= f_3(k, n, \mathrm{Re}) \end{aligned} \tag{8.23}$$

in which m_v is the ratio of the elastic parameter M_2 for phases A and B [see Eq. (8.19)], that is,

$$m_\mathrm{v} = (M_2)_B/(M_2)_A = \nu_B/\nu_A \tag{8.24}$$

From Eq. (8.22) it is seen that the sign of $(c_1)_\mathrm{i}$ depends on the sign of $-bd$, and therefore we define the stability parameter S_p as

$$S_p = -bd = f(k, n, m_\mathrm{v}, (M_2)_A, \mathrm{Re}, m_\mathrm{d}) \tag{8.25}$$

S_p depends on the layer thickness ratio n of the two fluids, the viscosity ratio k, the elasticity parameter ratio m_v, the density ratio m_d, the Reynolds

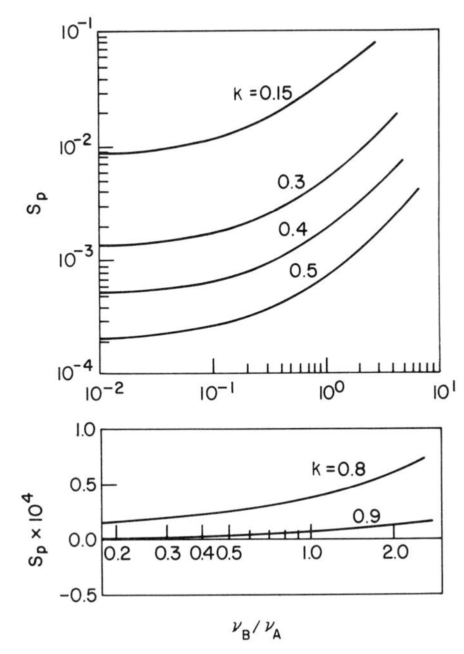

Fig. 8.19 Stability parameter in the x_1-x_2 plane versus elasticity parameter ratio, for various values of viscosity ratio [14]. Other system parameters: $n = 1.0$; $m_d = 1.0$; $\mathrm{Re} = 0.01$; $(M_2)_A = -0.2$. From A. A. Khan and C. D. Han, *Trans. Soc. Rheol.* **21**, 101. Copyright © 1977. Reprinted by permission of John Wiley & Sons, Inc.

number Re, and the magnitude of the elastic parameter M_2 in phase A. The effect of surface tension is assumed to be negligible, because, in polymer melt flows, the normal stress is of the magnitude of 10^4 N/m^2, whereas the surface tension is of the magnitude of 10^{-2} N/m (10 dyn/cm).

It is seen in Fig. 8.19 that, for an equal depth of the two fluids ($n = 1.0$), S_p is always positive, and the interface is always irregular for all values of viscosity ratio k other than unity, independent of the elasticity parameter ratio m_v of the two fluids. However, for unequal depths ($n \neq 1.0$) (see Fig. 8.20), S_p can be negative or positive (the interface being regular or irregular, respectively) depending on the viscosity ratio k of the fluid, again independent of the elasticity parameter ratio m_v. This indicates that interfacial irregularity in the $x_1 - x_2$ plane is controlled solely by the viscosity ratio of the two fluids.

It should be noted that the theoretical prediction presented above holds for small values of α, i.e., disturbances whose wavelength is large compared to the die opening ($\alpha \ll (d_A + d_B)$), and also for very low Reynolds numbers.

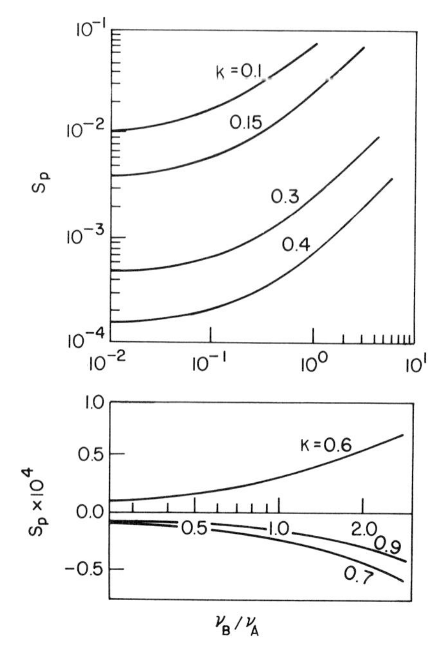

Fig. 8.20 Stability parameter in the $x_1 - x_2$ plane versus elasticity parameter ratio, for various values of viscosity ratio [14]. Other parameters: $n = 0.8$; $m_d = 1.0$; $Re = 0.01$; $(M_2)_A = -0.2$. From A. A. Khan and C. D. Han, *Trans. Soc. Rheol.* **21**, 101. Copyright © 1977. Reprinted by permission of John Wiley & Sons, Inc.

Let us now consider the situation where disturbances depend on the x_2 and x_3 coordinates, referring to Fig. 8.18, and express perturbation variables u, v, w, p', and σ_{ij} in terms of the dimensionless variables defined in Eq. (8.9) as follows:

$$\begin{gathered} u_1 = U(y) + u(x, y); \qquad u_2 = v(x, y); \qquad u_3 = w(x, y) \\ p = \bar{p} + p'(x, y); \qquad \tau_{ij} = \bar{\tau}_{ij} + \sigma_{ij}(x, y) \end{gathered} \tag{8.26}$$

The linearized equations of motion in terms of the perturbation variables become

$$\varepsilon_k \left(\frac{\partial u}{\partial \theta} + \frac{\partial U_k}{\partial y} v \right) = \frac{\partial \sigma_{12}}{\partial y} + \frac{\partial \sigma_{13}}{\partial x} \tag{8.27}$$

$$\varepsilon_k \frac{\partial v}{\partial \theta} = -\frac{\partial p'}{\partial y} + \frac{\partial \sigma_{22}}{\partial y} + \frac{\partial \sigma_{23}}{\partial x} \tag{8.28}$$

$$\varepsilon_k \frac{\partial w}{\partial \theta} = -\frac{\partial p'}{\partial x} + \frac{\partial \sigma_{23}}{\partial y} + \frac{\partial \sigma_{33}}{\partial x} \tag{8.29}$$

with ε_k as defined in Eq. (8.8).

Let us now again assume that all perturbation quantities follow an exponential type of time dependence and a periodic type of spatial dependence. Then, analogously to Eq. (8.10), we have:

$$q' = \hat{q}(y)\exp[i\alpha(x - c\theta)] \tag{8.30}$$

in which q' denotes an arbitrary perturbation variable, $\hat{q}(y)$ is the amplitude of a periodic function, α is a real positive number which is the reciprocal of wavelength, and $c = c_r + ic_i$, a complex number.

By employing the perturbation technique, Khan and Han [14] constructed regions of stability in terms of the flow properties, namely, the viscosity ratio k, the elasticity parameter ratio m_v, the layer thickness ratio n, the Reynolds number Re, the density ratio m_d, and the wave number α. Figure 8.21 shows the effect of layer thickness ratio n on the stability region. It is seen that as n decreases, the stability region becomes smaller. This is in agreement with the experimental observation that in the coextrusion of multilayer flat films, the interface (or interfaces) between the layers tends to become more irregular as the layer thicknesses deviate farther away from equality. Khan and Han concluded from their analysis that even in the *absence* of viscosity stratification (i.e., $k = 1.0$), the interface in the x_2-x_3 plane of two viscoelastic fluids, represented by the second-order fluid model, in plane Poiseuille flow can still be unstable for certain values of the elasticity parameter ratio, m_γ.

It should be pointed out that the details of a stability analysis of rheologically complex flow problems, such as the one discussed above, often depends

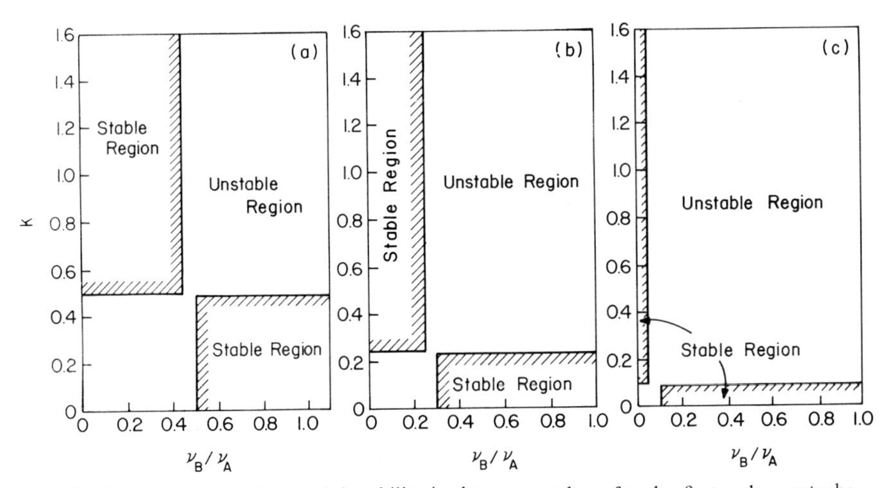

Fig. 8.21 Region of interfacial stability in the x_2-x_3 plane for the first-order perturbation, for various values of layer thickness ratio [14]: (a) $n = 0.7$; (b) $n = 0.5$; (c) $n = 0.3$. Other parameters: Re = 0.01; $(M_2)_A = -0.2$; $\alpha = 0.1$. From A. A. Khan and C. D. Han, *Trans. Soc. Rheol.* **21**, 101. Copyright © 1977. Reprinted by permission of John Wiley & Sons, Inc.

on the choice of a rheological model. Therefore the stability analysis presented above, which is based on the Coleman–Noll second-order fluid model, must be interpreted with caution, because it is possible for one to reach different conclusions by using other rheological models.

8.4 CONCLUDING REMARKS

In this chapter, emphasis was placed on the development of criteria for determining the processability of two or more thermoplastic polymeric materials for coextrusion operations. The criteria are based on both the rheological and physical properties points of view, which are affected by the molecular structure, molecular weight (and its distribution), and the compatibility of the polymer systems being coextruded.

When proper use of the information presented in this chapter is made, it may be possible for one to develop new products and/or process innovations. More specifically, one may be able to (1) select polymeric systems on the basis of their rheological properties, (2) select optimal processing conditions, and (3) design coextrusion dies intelligently, which will enable one to make products of high quality and to cut down, if not eliminate completely, the amount of waste product.

It was demonstrated in this chapter that theoretical analysis can help generalize the rather complicated relationships existing among many rheological/processing variables in the coextrusion processes, so that the occurrence of undesirable interfacial instability can be avoided and thus the possibility of having a poor quality product can be minimized.

Because of limitations of space, and the complexities involved in manipulating the mathematical equations, we have not discussed nonisothermal coextrusion operations. Often, in the industrial operation of coextrusion processes, because of a particular choice of polymer combinations (e.g., nylon and polyethylene; saran and polystyrene or polyethylene), two streams having different melt temperatures are fed to the same coextrusion die. Such a situation may arise from the necessity for minimizing the possible thermal degradation of one polymer, or for matching the melt viscosities of two or more streams. Hydrodynamic stability analysis of nonisothermal coextrusion processes requires consideration of the heat transferred between the layers across the phase interface, together with the continuity and momentum equations (see Problem 8.2).

For all coextruded products, adhesion between the layers is one of the most important problems. The mechanical properties (e.g., tensile strength, flexural strength) of the coextruded products are intimately related to this factor. In turn, the adhesion between the layers is related to the extent of compatibility of the polymers coextruded. After all, from the practical point of view, the choice of polymers to be coextruded will be based on the premise that they will have good, or at least acceptable, adhesion. Once the polymers are chosen for coextrusion, one must then find the optimal processing conditions that will give rise to those differences in their respective rheological properties that will guarantee a regular interface between the layers. Again, a discussion of the choice of polymer systems for coextrusion, and the mechanical properties of the coextruded products, is beyond the scope of this monograph.

Problems

8.1 On the basis of the particular physical properties desired, polymers A and B are chosen for making coextruded flat films. The rheological properties of these two polymers are very similar over the range of the processing conditions contemplated ($\tau_w = 10^4$–10^6 N/m^2 at 240°C), as given in Fig. 8.22. To all intents and purposes, you may assume that the two polymers have identical rheological properties. However, the

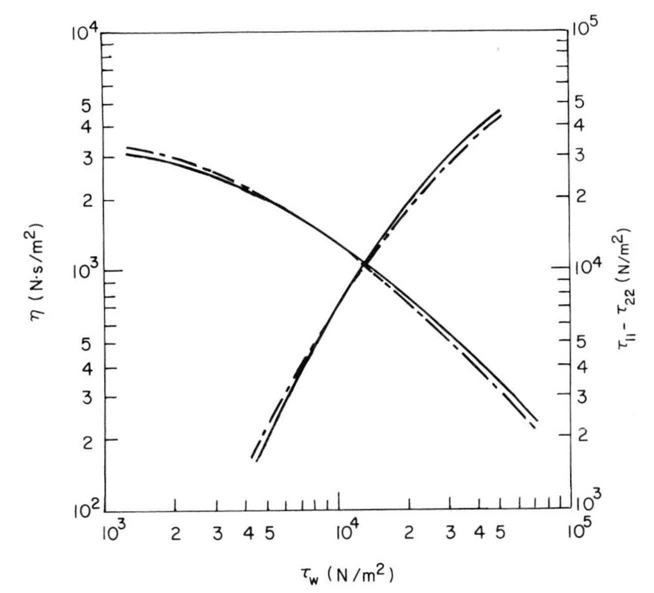

Fig. 8.22 Viscosity and first normal stress difference versus shear stress for two polymer melts.

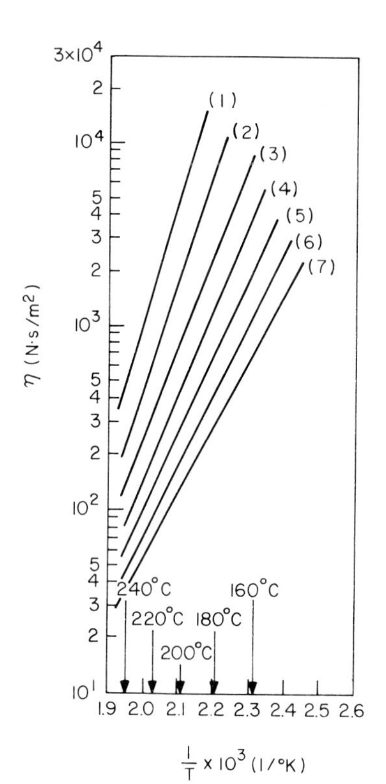

Fig. 8.23 Viscosity versus reciprocal temperature for polymer A at various shear stresses (N/m^2): (1) 2×10^4; (2) 3×10^4; (3) 4×10^4; (4) 5×10^4; (5) 6×10^4; (6) 7×10^4; (7) 8×10^4.

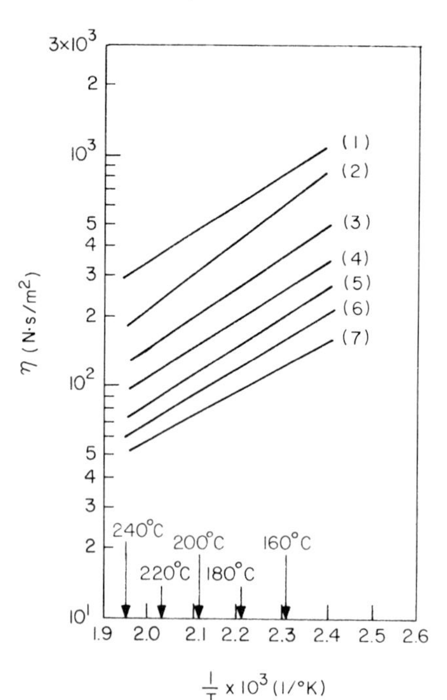

Fig. 8.24 Viscosity versus reciprocal temperature for polymer B at various shear stresses (N/m^2): (1) 2×10^4; (2) 3×10^4; (3) 4×10^4; (4) 5×10^4; (5) 6×10^4; (6) 7×10^4; (7) 8×10^4.

coextruded layers of the two polymers have poor adhesion and therefore a decision has been made to use a glue layer, which will be coextruded to make the layer configuration: polymer A/glue layer/polymer B.

The material to be used as the glue layer has a low molecular weight and may be considered as a Newtonian fluid ($\eta_0 = 50$ N sec/m^2 at $T = 240°$C).

(a) When the layer thicknesses inside the die are specified as: 0.05 mm of polymer A/0.10 mm of glue layer/0.05 mm of polymer B, plot the velocity and shear stress profiles inside the die, assuming that the layer arrangement is not perturbed during the flow inside the die. Can regular interface between the adjacent layers be produced at 240°C and $\tau_w = 10^5$ N/m^2?

(b) Assume that the layer thicknesses inside the die are specified as 0.10 mm of polymer A/0.05 mm of glue layer/0.15 mm of polymer B, and that interfacial instability occurs at the processing conditions: $T = 240°$C and $\tau_w = 10^5$ N/m^2. Which of the two interfaces, the interface between the polymer A and the glue layer, or the interface between the polymer B and the glue layer, will give rise to irregularity (i.e., instability) first?

8.2 Polymers A and B are to be coextruded to make two-layer flat films. The temperature dependence of the viscosities of the two polymers is given in Figs. 8.23 and 8.24. It is seen that the Arrhenius relationship

$$\eta = \eta_1 \exp\left[\frac{E}{R}\left(\frac{1}{T} - \frac{1}{T_1}\right)\right] \tag{8.31}$$

holds for both polymers, and that at a given temperature, both polymers follow a power law. It is suggested that in order to avoid the possibility of having interfacial instability during coextrusion, the two polymers be coextruded at different melt temperatures, namely, polymer A at 240°C and polymer B at 200°C. Can you indeed avoid the occurrence of interfacial instability during coextrusion? The following specifications are given for your evaluation: (i) Die opening is 0.2 cm; (ii) Die land length is 5 cm; (iii) Die width is 50 cm; (iv) Layer thicknesses inside the die are 0.15 cm of polymer A and 0.05 cm of polymer B. Carry out a theoretical analysis to answer the question.

8.3 Polymers A and B are to be coextruded at 200°C to produce three-layer flat films, with the layer arrangement $A/B/A$. The rheological properties of the two polymers are given in Figs. 8.23 and 8.24. Carry out a theoretical analysis to find the range of processing conditions (i.e., thickness ratio and wall shear stress) that will yield regular (i.e., stable) interfaces in coextrusion. Use the same die dimensions as given in Problem 8.2.

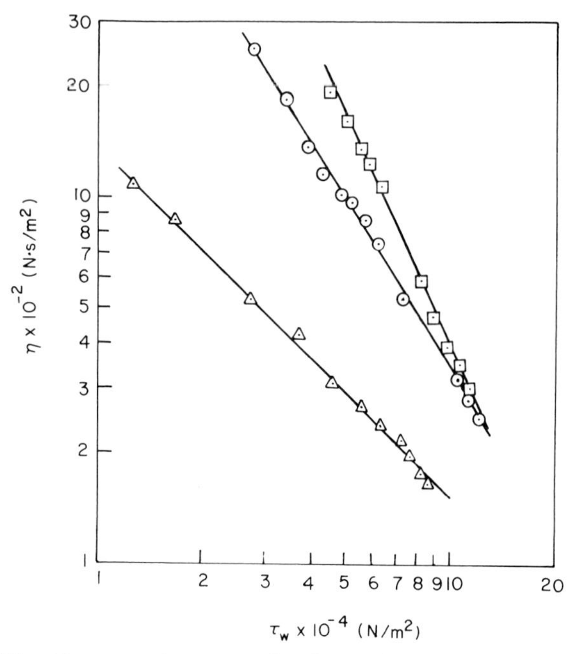

Fig. 8.25 Viscosity versus shear stress for three polymers at 200°C: (△) polymer A; (⊙) polymer B; (⊡) polymer C.

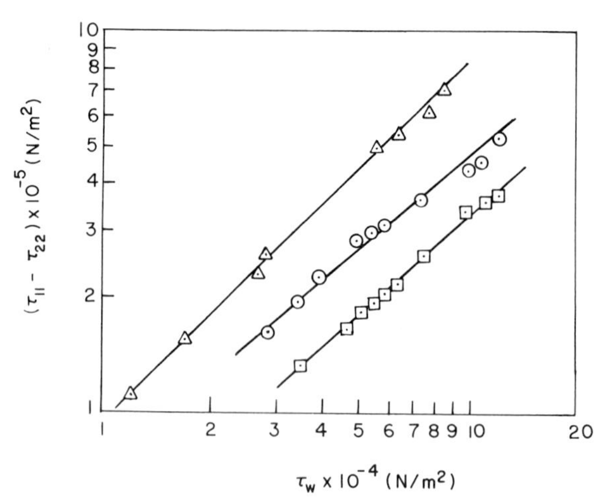

Fig. 8.26 First normal stress difference versus shear stress for three polymers at 200°C: (△) polymer A; (⊙) polymer B; (⊡) polymer C.

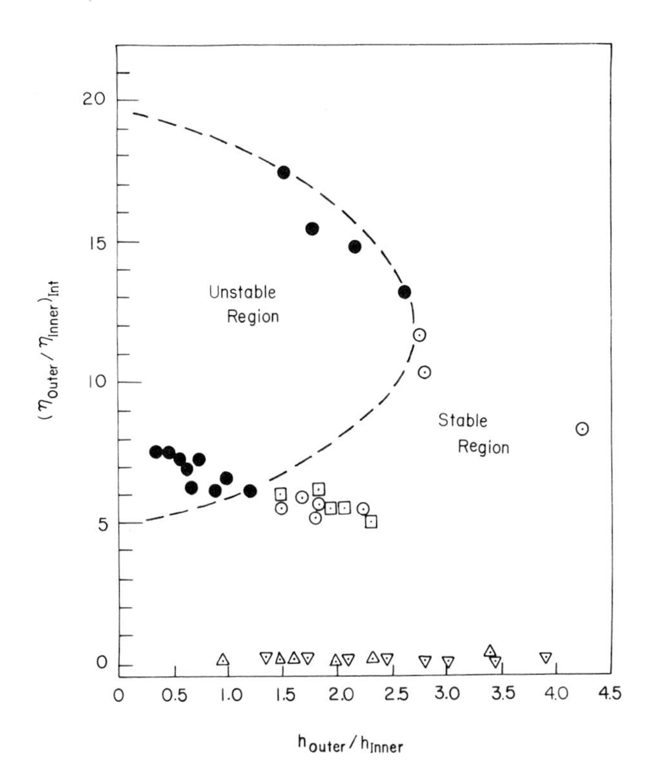

Fig. 8.27 Stability region, expressed in terms of the viscosity ratio at the interface and thickness ratio of the outer layer to the inner layer: (⊙, ●) $C/A/C$; (⊡) $B/A/B$; (△) $A/C/A$; (▽) $A/B/A$.

8.4 Polymers A, B, and C were used to produce three-layer flat films at 200°C with the various layer arrangements: $A/B/A$, $B/A/B$, $A/C/A$, and $C/A/C$. The rheological properties of the three polymers are given in Figs. 8.25 and 8.26.

The experimentally determined stability regions are given in Fig. 8.27 in terms of the ratio of the viscosity of the outer layer to that of the inner layer at the interface, $(\eta_{outer}/\eta_{inner})_{Int}$, and the layer thickness ratio, h_{outer}/h_{inner}. Obtain stability regions theoretically, in terms of $(\eta_{outer}/\eta_{inner})_{Int}$ and h_{outer}/h_{inner}, using (a) the Zaremba–DeWitt model [see Eq. (2.42) in Chapter 2]; (b) the convected Maxwell model [see Eq. (2.41) in Chapter 2].

REFERENCES

[1] C. D. Han and R. Shetty, *Polym. Eng. Sci.* **18**, 180 (1978).
[2] W. J. Schrenk, N. L. Bradley, T. Alfrey, and H. Maack, *Polym. Eng. Sci.* **18**, 620 (1978).

[3] J. H. Southern and R. L. Ballman, *J. Polym. Sci., Part A-2* **13**, 863 (1975).
[4] B. L. Lee and J. L. White, *Trans. Soc. Rheol.* **18**, 467 (1974).
[5] C. D. Han and D. Rao, *Polym. Eng. Sci.* **20**, 128 (1980).
[6] A. E. Everage, *Trans. Soc. Rheol.* **17**, 629 (1973).
[7] D. L. MacLean, *Trans. Soc. Rheol.* **17**, 385 (1973).
[8] H. S. Yu and E. M. Sparrow, *ASME Trans., Part C* **91**, 51 (1969).
[9] T. W. F. Russel, G. W. Hodgson, and G. W. Govier, *Can. J. Chem. Eng.* **37**, 9 (1959).
[10] C. S. Yih, *J. Fluid Mech.* **27**, 337 (1967).
[11] C. E. Hickox, *Phys. Fluids* **14**, 251 (1971).
[12] C. H. Li, *Phys. Fluids* **12**, 531 (1969).
[13] C. H. Li, *Phys. Fluids* **13**, 1701 (1970).
[14] A. A. Khan and C. D. Han, *Trans. Soc. Rheol.* **21**, 101 (1977).
[15] J. L. White, *Appl. Polym. Symp.* No 20, 155 (1973).
[16] C. D. Han, "Rheology in Polymer Processing." Academic Press, New York, 1976.
[17] H. B. Squire, *Proc. R. Soc. London, Ser. A* **142**, 621 (1933).
[18] F. J. Lockett, *Int. J. Eng. Sci.* **7**, 337 (1969).

Author Index

Numbers in italics refer to the pages on which complete references are listed.

E

F

G

H

I

J

K

N

O

P

R

Subject Index